Food Industry 4.0

Emerging Trends and Technologies in Sustainable Food Production and Consumption

Developments in Food Quality and Safety

Food Industry 4.0

Emerging Trends and Technologies in Sustainable Food Production and Consumption

Edited by

Abdo Hassoun

Sustainable AgriFoodtech Innovation & Research (SAFIR), Arras, France

Series Editor

Jose Manuel Lorenzo

ELSEVIER

ACADEMIC PRESS

An imprint of Elsevier

Academic Press is an imprint of Elsevier
125 London Wall, London EC2Y 5AS, United Kingdom
525 B Street, Suite 1650, San Diego, CA 92101, United States
50 Hampshire Street, 5th Floor, Cambridge, MA 02139, United States

ISBN: 978-0-443-15516-1

For Information on all Academic Press publications
visit our website at https://www.elsevier.com/books-and-journals

Publisher: Nikki Levy
Acquisitions Editor: Nina Bandeira
Editorial Project Manager: Dan Egan
Production Project Manager: Rashmi Manoharan
Cover Designer: Matthew Limbert

Typeset by MPS Limited, Chennai, India

Dedication

This book is dedicated to Gaza: Soul of the soul!

—Abdo Hassoun

Contents

List of contributors

Abderrahmane Aït-Kaddour Université Clermont Auvergne, INRAE, VetAgro Sup, UMRF, Lempdes, France; Department of Food Industrial Technology, Universitas Padjadjaran, Bandung, Indonesia

Farah Bader Goody Products Marketing Company, Jeddah, Saudi Arabia

Alaa El-Din A. Bekhit Department of Food Science, The University of Otago, Dunedin, New Zealand

Krzysztof B. Beć Institute of Analytical Chemistry and Radiochemistry, University of Innsbruck, Innsbruck, Austria

Zuhaib F. Bhat Division of Livestock Products Technology, SKUAST-J, Jammu and Kashmir, India

Giulio Maria Bianco Department of Civil Engineering and Computer Science Engineering, The University of Rome Tor Vergata, Rome, Italy

Barbara Bigliardi Department of Engineering and Architecture, University of Parma, Parma, Italy

Gonca Bilge Department of Food Engineering, Faculty of Engineering, Yeditepe University, Istanbul, Türkiye

Erik Bjørnerud Høgskolen i Østfold, Faculty of Teacher Education and Languages, Department of Natural Sciences, Practical-Aesthetic, Social and Religious Studies, Halden, Norway

Sofiane Boudalia Département d'Écologie et Génie de l'Environnement, Université 8 Mai 1945 Guelma, Guelma, Algeria; Laboratoire de Biologie, Eau et Environnement, Université 8 Mai 1945 Guelma, Guelma, Algeria

Fatma Boukid ClonBio Group, Ltd., Dublin, Ireland

Yana Cahyana Department of Food Industrial Technology, Universitas Padjadjaran, Bandung, Indonesia

Esra Capanoglu Department of Food Engineering, Faculty of Chemical and Metallurgical Engineering, Istanbul Technical University, Maslak, Istanbul, Turkey

Semra Çiçek Department of Agricultural Biotechnology, Faculty of Agriculture, Ataturk University, Erzurum, Turkey

John Colreavy Meat Technology Ireland, Teagasc, Ashtown, Dublin, Ireland

James Colwill Wolfson School, Loughborough University, Loughborough, Leicestershire, United Kingdom

Rui M.S. Cruz Department of Food Engineering, Institute of Engineering, Universidade do Algarve, Campus da Penha, Faro, Portugal; MED-Mediterranean Institute for Agriculture, Environment and Development and CHANGE-Global Change and Sustainability Institute, Faculty of Sciences and Technology, Campus de Gambelas, Universidade do Algarve, Faro, Portugal

José Antonio Entrenas Department of Animal Production, Faculty of Agriculture and Forestry Engineering, University of Cordoba, Campus Rabanales, Córdoba, Spain

Tuba Esatbeyoglu Department of Molecular Food Chemistry and Food Development, Institute of Food and One Health, Gottfried Wilhelm Leibniz University Hannover, Hannover, Germany

Eva Falch Department of Biotechnology and Food Science, Norwegian University of Science and Technology (NTNU), Trondheim, Norway

Alessandro Ferragina Food Quality and Sensory Science Department, Teagasc, Ashtown, Dublin, Ireland

Borja Ramis Ferrer The Institute of Electrical and Electronics Engineers, Madrid, Spain

Aberham Hailu Feyissa Research Group for Food Production Engineering, National Food Institute, Technical University of Denmark, Kgs Lyngby, Denmark

Serena Filippelli Department of Engineering and Architecture, University of Parma, Parma, Italy

Sigfredo Fuentes Digital Agriculture, Food and Wine Sciences Group, School of Agriculture, Food and Ecosystem Sciences, Faculty of Science, The University of Melbourne, Parkville, VIC, Australia; Tecnologico de Monterrey, School of Engineering and Sciences, Monterrey, Nuevo Leon, Mexico

Mohammed Gagaoua PEGASE, INRAE, Institut Agro, Saint-Gilles, France

Guillermo Garcia-Garcia Department of Agrifood System Economics, Institute of Agricultural and Fisheries Research & Training (IFAPA), Granada, Spain

Claudia Gonzalez Viejo Digital Agriculture, Food and Wine Sciences Group, School of Agriculture, Food and Ecosystem Sciences, Faculty of Science, The University of Melbourne, Parkville, VIC, Australia

Justyna Grabska Institute of Analytical Chemistry and Radiochemistry, University of Innsbruck, Innsbruck, Austria

Busra Gultekin Subasi Center for Innovative Food (CiFOOD), Department of Food Science, Aarhus University, Agro Food Park 48, Aarhus N 8200, Denmark

Cennet Pelin Boyaci Gunduz Department of Food Engineering, Adana Alparslan Turkes Science and Technology University, Adana, Turkey

Ruth M. Hamill Food Quality and Sensory Science Department, Teagasc, Ashtown, Dublin, Ireland

Abdo Hassoun Sustainable AgriFoodtech Innovation & Research (SAFIR), Arras, France

Mike Hibbett Irish Manufacturing Research, Mullingar, Co. Westmeath, Ireland

Christian W. Huck Institute of Analytical Chemistry and Radiochemistry, University of Innsbruck, Innsbruck, Austria

Sandeep Jagtap Sustainable Manufacturing Systems Centre, School of Aerospace, Transport and Manufacturing, Cranfield University, Cranfield, United Kingdom

Alan Kavanagh Irish Manufacturing Research, Mullingar, Co. Westmeath, Ireland

Yang Luo School of Intelligent Manufacturing Ecosystem, Xi'an Jiaotong-Liverpool University, Suzhou, P.R. China

Gaetano Marrocco Department of Civil Engineering and Computer Science Engineering, The University of Rome Tor Vergata, Rome, Italy

Jose L. Martinez-Lastra FAST-Lab, Faculty of Engineering and Natural Sciences, Tampere University, Tampere, Finland

Rizwan Matloob Department of Food Science, The University of Otago, Dunedin, New Zealand

Jyoti P. Mishra Food Quality and Sensory Science Department, Teagasc, Ashtown, Dublin, Ireland

Wael M. Mohammed FAST-Lab, Faculty of Engineering and Natural Sciences, Tampere University, Tampere, Finland

Cecilia Occhiuzzi Department of Civil Engineering and Computer Science Engineering, The University of Rome Tor Vergata, Rome, Italy

Gulay Ozkan Department of Food Engineering, Faculty of Chemical and Metallurgical Engineering, Istanbul Technical University, Maslak, Istanbul, Turkey

Fatih Özoğul Department of Seafood Processing Technology, Faculty of Fisheries, Cukurova University, Balcali, Adana, Turkey

Carlos Parra-López Department of Agrifood System Economics, Institute of Agricultural and Fisheries Research & Training (IFAPA), Granada, Spain

Mona Elisabeth Pedersen Nofima AS, Raw Materials and Process Optimization, Ås, Norway

Dolores Pérez-Marín Department of Animal Production, Faculty of Agriculture and Forestry Engineering, University of Cordoba, Campus Rabanales, Córdoba, Spain

Benedetta Pini Department of Engineering and Architecture, University of Parma, Parma, Italy

Ahmed Rady Food Quality and Sensory Science Department, Teagasc, Ashtown, Dublin, Ireland

Dele Raheem Artic Centre, University of Lapland, Rovaniemi, Finland

Sissel Beate Rønning Nofima AS, Raw Materials and Process Optimization, Ås, Norway

Eden Tongson Digital Agriculture, Food and Wine Sciences Group, School of Agriculture, Food and Ecosystem Sciences, Faculty of Science, The University of Melbourne, Parkville, VIC, Australia

Horst Treiblmaier School of International Management, Modul University Vienna, Vienna, Austria

Frank Trollman Trauma and Orthopaedics Department, University Hospitals of Derby and Burton, Derby, United Kingdom

Hana Trollman Department of Work, Employment, Management and Organisations, School of Business, University of Leicester, Brookfield, Leicester, United Kingdom

Sebahattin Serhat Turgut Department of Food Engineering, Faculty of Engineering and Natural Sciences, Suleyman Demirel University, Cunur, Isparta, Turkey; Research Group for Food Production Engineering, National Food Institute, Technical University of Denmark, Kgs Lyngby, Denmark

Lincoln C. Wood Management Department, The University of Otago, Dunedin, New Zealand; School of Management, Curtin University, Perth, Australia

Chapter 1

Industry 4.0 technologies: principles and applications in agriculture and the food industry

Abdo Hassoun[1] and Barbara Bigliardi[2]

[1]*Sustainable AgriFoodtech Innovation & Research (SAFIR), Arras, France,* [2]*Department of Engineering and Architecture, University of Parma, Parma, Italy*

1.1 Introduction

Over the last few years, several food sectors have been hit hard by a series of global challenges, such as pandemics (especially COVID-19), political conflicts (particularly the war in Ukraine), and climate shocks (such as floods or droughts). These extraordinary dilemmas have been added to those already facing food systems including climate change, increase in global population, plastic pollution, destruction of biodiversity, overuse of land and water resources, and food waste and loss (Jagtap et al., 2022; Peydayesh et al., 2023; Hamed et al., 2022).

Therefore food insecurity has been on the rise in recent years, but fortunately, so have been solutions and opportunities in agriculture and the food industry. For example, a wide range of green eco-friendly technologies is being developed to optimize food processing and create innovative pathways for the extraction and valorization of food waste and byproducts (Caldeira et al., 2020; Ozogul et al., 2021). Innovative technologies are transforming the agri-food sector, creating a new era of agriculture and food industry, which is characterized by enhanced productivity, increased efficiency, and improved quality and food safety (Ashraf et al., 2021; Ayed et al., 2022).

Development of innovative solutions has been accelerated with the advent of the fourth industrial revolution (labeled Industry 4.0) technologies. Although there is no common definition of Industry 4.0, it can be seen as the marriage of digital, physical, and biological innovations (Chapman et al., 2021; Hassoun, Aït-kaddour, et al., 2022). As for its definition, there is no general consensus on which technologies are being considered under the concept of Industry 4.0. However, in agriculture and the food sector, many references report artificial intelligence (AI), big data, smart sensors and the Internet of Things (IoT), robotics, blockchain, 3D printing, digital twins, cloud computing, among others, as being the main enablers of Industry 4.0 (Lezoche et al., 2020; Liu et al., 2021; Hassoun, Bekhit, et al., 2022; Sharma et al., 2022).

It should be underlined that the application of Industry 4.0 technologies in agriculture is often referred to as "agriculture 4.0" or "precision agriculture," or even "smart farming," whereas implementing Industry 4.0 innovations in the food industry is frequently labeled "food 4.0" or "smart food factory." The application of Industry 4.0 technologies in both agriculture and the food industry can be labeled "Agri-Food Industry 4.0" (Fig. 1.1).

Growing evidence shows that Industry 4.0 technologies have a huge potential to increase automation and enhance digitalization, save time, and decrease production costs. Consequently, increased adoption of Industry 4.0 technologies could offer tremendous opportunities to facilitate the achievement of food sustainability and the implementation of circular economy principles (Dantas et al., 2021; Hassoun, Prieto, et al., 2022; Javaid, Haleem, Singh, Suman & Gonzalez, 2022). For example, Bai et al. (2020) demonstrated that Industry 4.0 technologies have the potential to benefit all 17 sustainable development goals (SDGs), whereas Marvin et al. (2022) especially highlighted the role of digitalization and AI as key enablers of the transition toward sustainable food systems. Quiroz-Flores et al. (2023) showed the significant contribution of blockchain and IoT to enhancing transparency, traceability, process optimization, and food waste reduction.

Food Industry 4.0. DOI: https://doi.org/10.1016/B978-0-443-15516-1.00001-3

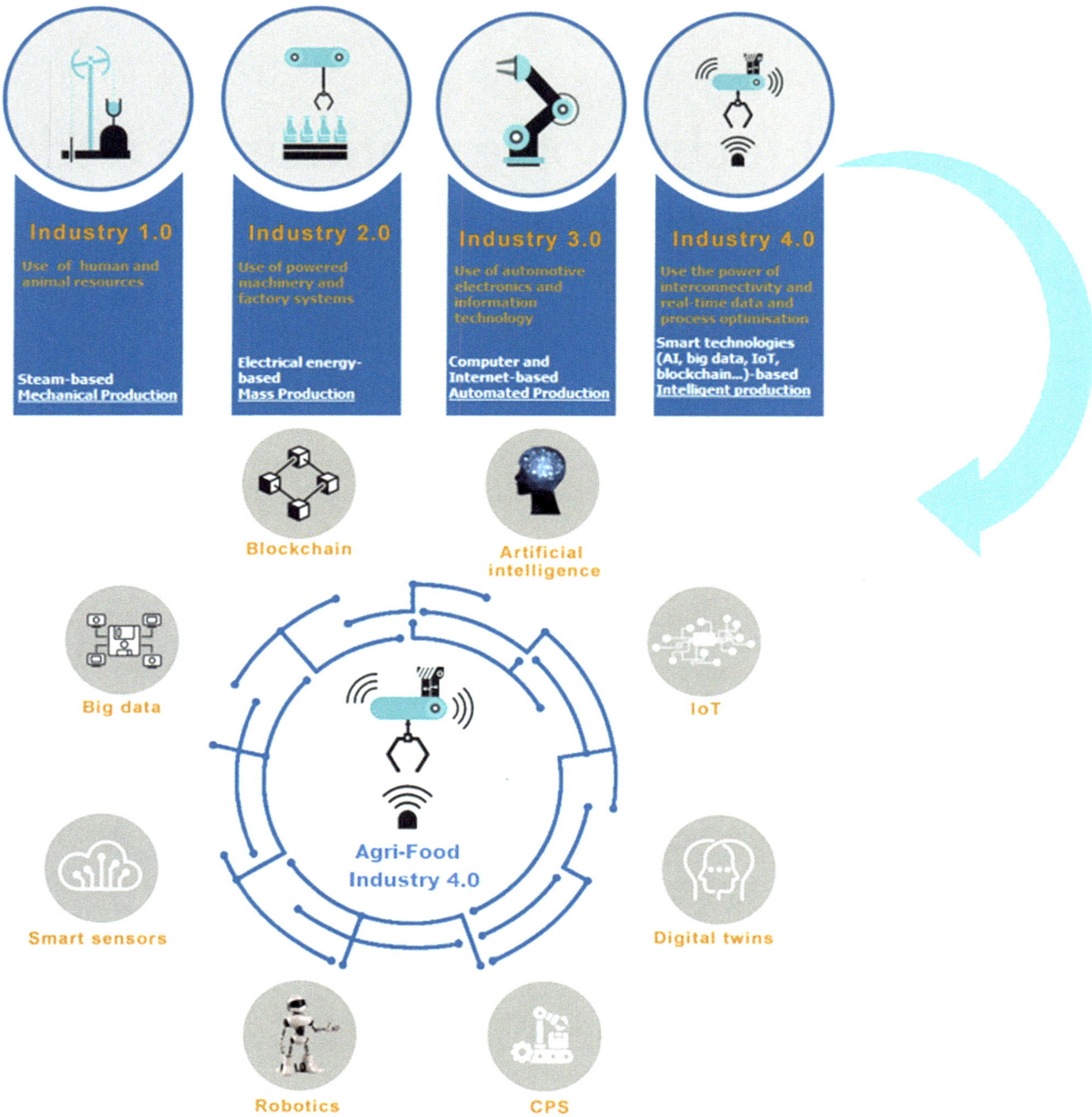

FIGURE 1.1 Overview of the industrial revolutions, displaying Agri-Food Industry 4.0 main technologies.

This chapter will first give a general overview of the four industrial revolutions, focusing on Industry 4.0 and its enabling technologies. Then, relevant examples on applications of Industry 4.0 innovations (such as AI, big data, IoT, and smart sensors) in agriculture (precision agriculture/smart farming) and the food industry (food 4.0/smart food factory) will be provided.

1.2 The fourth industrial revolution

Four industrial revolutions have taken place in the last two centuries, providing significant transformations to several production and consumption sectors, including agriculture and the food industry (Figs. 1.1 and 1.2).

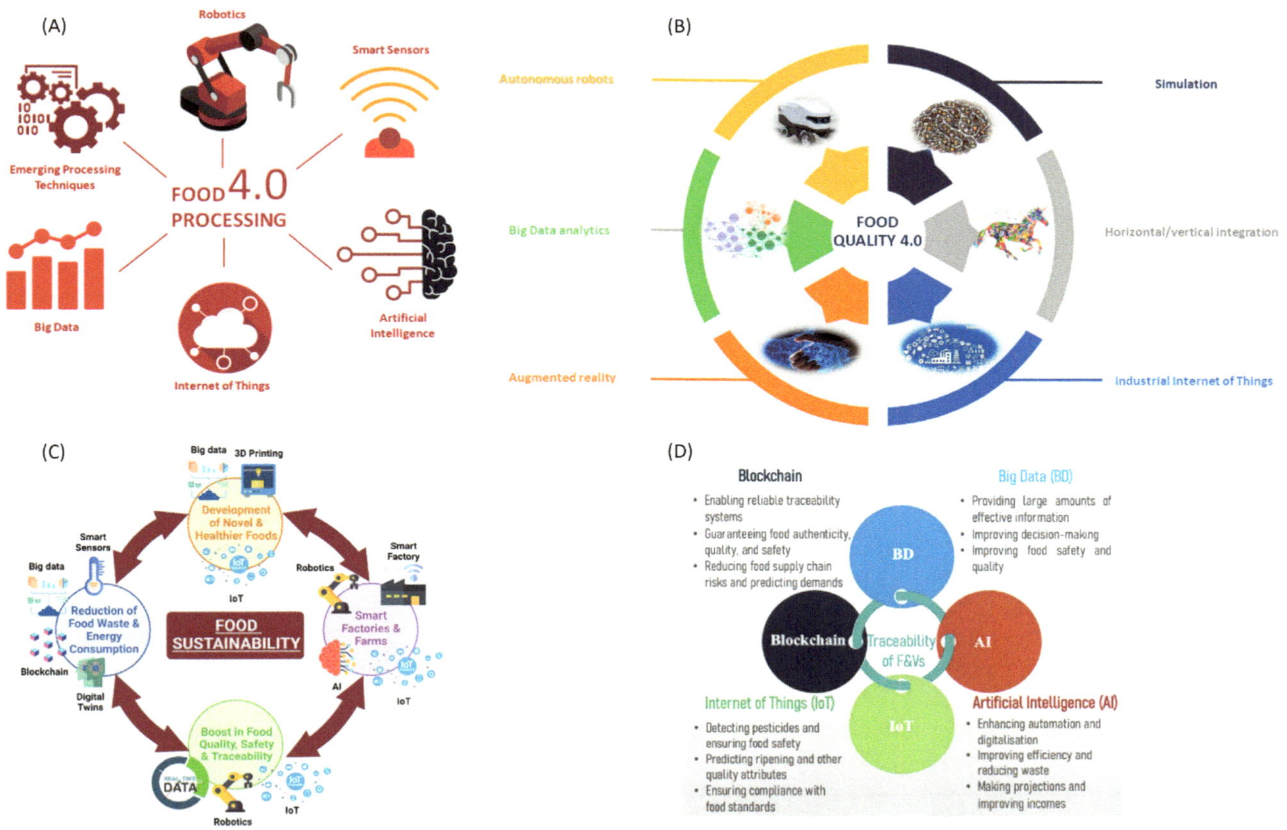

FIGURE 1.2 Main Industry 4.0 enablers in different sectors: food processing (A), food quality (B), food sustainability (C), and food traceability (D).

The first industrial revolution (Industry 1.0) started around the end of the 18th century and involved the use of coal, water, and steam, bringing with it the innovation of the steam engine that allowed the transition from manual artisanal production to mechanical production systems. However, the production was dominated by the use of human and animal resources, with textile and steel being the dominant industries (Bigliardi, 2021a, 2021b; Yang et al., 2021). The second industrial revolution (Industry 2.0) came about with the arrival of electricity, which enabled mass production around the late 1800s. This period was characterized by the invention of the internal combustion engine and the appearance of the first assembly line (Liu et al., 2021; Hassoun, Siddiqui, et al., 2022). The third industrial revolution (Industry 3.0) started around the 1970s and was driven by the development of information technology and electronics, which marked the beginning of the development of automated production systems. Industry 3.0 has led to the emergence of the first computers and digital systems, which enabled new ways of processing and sharing information (Bigliardi, 2021a, 2021b; Hassoun, Siddiqui, Smaoui, Ucak, Arshad, Bhat, et al., 2022).

The ongoing fourth industrial revolution (Industry 4.0), which started in an economic debate in Germany in 2011 and earned enormous research and industrial interests since 2015, refers to the alliance of several innovative technologies, linked to digital, physical, and biological disciplines. Industry 4.0 is based on harnessing the power of interconnectivity and real-time data and process optimization, allowing flexible intelligent production and manufacturing systems (Jagtap, Bader, Garcia-Garcia, Trollman, Fadiji, Salonitis 2021/2021; Karnik et al., 2022; Hassoun, Aït-kaddour, et al., 2022; Sharma et al., 2022).

Industry 4.0 enabling technologies can vary as a function of application fields. For example, our recent literature reviews showed that AI, big data, smart sensors, robotics, and IoT are widely used in food processing (Fig. 1.2A) (Hassoun, Jagtap, Trollman, et al., 2023), while big data, AI, robotics, augmented reality, and IoT are commonly used in food quality determination (Fig. 1.2B) (Hassoun, Jagtap, Garcia-Garcia, et al., 2023). A wide range of technological innovations (e.g., big data, 3D printing, robotics, IoT, blockchain, and digital twins) have been widely applied to study various topics related to food sustainability (Fig. 1.2C), while blockchain, big data,

AI, and IoT were the main Industry 4.0 enablers for food traceability (Fig. 1.2D) (Hassoun, Abdullah, et al., 2022; Hassoun, Kamiloglu, et al., 2023). Moreover, these technologies were found to be potential candidates for applications in the meat, labeled "Meat 4.0" (Echegaray et al., 2022) and the dairy; called "Dairy 4.0" (Hassoun, Garcia-Garcia, et al., 2023) sectors.

1.3 Precision agriculture and smart farming: application of Industry 4.0 technologies in agriculture

Agriculture plays a vital role in the economies of several countries worldwide, serving as a primary revenue source. Millions of individuals rely on agriculture to provide food, employment, and essential resources for human survival (United Nations, 2022). This sector holds a central position, influencing human lives and interconnecting industries across nations. Addressing the demands of the expanding world populace requires a substantial upsurge in annual cereal production by 3 billion tonnes and an over 200% surge in meat production by 2050 (Ayoub Shaikh et al., 2022; Trendov et al., 2019). This undertaking necessitates the expansion of crop yields and farming infrastructure, coupled with advanced technological approaches. The emergence of information technology introduces both challenges and prospects for the agricultural domain, particularly in achieving global food security (Alam et al., 2023; Kayikci et al., 2022).

The inception of precision farming dates back to the early 1980s when its primary objective was to enhance fertilizer application through the dynamic adjustment of rates and compositions tailored to specific sections within fields. This pioneering approach encompasses an intricate interplay of machinery, sensors, GPS technology, software applications, and remote sensing mechanisms. As articulated by Berger et al. (2020), the adoption of precision farming carries the potential to uplift the livelihoods of farmers while concurrently fostering the sustainability of food production systems. The benefits obtained from implementing precision farming are numerous, among which the increase in farmer profits, the increase in crop yields and animal performance, costs and labor reduction and process inputs optimization, the increase in occupational safety, and the reduction in the environmental impact of agricultural practices (Berger et al., 2020; Benyam et al., 2021).

Several contemporary technologies and breakthroughs have emerged in recent years under the umbrella of "Agriculture 4.0," or smart agriculture, with the potential to amplify crop production while concurrently minimizing water and energy consumption (Chandio et al., 2021; Rejeb, Rejeb, et al., 2022). The progression of Industry 4.0 principles has been extended to the agricultural sector, leading to the advent of "Farming 4.0" concept. Notably, various agricultural domains are experiencing an exceptionally rapid pace of technological evolution during the current Industrial Era 4.0 (Rose & Chilvers, 2018; da Silveira et al., 2021).

Precision agriculture, hailed as a novel global approach, seeks to amplify output, curtail labor duration, and ensure optimal administration of fertilization and irrigation protocols (Lu et al., 2022). The crux of this approach lies in the judicious harnessing of extensive datasets and insights to refine the allocation of agricultural resources, enhance yields, and elevate crop quality. Precision agriculture harnesses a rich tapestry of data from diverse sources, orchestrating an enhancement of crop yields while bolstering the efficiency and cost-effectiveness of pivotal cultivation strategies, spanning fertilizer input optimization, irrigation management, and targeted pesticide application (Ingram et al., 2022).

Specifically, as stated for example by Rijswijk et al. (2021), smart farming or smart agriculture encompasses the utilization of AI and IoT in the realm of cyber-physical farm management. This technological paradigm directly addresses a multitude of challenges associated with crop production by facilitating real-time monitoring of dynamic variables such as climate fluctuations, soil attributes, and moisture content. Smart agriculture is grounded on digital automation, data collection, data transmission, decision-making, data processing, and data analysis (Matei et al., 2017; Ayoub Shaikh et al., 2022; Javaid, Haleem, Singh, Suman, 2022, 2022). AI is progressively finding a broader array of applications within the realm of agriculture, driven by its ongoing advancement. Concurrently, the IoT, big data, and adoption of other Industry 4.0 technologies are fostering the creation of novel technologies and concepts (Ahmadzadeh et al., 2023; Sharma et al., 2021) (S.). Additionally, the concept of precision agriculture involves the use of cutting-edge sensor technology and sophisticated analytical tools to elevate both crop yields and managerial decision-making processes. Particularly, the use of smart sensors for remote sensing has found numerous applications in agriculture (Ullo & Sinha, 2021; Sishodia et al., 2020).

Precision agriculture and smart farming imply the use of other Industry 4.0 technologies such as the application of drones (Rejeb, Abdollahi, et al., 2022), blockchain technology (Torky & Hassanein, 2020), and robotics and other autonomous systems (Pearson et al., 2022; Oliveira et al., 2021).

It should be stressed that Industry 4.0 technologies are also increasingly used in animal production systems, leading to the emergence of the concept of precision livestock farming (Bahlo et al., 2019; Fuentes et al., 2022). For example, smart sensors and other advanced technologies are being increasingly used to track the animals and monitor their health, production efficiency, and welfare, providing opportunities to improve farm management and enhance productivity (Finger, 2023; Qiao et al., 2021; Monteiro et al., 2021).

Examples of the major available technologies adopted for the development of agriculture are summarized in Fig. 1.3, demonstrating the main benefits that can be obtained and the main references.

The advent of the "digital agricultural revolution" is thus poised to reshape the landscape of agriculture, yielding enhancements in efficiency, sustainability, inclusivity, and transparency. Nonetheless, the transformative potential of these advanced technologies can only be fully harnessed when integrated at a larger scale within the agricultural sector. Yet, within this promising horizon, lingering security challenges in agriculture demand resolution. The main intricacies stem from interoperability, heterogeneity, the management of vast data volumes, and the processing of immense datasets. The paramount challenge within the realm of Agriculture 4.0 is to ensure the accurate generation, seamless transfer, and secure processing of data, all while fortifying defenses against cyber threats (Duncan et al., 2019; Qazi et al., 2022; Zubaydi et al., 2023). Vital data-centric technologies, including analytics and intelligent systems, hinge on meticulous data integrity management. The amalgamation of diverse resources often gives rise to security predicaments, such as privacy erosion. In this intricate interplay of the IoT, cellular networks, and wireless technologies, lies the potential to surmount existing and emerging agricultural challenges. This necessitates a proactive approach to address facets like real-time data integrity, chronological precision, and device security, while also attending to security dimensions including encryption, data accuracy, and uninterrupted accessibility (Hassija et al., 2019; Abbasi et al., 2022; Rejeb, Rejeb, et al., 2022).

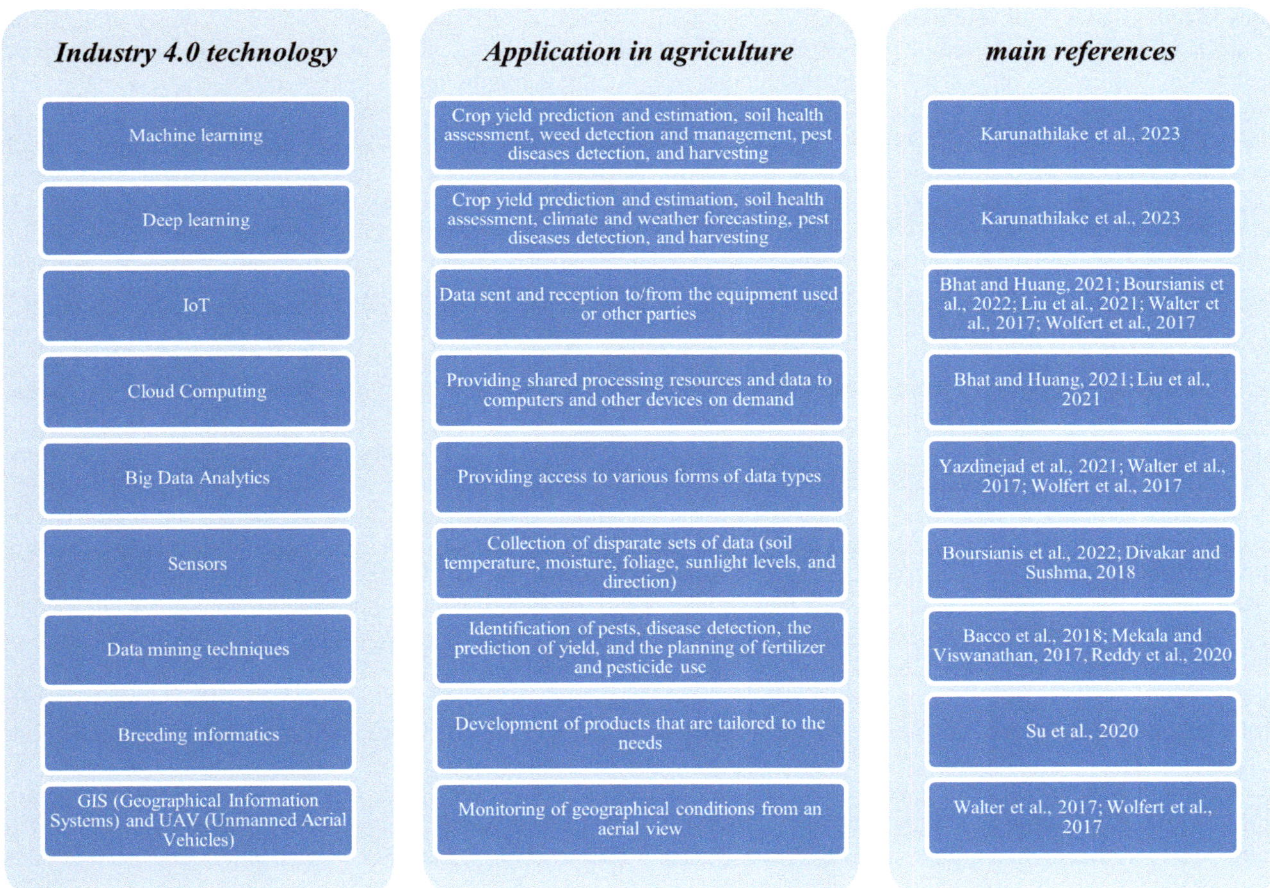

FIGURE 1.3 Examples of application of Industry 4.0 technologies in agriculture.

1.4 Smart factory: application of Industry 4.0 technologies in the food industry

The food industry encompasses entities engaged in the creation of food products, as well as those involved in the distribution and provision of food, beverages, and dietary supplements (Bigliardi & Galati, 2013; Luque et al., 2017). In essence, it includes the entire spectrum, spanning from the initial transformation of raw materials to the production of intermediary and final goods.

The pervasive Industry 4.0 revolution, which has left an indelible imprint on various sectors, holds a particularly notable sway over the food industry, transforming traditional food manufacturing into smart food factories. Harnessing technological resources becomes paramount for the food industry to harmonize nutritional requisites with human health considerations, food safety mandates, and prevailing regulations. Technology has traditionally found application in diverse stages of food and beverage fabrication, encompassing transportation, processing, packaging, and final product storage (Hassoun, Boukid, et al., 2022; Telukdarie et al., 2023).

Recent times have witnessed an incessant influx of novel food trends on the global stage, disseminated to consumers through the medium of social media. This dynamic landscape exerts additional pressure on the food industry, compelling it to adroitly embrace and leverage the aforementioned technological resources. The existing body of literature comprises various studies that underscore the profound influence of Industry 4.0 on the food industry, elucidating both its technical and economic ramifications. For example, Demir and Dincer (2020) spotlight digitalization, involving several facets like big data, cloud computing, virtualization, and cybersecurity, along with interaction dynamics like the IoT and cyber-physical systems, and the envisioned future of factories featuring additive manufacturing and automation as the primary enablers of Industry 4.0 in the food domain. Moreover, concerning the technical and economic merits of Industry 4.0, scholars such as Luque et al. (2017) emphasize a spectrum that includes technological advancement, operational flexibility, product personalization, and a discernible enhancement in consumer satisfaction.

Within the realm of food, the Industry 4.0 paradigm is commonly termed "Food 4.0," signifying a profound overhaul and revolutionary innovation in food manufacturing methodologies. This ongoing digital transformation encompasses diverse dimensions, including novel production techniques as exemplified by research studies conducted by various scholars such as Saucedo-Martínez et al. (2018), Kiel et al. (2017), and Yin et al. (2017). An integral facet involves the integration of external stakeholders, spanning enterprises, suppliers, and consumers, as demonstrated by various studies such as those by Burritt and Christ (2016), Preuveneers and Ilie-Zudor (2017), and Xu et al. (2018). This transformation further delves into lean production principles, as explored in works like Kolberg et al. (2017), while also encompassing the requalification of the workforce to realize multivendor and highly modular production systems, as highlighted by Weyer et al. (2015), along with pioneering managerial practices, as discussed by Hozdić (2015), among others.

In the extant literature, several papers delve into the application of the previously enumerated enabling technologies within the context of the food industry. Specifically, research concerning the Industry 4.0 domain has been carried out across various strata within the food sector, ranging from the overarching sector level to the intricate dynamics of supply chains, specific corporate entities, and distinct business domains. Notably, the examination of the available literature reveals an acute focus on the intersection of Industry 4.0 principles with sustainability concerns, reflecting a conscientious effort to align technological advancements with environmental and ethical imperatives.

Advanced robotic systems, for instance, were initially relegated to specific niches such as palletizing finished products, yet have now expanded to encompass diverse roles spanning from harvesting, cutting, processing, and packaging of food items, along with tasks like meat processing and automated quality assessment of baked food products. The merits of robotics within the food domain extend to facets like kinematics, dynamics, control, hygiene, productivity, and worker safety (Masey et al., 2010; Khan et al., 2018; Wang et al., 2022).

Automated control systems wield pivotal significance across the food industry's production continuum. For example, Cotrim et al. (2020) elaborated on the deployment of a Computational Vision System (CVS) based on Convolutional Neural Networks (CNNs) as a strategic technique in color analysis. This system's prowess was showcased in its capacity to categorize the degree of browning in bread crust during baking. Similarly, Bowler et al. (2020) expounded on the role of mixing in materials amalgamation, heat distribution enhancement in food products, and effecting material structure modifications. In these contexts, sensors become essential, particularly in critical processes like mixing, enabling real-time data acquisition. Various techniques, including ultrasound, are employed for real-time monitoring of industrial mixing processes. Ultrasonic sensors, characterized by their cost-effectiveness and applicability to opaque systems, utilize low-power, high-frequency sound waves to profile materials without altering their structures, relying on parameters like sound speed, wave attenuation, and material acoustic impedance (Hauptmann et al., 2002). Recently, various types of advanced smart sensors have been investigated for many applications in the food industry, such as reduction of food waste (Zhu et al., 2022), monitoring food quality and safety (Pereira et al., 2021; Kuswandi et al., 2022), and estimating food freshness (Das & Mishra, 2023), among others.

The IoT facilitates the collection and utilization of data from diverse sources, ushering in enhanced digitization and process automation. This technological advancement not only reduces production costs within the food industry but also augments production efficiency (Hong et al., 2014). Looking ahead, key tenets of the Food 4.0 trajectory encompass additive manufacturing via 3D printing, advanced robotic systems, and automated control systems, as evidenced by Lipton et al. (2015). 3D printing, synonymous with additive manufacturing, entails the layer-by-layer construction of food items. The technique encompasses diverse categories like bio-driven (meat-based), bottom-up (unconventional sources like algae and insects), and hybrid formulations blending insect derivatives with other printable foods like cheese (Tracey et al., 2022; Eswaran et al., 2023). Despite the manifold benefits, the widespread adoption of 3D food printing remains tempered by machinery costs and the relatively slow nature of the process when contrasted with incumbent techniques. Robotics and autonomous systems are rapidly evolving as promising technologies that hold the potential to enhance sustainable development while also bolstering the quality, productivity, and efficiency of the food supply chain (Hassoun, Prieto, et al., 2022). Within the realm of the food industry, intelligent sensors are experiencing growing adoption across various production processes, enabling intelligent control, real-time monitoring, and optimization of diverse manufacturing tasks. This adoption is accompanied by advancements in traceability and food quality (McVey et al., 2021; Hassoun, Prieto, et al., 2022). For instance, spectroscopy-based optical sensors are progressively employed to detect electromagnetic radiation frequency shifts. This application aids in real-time monitoring of food quality, ensuring authenticity, and overseeing food processing (Hassoun, Boukid, et al., 2022; Hassoun, Prieto, et al., 2022; Krause et al., 2021).

Incorporating Big Data yields tangible dividends, encompassing secure and accurate information storage, prediction of machinery malfunctions, streamlined maintenance management, and overall enhancement of production processes. Integral to this realm is cloud computing, which facilitates the storage of extensive datasets without relying on local memory or constant internet connectivity, enabling swift and secure information sharing with stakeholders (Chisenga et al., 2020; Misra et al., 2022; Rejeb, Keogh, et al., 2022).

Another important Industry 4.0 technology to be applied in the food industry is blockchain. Conventional food supply chains often grapple with deficiencies in traceability and the ability to track products, resulting in obscured labeling transparency, sluggish product innovation cycles, and intricate logistical challenges. Addressing these concerns within the food supply chain, blockchain technology emerges as a potential solution. Rooted in the principles of Industry 4.0, blockchain presents itself as a promising innovation characterized by its digital, decentralized, and distributed ledgers, maintained by a network of interconnected computers. This framework holds the potential to foster trust and transparency across the agri-food value chain. By enhancing traceability throughout the supply chain, blockchain seamlessly connects and tracks data from producers to consumers, enabling swifter and more accurate product recalls. Consequently, this technology mitigates risks, resulting in an elevated standard of food quality. Moreover, heightened traceability empowers the monitoring and authentication of claims such as "sustainable," "organic," and "halal" (Kayikci et al., 2022; Javaid, Haleem, Singh, Suman & Gonzalez, 2022). Remarkably, blockchain has demonstrated its efficacy in curbing global food losses within complex supply chains (Kayikci et al., 2022). Beyond this, blockchain serves as an integrated traceability tool capable of mitigating the disruption risk posed by pandemics like COVID-19 within the food system. For instance, in tandem with other emerging technologies such as radio frequency identification (RFID), blockchain has proven invaluable in maintaining the continuity of the food cold chain during the ongoing coronavirus crisis (Masudin et al., 2021). By ensuring a secure environment for real-time data aggregation and access, blockchain stands to revolutionize data management.

The transformative wave heralded by Industry 4.0 addresses pressing concerns such as labor scarcity and diversified demand. This transformation is exemplified by Matsumoto et al. (2020) study of the Japanese food industry, necessitating increased productivity and adaptability in the face of expanding markets and varied product offerings. Their case study of a dairy plant underscores the efficacy of a five-level horizontal model integrating automation technologies, manifesting improved efficiency and production standards compared to the preceding approach through cost and labor-hour comparisons.

Moreover, Industry 4.0 extends its influence beyond the production realm to ancillary activities linked to the food industry's value chain. These encompass applications like simulation-based decision-making for production line equilibrium and workstation design. Abd Rahman et al. (2020) explored the utility of production process data analysis as a decision support mechanism for waste reduction, outlining a model-driven decision support system (MD-DSS) merging simulation and data communication technologies for production process enhancement. Similarly, in the context of workplace design, the transition toward smaller batch production necessitates the evolution of flexible processes wherein human and automation systems collaboratively ensure safety, efficiency, and productivity. Papetti et al. (2019) emphasize the need to reconsider the human workstation's design amidst automated advancements, advocating for methods that evaluate worker well-being and shape operator-centric workspaces.

Given the myriad advantages and diverse applications that Industry 4.0 technological innovation holds for the food industry, investments are warranted to infuse these technologies more robustly into the sector. Concurrently, fostering a skilled workforce well-versed in these innovations is paramount to extract optimal value from this technological transformation.

1.5 Conclusion

Increased concerns about global food security have accelerated the need for next-generation innovations and technological advancements in both farms and food factories. This chapter has covered the main principles and applications of Industry 4.0 in agriculture and the food industry. According to the literature examined, there is a high demand for digitizing and automating various food production and manufacturing operations in different food sectors. Growing research shows that several Industry 4.0 technologies, such as AI, big data, the IoT, smart sensors, robotics, and blockchain have a great potential to enhance digitalization and automation, increase productivity due to efficient crop and livestock monitoring, reduce costs, improve food quality, safety and traceability, and foster sustainability in many food sectors. These transformative aspects have recently led to the emergence of innovative concepts including Agriculture 4.0, precision agriculture, smart farming, digital agriculture, and smart food factory.

Harnessing the full potential of emerging technologies means analyzing and addressing current challenges related to processing high amount of data, data security and privacy issues, high investment cost, and insufficient connectivity infrastructure, especially in less-developed regions where little or no advanced connectivity or equipment is available. However, collaboration between different actors in the agri-food supply chain can help overcome barriers that hamper the implementation of advanced technologies in various agriculture and food-related fields. Moreover, the application of metaverse, ChatGPT, and other emerging technologies is expected to greatly impact agriculture and the food industry in the near future.

References

Abbasi, R., Martinez, P., & Ahmad, R. (2022). The digitization of agricultural industry – A systematic literature review on agriculture 4.0. *Smart Agricultural Technology*, *2*, 100042. Available from https://doi.org/10.1016/J.ATECH.2022.100042.

Abd Rahman, M. S., Mohamad, E., & Abdul Rahman, A. A. (2020). Enhancement of overall equipment effectiveness (OEE) data by using simulation as decision making tools for line balancing. *Indonesian Journal of Electrical Engineering and Computer Science*, *18*(2), 1040–1047. Available from https://doi.org/10.11591/ijeecs.v18.i2.pp1040-1047.

Ahmadzadeh, S., Ajmal, T., Ramanathan, R., & Duan, Y. (2023). A comprehensive review on food waste reduction based on IoT and big data technologies. *Sustainability*, *15*(4), 3482. Available from https://doi.org/10.3390/su15043482.

Alam, Md. F. B., Tushar, S. , R., Zaman, S. Md, Gonzalez, E. D. R. Santibanez, Bari, A. B. M. M., & Karmaker, C. L. (2023). Analysis of the drivers of agriculture 4.0 implementation in the emerging economies: Implications towards sustainability and food security. *Green Technologies and Sustainability*, *1*(2). Available from https://doi.org/10.1016/j.grets.2023.100021.

Ashraf, S. A., Siddiqui, A. J., Elkhalifa, A. E. O., Khan, M. I., Patel, M., Alreshidi, M., Moin, A., Singh, R., Snoussi, M., & Adnan, M. (2021). Innovations in nanoscience for the sustainable development of food and agriculture with implications on health and environment. *Science of the Total Environment*, *768*, 144990. Available from https://doi.org/10.1016/j.scitotenv.2021.144990.

Ayed, R. B., Hanana, M., Ercisli, S., Karunakaran, R., Rebai, A., & Moreau, F. (2022). Integration of innovative technologies in the agri-food sector: The fundamentals and practical case of DNA-based traceability of olives from fruit to oil. *Plants*, *11*(9). Available from https://doi.org/10.3390/plants11091230, https://www.mdpi.com/2223-7747/11/9/1230/pdf?version = 1651481521.

Ayoub Shaikh, T., Rasool, T., & Rasheed Lone, F. (2022). Towards leveraging the role of machine learning and artificial intelligence in precision agriculture and smart farming. *Computers and Electronics in Agriculture*, *198*(May), 107119. Available from https://doi.org/10.1016/j.compag.2022.107119.

Bahlo, C., Dahlhaus, P., Thompson, H., & Trotter, M. (2019). The role of interoperable data standards in precision livestock farming in extensive livestock systems: A review. *Computers and Electronics in Agriculture*, *156*, 459–466. Available from https://doi.org/10.1016/j.compag.2018.12.007.

Bai, C., Dallasega, P., Orzes, G., & Sarkis, J. (2020). Industry 4.0 technologies assessment: A sustainability perspective. *International Journal of Production Economics*, *229*, 107776. Available from https://doi.org/10.1016/J.IJPE.2020.107776.

Benyam, A. (Addis), Soma, T., & Fraser, E. (2021). Digital agricultural technologies for food loss and waste prevention and reduction: Global trends, adoption opportunities and barriers. *Journal of Cleaner Production*, *323*, 129099. Available from https://doi.org/10.1016/j.jclepro.2021.129099.

Berger, K. ,, Verrelst, J., Féret, J.-B., Wang, Z., Wocher, M., Strathmann, M., Danner, M., Mauser, W., & Hank, T. (2020). Crop nitrogen monitoring: Recent progress and principal developments in the context of imaging spectroscopy missions. *Remote Sensing of Environment*, *242*(March), 111758. Available from https://doi.org/10.1016/j.rse.2020.111758.

Bigliardi, B. (2021a). Industry 4.0 applied to food. In C. M. Galanakis (Ed.), *Sustainable Food Processing and Engineering Challenges* (pp. 1–23). Academic Press.

Bigliardi, B. (2021b). *Industry 4.0 applied to food sustainable food processing and engineering challenges* (pp. 1–23). Italy: Elsevier. Available from https://www.elsevier.com/books/sustainable-food-processing-and-engineering-challenges/galanakis/978-0-12-822714-5, 10.1016/B978-0-12-822714-5.00001-2.

Bigliardi, B., & Galati, F. (2013). Innovation trends in the food industry: The case of functional foods. *Trends in Food Science & Technology*, *31*(2), 118–129. Available from https://doi.org/10.1016/J.TIFS.2013.03.006.

Bowler, A. L., Bakalis, S., & Watson, N. J. (2020). Monitoring mixing processes using ultrasonic sensors and machine learning. *Sensors*, *20*(7), 1813. Available from https://doi.org/10.3390/S20071813.

Burritt, R., & Christ, K. (2016). Industry 4.0 and environmental accounting: A new revolution? *Asian Journal of Sustainability and Social Responsibility*, *1*(1), 23–38. Available from https://doi.org/10.1186/s41180-016-0007-y.

Caldeira, C., Vlysidis, A., Fiore, G., De Laurentiis, V., Vignali, G., & Sala, S. (2020). Sustainability of food waste biorefinery: A review on valorisation pathways, techno-economic constraints, and environmental assessment. *Bioresource Technology*, *312*. Available from https://doi.org/10.1016/j.biortech.2020.123575.

Chandio, A. A., Jiang, Y., Rehman, A., Twumasi, M. A., Pathan, A. G., & Mohsin, M. (2021). Determinants of demand for credit by smallholder farmers': A farm level analysis based on survey in Sindh, Pakistan. *Journal of Asian Business and Economic Studies*, *28*(3), 225–240. Available from https://doi.org/10.1108/JABES-01-2020-0004.

Chapman, J., Power, A., Netzel, M. E., Sultanbawa, Y., Smyth, H. E., Truong, V. K., & Cozzolino, D. (2021). Challenges and opportunities of the fourth revolution: A brief insight into the future of food. *Critical Reviews in Food Science and Nutrition*, *62*(10), 2845–2853. Available from https://doi.org/10.1080/10408398.2020.1863328.

Chisenga, S. M., Tolesa, G. N., Workneh, T. S., & Owusu-Kwarteng, J. (2020). Biodegradable food packaging materials and prospects of the fourth industrial revolution for tomato fruit and product handling. *International Journal of Food Science*, *2020*, 1–17. Available from https://doi.org/10.1155/2020/8879101.

Cotrim, W., da, S., Minim, V. P. R., Felix, L. B., & Minim, L. A. (2020). Short convolutional neural networks applied to the recognition of the browning stages of bread crust. *Journal of Food Engineering*, *277*(January), 109916. Available from https://doi.org/10.1016/j.jfoodeng.2020.109916.

Dantas, T. E. T., de-Souza, E. D., Destro, I. R., Hammes, G., Rodriguez, C. M. T., & Soares, S. R. (2021). How the combination of circular economy and industry 4.0 can contribute towards achieving the sustainable development goals. *Sustainable Production and Consumption*, *26*, 213–227. Available from https://doi.org/10.1016/j.spc.2020.10.005.

Das, J., & Mishra, H. N. (2023). A comprehensive review of the spoilage of shrimp and advances in various indicators/sensors for shrimp spoilage monitoring. *Food Research International*, *173*. Available from https://doi.org/10.1016/j.foodres.2023.113270.

Demir, Y., & Dincer, F. I. (2020). The effects of industry 4.0 on the food and beverage industry. *Journal of Tourismology*, *6*(1), 133–145. Available from https://doi.org/10.26650/jot.2020.6.1.0006.

Duncan, S. E., Reinhard, R., Williams, R. C., Ramsey, F., Thomason, W., Lee, K., Dudek, N., Mostaghimi, S., Colbert, E., & Murch, R. (2019). Cyberbiosecurity: A new perspective on protecting U.S. food and agricultural system. *Frontiers in Bioengineering and Biotechnology*, *7*, 63. Available from https://doi.org/10.3389/fbioe.2019.00063, https://www.frontiersin.org/articles/10.3389/fbioe.2019.00063/full.

Echegaray, N., Hassoun, A., Jagtap, S., Tetteh-Caesar, M., Kumar, M., Tomasevic, I., Goksen, G., & Manuel Lorenzo, J. (2022). Meat 4.0: Principles and applications of industry 4.0 technologies in the meat industry. *Applied Sciences*, *12*(14). Available from https://doi.org/10.3390/app12146986.

Eswaran, H., Devi, R., & Pungampalayam, R. (2023). Annals of 3D printed medicine perspective approaches of 3D printed stuffs for personalized nutrition : A comprehensive review. *Annals of 3D Printed Medicine*, *12*(July), 100125. Available from https://doi.org/10.1016/j.stlm.2023.100125.

Finger, R. (2023). Digital innovations for sustainable and resilient agricultural systems. *European Review of Agricultural Economics*, *50*(4), 1277–1309. Available from https://doi.org/10.1093/erae/jbad021.

Fuentes, S., Gonzalez Viejo, C., Tongson, E., & Dunshea, F. R. (2022). The livestock farming digital transformation: Implementation of new and emerging technologies using artificial intelligence. *Animal Health Research Reviews*, *23*(1), 59–71. Available from https://doi.org/10.1017/S1466252321000177, http://journals.cambridge.org/action/displayJournal?jid = ahr.

Hamed, I., Jakobsen, A. N., & Lerfall, J. (2022). Sustainable edible packaging systems based on active compounds from food processing byproducts: A review. *Comprehensive Reviews in Food Science and Food Safety*, *21*(1), 198–226. Available from https://doi.org/10.1111/1541-4337.12870.

Hassija, V., Chamola, V., Saxena, V., Jain, D., Goyal, P., & Sikdar, B. (2019). A survey on IoT security: Application areas, security threats, and solution architectures. *IEEE Access*, *7*, 82721–82743. Available from https://doi.org/10.1109/ACCESS.2019.2924045.

Hassoun, A., Prieto, M. A., Carpena, M., Bouzembrak, Y., Marvin, H. J. P., Pallar, N., Barba., Hassoun, A., Siddiqui, S. A., Smaoui, S., Ucak, İ., Arshad, N., Bhat, Z. F., Bhat, H. F., Carpena, M., Miguel, A., Aït-kaddour, A., Pereira, J. A. M., Zacometti, C., Tata, A., Ibrahim, S. A., Ozogul, F., Camara, J. S., Arshad, N., Bhat, Z. F., Bhat, H. F., & Ibrahim, S. A. (2022). Emerging technological advances in improving the safety of muscle foods: Framing in the context of the food revolution 4.0. *Food Research International*, 0–42. Available from https://doi.org/10.1080/87559129.2022.2149776.

Hassoun, A., Siddiqui, S. A., Smaoui, S., Ucak, İ., Arshad, R. N., Garcia-Oliveira, P., Prieto, M. A., Aït-Kaddour, A., Perestrelo, R., Câmara, J. S., & Bono, G. (2022). Seafood processing, preservation, and analytical techniques in the age of industry 4.0. *Applied Sciences*, *12*(3), 1703. Available from https://doi.org/10.3390/APP12031703.

Hassoun, A., Prieto, M. A., Carpena, M., Bouzembrak, Y., Marvin, H. J. P., Pallarés, N., Barba, F. J., Punia Bangar, S., Chaudhary, V., Ibrahim, S., & Bono, G. (2022). Exploring the role of green and Industry 4.0 technologies in achieving sustainable development goals in food sectors. *Food Research International*, *162*. Available from https://doi.org/10.1016/j.foodres.2022.112068.

Hassoun, A., Abdullah, N. A., Aït-kaddour, A., Beşir, A., Zannou, O., Önal, B., Aadil, R. M., Lorenzo, J. M., Khaneghah, A. M., Regenstein, J. M., Beşir, A., Zannou, O., Önal, B., Aadil, R. M., & Lorenzo, J. M. (2022). Food traceability 4.0 as part of the fourth industrial revolution: Key enabling technologies. *Critical Reviews in Food Science and Nutrition*, *0*(0), 1–17. Available from https://doi.org/10.1080/10408398.2022.2110033.

Hassoun, A., Aït-kaddour, A., Abu-mahfouz, A. M., Rathod, N. B., Bader, F., Barba, F. J., Cropotova, J., Galanakis, C. M., Jambrak, A. R., Lorenzo, M., Måge, I., Ozogul, F., Regenstein, J., Rathod, B., Bader, F., Barba, F. J., Biancolillo, A., Cropotova, J., Galanakis, M., & Regenstein, J. (2022). The fourth industrial revolution in the food industry—Part I: Industry 4.0 technologies. *Critical Reviews in Food Science and Nutrition*, 1–17. Available from https://doi.org/10.1080/10408398.2022.2034735.

Hassoun, A., Bekhit, A. E., Jambrak, A. R., Regenstein, J. M., Chemat, F., Morton, J. D., Gudjónsdóttir, M., Carpena, M., Prieto, A., Varela, P., Arshad, R. N., Aadil, R. M., Ueland, Ø., Hassoun, A., Bekhit, A. E., Jambrak, A. R., Regenstein, J. M., Chemat, F., & Morton, J. D. The, Ø. U. (2022). The fourth industrial revolution in the food industry—part II : Emerging food trends trends. *Critical Reviews in Food Science and Nutrition*, *0*(0), 1–31. Available from https://doi.org/10.1080/10408398.2022.2106472.

Hassoun, A., Boukid, F., Pasqualone, A., Bryant, C. J., Jagtap, S., Trollman, H., Cropotova, J., García, G., Parra-l, C., & Barba, F. J. (2022). Current research in food science emerging trends in the agri-food sector: Digitalisation and shift to plant-based diets. *Current Research in Food Science*, *5*, 2261–2269. Available from https://doi.org/10.1016/j.crfs.2022.11.010.

Hassoun, A., Jagtap, S., Garcia-Garcia, G., Trollman, H., Pateiro, M., Lorenzo, M., Trif, M., Vasile, A., Muhammad, R., Simat, V., Cropotova, J, & C., S. (2023). Food quality 4.0: From traditional approaches to digitalized automated analysis. *Journal of Food Engineering*, *337*, 111216. Available from https://doi.org/10.1016/j.jfoodeng.2022.111216.

Hassoun, A., Jagtap, S., Trollman, H., Garcia-Garcia, G., Alhaj, N., Goksen, G., Bader, F., Ozogul, F., Barba, F. J., Cropotova, J., Munekata, P. E. S., & Lorenzo, M. (2023). Food processing 4.0: Current and future developments spurred by the fourth industrial revolution. *Food Control*, *145*, 109507. Available from https://doi.org/10.1016/j.foodcont.2022.109507.

Hassoun, A., Kamiloglu, S., Garcia-Garcia, G., Parra-Lopez, C., Trollman, H., Jagtap, S., Aadil, R. M., & Esatbeyoglu, T. (2023). Implementation of relevant fourth industrial revolution innovations across the supply chain of fruits and vegetables: A short update on Traceability 4.0. *Food Chemistry*, *409*, 135303. Available from https://doi.org/10.1016/j.foodchem.2022.135303.

Hassoun, A., Garcia-Garcia, G., Trollman, H., Jagtap, S., Cropotova, J., Bhat, Z., Centobelli, P., Parra-l, C., & Aït-kaddour, A. (2023). Birth of dairy 4.0: Opportunities and challenges in adoption of fourth industrial revolution technologies in the production of milk and its derivatives. *Current Research in Food Science*, *7*(May), 100535. Available from https://doi.org/10.1016/j.crfs.2023.100535.

Hauptmann, P., Hoppe, N., & Püttmer, A. (2002). Application of ultrasonic sensors in the process industry. *Measurement Science and Technology*, *13* (8), R73. Available from https://doi.org/10.1088/0957-0233/13/8/201.

Hong, I., Park, S., Lee, B., Lee, J., Jeong, D., & Park, S. (2014). IoT-Based Smart Garbage System for Efficient Food Waste Management. *Scientific World Journal*, 2014. Available from https://doi.org/10.1155/2014/646953, http://www.hindawi.com/journals/tswj/.

Hozdić, E. (2015). Smart factory for industry 4.0: A review. *International Journal of Modern Manufacturing Technologies*, *7*(1), 28–35. Available from http://modtech.ro/international-journal/vol7no12015/Hozdic_Elvis.pdf.

Ingram, J., Maye, D., Bailye, C., Barnes, A., Bear, C., Bell, M., Cutress, D., Davies, L., de Boon, A., Dinnie, L., Gairdner, J., Hafferty, C., Holloway, L., Kindred, D., Kirby, D., Leake, B., Manning, L., Marchant, B., Morse, A., ... Wilson, L. (2022). What are the priority research questions for digital agriculture? *Land use Policy*, *114*, 105962. Available from https://doi.org/10.1016/j.landusepol.2021.105962.

Jagtap, S., Bader, F., Garcia-Garcia, G., Trollman, H., Fadiji, T., & Salonitis, K. (2021). Food logistics 4.0: opportunities and challenges. *Logistics*, *5* (1). Available from https://doi.org/10.3390/LOGISTICS5010002.

Jagtap, S., Trollman, H., Trollman, F., Garcia-Garcia, G., Parra-López, C., Duong, L., Martindale, W., Munekata, P. E. S., Lorenzo, J. M., Hdaifeh, A., Hassoun, A., Salonitis, K., & Afy-Shararah, M. (2022). The Russia-Ukraine conflict: Its implications for the global food supply chains. *Foods*, *11*(14), 2098. Available from https://doi.org/10.3390/foods11142098.

Javaid, M., Haleem, A., Singh, R. P., Suman, R., & Gonzalez, E. S. (2022). Understanding the adoption of Industry 4.0 technologies in improving environmental sustainability. *Sustainable Operations and Computers*, *3*, 203–217. Available from https://doi.org/10.1016/j.susoc.2022.01.008, https://www.sciencedirect.com/journal/sustainable-operations-and-computers.

Javaid, M., Haleem, A., Singh, R. P., & Suman, R. (2022). Enhancing smart farming through the applications of Agriculture 4.0 technologies. *International Journal of Intelligent Networks*, *3*, 150–164. Available from https://doi.org/10.1016/j.ijin.2022.09.004.

Karnik, N., Bora, U., Bhadri, K., Kadambi, P., & Dhatrak, P. (2022). A comprehensive study on current and future trends towards the characteristics and enablers of industry 4.0. *Journal of Industrial Information Integration*, *27*(124), 100294. Available from https://doi.org/10.1016/j.jii.2021.100294.

Kayikci, Y., Subramanian, N., Dora, M., & Bhatia, M. S. (2022). Food supply chain in the era of Industry 4.0: Blockchain technology implementation opportunities and impediments from the perspective of people, process, performance, and technology. *Production Planning & Control*, *33*(2–3), 301–321. Available from https://doi.org/10.1080/09537287.2020.1810757.

Khan, Z. H., Khalid, A., & Iqbal, J. (2018). Towards realizing robotic potential in future intelligent food manufacturing systems. *Innovative Food Science & Emerging Technologies*, *48*, 11–24. Available from https://doi.org/10.1016/J.IFSET.2018.05.011.

Kiel, D., Arnold, C., & Voigt, K. I. (2017). The influence of the industrial internet of things on business models of established manufacturing companies – A business level perspective. *Technovation*, *68*(October), 4–19. Available from https://doi.org/10.1016/j.technovation.2017.09.003.

Kolberg, D., Knobloch, J., & Zühlke, D. (2017). Towards a lean automation interface for workstations. *International Journal of Production Research*, *55*(10), 2845–2856. Available from https://doi.org/10.1080/00207543.2016.1223384.

Krause, J., Grüger, H., Gebauer, L., Zheng, X., Knobbe, J., Pügner, T., Kicherer, A., Gruna, R., Längle, T., & Beyerer, J. (2021). Smart spectrometer—Embedded optical spectroscopy for applications in agriculture and industry. *Sensors*, *21*(13), 4476. Available from https://doi.org/10.3390/s21134476.

Kuswandi, B., Moradi, M., & Ezati, P. (2022). Food sensors: Off-package and on-package approaches. *Packaging Technology and Science*, *35*(12), 847–862. Available from https://doi.org/10.1002/pts.2683.

Lezoche, M., Panetto, H., Kacprzyk, J., Hernandez, J. E., & Díaz, M. M. E. A. (2020). Agri-food 4.0: A survey of the supply chains and technologies for the future agriculture. *Computers in Industry*, *117*, 103187. Available from https://doi.org/10.1016/j.compind.2020.103187, https://www.journals.elsevier.com/computers-in-industry.

Lipton, J. I., Cutler, M., Nigl, F., Cohen, D., & Lipson, H. (2015). Additive manufacturing for the food industry. *Trends in Food Science and Technology*, *43*(1), 114–123. Available from https://doi.org/10.1016/j.tifs.2015.02.004, http://www.elsevier.com/wps/find/journaldescription.cws_home/601278/description#description.

Liu, Y., Ma, X., Shu, L., Hancke, G. P., & Abu-Mahfouz, A. M. (2021). From Industry 4.0 to Agriculture 4.0: Current Status, Enabling Technologies, and Research Challenges. *IEEE Transactions on Industrial Informatics*, *17*(6), 4322–4334. Available from https://doi.org/10.1109/TII.2020.3003910, http://ieeexplore.ieee.org/xpl/RecentIssue.jsp?punumber = 9424.

Lu, Y., Liu, M., Li, C., Liu, X., Cao, C., Li, X., & Kan, Z. (2022). Precision Fertilization and Irrigation: Progress and Applications. *AgriEngineering*, *4*(3), 626–655. Available from https://doi.org/10.3390/agriengineering4030041.

Luque, A., Peralta, M. E., de las Heras, A., & Córdoba, A. (2017). State of the Industry 4.0 in the Andalusian food sector. *Procedia Manufacturing*, *13*, 1199–1205. Available from https://doi.org/10.1016/j.promfg.2017.09.195.

Marvin, H. J. P., Bouzembrak, Y., van der Fels-Klerx, H. J., Kempenaar, C., Veerkamp, R., Chauhan, A., Stroosnijder, S., Top, J., Simsek-Senel, G., Vrolijk, H., Knibbe, W. J., Zhang, L., Boom, R., & Tekinerdogan, B. (2022). Digitalisation and Artificial Intelligence for sustainable food systems. *Trends in Food Science & Technology*, *120*, 344–348. Available from https://doi.org/10.1016/J.TIFS.2022.01.020, http://www.elsevier.com/wps/find/journaldescription.cws_home/601278/description#description.

Masey, R. J. M., Gray, J. O., Dodd, T. J., & Caldwell, D. G. (2010). Guidelines for the design of low-cost robots for the food industry. *Industrial Robot*, *37*(6), 509–517. Available from https://doi.org/10.1108/01439911011081650.

Masudin, I., Ramadhani, A., Restuputri, D. P., & Amallynda, I. (2021). The effect of traceability system and managerial initiative on Indonesian food cold chain performance: A Covid-19 pandemic perspective. *Global Journal of Flexible Systems Management*, *22*(4), 331–356. Available from https://doi.org/10.1007/s40171-021-00281-x.

Matei, O., Rusu, T., Petrovan, A., & Mihuţ, G. (2017). A data mining system for real time soil moisture prediction. *Procedia Engineering*, *181*, 837–844. Available from https://doi.org/10.1016/j.proeng.2017.02.475.

Matsumoto, T., Chen, Y., Nakatsuka, A., & Wang, Q. (2020). Research on horizontal system model for food factories: A case study of process cheese manufacturer. *International Journal of Production Economics*, *226*(August 2018), 107616. Available from https://doi.org/10.1016/j.ijpe.2020.107616.

McVey, C., Elliott, C. T., Cannavan, A., Kelly, S. D., Petchkongkaew, A., & Haughey, S. A. (2021). Portable spectroscopy for high throughput food authenticity screening: Advancements in technology and integration into digital traceability systems. *Trends in Food Science and Technology*, *118*, 777–790.

Misra, N. N., Dixit, Y., Al-Mallahi, A., Bhullar, M. S., Upadhyay, R., & Martynenko, A. (2022). IoT, big data and artificial intelligence in agriculture and food industry. *IEEE Internet of Things Journal*. Available from https://doi.org/10.1109/jiot.2020.2998584, 1–1.

Monteiro, A., Santos, S., & Gonçalves, P. (2021). Precision agriculture for crop and livestock farming—Brief review. *Animals*, *11*(8), 2345. Available from https://doi.org/10.3390/ANI11082345.

Oliveira, L. F. P., Moreira, A. P., & Silva, M. F. (2021). Advances in agriculture robotics: A state-of-the-art review and challenges ahead. *Robotics*, *10*, 52.

Ozogul, F., Cagalj, M., Šimat, V., Ozogul, Y., Tkaczewska, J., Hassoun, A., Kaddour, A. A., Kuley, E., Rathod, N. B., & Phadke, G. G. (2021). Recent developments in valorisation of bioactive ingredients in discard/seafood processing by-products. *Trends in Food Science and Technology*, *116*, 559–582. Available from https://doi.org/10.1016/j.tifs.2021.08.007, http://www.elsevier.com/wps/find/journaldescription.cws_home/601278/description#description.

Papetti, A., Scafà, M., Brunzini, A., & Mandolini, M. (2019). Multiperspective ergonomic assessment approach for human centered workplace design. *Springer Science and Business Media LLC*, 675–685. Available from https://doi.org/10.1007/978-3-030-31154-4_57.

Pearson, S., Tania Camacho-Villa, C., Valluru, Ravi, Oorbessy, Gaju, Mini, Rai, C., Gould, I., Brewer, S., & Sklar, E. (2022). Robotics and autonomous systems for net zero agriculture. *Current Robotics Reports*, *3*(2), 57–64. Available from https://doi.org/10.1007/S43154-022-00077-6.

Pereira, P. F. M., de Sousa Picciani, P. H., Calado, V., & Tonon, R. V. (2021). Electrical gas sensors for meat freshness assessment and quality monitoring: A review. *Trends in Food Science and Technology*, *118*(November 2020), 36–44. Available from https://doi.org/10.1016/j.tifs.2021.08.036, http://www.elsevier.com/wps/find/journaldescription.cws_home/601278/description#description.

Peydayesh, M., Bagnani, M., Soon, W. L., & Mezzenga, R. (2023). Turning food protein waste into sustainable technologies. *Chemical Reviews*, *123*(5), 2112–2154. Available from https://doi.org/10.1021/acs.chemrev.2c00236, http://pubs.acs.org/journal/chreay.

Preuveneers, D., & Ilie-Zudor, E. (2017). The intelligent industry of the future: A survey on emerging trends, research challenges and opportunities in Industry 4.0. *Journal of Ambient Intelligence and Smart Environments*, *9*(3), 287–298. Available from https://doi.org/10.3233/AIS-170432.

Qazi, S., Khawaja, B. A., & Farooq, Q. U. (2022). IoT-equipped and AI-enabled next generation smart agriculture: A critical review, current challenges and future trends. *IEEE Access*, *10*, 21219–21235. Available from https://doi.org/10.1109/ACCESS.2022.3152544.

Qiao, Y., Kong, H., Clark, C., Lomax, S., Su, D., Eiffert, S., & Sukkarieh, S. (2021). Intelligent perception for cattle monitoring: A review for cattle identification, body condition score evaluation, and weight estimation. *Computers and Electronics in Agriculture*, *185*(April). Available from https://doi.org/10.1016/j.compag.2021.106143.

Quiroz-flores, J. C., Aguado-rodriguez, R. J., Zegarra-aguinaga, E. A., Collao-diaz, M. F., & Flores-perez, A. E. (2023). Industry 4.0, circular economy and sustainability in the food industry: a literature review. *International Journal of Industrial Engineering and Operations Management*. Available from https://doi.org/10.1108/IJIEOM-12-2022-0071.

Rejeb, A., Abdollahi, A., Rejeb, K., & Treiblmaier, H. (2022). Drones in agriculture: A review and bibliometric analysis. *Computers and Electronics in Agriculture*, *198*, 107017. Available from https://doi.org/10.1016/j.compag.2022.107017.

Rejeb, A., Keogh, J. G., & Rejeb, Karim (2022). Big data in the food supply chain: A literature review. *Journal of Data, Information and Management*, *4*(1), 33–47. Available from https://doi.org/10.1007/S42488-021-00064-0.

Rejeb, A., Rejeb, K., Abdollahi, A., Al-Turjman, F., & Treiblmaier, H. (2022). The Interplay between the Internet of Things and agriculture: A bibliometric analysis and research agenda. *Internet of Things (Netherlands)*, *19*(July), 100580. Available from https://doi.org/10.1016/j.iot.2022.100580.

Rijswijk, K., Klerkx, L., Bacco, M., Bartolini, F., Bulten, E., Debruyne, L., Dessein, J., Scotti, I., & Brunori, G. (2021). Digital transformation of agriculture and rural areas: A socio-cyber-physical system framework to support responsibilisation. *Journal of Rural Studies*, *85*, 79–90. Available from https://doi.org/10.1016/J.JRURSTUD.2021.05.003.

Rose, D. C., & Chilvers, J. (2018). Agriculture 4.0: Broadening responsible innovation in an era of smart farming. *Frontiers in Sustainable Food Systems*, *2*. Available from https://doi.org/10.3389/fsufs.2018.00087.

Saucedo-Martínez, J. A., Pérez-Lara, M., Marmolejo-Saucedo, J. A., Salais-Fierro, T. E., & Vasant, P. (2018). Industry 4.0 framework for management and operations: A review. *Journal of Ambient Intelligence and Humanized Computing*, *9*(3), 789–801. Available from https://doi.org/10.1007/s12652-017-0533-1, http://www.springer.com/engineering/journal/12652.

Sharma, S., Gahlawat, V. K., Rahul, K., Mor, R. S., & Malik, M. (2021). Sustainable innovations in the food industry through artificial intelligence and big data analytics. *Logistics*, *5*(4), 66. Available from https://doi.org/10.3390/logistics5040066.

Sharma, V., Tripathi, A. K., & Mittal, H. (2022). Technological revolutions in smart farming: Current trends, challenges & future directions. *Computers and Electronics in Agriculture*, *201*(August), 107217. Available from https://doi.org/10.1016/j.compag.2022.107217.

da Silveira, F., Lermen, F. H., & Amaral, F. G. (2021). An overview of agriculture 4.0 development: Systematic review of descriptions, technologies, barriers, advantages, and disadvantages. *Computers and Electronics in Agriculture*, *189*(August), 106405. Available from https://doi.org/10.1016/j.compag.2021.106405.

Sishodia, R. P., Ray, R. L., & Singh, S. K. (2020). Applications of remote sensing in precision agriculture: A review. *Remote Sensing*, *12*(19), 3136. Available from https://doi.org/10.3390/RS12193136.

Telukdarie, A., Munsamy, M., Katsumbe, T. H., Maphisa, X., & Philbin, S. P. (2023). Industry 4.0 technological advancement in the food and beverage manufacturing industry in South Africa—bibliometric analysis via natural language processing. *Information*, *14*(8), 454. Available from https://doi.org/10.3390/info14080454.

Torky, M., & Hassanein, A. E. (2020). Integrating blockchain and the internet of things in precision agriculture : Analysis, opportunities, and challenges. *Computers and Electronics in Agriculture*, *178*(November 2019), 105476. Available from https://doi.org/10.1016/j.compag.2020.105476.

Tracey, C. T., Predeina, A. L., Krivoshapkina, E. F., & Kumacheva, E. (2022). A 3D printing approach to intelligent food packaging. *Trends in Food Science and Technology*, *127*(May), 87–98. Available from https://doi.org/10.1016/j.tifs.2022.05.003, http://www.elsevier.com/wps/find/journaldescription.cws_home/601278/description#description.

Trendov, N. M., Varas, S., & Zeng, M. (2019). Digital technologies in agriculture and rural areas. *Briefing Paper FAO (Issue June 2018)*. Available from http://www.fao.org/3/ca4887en/ca4887en.pdf.

United Nations. (2022). The Sustainable Development Goals Report 2022. https://unstats.un.org/sdgs/report/2022/The-Sustainable-Development-Goals-Report-2022.pdf.

Ullo, Silvia Liberata, & Sinha, G. R. (2021). Advances in IoT and smart sensors for remote sensing and agriculture applications. *Remote Sensing*, *13*(13), 2585. Available from https://doi.org/10.3390/rs13132585.

Wang, Z., Hirai, S., & Kawamura, S. (2022). Challenges and opportunities in robotic food handling: A review. *Frontiers in Robotics and AI*, *8*, 433. Available from https://doi.org/10.3389/FROBT.2021.789107/BIBTEX.

Weyer, Stephan, Schmitt, Mathias, Ohmer, Moritz, & Gorecky, Dominic (2015). Towards Industry 4.0 - Standardization as the crucial challenge for highly modular, multi-vendor production systems. *IFAC-PapersOnLine*, *48*(3), 579–584. Available from https://doi.org/10.1016/j.ifacol.2015.06.143.

Xu, L., Da, Xu, E. L., & Li, L. (2018). Industry 4.0: state of the art and future trends. *International Journal of Production Research*, *56*(8), 2941–2962. Available from https://doi.org/10.1080/00207543.2018.1444806.

Yang, X., Shu, L., Chen, J., Ferrag, M. A., Wu, J., Nurellari, E., & Huang, K. (2021). A survey on smart agriculture: Development modes, technologies, and security and privacy Challenges. *IEEE/CAA Journal of Automatica Sinica*, *8*(2), 273–302. Available from https://doi.org/10.1109/JAS.2020.1003536, https://www.ieee.org/membership-catalog/productdetail/showProductDetailPage.html?product = PER284-EPC.

Yin, Y., Stecke, K. E., & Li, D. (2017). The evolution of production systems from Industry 2.0 through Industry 4.0. *International Journal of Production Research*, *56*(1–2), 848–861. Available from https://doi.org/10.1080/00207543.2017.1403664.

Zhu, J., Luo, Z., Liu, Y., Tong, H., & Yin, K. (2022). Environmental perspectives for food loss reduction via smart sensors: A global life cycle assessment. *Journal of Cleaner Production*, *374*(April), 133852. Available from https://doi.org/10.1016/j.jclepro.2022.133852.

Zubaydi, H. D., Varga, P., & Molnár, S. (2023). Leveraging blockchain technology for ensuring security and privacy aspects in internet of things: A systematic literature review. *Sensors*, *23*(2), 788. Available from https://doi.org/10.3390/S23020788.

Further reading

Bhat, S. A., & Huang, N.-F. (2021). Big data and AI revolution in precision agriculture: Survey and challenges. *IEEE Access*, *9*, 110209–110222.

Boursianis, A. D., Papadopoulou, M. S., Diamantoulakis, P., Liopa-Tsakalidi, A., Barouchas, P., Salahas, G., & Goudos, S. K. (2022). Internet of things (IoT) and agricultural unmanned aerial vehicles (UAVs) in smart farming: A comprehensive review. *Internet of Things*, *18*, 100187.

Karunathilake, E. M. B. M., Le, A. T., Heo, S., Chung, Y. S., & Mansoor, S. (2023). The path to smart farming: Innovations and opportunities in precision agriculture. *Agriculture*, *13*(8), 1593.

Su, J., Yi, D., Su, B., Mi, Z., Liu, C., Hu, X., & Chen, W. H. (2020).). Aerial visual perception in smart farming: Field study of wheat yellow rust monitoring. *IEEE Transactions on Industrial Informatics*, *17*(3), 2242–2249.

Walter, A., Finger, R., Huber, R., & Buchmann, N. (2017). Smart farming is key to developing sustainable agriculture. *Proceedings of the National Academy of Sciences*, *114*(24), 6148–6150.

Wolfert, S., Ge, L., Verdouw, C., & Bogaardt, M. J. (2017). Big data in smart farming—a review. *Agricultural Systems*, *153*, 69–80.

Chapter 2

Industry 4.0 and food sustainability: role of automation, digitalization, and green technologies

Barbara Bigliardi[1], Serena Filippelli[1], Benedetta Pini[1], Eva Falch[2], Cennet Pelin Boyaci Gunduz[3] and Abdo Hassoun[4]

[1]*Department of Engineering and Architecture, University of Parma, Parma, Italy,* [2]*Department of Biotechnology and Food Science, Norwegian University of Science and Technology (NTNU), Trondheim, Norway,* [3]*Department of Food Engineering, Adana Alparslan Turkes Science and Technology University, Adana, Turkey,* [4]*Sustainable AgriFoodtech Innovation & Research (SAFIR), Arras, France*

2.1 Introduction

The agrifood systems are currently facing an unprecedented convergence of pressures due to climate change and resource scarcity, the rapid increase in world population, the outbreak of pandemics and conflicts, just to mention a few. These global extraordinary issues highlight the imperative need to explore and find new and alternative solutions to address or, at least, mitigate these devastating environmental, social, and economic challenges. Therefore the last few years have seen a continuous commitment to research and development, so that the speed of current technological advances and innovations has no historical precedent (Galanakis, 2021; Galanakis et al., 2021; Jung, 2021; Echegaray, 2022). Indeed, the adoption of digital technologies and other fourth industrial revolution (Industry 4.0) innovations has been accelerated in recent years, aiming at delivering opportunities to reduce the environmental impact of food systems, increase productivity and profitability, ensure high-food quality, and enhance food safety and traceability (Hassoun & Abdullah, 2022; Hassoun & Siddiqui, 2022; Hassoun & Jagtap, 2023; Hassoun & Kamiloglu, 2023).

Industry 4.0 and its enabling technologies have attracted tremendous attention from researchers and the academic community in recent times. According to data obtained using the database Scopus, the number of publications dealing with food Industry 4.0 has increased significantly since 2015 (Fig. 2.1). Growing research shows that agriculture and the food industry are being revolutionized and drastically changed through the increased use and harnessing of artificial intelligence, Big data, blockchain, the Internet of Things, smart sensors, and robotics, among other breakthroughs. These advanced technologies are enabling smart solutions in the context of precision agriculture and smart food factories, leading to the emergence of a number of new food trends (Hassoun & Aït-kaddour, 2022; Hassoun & Bekhit, 2022; Hassoun & Cropotova, 2022; Silva, 2023).

Publication trend on Industry 4.0 since 2015. Data were obtained from Scopus in June 2023 with the following search criteria: TITLE-ABS-KEY (Food Industry 4.0) OR (Fourth Industrial Revolution) AND (Food Industry) OR (Agriculture).

Industry 4.0 technologies are digitally transforming many industrial sectors, including agriculture and the food industry to become highly interconnected and controlled remotely. Thus digitalization is becoming one of the megatrends in the food sector, and along with Industry 4.0, digitalization and automation are changing the way food is produced, processed, packaged, transported, and consumed (Appelqvist, 2022; Hassoun & Boukid, 2022; Hassoun & Jagtap, 2023).

In addition, substantial research effort has been devoted to developing green and novel technologies (such as ultrasound, high-pressure processing, pulsed electric field, and microwave, among others) that can be used in food preservation, extraction, and processing (Pinheiro, 2021; Ulug et al., 2021; Hassoun & Prieto, 2022). Although fermentation is an ancient technology, it is currently undergoing significant transformations, leading to the emergence of precision fermentation, which employs genetically engineered microorganisms as starters for the production of target molecules

Food Industry 4.0. DOI: https://doi.org/10.1016/B978-0-443-15516-1.00002-5

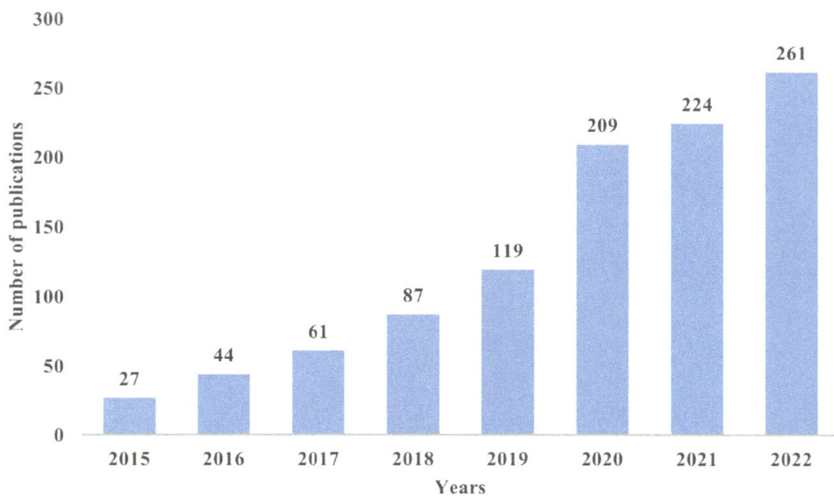

FIGURE 2.1 Publication trend on Industry 4.0 since 2015. Data were obtained from Scopus in June 2023 with the following search criteria: TITLE-ABS-KEY (Food Industry 4.0) OR (Fourth Industrial Revolution) AND (Food Industry) OR (Agriculture).

(Terefe, 2022). Among other applications of fermentation, a recent research study has highlighted the fact that fermentation could be a promising strategy for valorization of food residues (Carpena, 2022).

The interplay between Industry 4.0 and green technologies and food sustainability has been highlighted in many publications. Several issues related to environmental, economic, and social sustainability and sustainable development goals can be addressed by harnessing opportunities offered by these technologies (Bigliardi, 2021; Sun, 2021; Bradu, 2022; Hassoun & Prieto, 2022; Abbate et al., 2023). As food systems have recently been plagued by the grand sustainability challenge of food loss and food waste (Ciccullo, 2021; Peydayesh et al., 2023), this topic has received enormous attention from academia and practitioners. Many recent publications have shown successful applications of Industry 4.0 and green technologies in reducing food loss and waste (Ruggieri, 2020; Lopes de Sousa Jabbour, 2021; Hassoun & Cropotova, 2023).

This chapter is intended to provide a general overview of recent applications of Industry 4.0 and green technologies in different food sectors and to show how these technological innovations can be used to foster automation and digitalization. A special focus will be put on the role of these technologies in providing innovative solutions to enhance food sustainability and circular economy.

2.2 Interplay between Industry 4.0 technologies and food sustainability

This section discusses the relationship between Industry 4.0 digital technologies and sustainability in the food industry. To study this relationship in the best possible way, a two-stage methodology was applied, which firstly allowed us to identify the main theoretical roots on which this research topic is based, and secondly the main areas of interest on which the scientific literature has focused from 2011 (i.e., the year in which the concept of Industry 4.0 emerged in Germany) to date. For the first part, reference cocitation analysis (RCCA) was applied through the VOSviewer software, following the methodological indications of Daniel et al. (2022). It allows the discovery of the intellectual roots of the field of study through the identification of the most cited publications. The basic idea underlying this technique is that the higher the frequency with which two documents are cited together, the greater the probability that they belong to the same research domain (Anand & Dumazert, 2022). The second part, on the other hand, consisting of the identification of the main research fronts, was done by means of keyword cooccurrence analysis, that is, a bibliometric tool with which the relationship between two or more keywords is studied by considering their cooccurrence in a sample of documents (Bigliardi & Filippelli, 2022). To perform both analyses, we started from a sample of articles published on the Scopus database identified through the following search query (TITLE-ABS-KEY) ("Industry 4.0" OR "Fourth industrial revolution" OR "Industrie 4.0" OR "Cyber physical systems" OR "Industrial internet" OR "smart manufacturing" OR "cyberphysical systems" OR "enabling technolog*" OR "digital technolog*") AND TITLE-ABS-KEY ("food") AND TITLE-ABS-KEY ("sustainab*" OR "triple bottom line") AND (LIMIT-TO (LANGUAGE, "English")). In particular, the string consists of "Industry 4.0-related keywords," "sustainability-related keywords," as suggested by Ghobakhloo (2020), and the term food. Furthermore, only English-language contributions were considered. The search returned 386 documents. In the following paragraphs, the results of the RCCA and the keywords cooccurrence analysis will be presented.

2.2.1 Main theoretical foundations

The main intellectual roots were identified through the RCCA. Fig. 2.2 displays the RCCA map for the sample of scientific documents included in the study, depicting interrelationships among the most cocited references. This analysis, conducted using the VOSviewer software, only considered papers in the sample with a minimum of three citations, and highlights three main clusters of articles that give rise to the three main lines of research on the relationship between Industry 4.0 and food sustainability: Industry 4.0 and sustainability (blue cluster), smart agriculture (green cluster), and food monitoring (red cluster).

The Industry 4.0 paradigm is closely intertwined with the concept of sustainability and sustainable development. In particular, the 17 sustainable development goals (SDGs) identified by the United Nations for countries, industries, and companies to achieve sustainable development are partly achievable through Industry 4.0 technologies. Generally speaking, Industry 4.0 technologies have an effect on each of the three fundamental pillars of sustainability, which, according to the perspective of the Triple Bottom Line, are the environmental, economic, and social dimensions. These dimensions are reflected in the aforementioned SDGs. The goals that point back to the environmental dimension refer to the need to have sustainable cities, cleaner and more affordable energy, and sensible reduction of environmental impact (Nascimento et al., 2018). In this regard, the adoption of Industry 4.0 technologies enables the reduction of energy and resource consumption. Indeed, they can detect and analyze data and information from production processes and the entire supply chain in real time, ensuring a decrease in waste materials and CO_2 emissions (Sarkis & Zhu, 2018). The digital transformation undertaken by the fourth industrial revolution supports environmental sustainability through the sustainable transformation of energy and resources. For instance, the digitalization of energy systems has been an important step in the direction of environmental sustainability (Ghobakhloo, 2020). In addition, digital technologies make production systems more efficient, facilitate resource planning and allocation, and contribute positively to energy and material savings (Jose & Ramakrishna, 2018).

Among the economic goals, we find fostering economic growth and reducing inequality, which is reflected in the right to decent work and stimulating innovation. In this context, Industry 4.0 technologies allow for shorter times for both setup and delivery, reduced direct costs, greater production flexibility, and consequently, higher productivity (Dalenogare et al., 2018; Bai et al., 2020). In addition, smart technologies allow for increased transparency of processes, enabling a reduction in associated costs (Hofmann & Rüsch, 2017). Another relevant aspect pertaining to the economic dimension of sustainability is related to the introduction of new business models enabled by Industry 4.0 technologies (Rudtsch et al., 2014). As highlighted by Lopes de Sousa Jabbour et al. (2018), from an economic point of view, digital transformation allows companies to acquire relevant information from the market, which can be used to identify and exploit opportunities for environmental sustainability. Finally, goals that invoke the social dimension relate to ensuring

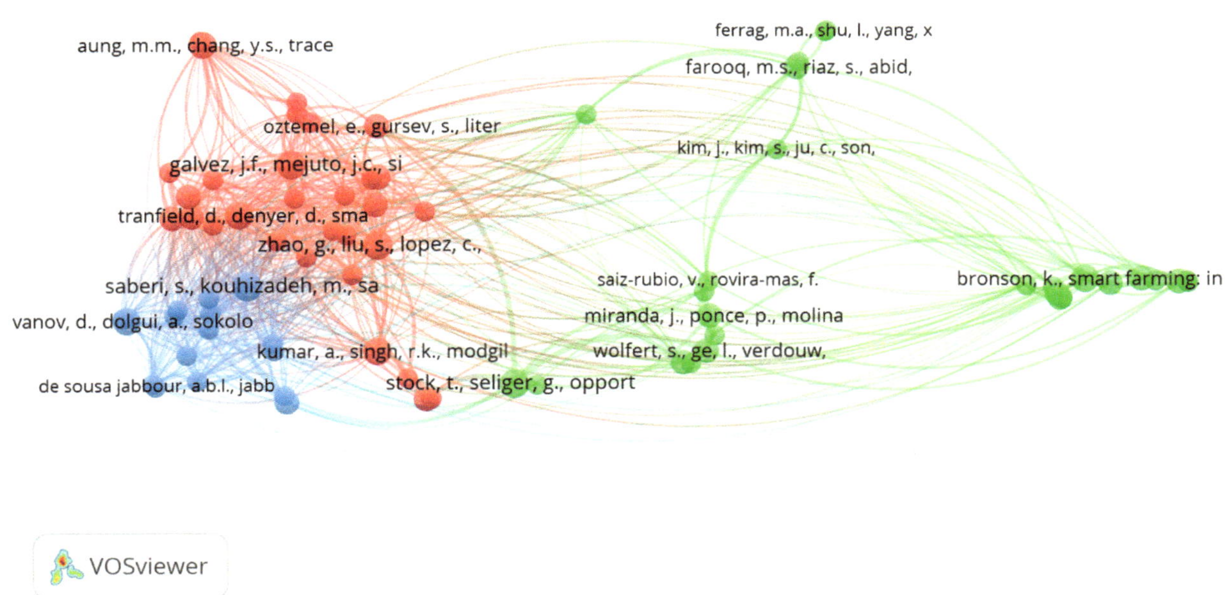

FIGURE 2.2 Reference cocitation map (implemented through VOSviewer).

health and well-being, ending hunger, establishing gender equality, and ensuring peace and justice. Here, Industry 4.0 technologies provide opportunities to improve working conditions through, for example, the automation of repetitive tasks (Müller et al., 2018; Bai et al., 2020).

The concept of smart agriculture is intertwined with that of sustainability: digital technologies and all precision farming techniques help agribusinesses become more sustainable (Deichmann et al., 2016). This strategy offers the advantage, among others, of better meeting the demands of consumers, who are now more sensitive to sustainability issues, especially environmental and social ones. Moreover, achieving definite sustainability milestones is one of the key points of the policy initiated by the European Union with the "Farm to Fork" strategy. This strategy is designed to transform the European food system to reduce its impact on third countries (Nukala et al., 2016). It touches many aspects of the food supply chain, from farming to the way food is labeled.

In addition, farms are increasingly expressing the need to deal effectively with highly challenging production conditions caused by climate change (Beckman et al., 2021). In this regard, the digital technologies of Industry 4.0 are a valuable tool. As a result of the application of these technologies in the field of food, the term "Agriculture 4.0" or "Smart Agriculture" was born, which is considered as an upgrade of precision agriculture. The latter merely takes into account the variability encountered, while Smart Agriculture goes further, basing its activities not only on location but on information derived from real-time data (Wolfert et al., 2014; da Silveira et al., 2021). In fact, Industry 4.0 technologies are useful to support the farmer both in his daily activities and in planning strategies for his business, including relations with all links in the supply chain, generating a virtuous circulus that can create value for the individual farm and its partners (Ibrahim et al., 2019).

There are many technologies used, and numerous examples of virtuous applications are reported in the literature. For example, Wolfert et al. (2017) highlight the importance of Big data as an important pillar of Industry 4.0. Machines are equipped with sensors that can detect and collect data from their surroundings, which is used to modify their behavior. They can collect multiple types of information, from meteorological data to market data. At the farm level, the authors identify two main areas of application: farm processes (e.g., those related to product creation and marketing) and farm management (e.g., to measure farm performance and compare it with previous years' or other farms' ones). Tzounis et al. (2017) conducted a study to identify the different applications of the Internet of Things (IoT) technology in agriculture, finding a considerable increase in its use since 2015. Barcelo-Ordinas et al. (2013) divided the potential applications of IoT in agriculture into different types: scaled sensor networks for monitoring agricultural infrastructure, tag-based networks, for example, radio frequency identification (RFID) for product tracking, and multimedia sensor networks for detecting diseases or insects that damage plantations. The issue of traceability is particularly relevant, both to ensure consistent product quality throughout the supply chain and to address privacy and security issues. In this regard, IoT technology can be combined with blockchain technology (Ferrag et al., 2020). Specifically, blockchain is used for encrypted data sharing, and can consequently be applied in the field of IoT-based agriculture as a digital ledger containing all relevant messages. IoT is also applied in specific areas. For example, in the study conducted by Jagtap et al. (2019), IoT is used to monitor potato waste in food manufacturing processes.

Globalization has been a turning point for the food industry, providing consumers with a wide range of products at competitive prices. However, this phenomenon has increased the distance traveled by food, that is, the distance from producer to consumer. It has caused the emergence of a new issue of interest related to maintaining food quality and ensuring safety along the supply chain (Aung & Chang, 2014). This challenge is corroborated by a growing consumer interest in food quality and safety. In fact, consumers' attention threshold on these two issues has significantly increased in recent years, especially in the face of ongoing food crises and incidents. To meet these pressing needs, laws have been enacted, policies developed, and standards instituted related to quality and safety management in the food industry (Creydt & Fischer, 2019). To meet these requirements, the tool of traceability has been used. This system benefits both supply chain operators and end consumers by satisfying the need for information about the origin and quality of purchased food products. In addition, consumers demand the verifiability of such information to make informed and risk-free food choices (Galvez et al., 2018). Food traceability, in fact, consists of a system of monitoring quality and safety through the ability to follow each stage of a food's production pathway, from the raw materials used to the finished product, ready to be consumed (Feng et al., 2020). Numerous definitions of traceability exist in the literature. In the specific context of the food supply chain, it is defined by Bosona and Gebresenbet (2013) as a function of logistics management devoted to acquiring, storing, and transmitting all relevant information about a food, feed, substance, or animal so that the product can be checked for quality and safety, followed every step of the way, and traced if needed. Through this continuous monitoring activity, timely measures, such as recall, can be put in place if a food is deemed to be at risk (Feng et al., 2020). As evidence of the importance of food traceability, it has become mandatory throughout the European Union since the beginning of 2005. All food companies operating in the European territory must have a system of traceability of products in the path from farm to fork (Walaszczyk & Galińska, 2020). In addition to being a tool for ensuring food safety and quality, traceability also allows companies to improve the quality of incoming raw materials and inventory management (Aung & Chang, 2014). Existing food traceability systems rely on the use of objective analytical methods to monitor food safety parameters, such as the presence of contaminants, and derive information about the geographic origin and biological and chemical identity of production methods (Creydt & Fischer, 2019). Different technologies are employed to enhance food traceability, but the use of blockchain stands out among them all. There are numerous examples in the literature of the application of blockchain technology in the food industry to meet the need for traceability. Yiannas (2018) highlighted suitability for ensuring transparency to the end consumer. Tian (2016) and Mondal et al. (2019) presented examples of food supply chain traceability systems based on the combination of blockchain technology and RFID. Antonucci et al. (2019) reviewed all applications of blockchain technology in the food supply chain, emphasizing its usefulness in tracking systems.

2.2.2 Hottest research topics

Through keyword cooccurrence analysis, it has been possible to identify the main research themes, displayed in Fig. 2.3: Industry 4.0 to address climate change (yellow cluster), Big data for circular economy in the food industry (purple cluster), agricultural robots (blue cluster), food safety supervision (green cluster), and Industry 4.0 in the food industry to face COVID-19 (red cluster).

2.2.2.1 Industry 4.0 to address climate change

The manufacturing industry is a major emitter of greenhouse gases worldwide. Specifically, the food sector is a major contributor to these emissions, determining food production as one of the main causes of climate change. Indeed, in

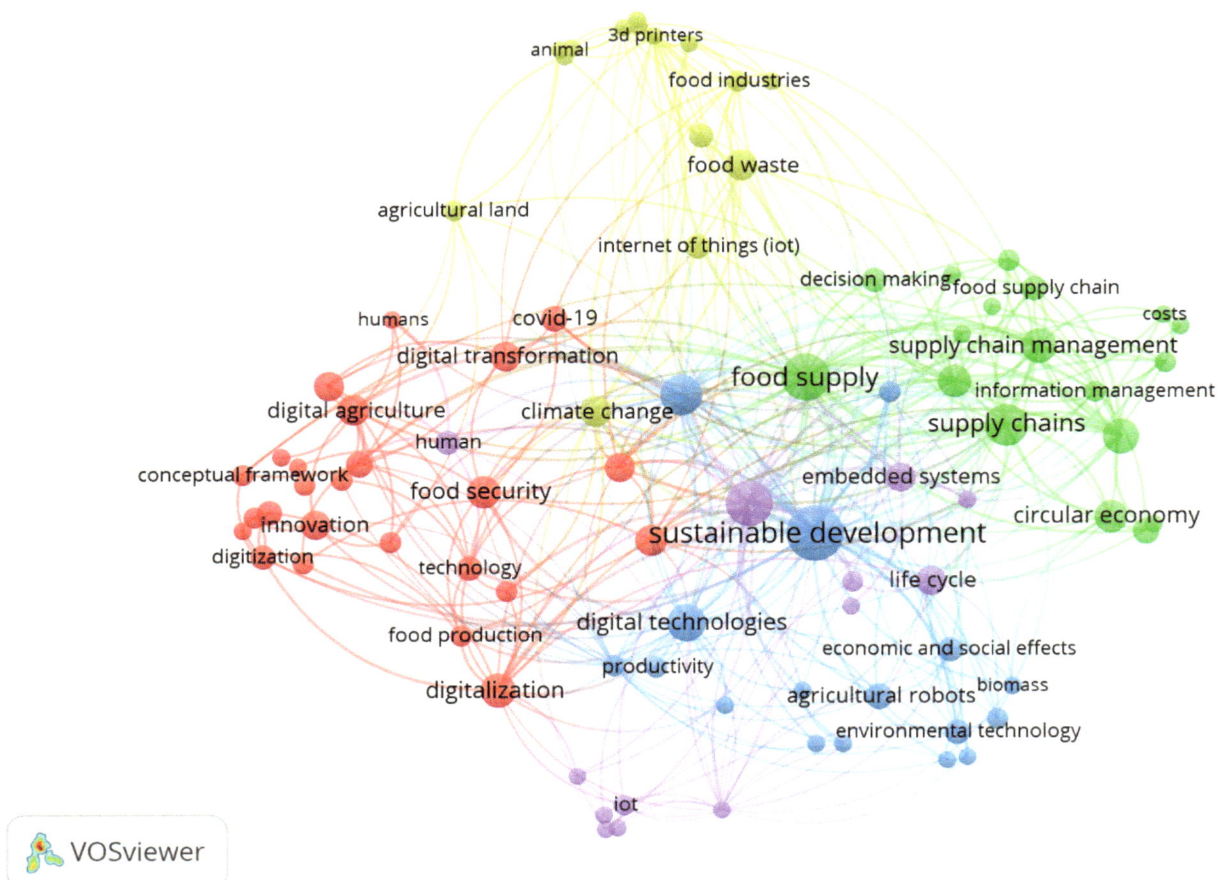

FIGURE 2.3 The hottest research topics.

recent years, the food industry has faced numerous challenges, such as population growth and nutrition crises, which have made it imperative to resort to innovative solutions to make the food industry more sustainable (Hassoun & Siddiqui, 2022). In addition, climate change mitigation is one of the SDGs set by the United Nations in the 2030 Agenda, while the Paris Agreement emphasizes the key role of technology in achieving this goal (Fritzsche et al., 2018). In this regard, Industry 4.0 digital technologies can make a strong contribution to the transformation of the manufacturing industry, and specifically food, moving it toward more sustainable and climate-friendly modes of production. Indeed, Industry 4.0 innovations enable increased energy and resource efficiency, becoming in effect enablers of sustainable practices (Yu et al., 2022).

2.2.2.2 Big data for circular economy in food industry

The objectives of the circular economy include reducing waste and extending the life cycle of products. Applying the concept of circular economy to the food industry, it is possible to refer to circular food supply chains as the result of the transition to smarter, more efficient, and resilient supply chains based on circular economy principles. Percin (2022) defines circular food supply chains as supply chains based on the restorative, regenerative, and circularity perspectives of circular economy, in which end-of-life products are transformed through restoration, the use of renewable energy is encouraged, and food waste, losses, and inefficiencies that may arise in food production are eliminated. However, due to the need for data, these supply chains are supported by digitization and, more specifically, by the digital technologies of Industry 4.0. Among the different technologies, Big data plays a key role. Big data contains a great variety, arriving in increasing volumes and with great speed. The data can be recorded in real time, creating economic, environmental, and social value at every stage of the supply chain in the food industry. One example is the use of large volumes of agricultural data for more efficient water use. Basing decisions along the food supply chain on Big data facilitates risk management and improves the development of circular strategies from farm to fork (Cheng et al., 2017; Bag et al., 2020).

2.2.2.3 Agricultural robots

The third promising research stream focuses on the digitization of agriculture and in particular on the use of agricultural robots. As expressed earlier, the concept of Industry 4.0 has been embraced by Agriculture 4.0. It encompasses the management of the agricultural ecosystem in real time based on a high degree of automation. Agricultural robots are the result of the integration of several emerging technologies, including artificial intelligence and robotics. They are widely used in industrial production to increase productivity and decrease risks related to product defects. In the agricultural context, these robots are capable of transforming the agricultural production process, finding application in animal husbandry, aerial monitoring, or as intelligent tractors capable of generating paths that ensure the safety of farmland (Liu, Ma, et al., 2020; Liu, Ramakrishnan, et al., 2020).

2.2.2.4 Food safety supervision

Among the theoretical foundations identified with the RCCA, we can find the topic of food monitoring to ensure food safety and food security. In particular, food safety and its supervision have become a hot topic in recent years. Indeed, food, being a basic necessity, significantly influences the social and economic development of a country. Due to the multiple food crises that have affected several countries around the world, ensuring food security has become a top priority (Maksimovic et al., 2015; Qureshi et al., 2022). In particular, IoT technology has been helpful to food safety supervisors as it allows for the tracking of food and the sharing of large amounts of information. In fact, it allows information from the food supply chain to be integrated, facilitating the monitoring of processes and preventing the occurrence of food safety problems (Bouzembrak et al., 2019).

2.2.2.5 Industry 4.0 in the agrifood industry to face COVID-19

The COVID-19 pandemic drastically changed the reality we live in, marking a clear distinction between the period before and after the outbreak. The food industry was no exception, experiencing marked consequences of this health-related economic event. From the consumer's point of view, the way in which food is purchased and consumed has changed. From the industry's point of view, the pandemic caused problems from the moment of its occurrence, taking away the workforce due to containment measures and forcing numerous factories to close due to a lack of personnel (Nakat & Bou-Mitri, 2021). According to Galanakis (2020), the main areas of the food sector that have been affected by the pandemic are food safety, bioactive food ingredients, food security, and sustainability. As far as food safety and food security are concerned, these were the first issues to be promptly addressed. In fact, it was necessary to implement numerous precautionary measures, in addition to the existing ones, along the entire process from farm to fork while guaranteeing the supply of food in a reality based on social distancing (Galanakis, 2021; Galanakis et al., 2021). In this context, the digital technologies of Industry 4.0 have provided valuable support. Another trend that arose as a result of the pandemic concerns the new consumers' need to strengthen their immune systems as much as possible. Already before this crisis, there had been an increased awareness of healthier foods, slowly changing the food landscape to make room for organic products or functional foods (Bigliardi & Galati, 2013). This trend has become even more pronounced during and post-COVID-19 crisis, recording an increase in consumer interest in healthy and sustainable foods (Galanakis, 2021; Galanakis et al., 2021). However, the pandemic has caused a sudden lack of information regarding consumer preferences, highlighting the need to address this critical issue. For this reason, the use of Industry 4.0 technologies has been vital. Big data and artificial intelligence have allowed consumers' decisions to be monitored in real time, without being physically present at the place of purchase (Hasnan & Yusoff, 2018). Finally, another important trend accentuated by the pandemic is that of sustainability, closely linked to the concept of food waste, which will be tackled in Section 2.4 (Galanakis, 2021; Galanakis et al., 2021).

2.3 Green food processing for reducing, preventing, and valorizing processing side streams

Food and nutrients are lost and wasted along food production and consumption chains, from primary production to the consumer. It is estimated that 1.3 billion tons of food are annually lost globally, which makes up around 1/3 of all food (FAO, 2015) and 6% of greenhouse gas emissions (Poore & Nemecek, 2018). These large volumes comprise major nutrients, such as proteins, lipids, and carbohydrates, and minor components that are important for human nutrition. Better utilization of these food resources is important for food security, reduction of environmental footprint, and for gaining economy in food production. Although waste prevention is the preferred approach in the waste hierarchy (Usmani et al., 2021; Teigiserova et al., 2020), food processing can enable the reuse, recovery, or conversion of lost biomolecules into value-added products with a circular bioeconomy approach (Usman et al., 2021; Arshad et al., 2021).

There are different terms in use to describe the potential food resources. According to the Food and Agriculture Organization of the United Nations (FAO) definition (FAO, 2019) "food loss" is the decrease in the quantity or quality of food resulting from decisions and actions by food suppliers in the chain, excluding retail, food service providers, and consumers. "Food waste" is generated later in the value chain and is defined as the decrease in quality and quantity of food resulting from decisions and actions by retailers, food services, and consumers. As the food loss and waste are calculated based on the resources that were intended for food in the first place, other potential nutrients are not a part of this calculation. These potential food resources can be defined as the parts that are not regarded as the main product categorized as processing side streams, residual raw materials, coproducts, plus-products, secondary raw materials, by-products, or waste streams. These raw materials can comprise edible and nonedible (Teigiserova et al., 2019) parts, such as peels from fruits and vegetables, press cakes from edible oils, spent grains from breweries, whey from cheese production, shells from shellfish and cut-offs, and trimmings and viscera from animal and fish processing. Additionally, it can comprise lost food parts due to, for example, quality deviation, short shelf life, or consumer rejections. Food processing is essential to transform these bioresources into edible, palatable, functional, safe, and available food, or health products. As foods are complex with a coexistence between major and minor nutrients, high-value application often includes extraction of the valuable nutrients, such as proteins or their hydrolysates, oils, carbohydrates, or minor bioactives, such as smaller organic molecules. Processing can also be used to extend the shelf life of food and thereby prevent or minimize the generation of food loss and waste.

All food processing should be effective and green (Arshad et al., 2022; Tsironi et al., 2021). The term "green food processing" has been widely used in recent research studies (Boye and Arcand, 2013; Chemat et al., 2017; Hassoun & Aït-Kaddour, et al., 2022) and describes the techniques that not only minimize environmental footprints but also protect the product from harsh conditions that can lead to safety or nutritional concerns. The green processes represent better alternatives than traditional processes characterized by the use of elevated temperatures, use of hazardous chemicals, use of scarce natural resources, and so on. In some foods, elevated temperatures can lead to the formation of potentially health-hazardous components, such as acrylamide, heterocyclic amines, nitrosamines, furans, cyclic compounds, and harmful lipid oxidation products (Andres et al., 2017; Pedreschi et al., 2017; Zhang et al., 2012). Moreover, elevated temperatures can change characteristics and reduce the functional, bioactive, or sensory properties of some of

TABLE 2.1 Some characteristics of emerging green food process techniques for improving food resource utilization.

Processing technique	Principle	Green advantages	Applications
Enzymatic hydrolysis	The use of enzymes to hydrolysis of molecules to ease extraction or produce functional or bioactive ingredients	Nonthermal, no harmful solvents, low energy use, Specific to certain molecules	Extraction of different molecules by hydrolysis or to hydrolyze molecules to produce specific functional, bioactive, or sensory properties
Pulse electric fields	The use of electric field pulses to disrupt cells and extract their content or to reduce bacterial load	Nonthermal, no harmful solvents, low energy use, selective to certain molecules	Extraction of different molecules and extended shelf life
Ultrasound	The use of acoustic cavitation to disrupt cells and extract their content or to reduce bacterial load (inactivate enzymes and microorganisms)	No harmful solvents and no traditional heat treatment. Low energy use	Extraction, shelf life extension, and preparation of functional properties. Often used in combination with other techniques. Pretreatment before enzymatic hydrolysis
High hydrostatic pressure	Using high isostatic pressure that will disrupt membranes and walls and enable extraction. Additionally, it can reduce bacterial load	Nonthermal, no harmful solvent use	Extraction, texture, and extended shelf life
Supercritical fluids	Using (gentle) supercritical solvents at a stage between a gas and a liquid to extract compounds	Nonthermal, no harmful solvent use, Low energy use, specific to certain molecules	Primarily used for unipolar compounds. High investment costs
Microwaves	Using microwaves to heat and increase pressure inside the cells that will disrupt and enable extraction	No use of harmful solvents and low in energy use	Extraction of organic molecules from food. The main drawback is the uneven heat distribution
Membrane filtration	Physical process with the use of pressure through a membrane fractionating into different molecule sizes	Nonthermal, no use of harmful solvents, low energy use	For liquids. Can be used for utilization, purification, and fractionating into different bioactive and functional properties

the heat-sensitive molecules (Struder et al., 2021). The following paragraphs will discuss a selection of the emerging green processes that can be used for reducing, preventing, and valorizing processing side streams. Table 2.1 provides a summary of the processes detailed below.

2.3.1 Enzyme-assisted hydrolysis

Enzyme-assisted hydrolysis is a very promising technological approach that can be used to reduce the environmental footprint in food processing (Boye and Arcand, 2013). It is a mild and gentle process where enzymes are used as catalysts to hydrolyze molecules, and there is no need for elevated temperatures or the addition of harmful chemicals. This process can help recover nutrients from processing side streams and convert them into proteins, peptides and amino acids, oils, and fatty acids, saccharides and sugar molecules, and so on. Another advantage is that selected enzymes at optimal conditions can act specifically on certain molecules. Consequently, this enables controlled hydrolysis of certain biomolecules, such as peptides, with targeted functional, sensory, or bioactive properties, and fatty components enriched with unsaturated fatty acids, such as omega-3 fatty acids. When using proteases to hydrolyze processing side streams from seafood or meat, the proteins are hydrolyzed into peptides, but, at the same time, lipids are released and can be separated by centrifugation instead of using chemical extraction or high-temperature treatment (Araujo et al., 2021; Liu, Ma, et al., 2020; Liu, Ramakrishnan, et al., 2020). Similarly, enzymes can be used for plant-based side streams either to hydrolyze the targeted nutrients for extraction purposes or act on other compounds that can help release the targeted nutrients from the matrix (Radenkovs et al., 2018). Choosing the right enzymes can help hydrolyze, recover, and accommodate a range of different molecules. Some studies highlight the advantages of pretreating the raw material before the enzyme hydrolysis to increase the efficiency and yield (Mora & Toldrá, 2023).

2.3.2 Pulse electric field

Pulse electric field (PEF) is another emerging green process, utilized for the disruption of cells that leads to the release of their content. This innovative technique can be used both for the extraction of valuable compounds or to reduce bacterial load and thereby extend the shelf life of food. The food material is placed between two electrodes and exposed to short-time high electric field pulses (Gómez et al., 2019). This technique has been used alone to valorize processing side streams from meat and seafood (Ghosh et al., 2019; Al Khawli et al., 2019; He et al., 2017; Li et al., 2016; Wang et al., 2021; Franco et al., 2020) or combined with other methods to extract valuable components from these materials. Arshad et al. (2021) recently reviewed the application of this technique and highlighted its potential for the effective valorization of food waste and by-products. These authors considered this method among the green extraction techniques due to, for example, nonthermal running conditions and no use of harmful solvents. They further reported that PEF can be selective to certain biomolecules. However, one of the reported limitations of this technique is that it might lead to higher levels of lipid oxidation (Arroyo et al., 2015).

2.3.3 Ultrasound

Ultrasound is used to extract valuable compounds from food side streams (Baiano, 2014; Paini et al., 2022) and to preserve food. During ultrasound treatment, the acoustic cavitation leads to the breakage of the cell walls, thus extractable compounds are released. Ultrasound has been used to extract gelatin and other protein compounds from animal residual raw material (Huang et al., 2013), oil from fish head (Bruno et al., 2019), and to deactivate proteases and microorganisms in animal residual raw materials (Li et al., 2021). Ultrasound is often used in combination with other technologies and as a pretreatment before other extraction methods (Singla & Sit, 2021; Li et al., 2019; Nitayavardhana et al., 2008). Ultrasound pretreatment before enzymatic hydrolysis is shown to reduce the particle size of the material and thereby contribute to a better interaction between enzyme and substrate (Wang et al., 2020; Li et al., 2019). As described by Singla and Sit (2021), the ultrasound technique is a physical technique that produces green extracts with no residual solvents, contaminants, or artifacts. Higuera-Barraza et al. (2016) reviewed the effect of ultrasound on properties, such as water holding, gelling, emulsification, and foaming, which are important properties when the treated materials are used as food ingredients.

2.3.4 High-pressure processing

High-pressure processing (HPP) has been efficiently used to extract valuable compounds from food side streams (Khan et al., 2019). The bioactive compounds in the extract maintain their natural quality and functionality due to the lower temperature used when applying this technique. During the HPP treatment, the product is placed in a sealed vessel and subjected to ultra-high pressure (100−600 MPa) that is transmitted by a liquid medium (usually water). This provides a pasteurization effect (Huang et al., 2020) at low temperatures, hence its name "cold pasteurization." Khan et al. (2019) also highlighted several advantages, such as time and energy efficiency, low use of solvents, and safety of this technique compared to traditional extraction methods.

2.3.5 Supercritical solvent extraction

This technique allows solvent extraction at low temperatures and is therefore suitable for the extraction of heat-sensitive biomolecules. According to Usman et al. (2022), it is one of the most advanced techniques that can be used to replace organic solvents traditionally used in extraction. This technique can be used to extract a range of natural nonpolar compounds, such as unsaturated oils (Gallego et al., 2013; Paini et al., 2022; Baiano, 2014) where its ability to reduce lipid oxidation during extraction (Rubio-Rodríguez et al., 2010) is of particular importance. However, additional solvents are needed when polar components are the target for extraction (Paini et al., 2022). A supercritical solvent possesses physical properties of both gas and liquid states under certain temperatures and pressures. Indeed, the density is like a liquid and the viscosity is similar to a gas while its diffusivity is placed between gas and liquid (Gallego et al., 2013). The most used supercritical solvent is CO_2. However, the high initial investment cost has been suggested as one of the reasons for the limited use of this technology in the industry (Baiano, 2014; Alfio et al., 2021).

2.3.6 Microwave-assisted extraction

Microwave-assisted extraction (MAE) is an emerging technology that can be used to extract bioactives such as antioxidants, aromas, essential oils pigments, and other organic molecules from food materials. Alvi et al. (2022) have recently explored the potential of microwave-based extraction as a green and eco-friendly technique with no use of solvent and with reduced energy consumption for the extraction of biomolecules. The principle behind MAE is that electromagnetic radiation generates an electric field resulting in heating and an increased pressure inside cells, leading to cell rupture. Food generally contains water, which acts as a solvent for the bioactives released from the cells. However, omega-3 fatty acids and other hydrophobic compounds can also be extracted from food side streams by using MAE. Alvi et al. (2022) gave an overview of MAE studies on different biomolecules and explained how this technique can be coupled with other more traditional extraction techniques. Baiano (2014) highlighted several modified microwave extraction systems, such as the use of vacuum and nitrogen for particularly thermal sensitive compounds and to avoid oxidation, respectively. Although this technique is widely used in analytical laboratories, industrial use is still limited (Paini et al., 2022). The main limitation is the uneven heat distribution in the samples treated (Chemat et al., 2017).

2.3.7 Membrane filtration

Membrane filtration is a pressure-driven membrane process that can be used to fraction molecules into different sizes based on different properties and purities. This technique has been used to recover a range of valuable compounds from food side streams (Castro-Muñoz & Barragán-Huerta, Fíla, et al., 2018; Castro-Muñoz & Fíla, 2018; Alavi & Ciftci, 2023; Zagklis & Paraskeva, 2018). For example, membrane filtration has been applied to whey proteins from cheese production, spent grains from breweries, press liqueurs from fruit, fish by-products and mill waste, and so on (Damar et al., 2020; Chen et al., 2023; Ambrosi et al., 2014; Figueroa et al., 2011; Ghalamara et al., 2020; Hjellnes et al., 2021; Benamar et al., 2020).

2.4 Fermentation and food sustainability

2.4.1 Sustainability in food systems

Production in many industries to meet the current needs of the growing population results in wasteful resource consumption. As sustainability is meeting the needs of the present population without compromising the ability of future generations to meet their own needs, major transformations in the food systems are necessary to sustainably feed the growing population with minimizing the use of natural resources (Bansard & Schröder, 2021; Doran, 2021). Especially, the COVID-19 pandemic has shown that our food systems are fragile, and it is very urgent to transform food systems to become more sustainable and resilient to future challenges (Boyaci-Gunduz et al., 2021).

In that context, more productive, environmentally friendly, and less wasteful production and processing methods have been investigated from production to consumption for sustainable food systems, such as reducing greenhouse gas (GHG) emissions, less and more efficient natural resources usage, decreasing loss and waste and utilization of waste and by-products, production of food products with reduced environmental impact and carbon footprint, using ingredients and materials produced from renewable sources, and also, production of bio-based ingredients and products from microorganisms through fermentation. In the food industry, fermentation could be an environmentally friendly alternative from a food sustainability point of view.

2.4.2 Application of fermentation for sustainable food systems

Fermentation has been used as a production and preservation method since ancient times. In food technology, fermented foods and beverages are mostly made through the controlled growth of microorganisms and the conversion of major and minor food components enzymatically (Marco et al., 2017). A broad range of microorganisms plays important roles in food and biotechnology industries not only for the production of fermented foods and beverages but also for the production of aroma compounds, vitamins, pigments, antimicrobial compounds, and some food additives (Erten et al., 2021). In that context, fermentation could be used as a sustainable alternative in the food industry for more eco-friendly food products and ingredients production.

Meat and dairy-based products are important sources of protein. However, animal-based food products result in twice the global GHG emissions compared to plant-based foods (Gaillac & Marbach, 2021; Xu et al., 2021). High amounts of GHG emissions and land usage result in the high carbon, water, and land footprints of animal-based products, and therefore, interest in sustainable alternatives is increasing among consumers and producers (McClements et al., 2021). Therefore the production of meat and dairy-based food products and ingredients through alternative production methods instead of traditional agricultural systems has been gaining importance for sustainable food systems.

Alternative proteins have been gaining attention for the development of sustainable food products and ingredients. The protein alternatives are produced from plants by cultivation directly from animal cells or by way of fermentation from microorganisms (Grossmann & Weiss, 2021; Nielson-Stowell, 2021). According to a recent survey study, alternative proteins could be used more by 2040 than traditional ones (Cervera, 2023; GEA, 2023). Production of alternative proteins via fermentation is categorized as traditional fermentation, biomass fermentation, and precision fermentation, as shown in Fig. 2.4 (GFI, 2021).

Traditional fermentation has been used since ancient times for the production of many fermented foods and beverages including bread, yogurt, cheese, beer, wine, pickles, and kimchi. Different microorganisms including lactic acid bacteria, yeast, molds, and some other groups of bacteria are responsible for the traditional fermentation of foods and beverages. During fermentation, microbial activities result in biochemical transformations, which improve some sensory characteristics and functional properties, such as aroma, texture, alcohol production, organic acids generation, shelf life, and nutritional attributes of the final food products (Boyaci-Gunduz et al., 2022). Traditional fermentation is a significant production methodology as it can improve the sensory, functional, and nutritional attributes of many ingredients (GFI, 2021). As reported by the Good Food Institute-State of the Industry Report, the application of traditional fermentation accounts for 7% of fermented alternative protein production (GFI, 2021). On the other hand, biomass fermentation (49%) and precision fermentation (44%) are commonly applied by more companies in the production of fermented alternative proteins for animal-free meat, seafood, eggs, and dairy.

Biomass fermentation relies on protein-dense and fast-growing microorganisms like algae, filamentous fungi, and yeast to efficiently produce high amounts of protein (GFI, 2021). Microorganisms rapidly grow under optimum conditions and generate a high amount of protein content (more than 50% dry weight). The resulting whole-cell biomass can be served as the main ingredient of a food product or blended with other food ingredients (Teng et al., 2021). Precision fermentation leverages microbial hosts to act as cell factories for the production of high-value functional food ingredients such as enzymes, flavoring agents, proteins, vitamins, natural pigments, and fats at high yields and purity. Those high-value ingredients improve the sensory and functional characteristics of plant-based products or cultivated meat (GFI, 2021).

Research and innovations in the fermentation-based production of meat, seafood, dairy, or supporting technologies field have been proceeding (Santo et al., 2020). Fermentation technology has been started to be used by companies for the production of many food products and ingredients including meat products, eggs, seafood, milk, cheese, ice cream, fats, oils, honey, infant formula, collagen, gelatin, and pet food (GFI, 2021). During the production of sustainable alternatives, it is critically important to produce products with similar sensory and nutritional properties. Biotechnology companies have already started to develop ingredients through fermentation. For example, biomass fermentation can produce protein alternatives and also, convert waste products into value-added products by microorganisms (Teng et al., 2021; Caporusso et al., 2021). Companies are trying to develop products and processes with zero-waste technology by transforming unavoidable by-products. For example, a start-up company reported the cycling of vegetable oil, brewing, and distilling industries wastes into additional protein through solid-state fermentation by fungi (Planetarians, 2021; Askew, 2021).

Fermentation can be applied effectively for the production of specific compounds such as enzymes, vitamins, natural pigments, flavors, and also, fats to improve sensory characteristics and functional attributes of plant-based products (Specht, 2021). For example, Biospringer has produced a new protein through yeast fermentation with patented technology for cheese and meat substitutes (Green, 2020). The yeast protein that is produced through fermentation could be

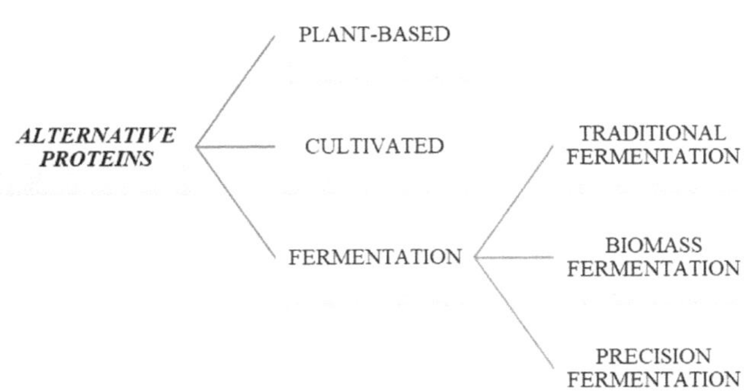

FIGURE 2.4 Alternative proteins production. *Adapted from GFI. (2021). Good Food Institute.* Fermentation. *https://gfi.org/fermentation. Accessed 18.10.22.*

a sustainable and controlled alternative as it is less dependent on harvest or seasonality. In addition, it limits land use. Moreover, the natural pigments obtained from microorganisms could be sustainable alternatives to synthetic food colorants as the production of color extracts from fruit and vegetables requires water and land resources and this can result in higher land, carbon, and water footprints. In addition, microbial pigments are unaffected by seasonal variations compared to fruits and vegetables. Moreover, the usage of microbial culture for pigment production could be sustainable as it does not require animals or fields and large-scale production will cause little or no impact on biodiversity and the environment (Rana et al., 2021).

Research and innovative applications in the fermentation-based production of foods and food ingredients are proceeding rapidly because fermentation could produce various foods and food ingredients as a sustainable, cost-effective, and environmentally friendly technology.

2.5 Industry 4.0 technologies and food waste

Combating food waste is one of the SDGs of the United Nations (UN) 2030 Agenda. To realize these goals, different states have set up specific projects. In 2011 FAO calculated food waste and found that it accounted for about one-third of the total food produced worldwide. After more than a decade, although attention on the issue has increased, the problem is still present, so much so that the FAO has created two indices to measure it according to different aspects. The food loss index concerns agricultural or livestock products that are discarded before they even reach restaurant kitchens or the end consumer. This is not because the products are of poor quality, but due to other reasons, such as size or appearance or other inconveniences that occur along the supply chain, from harvest to retail (inadequate harvesting times or practices, incorrect storage conditions, technical malfunctions or human errors in packaging and transport) (Koester & Galaktionova, 2021). The food waste index, on the other hand, measures the food (especially fresh) that is wasted when it reaches the final stages of the supply chain. This is the type of waste that occurs in shops, restaurants, or private homes due to a variety of causes, ranging from variability in demand, leftovers that remain on plates or buffets, or, in the case of households, due to poor meal planning, or incorrect storage and misinterpretation of labels (Ferreira et al., 2013). According to the Food Waste Index Report 2021, 931 million tonnes of food was wasted in 2019, equivalent to about 17% of all food available to the world's population. Most of this food waste was found to come from household (11%), especially from singles, who throw away an average of 50% more than large families, followed by catering and, lastly, trade (Zhongming et al., 2021) to recognize the centrality of the issue of food waste and the urgency of a sustainable solution to address it. Firstly, it is necessary to understand where food waste originates from. For instance, in the retail sector, food waste can result from mismanagement of replenishment operations or from problems caused by improper handling of the food product (e.g., poor refrigeration). In the literature, several studies have highlighted the differences between different types of retailers (e.g., Filimonau & Gherbin, 2017; Filimonau et al., 2019). Supermarkets, for example, prioritize the search for bulk products, generating problems related to the size and frequency of restocking (Filimonau & Gherbin, 2017). Table 2.2 summarizes the main causes of food waste at different stages of the supply chain.

The emergence of Industry 4.0 has offered a promising and safe approach for the agro-food industry to address food waste. By harnessing advanced digital technologies, it is possible to promote the circular economy, while reducing the

TABLE 2.2 The main causes of food waste at different stages of the supply chain (Storup et al., 2016).

Stage	Main causes of food waste
Production (postharvest, handling, and storage)	Animal mortality, both during transport and on-farm; milk loss due to animal problems; crops harvested early or late; damage during harvest, storage, or transport
Processing	Losses in the various manufacturing processes, such as washing, slicing, and peeling; unplanned or planned plant stoppages; damage to product during storage; bad relations with suppliers
Retail	Expiry date reached in store; unsold products; products damaged during storage; products not sold because they do not meet aesthetic standards
Consumption	Waste generated during storage; poor demand forecasting (e.g., at restaurant level); food preparation waste; household waste

Adapted from Storup, K., Mattfolk, K., Voinea, D., Jakobsen, B., Bain, M., Reverté Casas, M., & Oliveira, P. (2016). Combating food waste: An opportunity for the EU to improve the resource-efficiency of the food supply chain. *European Court of Auditors Special Edition, 34*, 6–7.

environmental impact of food loss and waste along the food supply chain (Ruggieri, 2020). These technologies enable the acquisition and sharing of data, promoting the creation and coordination of networks in the food supply chain. In this way, the alignment and coordination of people, information, knowledge, and processes are fostered, reaching the issues that can lead to food waste (Annosi et al., 2021). In the literature, there are numerous examples of efficient use of 4.0 technologies to tackle the problem of food waste. At the production level, food production can be based on artificial intelligence algorithms and machine learning methods that can optimize sales and production planning, as well as process and plant control systems (Hasnan & Yusoff, 2018). This is done by applying artificial intelligence to the entire value chain, especially in manufacturing plants. Indeed, with artificial intelligence, it is possible to make targeted predictions of food requirements, improving the predictability and controllability of value-creation processes and reducing quality-related food losses. Artificial intelligence can also be exploited for specific purposes: for instance, to monitor food storage temperature. Mena et al. (2014) explained how the use of digital solutions allows for the effective management of the timely transfer of foodstuffs along the supply chain, ensuring proper storage. In this way, the percentage of products that are not suitable to be sold can be significantly decreased. In addition, digital technologies help to prevent food surplus, from which waste is generated, caused by a mismatch between supply and demand. Indeed, Industry 4.0 technologies help to correctly size the amount of products to be placed on the market, preventing food overproduction (Tromp et al., 2016). In this case, Big data analysis is of paramount importance as it improves sales forecasts by making them more accurate and ensuring proper alignment between retailers and suppliers (Bradlow et al., 2017). Another important contribution is made by digital platforms that contribute to the recovery of food waste following the principles of the circular economy. These online platforms facilitate the creation of a network of actors, acting as intermediaries between them (Jedermann et al., 2014). Recently, it was argued that Industry 4.0 technologies (e.g., artificial intelligence, Big data, smart sensors, and IoT) have a huge potential to reduce and valorize seafood waste and by-products (Hassoun & Cropotova, 2023).

2.6 Conclusion

The developments that occurred in production processes and their automation in the last decade have led to the Fourth Industrial Revolution, the so-called "Industry 4.0." Specifically, Industry 4.0 refers to a trend in industrial automation integrating new production technologies and has a threefold aim, namely: (1) to improve working conditions, (2) to increase productivity, and (3) to improve the quality of plant production.

In the last years, more and more people, both researchers and practitioners, are increasingly paying attention to this topic, as shown by the number of results obtained searching on Google or on scientific databases (such as Scopus) with the keyword "Industry 4.0."

The available literature provides several studies stressing that the Industry 4.0 revolution is influencing almost all industries, and in particular, it has a significant influence on the food industry. Indeed, it has been proven that Industry 4.0 technologies have improved not only productivity (i.e., quantity) but also food quality, traceability, and safety.

This chapter, after presenting the results from an analysis of the available scientific literature, and in particular the hottest topics to be further investigated, shows that an important issue related to Industry 4.0 is that of sustainability. Indeed, the industry has significant responsibility for pollution in general, depletion of fossil fuels, and carbon dioxide emissions. In this context, a large amount of energy (heat) is required, and as a consequence, clean energy production techniques and energy recovery are of great importance. Industry 4.0 technologies can help to address this issue by optimizing energy and resource utilization. Digitization, from an Industry 4.0 point of view represents the true power of Industry 4.0, while from the sustainability point of view, it may lead to higher resource and energy consumption.

Based on the results presented in this chapter, we can conclude that the fourth industrial revolution and, in particular, its technological innovations have brought numerous advantages in terms of efficiency and sustainability to the food industry. However, this has to be considered not as a mere point of arrival, but as a starting point for continuous improvement.

References

Abbate, S., Centobelli, P., & Cerchione, R. (2023). The digital and sustainable transition of the agri-food sector. *Technological Forecasting and Social Change*, *187*, 122222. Available from https://doi.org/10.1016/j.techfore.2022.122222.

Alavi, F., & Ciftci, O. N. (2023). Purification and fractionation of bioactive peptides through membrane filtration: A critical and application review. *Trends in Food Science & Technology*, *131*, 118–128. Available from https://doi.org/10.1016/j.tifs.2022.11.024.

Alfio, V. G., Manzo, C., & Micillo, R. (2021). From fish waste to value: An overview of the sustainable recovery of omega-3 for food supplements. *Molecules (Basel, Switzerland)*, *26*(4), 1002. Available from https://doi.org/10.3390/molecules26041002.

Al Khawli, F., Pateiro, M., Domínguez, R., Lorenzo, J. M., Gullón, P., Kousoulaki, K., Ferrer, E., Berrada, H., & Barba, F. J. (2019). Innovative green technologies of intensification for valorization of seafood and their by-products. *Marine Drugs*, *17*, 689.

Alvi, T., Asif, Z., & Iqbal Khan, M. K. (2022). Clean label extraction of bioactive compounds from food waste through microwave-assisted extraction technique—A review. *Food Bioscience*, *46*, 101580. Available from https://doi.org/10.1016/j.fbio.2022.101580.

Ambrosi, A., Cardozo, N. S. M., & Tessaro, I. C. (2014). Membrane separation processes for the beer industry: A review and state of the art. *Food and Bioprocess Technology*, *7*, 921–936.

Anand, A., & Dumazert, J. P. (2022). Knowledge sharing in organization: Reviewing the foundations of the field and current themes using bibliometrics. *Knowledge and Process Management*, *29*(3), 270–283.

Andres, S., Schultrich, K., Monien, B., Buhrke, T., Bakhiya, N., Frenzel, F., Abraham, K., Schäfer, B., & Lampen, A. (2017). Erhitzungsbedingte Kontaminanten in Lebensmitteln: Acrylamid, Furan und Fettsäureester von Monochlorpropandiolen und Glycidol [Heat-induced contaminants in foodstuffs: Acrylamide, furan, and fatty acid esters of monochloropropanediols and glycidol]. *Bundesgesundheitsblatt Gesundheitsforschung Gesundheitsschutz*, *60*(7), 737–744. Available from https://doi.org/10.1007/s00103-017-2564-3, PMID: 28523455.

Annosi, M. C., Brunetta, F., Bimbo, F., & Kostoula, M. (2021). Digitalization within food supply chains to prevent food waste. Drivers, barriers and collaboration practices. *Industrial Marketing Management*, *93*, 208–220.

Antonucci, F., Figorilli, S., Costa, C., Pallottino, F., Raso, L., & Menesatti, P. (2019). A review on blockchain applications in the agri-food sector. *Journal of the Science of Food and Agriculture*, *99*(14), 6129–6138.

Appelqvist, I., De Wet, H., Marusic, S., Janakievski, P., et al. (2022). 'Industry 4.0 and the impact on the agrifood industry'. In R. B. Pablo Juliano, J. Sellahewa, K. Knoerzer, & M. Nguyen (Eds.), *Food engineering innovations across the food supply chain* (pp. 343–356). London, UK: Academic Press. Available from https://doi.org/10.1016/B978-0-12-821292-9.00001-7.

Araujo, J., Sica, P., Costa, C., & Márquez, M. C. (2021). Enzymatic hydrolysis of fish waste as an alternative to produce high value-added products. *Waste and Biomass Valorization*, *12*(2), 847–855. Available from https://doi.org/10.1007/s12649-020-01029-x.

Arroyo, C., Eslami, S., Brunton, N. P., Arimi, J. M., Noci, F., & Lyng, J. G. (2015). An assessment of the impact of pulsed electric fields processing factors on oxidation, colour, texture, and sensory attributes of turkey breast meat. *Poultry Science*, *94*, 1088–1095.

Arshad, R. N., Abdul-Malek, Z., Roobab, U., Ranjha, M. M. A. N., Režek Jambrak, A., Qureshi, M. I., Khan, N., Manuel Lorenzo, J., & Aadil, R. M. (2022). Nonthermal food processing: A step towards a circular economy to meet the sustainable development goals. *Food Chemistry: X*, *16*, 100516–100516.

Arshad, R. N., Abdul-Malek, Z., Roobab, U., Qureshi, M. I., Khan, N., Ahmad, M. H., Liu, Z., & Aadil, R. M. (2021). Effective valorization of food wastes and by-products through pulsed electric field: A systematic review. *Journal of Food Process Engineering*, *44*(3). Available from https://doi.org/10.1111/jfpe.13629, n/a–n/a.

Askew, K. (2021). *Single ingredient plant-based meat start-up launches fund raise: 'We have done the hard work. Now we feel ready to move further'*. Available from https://www.foodnavigator.cm/Article/2021/09/01/Single-ingredient-plant-based-meat-start-up-launches-fund-raise-We-have-done-the-hard-work-Now-we-feel-ready-to-move-further. Accessed 23.10.21.

Aung, M. M., & Chang, Y. S. (2014). Traceability in a food supply chain: Safety and quality perspectives. *Food Control*, *39*, 172–184.

Bag, S., Wood, L. C., Xu, L., Dhamija, P., & Kayikci, Y. (2020). Big data analytics as an operational excellence approach to enhance sustainable supply chain performance. *Resources, Conservation and Recycling*, *153*, 104559.

Bai, C., Dallasega, P., Orzes, G., & Sarkis, J. (2020). Industry 4.0 technologies assessment: A sustainability perspective. *International Journal of Production Economics*, *229*, 107776.

Baiano, A. (2014). Recovery of biomolecules from food wastes – A review. *Molecules (Basel, Switzerland)*, *19*(9), 14821–14842. Available from https://doi.org/10.3390/molecules190914821.

Bansard, J., & Schröder, M. (2021). *The sustainable use of natural resources: The governance challenge*. Available from https://www.iisd.org/articles/deep-dive/sustainable-use-natural-resources-governance-challenge. Accessed 09.12.22.

Barcelo-Ordinas, J. M., Chanet, J. P., Hou, K. M., & García-Vidal, J. (2013). *A survey of wireless sensor technologies applied to precision agriculture. Precision agriculture'13* (pp. 801–808). Wageningen: Wageningen Academic Publishers.

Beckman, J., Ivanic, M., & Jelliffe, J. (2021). Market impacts of farm to fork: Reducing agricultural input usage. *Applied Economic Perspectives and Policy*.

Benamar, A., Mahjoubi, F. Z., Barka, N., Kzaiber, F., Boutoial, K., Ali, G. A., & Oussama, A. (2020). Olive mill wastewater treatment using infiltration percolation in column followed by aerobic biological treatment. *SN Applied Sciences*, *2*, 1–12.

Bigliardi, B. (2021). Industry 4.0 applied to food. In C. M. Galanakis (Ed.), *Sustainable food processing and engineering challenges* (pp. 1–23). London, UK: Academic Press.

Bigliardi, B., & Filippelli, S. (2022). Sustainability and open innovation: Main themes and research trajectories. *Sustainability*, *14*(11), 6763.

Bigliardi, B., & Galati, F. (2013). Innovation trends in the food industry: The case of functional foods. *Trends in Food Science & Technology*, *31*(2), 118–129.

Bosona, T., & Gebresenbet, G. (2013). Food traceability as an integral part of logistics management in food and agricultural supply chain. *Food Control*, *33*(1), 32–48.

Bouzembrak, Y., Klüche, M., Gavai, A., & Marvin, H. J. (2019). Internet of things in food safety: Literature review and a bibliometric analysis. *Trends in Food Science & Technology*, *94*, 54–64.

Boyaci-Gunduz, C. P., Agirman, B., Gaglio, R., Franciosi, E., Francesca, N., Settanni, L., & Erten, H. (2022). Evaluation of the variations in chemical and microbiological properties of the sourdoughs produced with selected lactic acid bacteria strains during fermentation. *Food Chemistry X*, *14*, 100357. Available from https://doi.org/10.1016/j.fochx.2022.100357.

Boyaci-Gunduz, C. P., Ibrahim, S. A., Wei, O. C., & Galanakis, C. M. (2021). Transformation of the food sector: Security and resilience during the COVID-19 Pandemic. *Foods*, *10*(3), 497.

Bradlow, E. T., Gangwar, M., Kopalle, P., & Voleti, S. (2017). The role of big data and predictive analytics in retailing. *Journal of Retailing*, *93*(1), 79–95.

Bradu, P., et al. (2022). Recent advances in green technology and Industrial Revolution 4.0 for a sustainable future. *Environmental Science and Pollution Research [Preprint]*. Available from https://doi.org/10.1007/s11356-022-20024-4.

Bruno, S. F., Kudre, T. G., & Bhaskar, N. (2019). Effects of different pretreatments and proteases on recovery, umami taste compound contents and antioxidant potentials of *Labeo rohita* head protein hydrolysates. *Journal of Food Science and Technology*, *56*, 1966–1977. Available from https://doi.org/10.1007/s13197-019-03663-3.

Caporusso, A., Capece, A., & De Bari, I. (2021). Oleaginous yeasts as cell factories for the sustainable production of microbial lipids by the valorization of agri-food wastes. *Fermentation*, *7*(2), 50.

Carpena, M., et al. (2022). 'Application of fermentation for the valorization of residues from cactaceae family'. *Food Chemistry*, 135369. Available from https://doi.org/10.1016/J.FOODCHEM.2022.135369.

Castro-Muñoz, R., Barragán-Huerta, B. E., Fíla, V., Denis, P. C., & Ruby-Figueroa, R. (2018). Current role of membrane technology: From the treatment of agro-industrial by-products up to the valorization of valuable compounds. *Waste and Biomass Valorization*, *9*(4), 513–529. Available from https://doi.org/10.1007/s12649-017-0003-1.

Castro-Muñoz, R., & Fíla, V. (2018). Membrane-based technologies as an emerging tool for separating high added-value compounds from natural products – Review. *Trends in Food Science & Technology*, *82*, 8–20.

Cervera, M. (2023). *Alternative proteins will constitute a quarter of all foods by 2040, says chef survey*. Available from https://www.foodingredientsfirst.com/news/alternative-proteins-will-constitute-a-quarter-of-all-foods-by-2040-says-chef-survey.html. Accessed 14.01.23.

Chemat, F., Rombaut, N., Meullemiestre, A., Turk, M., Perino, S., Fabiano-Tixier, A.-S., & Abert-Vian, M. (2017). Review of green food processing techniques. Preservation, transformation, and extraction. *Innovative Food Science & Emerging Technologies*, *41*, 357–377. Available from https://doi.org/10.1016/j.ifset.2017.04.016.

Chen, G. Q., Qu, Y., Gras, S. L., & Kentish, S. E. (2023). Separation technologies for whey protein fractionation. *Food Engineering Reviews*, 1–28.

Cheng, H., Chen, C., Wu, S., Mirza, Z. A., & Liu, Z. (2017). Emergy evaluation of cropping, poultry rearing, and fish raising systems in the drawdown zone of Three Gorges Reservoir of China. *Journal of Cleaner Production*, *144*, 559–571.

Cicculo, F., et al. (2021). Implementing the circular economy paradigm in the agri-food supply chain: The role of food waste prevention technologies. *Resources, Conservation and Recycling*, *164*(September 2020), 105114. Available from https://doi.org/10.1016/j.resconrec.2020.105114.

Creydt, M., & Fischer, M. (2019). Blockchain and more-algorithm driven food traceability. *Food Control*, *105*, 45–51.

Dalenogare, L. S., Benitez, G. B., Ayala, N. F., & Frank, A. G. (2018). The expected contribution of Industry 4.0 technologies for industrial performance. *International Journal of Production Economics*, *204*, 383–394.

Damar, I., Cinar, K., & Gulec, H. A. (2020). Concentration of whey proteins by ultrafiltration: Comparative evaluation of process effectiveness based on physicochemical properties of membranes. *International Dairy Journal*, *111*, 104823.

Daniel, C., Walsh, I., & Mesmer-Magnus, J. (2022). Mindfulness: Unpacking its three shades and illuminating integrative ways to understand the construct. *International Journal of Management Reviews*.

Deichmann, U., Goyal, A., & Mishra, D. (2016). Will digital technologies transform agriculture in developing countries? *Agricultural Economics*, *47*(S1), 21–33.

Doran, P. (2021). *Doing more with less: Ensuring sustainable consumption and production*. Available from https://www.iisd.org/articles/deep-dive/doing-more-less-ensuring-sustainable-consumption-and-production. Accessed 09.12.22.

Echegaray, N., et al. (2022). 'Meat 4.0: Principles and applications of Industry 4.0 technologies in the meat industry'. *Applied Sciences*, *12*(14), 6986. Available from https://doi.org/10.3390/app12146986.

Erten, H., Boyaci Gunduz, C. P., & Agirman, B. (2021). Introduction to yeasts genera important in food biotechnology. In Q.-S. Wu, Y.-N. Zou, F. Zhang, & B. Shu (Eds.), *An introduction to microorganisms* (pp. 33–101). New York, USA: Nova Science Publishers.

FAO (2015). WFP. The state of food insecurity in the world. In *Meeting the 2015 international hunger targets: Taking stock of uneven progress*, Rome, Italy.

FAO (2019). *The State of Food and Agriculture 2019. Moving forward on food loss and waste reduction*. Rome. Licence: CC BY-NC-SA 3.0 IGO.

Feng, H., Wang, X., Duan, Y., Zhang, J., & Zhang, X. (2020). Applying blockchain technology to improve agri-food traceability: A review of development methods, benefits and challenges. *Journal of Cleaner Production*, *260*, 121031.

Ferrag, M. A., Shu, L., Yang, X., Derhab, A., & Maglaras, L. (2020). Security and privacy for green IoT-based agriculture: Review, blockchain solutions, and challenges. *IEEE Access*, *8*, 32031–32053.

Ferreira, M., Martins, M. L., & Rocha, A. (2013). Food waste as an index of foodservice quality. *British Food Journal*, *115*, 1628–1637.

Figueroa, R. A. R., Cassano, A., & Drioli, E. (2011). Ultrafiltration of orange press liquor: Optimization for permeate flux and fouling index by response surface methodology. *Separation and Purification Technology*, *80*(1), 1–10.

Filimonau, V., & Gherbin, A. (2017). An exploratory study of food waste management practices in the UK grocery retail sector. *Journal of Cleaner Production*, *167*, 1184–1194.

Filimonau, V., Krivcova, M., & Pettit, F. (2019). An exploratory study of managerial approaches to food waste mitigation in coffee shops. *International Journal of Hospitality Management*, *76*, 48–57.

Franco, D., Munekata, P. E. S., Agregán, R., Bermúdez, R., López Pedrouso, M., Pateiro, M., & Lorenzo, J. M. (2020). Applications of pulsed electric fields for obtaining antioxidant extracts from fish residues. *Antioxidants*, *9*(2), 90.

Fritzsche, K., Niehoff, S., & Beier, G. (2018). Industry 4.0 and climate change—Exploring the science-policy gap. *Sustainability*, *10*(12), 4511.

Gaillac, R., & Marbach, S. (2021). The carbon footprint of meat and dairy proteins: A practical perspective to guide low carbon footprint dietary choices. *Journal of Cleaner Production*, *321*, 128766.

Galanakis, C. M. (2020). The food systems in the era of the coronavirus (COVID-19) pandemic crisis. *Foods*, *9*(4), 523.

Galanakis, C. M. (2021). Functionality of food components and emerging technologies. *Foods*, *10*(1), 128.

Galanakis, C. M., et al. (2021). 'Innovations and technology disruptions in the food sector within the COVID-19 pandemic and post-lockdown era'. *Trends in Food Science & Technology*, *110*(February), 193–200. Available from https://doi.org/10.1016/j.tifs.2021.02.002.

Gallego, R., Jose A.M., Miguel H., M.A. Castro-Puyana, & Elena I. (2013). Supercritical fluid extraction. In J. Prado & M. Rostagno (Eds.) *Natural product extraction: Principles and applications*. Royal Society of Chemistry. MLA 9th Edition (Modern Language Assoc.).

Galvez, J. F., Mejuto, J. C., & Simal-Gandara, J. (2018). Future challenges on the use of blockchain for food traceability analysis. *TrAC Trends in Analytical Chemistry*, *107*, 222–232.

GEA. (2023). *GEA International Chef Survey: Alternative Proteins Constitute at Least a Quarter of All Food by 2040*. Available from https://www.gea.com/en/news/corporate/2023/chef-survey.jsp. Accessed 14.01.23.

GFI. (2021). Good Food Institute. *Fermentation*. Available from https://gfi.org/fermentation. Accessed 18.10.22.

Ghalamara, S., Silva, S., Brazinha, C., & Pintado, M. (2020). Valorization of fish by-products: Purification of bioactive peptides from codfish blood and sardine cooking wastewaters by membrane processing. *Membranes*, *10*(3), 44. Available from https://doi.org/10.3390/membranes10030044.

Ghobakhloo, M. (2020). Industry 4.0, digitization, and opportunities for sustainability. *Journal of Cleaner Production*, *252*, 119869.

Ghosh, S., Gillis, A., Sheviryov, J., Levkov, K., & Golberg, A. (2019). Towards waste meat biorefinery: Extraction of proteins from waste chicken meat with non-thermal pulsed electric fields and mechanical pressing. *Journal of Cleaner Production*, *208*, 220–223.

Gómez, B., Munekata, P. E. S., Gavahian, M., Barba, F. J., Martí-Quijal, F. J., Bolumar, T., Campagnol, P. C. B., Tomasevic, I., & Lorenzo, J. M. (2019). Application of pulsed electric fields in meat and fish processing industries: An overview. *Food Research International*, *123*, 95–105.

Green, M. (2020). *Biospringer launches "revolutionary" yeast protein eliminating off-notes from plant-based foods*. Available from https://www.foodingredientsfirst.com/news/biospringer-launches-revolutionary-yeast-protein-eliminating-off-notes-from-plant-based-foods.html. Accessed 09.12.22.

Grossmann, L., & Weiss, J. (2021). Alternative protein sources as technofunctional food ingredients. *Annual Review Food Science and Technology*, *12*, 93–117.

Hasnan, N. Z. N., & Yusoff, Y. M. (2018). Short review: Application areas of industry 4.0 technologies in food processing sector. *2018 IEEE student conference on research and development (SCOReD)*, 1–6, IEEE.

Hassoun, A., Abdullah, N. A., et al. (2022). 'Food traceability 4.0 as part of the fourth industrial revolution: Key enabling technologies'. *Critical Reviews in Food Science and Nutrition*, *0*(0), 1–17. Available from https://doi.org/10.1080/10408398.2022.2110033.

Hassoun, A., Aït-kaddour, A., et al. (2022). The fourth industrial revolution in the food industry—Part I: Industry 4.0 technologies. *Critical Reviews in Food Science and Nutrition*, 1–17. Available from https://doi.org/10.1080/10408398.2022.2034735.

Hassoun, A., Bekhit, A. E., et al. (2022). The fourth industrial revolution in the food industry—Part II: Emerging food trends trends. *Critical Reviews in Food Science and Nutrition*, *0*(0), 1–31. Available from https://doi.org/10.1080/10408398.2022.2106472.

Hassoun, A., Boukid, F., et al. (2022). Current research in food science emerging trends in the agri-food sector: Digitalisation and shift to plant-based diets. *Current Research in Food Science*, *5*, 2261–2269. Available from https://doi.org/10.1016/j.crfs.2022.11.010.

Hassoun, A., Cropotova, J., et al. (2022). 'Consumer acceptance of new food trends resulting from the fourth industrial revolution technologies: A narrative review of literature and future perspectives'. *Frontiers in Nutrition*, *9*, 972154. Available from https://doi.org/10.3389/fnut.2022.972154.

Hassoun, A., Cropotova, J., et al. (2023). 'Use of industry 4.0 technologies to reduce and valorize seafood waste and by-products: A narrative review on current knowledge'. *Current Research in Food Science*, *6*, 100505. Available from https://doi.org/10.1016/j.crfs.2023.100505.

Hassoun, A., Jagtap, S., et al. (2023). 'Food quality 4.0: From traditional approaches to digitalized automated analysis'. *Journal of Food Engineering*, *337*, 111216. Available from https://doi.org/10.1016/j.jfoodeng.2022.111216.

Hassoun, A., Kamiloglu, S., et al. (2023). 'Implementation of relevant fourth industrial revolution innovations across the supply chain of fruits and vegetables: A short update on Traceability 4.0'. *Food Chemistry*, *409*, 135303. Available from https://doi.org/10.1016/j.foodchem.2022.135303.

Hassoun, A., Prieto, M. A., et al. (2022). 'Exploring the role of green and Industry 4.0 technologies in achieving sustainable development goals in food sectors'. *Food Research International*, *162*, 112068. Available from https://doi.org/10.1016/j.foodres.2022.112068.

Hassoun, A., Siddiqui, S. A., et al. (2022). 'Emerging technological advances in improving the safety of muscle foods: Framing in the context of the food revolution 4.0'. *Food Research International*, 0–42. Available from https://doi.org/10.1080/87559129.2022.2149776.

He, G., Yin, Y., Yan, X., & Wang, Y. (2017). Semi-bionic extraction of effective ingredient from fishbone by high intensity pulsed electric fields. *Journal of Food Process Engineering*, *40*(2).

Higuera-Barraza., Del Toro-Sanchez, C. L., Ruiz-Cruz, S., & Márquez-Ríos, E. (2016). Effects of high-energy ultrasound on the functional properties of proteins. *Ultrasonics Sonochemistry*, *31*, 558–562. Available from https://doi.org/10.1016/j.ultsonch.2016.02.007.

Hjellnes, V., Rustad, T., Jensen, I. J., Eiken, E., Pettersen, S. M., & Falch, E. (2021). Ultrafiltration of saithe (*Pollachius virens*) protein hydrolysates and its effect on antioxidative activity. *Catalysts*, *11*(9), 1053.

Hofmann, E., & Rüsch, M. (2017). Industry 4.0 and the current status as well as future prospects on logistics. *Computers in Industry*, *89*, 23–34.

Huang., Hsu, C.-P., Yang, B. B., & Wang, C.-Y. (2013). Advances in the extraction of natural ingredients by high pressure extraction technology. *Trends in Food Science & Technology*, *33*(1), 54–62. Available from https://doi.org/10.1016/j.tifs.2013.07.001.

Huang, H.-W., Hsu, C.-P., & Wang, C.-Y. (2020). Healthy expectations of high hydrostatic pressure treatment in food processing industry. *Journal of Food and Drug Analysis*, *28*(1), 1–13.

Ibrahim, H., Mostafa, N., Halawa, H., Elsalamouny, M., Daoud, R., Amer, H., & ElSayed, H. (2019). A layered IoT architecture for greenhouse monitoring and remote control. *SN Applied Sciences*, *1*(3), 1–12.

Jagtap, S., Bhatt, C., Thik, J., & Rahimifard, S. (2019). Monitoring potato waste in food manufacturing using image processing and internet of things approach. *Sustainability*, *11*(11), 3173.

Jedermann, R., Nicometo, M., Uysal, I., & Lang, W. (2014). Reducing food losses by intelligent food logistics. Philosophical transactions of the Royal Society A: Mathematical. *Physical and Engineering Sciences*, *372*(2017), 20130302.

Jose, R., & Ramakrishna, S. (2018). Materials 4.0: Materials big data enabled materials discovery. *Applied Materials Today*, *10*, 127–132.

Jung, J., et al. (2021). 'The potential of remote sensing and artificial intelligence as tools to improve the resilience of agriculture production systems'. *Current Opinion in Biotechnology*, *70*, 15–22. Available from https://doi.org/10.1016/J.COPBIO.2020.09.003.

Khan, S. A., Aslam, R., & Makroo, H. A. (2019). High pressure extraction and its application in the extraction of bio-active compounds: A review. *Journal of Food Process Engineering*, *42*(1). Available from https://doi.org/10.1111/jfpe.12896, e12896–n/a.

Koester, U., & Galaktionova, E. (2021). FAO food loss index methodology and policy implications. *Studies in Agricultural Economics*, *123*(1), 1–7.

Li, X., Liu, Y., Wang, Y., Wang, J., Xu, Y., Yi, S., Zhu, W., Mi, H., Li, T., & Li, J. (2021). Combined ultrasound and heat pretreatment improve the enzymatic hydrolysis of clam (Aloididae aloidi) and the flavor of hydrolysates. *Innovative Food Science & Emerging Technologies*, *67*, 102596. Available from https://doi.org/10.1016/j.ifset.2020.102596.

Li, X., Mettu, S., Martin, G. J. O., Ashokkumar, M., & Lin, C. S. K. (2019). Ultrasonic pretreatment of food waste to accelerate enzymatic hydrolysis for glucose production. *Ultrasonics Sonochemistry*, *53*, 77–82. Available from https://doi.org/10.1016/j.ultsonch.2018.12.035.

Li, M. J., Lin, J., & Chen, T. (2016). Fang pulsed electric field-assisted enzymatic extraction of protein from abalone (Haliotis Discus Hannai Ino) viscera. *Journal of Food Process Engineering*, *39*(6), 702–710.

Liu, Y., Ma, X., Shu, L., Hancke, G. P., & Abu-Mahfouz, A. M. (2020). From Industry 4.0 to Agriculture 4.0: Current status, enabling technologies, and research challenges. *IEEE Transactions on Industrial Informatics*, *17*(6), 4322–4334.

Liu, Y., Ramakrishnan, V. V., & Dave, D. (2020). Lipid class and fatty acid composition of oil extracted from Atlantic salmon by-products under different optimization parameters of enzymatic hydrolysis. *Biocatalysis and Agricultural Biotechnology*, *30*, 101866.

Lopes de Sousa Jabbour, A. B., et al. (2021). 'Are food supply chains taking advantage of the circular economy? A research agenda on tackling food waste based on Industry 4.0 technologies'. *Production Planning & Control [Preprint]*. Available from https://doi.org/10.1080/09537287.2021.1980903.

Maksimovic, M., Vujovic, V., & Omanovic-Miklicanin, E. (2015). A low cost internet of things solution for traceability and monitoring food safety during transportation. *HAICTA*, 583–593.

Marco, M. L., Heeney, D., Binda, S., Cifelli, C. J., Cotter, P. D., Foligné, B., Gänzle, M., Kort, R., Pasin, G., Pihlanto, A., Smid, E. J., & Hutkins, R. (2017). Health benefits of fermented foods: Microbiota and beyond. *Current Opinion in Biotechnology*, *44*, 94–102. Available from https://doi.org/10.1016/j.copbio.2016.11.010.

McClements, D. J., Weiss, J., Kinchla, A. J., Nolden, A. A., & Grossmann, L. (2021). Methods for testing the quality attributes of plant-based foods: Meat- and processed-meat analogs. *Foods*, *10*(2), 260.

Mena, C., Terry, L. A., Williams, A., & Ellram, L. (2014). Causes of waste across multi-tier supply networks: Cases in the UK food sector. *International Journal of Production Economics*, *152*, 144–158.

Mondal, S., Wijewardena, K. P., Karuppuswami, S., Kriti, N., Kumar, D., & Chahal, P. (2019). Blockchain inspired RFID-based information architecture for food supply chain. *IEEE Internet of Things Journal*, *6*(3), 5803–5813.

Mora, L., & Toldrá, F. (2023). Advanced enzymatic hydrolysis of food proteins for the production of bioactive peptides. *Current Opinion in Food Science*, 49. Available from https://doi.org/10.1016/j.cofs.2022.100973.

Müller, J. M., Kiel, D., & Voigt, K. I. (2018). What drives the implementation of Industry 4.0? The role of opportunities and challenges in the context of sustainability. *Sustainability*, *10*(1), 247.

Nakat, Z., & Bou-Mitri, C. (2021). COVID-19 and the food industry: Readiness assessment. *Food Control*, *121*, 107661.

Nascimento, D. L. M., Alencastro, V., Quelhas, O. L. G., Caiado, R. G. G., Garza-Reyes, J. A., Rocha-Lona, L., & Tortorella, G. (2018). Exploring Industry 4.0 technologies to enable circular economy practices in a manufacturing context: A business model proposal. *Journal of Manufacturing Technology Management*, *30*(3), 607–627.

Nielson-Stowell, A. (2021). *Fermentation for alternative proteins*. The Fermentation Association. Available from https://fermentationassociation.org/fermentation-for-alternative-proteins/. Accessed 12.12.22.

Nitayavardhana, S., Rakshit, S. K., Grewell, D., van Leeuwen, J., & Khanal, S. K. (2008). Ultrasound pretreatment of cassava chip slurry to enhance sugar release for subsequent ethanol production. *Biotechnology and Bioengineering*, *101*, 487–496.

Nukala, R., Panduru, K., Shields, A., Riordan, D., Doody, P., & Walsh, J. (2016). Internet of things: A review from 'Farm to Fork'. *2016 27th Irish signals and systems conference (ISSC)*, 1–6, IEEE.

Paini, J., Benedetti, V., Ail, S. S., Castaldi, M. J., Baratieri, M., & Patuzzi, F. (2022). Valorization of wastes from the food production industry: A review towards an integrated agri-food processing biorefinery. *Waste and Biomass Valorization*, *13*(1), 31–50. Available from https://doi.org/10.1007/s12649-021-01467-1.

Pedreschi, F., Murkovic, M., & Murkovic, M. (2017). Potentially toxic food components formed by excessive heat processing. *Dietary Ages and their Role in Health and Disease*. Available from https://doi.org/10.1201/9781315120041-9.

Percin, S. (2022). Evaluating the circular economy–based big data analytics capabilities of circular agri-food supply chains: The context of Turkey. *Environmental Science and Pollution Research*, *29*(55), 83220–83233.

Peydayesh, M., Bagnani, M., & Soon, W. L. (2023). Turning food protein waste into sustainable technologies. *Chemical Reviews*, *123*(5), 2112–2154. Available from https://doi.org/10.1021/acs.chemrev.2c00236.

Pinheiro, A. C. D. A. S., et al. (2021). 'Innovative non-thermal technologies for recovery and valorization of value-added products from crustacean processing by-products—An opportunity for a circular economy approach'. *Foods*, *10*(9), 2030. Available from https://doi.org/10.3390/FOODS10092030.

Planetarians. (2021). *Next gen plant based meat*. Available from https://www.planetarians.com/. Accessed 25.10.21.

Poore, J., & Nemecek, T. (2018). Reducing foods environmental impacts through producers and consumers. *Science (New York, N.Y.)*, *360*(3692), 987–992.

Qureshi, T., Saeed, M., Ahsan, K., Malik, A. A., Muhammad, E. S., & Touheed, N. (2022). Smart agriculture for sustainable food security using internet of things (IoT). *Wireless Communications and Mobile Computing*, *2022*, 1–10.

Radenkovs, V., Juhnevica-Radenkova, K., Górnaś, P., & Seglina, D. (2018). Non-waste technology through the enzymatic hydrolysis of agro-industrial by-products. *Trends in Food Science & Technology*, *77*, 64–76. Available from https://doi.org/10.1016/j.tifs.2018.05.013.

Rana, B., Bhattacharyya, M., Patni, B., Arya, M., & Joshi, G. K. (2021). The realm of microbial pigments in the food color market. *Frontiers in Sustainable Food Systems*, *5*, 54.

Rubio-Rodríguez, N., Beltrán, S., Jaime, I., de Diego, S. M., Sanz, M. T., & Carballido, J. R. (2010). Production of omega-3 polyunsaturated fatty acid concentrates: A review. *Innovative Food Science and Emerging Technology*, *11*, 1–12.

Rudtsch, V., Gausemeier, J., Gesing, J., Mittag, T., & Peter, S. (2014). Pattern-based business model development for cyber-physical production systems. *Procedia CIRP*, *25*, 313–319.

Ruggieri, R., et al. (2020). Food losses and food waste: The Industry 4.0 opportunity for the sustainability challenge. *Rivista di Studi Sulla Sostenibilita*, *2020*(1), 159–177. Available from https://doi.org/10.3280/RISS2020-001009.

Santo, R., Kim, B. F., Goldman, S. E., Dutkiewicz, J. A. N., Biehl, E., Bloem, M. W., et al. (2020). Considering plant-based meat substitutes and cell-based meats: A public health and food systems perspective. *Frontiers Sustainable Food Systems*, *4*, 134.

Sarkis, J., & Zhu, Q. (2018). Environmental sustainability and production: Taking the road less travelled. *International Journal of Production Research*, *56*(1–2), 743–759.

da Silva, F. T., et al. (2023). Open innovation in agribusiness: Barriers and challenges in the transition to agriculture 4.0. *Sustainability*, *15*(11), 8562. Available from https://doi.org/10.3390/SU15118562.

da Silveira, F., Lermen, F. H., & Amaral, F. G. (2021). An overview of agriculture 4.0 development: Systematic review of descriptions, technologies, barriers, advantages, and disadvantages. *Computers and Electronics in Agriculture*, *189*, 106405.

Singla, M., & Sit, N. (2021). Application of ultrasound in combination with other technologies in food processing: A review. *Ultrasonic Sonochemistry*, *73*(2021), 105506.

de Sousa Jabbour, A. B. L., Jabbour, C. J. C., Foropon, C., & Godinho Filho, M. (2018). When titans meet–Can industry 4.0 revolutionise the environmentally-sustainable manufacturing wave? The role of critical success factors. *Technological Forecasting and Social Change*, *132*, 18–25.

Specht, L. (2021). *Good Food Institute—The science of fermentation*. Available from https://gfi.org/science/the-science-of-fermentation/. Accessed 07.12.21.

Storup, K., Mattfolk, K., Voinea, D., Jakobsen, B., Bain, M., Reverté Casas, M., & Oliveira, P. (2016). Combating food waste: An opportunity for the EU to improve the resource-efficiency of the food supply chain. *European Court of Auditors Special Edition*, *34*, 6–7.

Sun, X., et al. (2021). The application of Industry 4.0 technologies in sustainable logistics: A systematic literature review (2012–2020) to explore future research opportunities. *Environmental Science and Pollution Research*, *2021*(1), 1–32. Available from https://doi.org/10.1007/S11356-021-17693-Y.

Teigiserova, D. A., Hamelin, L., & Thomsen, M. (2019). Review of high-value food waste and food residues biorefineries with focus on unavoidable wastes from processing. *Resources, Conservation and Recycling*, *149*, 413–426. Available from https://doi.org/10.1016/j.resconrec.2019.05.003.

Teigiserova, D. A., Hamelin, L., & Thomsen, M. (2020). Towards transparent valorization of food surplus, waste and loss: Clarifying definitions, food waste hierarchy, and role in the circular economy. *The Science of the Total Environment*, *706*. Available from https://doi.org/10.1016/j.scitotenv.2019.136033, 136033–136033.

Teng, T. S., Chin, Y. L., Chai, K. F., & Chen, W. N. (2021). Fermentation for future food systems: Precision fermentation can complement the scope and applications of traditional fermentation. *EMBO Reports*, *22*(5), e52680.

Terefe, N. S. (2022). *Recent developments in fermentation technology: Toward the next revolution in food production. Food Engineering Innovations Across the Food Supply Chain* (pp. 89–106). Elsevier Inc. Available from https://doi.org/10.1016/b978-0-12–821292-9.00026-1.

Tian, F. (2016, June). An agri-food supply chain traceability system for China based on RFID & blockchain technology. In *2016 13th international conference on service systems and service management (ICSSSM)* (pp. 1–6). IEEE.

Tromp, S. O., Haijema, R., Rijgersberg, H., & van der Vorst, J. G. (2016). A systematic approach to preventing chilled-food waste at the retail outlet. *International Journal of Production Economics*, *182*, 508–518.

Tsironi, T., Koutinas, A., Mandala, I., & Stoforos, N. G. (2021). Current and new green deal solutions for sustainable food processing. *Current Opinion in Environmental Science & Health*, *21*, 100244.

Tzounis, A., Katsoulas, N., Bartzanas, T., & Kittas, C. (2017). Internet of things in agriculture, recent advances and future challenges. *Biosystems Engineering*, *164*, 31–48.

Ulug, S. K., Jahandideh, F., & Wu, J. (2021). 'Novel technologies for the production of bioactive peptides'. *Trends in Food Science & Technology*, *108*, 27–39. Available from https://doi.org/10.1016/J.TIFS.2020.12.002.

Usman, I., Hussain, M., Imran, A., Afzaal, M., Saeed, F., Javed, M., Afzal, A., Ashfaqc, I., Al Jbawi, I., & Saewane, S. A. (2022). Traditional and innovative approaches for the extraction of bioactive compounds. *International Journal of Food Properties*, *25*(1), 1215–1233. Available from https://doi.org/10.1080/10942912.2022.2074030.

Usman, I., Sharma, M., Awasthi, A. K., Sharma, G. D., Cysneiros, D., Nayak, S. C., Thakur, V. K., Naidu, R., Pandey, A., & Gupta, V. K. (2021). Minimizing hazardous impact of food waste in a circular economy – Advances in resource recovery through green strategies. *Journal of Hazardous Materials*, *416*, 126154. Available from https://doi.org/10.1016/j.jhazmat.2021.126154.

Walaszczyk, A., & Galińska, B. (2020). Food origin traceability from a consumer's perspective. *Sustainability*, *12*(5), 1872.

Wang, D., Hou, F., Ma, X., Chen, W., Yan, L., Ding, T., Ye, X., & Liu, D. (2020). Study on the mechanism of ultrasound-accelerated enzymatic hydrolysis of starch: Analysis of ultrasound effect on different objects. *International Journal of Biological Macromolecules*, *148*, 493–500. Available from https://doi.org/10.1016/j.ijbiomac.2020.01.064.

Wang, M., Zhou, J., Collado, M. C., & Barba, F. J. (2021). Accelerated solvent extraction and pulsed electric fields for valorization of rainbow trout (*Oncorhynchus mykiss*) and sole (Dover sole) by-products: Protein content, molecular weight distribution and antioxidant potential of the extracts. *Marine Drugs*, *19*, 207.

Wolfert, S., Ge, L., Verdouw, C., & Bogaardt, M. J. (2017). Big data in smart farming—a review. *Agricultural Systems*, *153*, 69–80.

Wolfert, S., Goense, D., & Sørensen, C. A. G. (2014). A future internet collaboration platform for safe and healthy food from farm to fork. *2014 annual SRII global conference*, 266–273, IEEE.

Xu, X., Sharma, P., Shu, S., Lin, T.-S., Ciais, P., Tubiello, F. N., et al. (2021). Global greenhouse gas emissions from animal-based foods are twice those of plant-based foods. *National Food*, *2*(9), 724–732.

Yiannas, F. (2018). A new era of food transparency powered by blockchain. *Innovations: Technology, Governance, Globalization*, *12*(1-2), 46–56.

Yu, M., Kubiczek, J., Ding, K., Jahanzeb, A., & Iqbal, N. (2022). Revisiting SDG-7 under energy efficiency vision 2050: The role of new economic models and mass digitalization in OECD. *Energy Efficiency*, *15*(1), 1–20.

Zagklis, D., & Paraskeva, C. A. (2018). Isolation of organic compounds with high added values from agro-industrial solid wastes. *Journal of Environmental Management*, *216*, 183–191. Available from https://doi.org/10.1016/j.jenvman.2017.04.083.

Zhang, Q., Saleh, A. S. M., Chen, J., & Shen, Q. (2012). Chemical alterations taken place during deep-fat frying based on certain reaction products: A review. *Chemistry and Physics of Lipids*, *165*(6), 662–681. Available from https://doi.org/10.1016/j.chemphyslip.2012.07.002.

Zhongming, Z., Linong, L., Xiaona, Y., Wangqiang, Z., & Wei, L. (2021). *UNEP Food Waste Index Report*, 2021.

Chapter 3

Food Quality 4.0: contribution to sustainability

Abderrahmane Aït-Kaddour[1,2], Sofiane Boudalia[3,4], Fatih Özoğul[5], Semra Çiçek[6], Yana Cahyana[2] and Zuhaib F. Bhat[7]

[1]Université Clermont Auvergne, INRAE, VetAgro Sup, UMRF, Lempdes, France, [2]Department of Food Industrial Technology, Universitas Padjadjaran, Bandung, Indonesia, [3]Département d'Écologie et Génie de l'Environnement, Université 8 Mai 1945 Guelma, Guelma, Algeria, [4]Laboratoire de Biologie, Eau et Environnement, Université 8 Mai 1945 Guelma, Guelma, Algeria, [5]Department of Seafood Processing Technology, Faculty of Fisheries, Cukurova University, Balcali, Adana, Turkey, [6]Department of Agricultural Biotechnology, Faculty of Agriculture, Ataturk University, Erzurum, Turkey, [7]Division of Livestock Products Technology, SKUAST-J, Jammu and Kashmir, India

3.1 Introduction

The food industry is a heavyweight in economic terms and can be considered as a complex network of farmers and diverse businesses that together supply much of the food consumed by the world population. This supply capacity is principally associated with the possibility to safely preserve, package, and transport foods from across the globe until reaching the consumers' homes. By 2022, the global food market was estimated to be worth $2540.05 billion and is forecasted to grow by 6% per year to reach a value of $5194.60 billion by 2029 (http://www.fortunebusinessinsights.com). The rapid growth of this sector provides economic opportunities but the associated risks, such as dysfunctions of quality and safety in supply chains, can lead to traceability issues, health issues, and negative environmental impact, food wastage and energy loss. In addition, the food industry must cater to new dietary trends and consider the new consumers' requirements who are increasingly oriented toward specific requests, like, for example, zero defects, health benefits, limited environmental impact, and food product history through the supply chain, all demanding the implementation of the fourth industrial revolution.

The fourth industrial revolution, also known as Industry 4.0, is about automation and data exchange within manufacturing practices and technologies (Jagtap et al., 2021). It involves connected systems, also termed as Cyber-Physical Systems, leading to increased automation, better communication, self-monitoring, and the deployment of intelligent machines, which forms the basis for intelligent manufacturing or smart factory (Hassoun, Jagtap, Garcia-Garcia, Trollman, Pateiro, Lorenzo, Trif, Rusu, Aadil, Šimat, Cropotova, Câmara 2023; Jagtap et al., 2021). This new industry generation and digitalization provide a new opportunity to improve productivity, efficiency, safety, traceability, and quality of food without neglecting the great sustainable goal. To reach this target, the most important point is to manage the cornerstone of food quality. Food industry 4.0 technologies (e.g., artificial intelligence or AI, Big Data or BD, smart sensors, etc.) can be used for this purpose and give birth to the concept of Food Quality 4.0. The concept of Food Quality 4.0 has been recently defined (Hassoun, Jagtap, Garcia-Garcia, Trollman, Pateiro, Lorenzo, Trif, Rusu, Aadil, Šimat, Cropotova, Câmara 2023). Food Quality 4.0 is considered as the outcome of the use of Industry 4.0 technologies to determine food quality in the most efficient, rapid, and reliable manner through the supply chain (Hassoun, Jagtap, Garcia-Garcia, Trollman, Pateiro, Lorenzo, Trif, Rusu, Aadil, Šimat, Cropotova, Câmara 2023).

3.2 Definition and concepts

The Food Industry is a sector that encompasses all stages of the food chain such as production, harvesting, processing, food supply, packaging, transportation, distribution, consumption, and waste management (Sadiku et al., 2019). Under the influence of new mass markets, consumer demands, and requirements of the authorities, the Food Industry showed

Food Industry 4.0. DOI: https://doi.org/10.1016/B978-0-443-15516-1.00003-7

FIGURE 3.1 Food Industry 4.0, Food Quality 4.0, and food sustainability concepts.

serious improvement in processing, food storage, packaging, labeling, and transportation stages (Silva et al., 2018). Inspired by Industry 4.0, the concept of Food Industry 4.0 has emerged (Akyazi et al., 2020; Hassoun, Bekhit, et al., 2022). The Food Industry 4.0 has been characterized by new concepts including data analysis, cyber security, AI, basic digital skills, the Internet of Things (IoT) technology, blockchain, digital twins, smart sensors technology, BD, machine learning, collaborative/autonomous robots, traceability for food production, and consumption patterns (Fig. 3.1) (Hassoun, Aït-Kaddour, et al., 2022; Jambrak et al., 2021; Kayikci et al., 2022). Despite its evolution toward Industry 4.0, the key point of the agrifood sector is the management of both the quality and the safety of food.

Food quality can be defined as the combination of measurable and controllable characteristics of foods acceptable to the consumer, such as specific nutritional value, color, taste, odor, appearance, physical, chemical, and microbiological quality (García-Oliveira et al., 2022; Hassoun, Jagtap, Garcia-Garcia, Trollman, Pateiro, Lorenzo, Trif, Rusu, Aadil, Šimat, Cropotova, Câmara, 2023). Nowadays, a multistep, destructive, high-cost, low performance, using environmentally hazardous chemicals and time-consuming methods are used to measure these food characteristics, for example, Soxhlet solvent extraction, Kjeldahl and Dumas methods, CIELAB system, and total viable counts (TVCs) or polymerase chain reaction (PCR) (Brühl & Unbehend, 2021; Hassoun et al., 2020; Hassoun, Jagtap, Garcia-Garcia, Trollman, Pateiro, Lorenzo, Trif, Rusu, Aadil, Šimat, Cropotova, Câmara, 2023). In the past decade, these techniques changed rapidly due to the benefits of Industry 4.0 technologies and led to the rise of a new concept, Food Quality 4.0 (Sarkar et al., 2022). Food Quality 4.0 includes the use of Industry 4.0 technologies such as digital methods and automations, fingerprint technologies, smart sensors, omics technologies, bioinformatics, AI, and BD in the evaluation of food quality and can create fast, reliable, and objective data (Fig. 3.1) (Bouzembrak et al., 2019; Jeevanandam et al., 2021; Kakani et al., 2020; Marvin et al., 2022; Valdés et al., 2022) that can be used for a better food quality management matching higher sustainability.

3.3 Conventional methods and technologies

Before Industry 4.0, food systems and the food industry used physicochemical, microbiological, toxicological and sensory methods or techniques (Fig. 3.2) to assess the quality of foods (Hassoun, Jagtap, Garcia-Garcia, Trollman, Pateiro, Lorenzo, Trif, Rusu, Aadil, Šimat, Cropotova, Câmara, 2023; Hassoun, Jagtap, Trollman, et al., 2023; Romanello & Veglio, 2022). These methods are both used on the upstream (production stage: agriculture) and on the downstream part (transformation stage: food processing) of the agrifood supply chain. However, most of them remain expensive, destructive, time- and energy-consuming, and laborious and can generate toxic waste (Melini & Melini, 2020). Moreover, the use of these methods alone cannot meet the new consumer's expectations to achieve the sustainability of our food systems (reduce production waste and carbon footprint and promote livestock animal welfare) (Aprile & Punzo, 2022; Chen et al., 2022; Holotová et al., 2021). Consequently, to drive quality implementation in the agrifood sector, other innovative technologies, driven by Food Quality 4.0 are being developed and combined with the traditional methods when necessary.

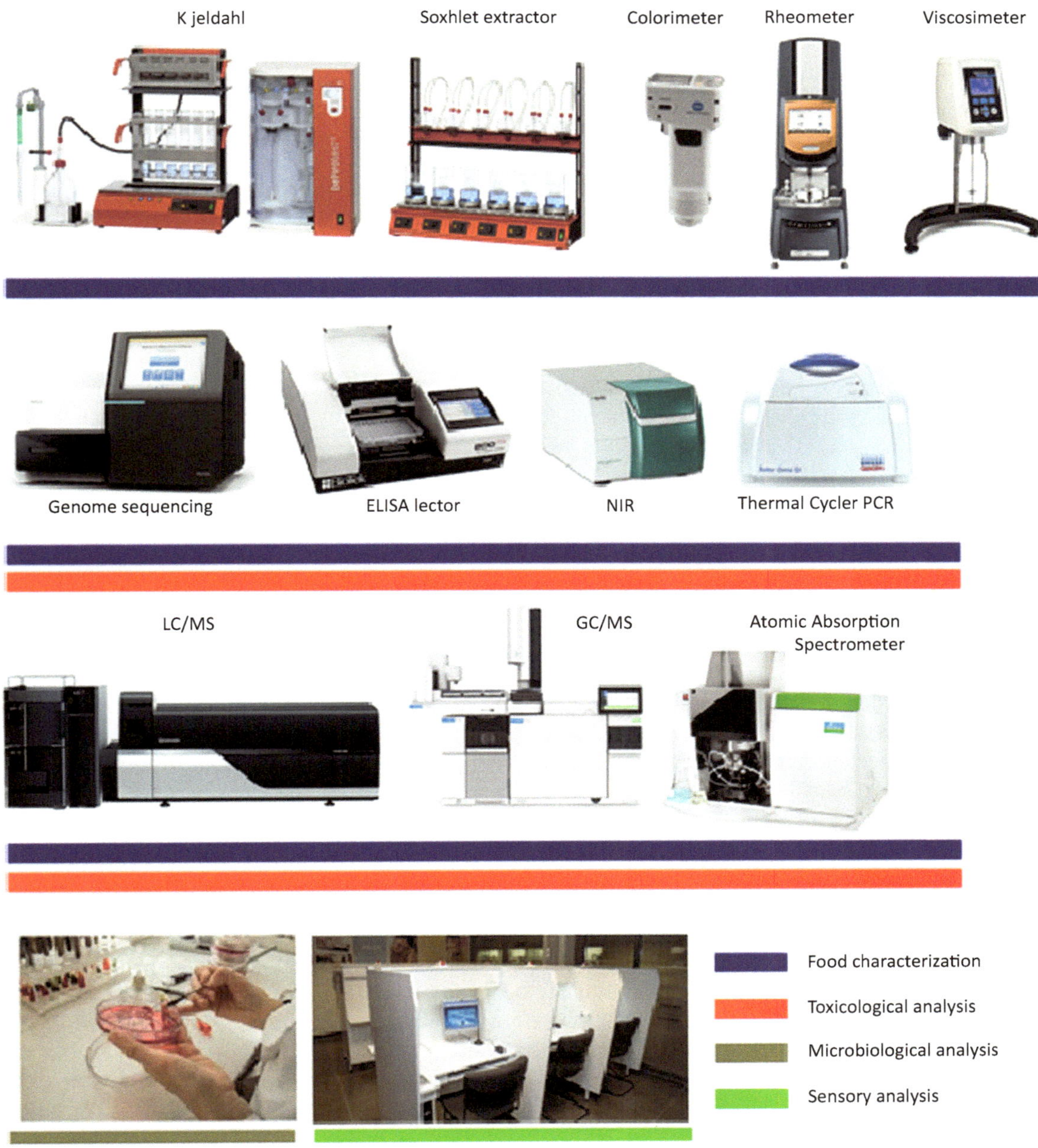

FIGURE 3.2 Conventional methods used in food quality assessment.

Among these conventional methods, the physicochemical analysis seems to be the most used to determine food proximate composition (moisture, ash, lipids, proteins, and carbohydrates) and other food compounds (e.g., vitamins and carotenoids) (Kamiloglu et al., 2020; Melini & Melini, 2020; Wu et al., 2022). Most of them are supported by international organizations (AACC, 2000; AOAC, 1995; AOCS, 1996).

Evaluation of the moisture and ash levels in most food matrices is done physically using oven drying, vacuum oven, freeze drying, or microwave oven. These techniques are based on direct or indirect measurements of water removed from the food. Furthermore, chemical methods are applicable such as Karl Fischer used in foods with low moisture content (Nielsen, 2006).

In the case of fat content, several techniques, including the Soxhlet, Goldfish, and Mojonnier methods, are developed. The extraction phase using an organic solvent, such as diethyl ether, petroleum ether, or trichloroethylene is required. The weight loss of the sample or the weight of fat extracted is used to determine the amount of fat present (Aravind et al., 2021). In some food matrices, such as dairy products, the Gerber method is developed using acid treatment rather than organic solvent. The addition of sulfuric acid and amyl alcohol to the sample initiates the digestion of protein, produces heat, and releases fat. Centrifugation and the additional hot water treatment cause the rising of fat to the graded portion of the test bottle where it can be measured by volume (Min & Ellefson, 2010). Concerning protein content evaluation, Kjeldahl and Dumas methods are both based on determining the nitrogen level that is converted to proteins using a specific conversion factor (6.25 for most foods). Proteins and other organic food components are digested with sulfuric acid in the presence of catalysts (mercury, copper, or selenium). Organic nitrogen is converted to ammonia, which is measured by titration or colorimetrically (Chang & Zhang, 2017). In the case of total carbohydrate analysis, physical measurements based on specific gravity or refractive index, reductiometric (e.g., Somogyi–Nelson method), and colorimetric (phenol–sulfuric acid method) are the oldest used methods. Moreover, specific and sensitive enzymatic methods (e.g., glucose-oxidase method) can be used to determine only a single component (BeMiller, 2017). Atomic emission spectroscopy (AAS), flame atomic emission spectroscopy (FAAS), inductively coupled plasma mass spectrophotometer (ICP-MS), inductively coupled plasma-optical emission spectrometry (ICP-OES), and inductively coupled plasma atomic emission spectrometry (ICP-AES) are among the most widely used techniques to assess the composition of minerals in different food matrices (Boudebbouz et al., 2021). Moreover, energy-dispersive X-ray fluorescence (ED-XRF) and wavelength-dispersive X-ray fluorescence (WD-XRF) can also be used to determine elemental food profiles (Herreros-Chavez et al., 2019). These analytical methods are also applicable to assess food contamination by heavy metals (Pb, Cd, and Cr) (Boudebbouz et al., 2022).

Rapid instrumental methods are now available and widely used to quantify food compounds. Hydrometry and refractometry are applicable in moisture quantification without water removal (Nielsen, 2006). Nuclear magnetic resonance, mass spectrometry, and chromatographic methods, such as gas–liquid chromatography and gas–solid chromatography, thin-layer chromatography, and high-performance liquid chromatography, are used to separate mixtures into their components, identify each component by retention time, and provide a measurement of the quantity of each component. These instrumental analytical methods are also used to monitor food authenticity (Esteki et al., 2021). These methods are unrestrictive and present many advantages, such as high sensitivity, good stability, and reproducibility, but need high acquisition and maintenance costs and rigorous calibration.

Microbiological analysis is widely used to assess food pathogens, spoilage microorganisms, and food-grade microorganisms (e.g., *Lactobacillus* spp. and *Bifidobacterium* spp.) (Moradi et al., 2021). The most common technique is still TVC after inoculation and growth monitoring of microorganisms on selective plates (Jasson et al., 2010). Other methodological approaches such as the enzyme-linked immunosorbent assays (ELISAs) and PCR characterized by their rapidity, mobility, high sensitivity, and low detection limit can be used extensively in food analysis (Tramuta et al., 2022; Xu et al., 2022).

Another approach to evaluate food quality is sensory analysis. However, the effectiveness of sensory methods, which combine the senses of sight, smell, taste, and touch, depends on the panelists' experience and training, and the results are often difficult to quantify (Wu et al., 2022). Traditionally, descriptive sensory analyses are applied but they are associated with disadvantages such as time consumption and high costs. Other instrumental methods can be used to evaluate textural parameters related to rheology (e.g., viscosity and elasticity) (Chen & Opara, 2013). These methods study the viscoelastic behavior of food, and the relationship between stress and material structure/deformation under different chemical, physical, or other conditions (Wang et al., 2019).

Besides texture, another parameter that can significantly affect the quality and acceptability of food is color (Brühl & Unbehend, 2021). Instrumental evaluation uses a colorimeter based on the CIELAB color space system, which determines the exact color of the food product in a three-dimensional space using three coordinates defined as L^*, a^*, and b^*. Additionally, other parameters such as chroma (C^*) and hue (h^*) can also be obtained from a^* and b^* values (Milovanovic et al., 2020). Color evaluation can also be performed by trained and qualified assessors or consumers (Boudalia et al., 2020). They provide a subjective hedonic score for color, which is often affected by several factors, such as testing conditions, lighting, color tones, training of assessors, cultural characteristics, and difficulty in finding appropriate matches between standards and tested samples (Hassoun, Jagtap, Garcia-Garcia, Trollman, Pateiro, Lorenzo, Trif, Rusu, Aadil, Šimat, Cropotova, Câmara, 2023; Tomasevic et al., 2021).

To analyze traceability and prevent adulteration, ELISA and DNA-based methods (DNA sequencing and PCR) are widely used to deliver the necessary information on the origin and the transport chains of food during its movement through the supply chain to reach food industries, retailers, and consumers (Guldiken et al., 2020). For example, the determination of genetically modified glyphosate-resistant transgenic soybeans was investigated by Qiang et al. (2016) using the PCR-ELISA method. Nevertheless, due to the low stability of protein-based methods (e.g., ELISA),

DNA-based methods offer more accurate results (Demeke & Dobnik, 2018; Qian et al., 2018). Although certain contaminants, such as mycotoxins in cereals, can be detected by the ELISA method, other analytical methods, such as ultra-high performance liquid chromatography-mass spectroscopy/mass spectroscopy (UHPLC-MS/MS) remain more sensitive (Zhang et al., 2023).

These conventional methods used to assess food quality are directly associated with the intrinsic characteristics of the product. During the last decade, other methods have been introduced to assess the quality of food that relates to its impact on our environment. Life cycle assessment (LCA) is one of the most influential methods that can affect decision-making regarding food consumption (Notarnicola et al., 2017). It is a powerful tool used to analyze the environmental impacts and resources used throughout a product's life, from raw material extraction to production (e.g., livestock, agriculture, fisheries, and aquaculture and packaging production), the product use, and finally waste disposal. The standardized methodology of LCA is described in ISO standards 14040−44 [International Organization for Standardization (ISO), 2006a,b] and results quantify the multiple environmental impacts using several indicators such as the global warming potential (GWP), Water Footprint Indicator, and waste production (Del Borghi et al., 2020). Several LCA studies have been conducted and published in agrifood sectors, defining the impact assessment of food products, such as breakfast cereals and snacks (Jeswani et al., 2015), dairy products (Berton et al., 2021; Famiglietti et al., 2019), meat (Ferronato et al., 2021), and pasta (Ruini et al., 2013).

3.4 Food quality sensors

Measurement of food quality and safety is of importance for the safe delivery of food products to consumers. Conventional analytical techniques used to measure food component stability are time consuming and laborious. This led to the development of different kinds of sensors capable of detecting the change in food components, such as vitamins, ethylene and other volatile gases, biogenic amines, or the presence of pesticide residues, toxins and pathogens, drugs, allergens, and undesirable food additives (sunset yellow, tartrazine, orange II, rhodamine B).

A sensor is a device capable of detecting, locating, or quantifying energy or matter followed by producing a signal for the detection or measurement of a physical or chemical property to which the device responds (Kerry et al., 2006; Müller & Schmid, 2019). A sensor typically consists of two components, namely receptor and transducer. The first component serves to detect the presence, activity, composition, or concentration of certain chemical or physical analytes. This physical or chemical information is then converted into a form of energy which is then measured by the transducer (Mlalila et al., 2016). The measured signal is further converted into a useful analytic signal, which can be an electrical, chemical, optical, or thermal signal.

Different kinds of sensors have been developed to analyze food safety, freshness, and spoilage in food products. Food quality can be determined indirectly by monitoring environmental indicators, such as time, temperature, and radio frequency identification, or directly by measuring chemicals, toxins, or unwanted pathogens present in food. Food spoilage can be detected from the presence of gases such as CO_2 or H_2S, which can be detected using nondispersive infrared sensors or chemical sensors (Pandey & Kim, 2007; Park et al., 2015). A chemical sensor can react to a specific target chemical (analyte) and transform the chemical data obtained by using the target substance concentration in a sample into an analytically useful signal.

Another type of sensor is the biosensor. This type of sensor comprises recognition elements, which can be biological materials (e.g., enzymes, antibodies, nucleic acids, and cell receptors), biologically derived material (e.g., aptamers and recombinant antibodies), or biomimetic material (e.g., imprinted polymers and synthetic catalysts). This sensor is coupled with transducing elements that measure physical activity, such as temperature, humidity, acceleration, and distance, after the alteration in chemical flow produced due to interaction with the analyte, and convert it into an electrical current (Ayankojo et al., 2020; El-Moghazy et al., 2018; Zhou et al., 2021).

In this section, two types of sensors, namely spectral and electrochemical are discussed. These two sensors are the most commonly used in food quality analysis. Furthermore, the application of different kinds of sensors in food intelligent packaging is also explored.

3.4.1 Spectral sensors

Recently, the need for in situ and on-the-field sensing has been driving tremendous progress for spectral sensors with reduced footprint, lower cost, increased portability, and consumer-friendliness, enabling them to be embedded into smartphones (Crocombe, 2018; Giussani et al., 2022; Yang, Albrow-Owen, et al., 2021). Spectroscopy-based sensors are being developed and applied in many applications. Spectroscopic methods are based on the interaction between electromagnetic radiation and matter at various wavelengths. Among the range of spectroscopy techniques, NIR spectrometers have become available in a miniaturized form and lower cost than traditional ones (Beć et al., 2021; Giussani et al., 2022).

The NIR spectrum particularly in the region 1000−1700 nm presents an absorption band arising from overtones and combinations of vibrational modes of O−H, C−H, and N−H bonds, providing interesting opportunities for organic material analysis. Along with its higher sensitivity compared to visible spectroscopy, NIR spectroscopy has wider application fields in agrifood and healthcare (Osborne, 2000; Pasquini, 2018).

The NIR analysis includes collecting and processing NIR spectra, eliminating abnormal samples, establishing (validating) the model using chemometric methods, and predicting the target parameters of the unknown sample to characterize its properties. To overcome the drawback of the NIR spectrum, the application of chemometrics is necessary. This technique allows for extracting chemically relevant information from NIR spectra and constructing calibration models to correlate spectral features with, for example, quality and safety parameters of food (Anjos et al., 2015; Daszykowski et al., 2008; Qu et al., 2015).

NIR-based sensors have been proved to be efficient in several applications. For example, the NIR spectroscopy method has been applied to determine the quality of oil seed and edible oil (Li et al., 2020). NIR application to identify the quality of beef, citrus fruits, cereals, and the adulteration paprika powder combined with chemometrics has also been successfully employed (De Marchi, 2013; Jespersen & Munck, 2009; Magwaza et al., 2012; Monago-Maraña et al., 2022).

Another promising method is hyperspectral imaging (HSI), which combines spectroscopy (commonly used vis/NIR, fluorescence, and Raman) and imaging methods to simultaneously obtain spectral and spatial information (Özdoğan et al., 2021; Qin et al., 2020). Features of HSI display desirable characteristics such as accuracy, high speed, automation, and real-time monitoring, which could be suitable for monitoring food quality and safety inspection of large sample sets, aligning well with the principles of Food Quality 4.0 (Hassoun, Jagtap, Trollman, et al., 2023). HSI has been applied to evaluate the sensory and freshness parameters (Özdoğan et al., 2021), authentication (Qin et al., 2020), and quality of egg (Yao, Sun, et al., 2022) and meat (Fu & Chen, 2019). In general, HSI is widely combined with the Vis/NIR range of 400−1000 nm. Using this method, several quality indicators such as TBARS, texture profile analysis (TPA), TVB-N and color can be predicted (Hassoun, Jagtap, Garcia-Garcia, Trollman, Pateiro, Lorenzo, Trif, Rusu, Aadil, Šimat, Cropotova, Câmara 2023).

For example, the prediction of TVB-N in pacific white shrimp using HSI at spectral range 860−1700 nm has been successfully determined using least-squares support-vector machine (LS-SVM) and partial least squares regression (PLSR) models (Guo et al., 2021; Yu et al., 2019). The PLSR model in HSI has also been used to predict the proximate composition and α-galactosides content in plant-based meat analogs using a spectral range of 950−1654 nm (Squeo et al., 2022).

3.4.2 Electrochemical sensors

Electrochemical sensors can be classified into amperometric, potentiometric, photoelectrochemical, impedimetric, and electrogenerated chemiluminescence. Amperometric sensors estimate the analyte's concentration using a voltage placed between electrodes and a reference. This voltage initiates electrochemical oxidation−reduction, and the resulting current, which is a quantitative indicator of the analyte's concentration, is then measured. Conductometric sensors, also known as impedimetric sensors, measure changes in the surface impedance to quantify analyte-specific recognition events on the electrode (Baranwal et al., 2022).

Current sensor technology is characterized by the development of better sensor materials to improve selectivity, sensitivity, and stability. Other trends in sensor technology include the development of fast sensors, an extension of sensor applications adapted to different pressures, temperatures, and aggressive media, and the development of biological principle-based sensors. Miniaturization of sensors is also an interesting trend in sensor technology because it allows the direct application of sensors in small volumes of media such as in biological media (Guth et al., 2009).

Electrochemical sensors in food particularly related to food safety or quality have been employed to detect different components such as the presence of heavy metals, pesticide residues, illegal additives, veterinary drugs, and microbial pathogens. The application of electrochemical sensors is tabulated in Table 3.1.

Heavy metal ions including Cd^{2+} and Pb^{2+} pose serious problems to health and the environment due to their prolonged half-life, nonbiodegradability, progressive concentration in food, and potential accumulation in the human body. Detecting the presence of these ions by conventional methods, such as atomic absorption spectroscopy (AAS), inductively coupled plasma mass spectrometry (ICPMS), high-performance liquid chromatography (HPLC-ICPMS), and surface-enhanced raman spectroscopy (SERS), is considered to be expensive and time consuming. The development of the electrochemical method overcomes such problems. The application of

TABLE 3.1 Electrochemical application in food quality and safety.

Detected components	Detecting electrode	Limit of detection or range of detection	References
Heavy metal: Cd^{2+} and Pb^{2+}	Nanostructured bismuth, antimony nanoparticle	0.4 μg/L (Cd^{2+}) 0.1 μg/L (Pb^{2+})	Serrano et al. (2016), Zidarič et al. (2017)
Pesticide residues: – Organophosphate – Methylcarbamate	– Coomassie brilliant blue dye surface–confined carbon nanoblack (GCE/CBnano@CoomBB)–modified electrode – Boron-doped diamond thin-film electrode	20 μL 5–20 ng/mL	Rao et al. (2002), Shalini Devi et al. (2018)
Illegal additive: Sudan I in ketchup sauce and chili powder	Gold nanoparticles (AuNPs)–decorated reduced graphene oxide (RGO)–modified electrode	–	Li et al. (2015)
Veterinary drug: β-lactam antibiotic residues in milk	Co^{2+}-tetradentate nitrilotriacetic acid (NTA)–modified screen-printed carbon electrodes combined with recombinant bacterial penicillin-binding protein (PBP) tagged by an N-terminal hexahistidine tail	–	Conzuelo et al. (2013)
Microbial pathogen: *Listeria monocytogenes*	Mouse monoclonal antibody immobilized on self-assembled monolayer (SAM)-modified gold (Au) electrodes	10^2 to 10^6 CFU/mL	Cheng et al. (2014)

electrochemical sensors using a nanostructured bismuth film electrode (nsBiFE) has been successfully employed to detect multiple heavy metals including Cd^{2+} and Pb^{2+}. The electrode is an anodic stripping voltammetry and mercury free (Serrano et al., 2016; Zidarič et al., 2017).

Pesticides, including organophosphate and carbamate, which are initially used to prevent pests can be harmful to health due to the presence of their residues in food and drinking water. Detection of these compounds using the electrochemical method has been applied. Acetylcholinesterase has been widely used for the detection of pesticides. Recently, organophosphorus hydrolase has also been examined and has shown promising results due to its high turnover rate and its potential to be reused hence suitability for continuous monitoring (Stoytcheva et al., 2016). A recent study on coomassie brilliant blue dye surface–confined carbon nanoblack (GCE/CBnano@CoomBB)-modified electrode shows promising results in the detection of pesticide residues particularly organophosphate. Similarly, the application of a boron-doped diamond thin-film electrode seems to be promising to detect methyl carbamate with a lower detection limit of 5–20 ng/mL (Rao et al., 2002; Shalini Devi et al., 2018).

Electrochemical sensors have also been applied to detect illegal additives, veterinary drugs, and microbial pathogens. Illegal additives such as Sudan I can be detected in ketchup sauce and chili powder using gold nanoparticles (AuNPs)-decorated reduced graphene oxide modified electrode. The presence of β-lactam antibiotic residues in milk can be detected using an electrochemical sensor (Conzuelo et al., 2013; Li et al., 2015). Microbial pathogens such as *Listeria monocytogenes* can be detected using mouse monoclonal antibodies immobilized on self-assembled monolayer (SAM)-modified gold (Au) electrodes. This method can even detect the microbial pathogen without sample pretreatment hence straightforward, reliable, simple, sensitive, and cheap method (Cheng et al., 2014).

Most recently, nitrite, which is not only naturally present in various vegetables but also commonly used as a food additive, has been studied using electrochemical sensors. Using Cu–Co Prussian Blue analog (CuII3[CoIII(CN)6]2)/ multiwalled carbon nanotubes (Cu–Co PBA/MWCNTs) composite, nitrite can be detected with a detection limit of 0.5 μM (Liu et al., 2019). Nitrite in food is also reported to be successfully detected using a Pt-Nafion-based nitric oxide sensor (Gill et al., 2019). $SiO_2/Al_2O_3/C$ grafted 3-n propylpyridinium silsesquioxane chloride-based nonenzymatic electrochemical sensor is another method showing encouraging results on nitrite detection (low limit of detection 0.01 μM) with high sensitivity, chemical stability, and insignificant interference from coexisting species present in sausage meat and food industry discharges (Alsaiari et al., 2022).

3.4.3 Sensors used in intelligent packaging

Different objectives of the food industry, such as protecting consumers from contaminated food products, increasing food productivity, and reducing spoilage-related losses, can be successfully achieved by monitoring food quality. Recent advance in packaging technology allows to monitor packaged food in real time, thanks to the development of sensor technology. This type of packaging is not only able to contain the food and protect it from the environment but also communicates the quality of packaged food with the consumers during storage (Becerril et al., 2021; Poyatos-Racionero et al., 2018).

Intelligent packaging may consist of three main parts, which are data carriers, indicators, and sensors (Ghaani et al., 2016). Data carriers, commonly placed on tertiary packaging, serve to store and transmit information about distribution and storage hence improving traceability, automatization, protection, or counterfeit protection (McFarlane & Sheffi, 2003). Barcode labels and radio frequency identification (RFID) tags are the most frequently used data carriers (Ghaani et al., 2016).

Indicators serve to determine the presence or absence of a substance, the extent of a reaction, or the concentration of a particular substance. Different indicators have been used in intelligent packaging including time–temperature indicators (TTIs), freshness indicators, and gas indicators (Müller & Schmid, 2019).

TTIs function is based on the detection of time and temperature-dependent chemical, mechanical, electrochemical, microbiological, and enzymatic changes in packaged food. Three types of TTIs are commonly found viz. critical temperature indicators, partial history indicators, and full history indicators. Critical temperature indicators can record the history indicating whether the products have been treated or stored at higher or lower than a permitted temperature. Partial history indicators can indicate whether a packaged food has been subjected to a temperature that can affect the food quality. Meanwhile, a full history indicator is capable of recording the complete temperature profile throughout the food supply chains (Lee & Rahman, 2013).

Freshness indicators are commonly placed inside the packaging to provide information on the loss of freshness due to microbiological growth, the presence of microbiological metabolites, or chemical changes (Kuswandi et al., 2013). Metabolites, such as glucose, carbon dioxide, organic acids, ethanol, biogenic amines, adenosine triphosphate (ATP) degradation products, volatile nitrogen compounds, and sulfuric compounds, are frequently used to monitor the quality of packaged food (Fang et al., 2017; Nopwinyuwong et al., 2010). Meanwhile, gas indicators provide information about food quality triggered by the change in the atmosphere inside the packaging due to enzymatic or chemical reactions or gas-generating microorganism metabolism. The concentration of gases, such as O_2, CO_2, water vapor, hydrogen sulfide, and ethanol, is believed to correlate with the extent of spoilage (Fang et al., 2017; Meng et al., 2014).

The third component in intelligent packaging is a sensor. Types of sensors commonly used in intelligent packaging are gas sensors and biosensors. Broadly speaking, applications of intelligent packaging in various food products have been widely studied, showing interesting results.

The application of TTIs in food is based on the change in irreversible discoloration of the indicator. The indicator is made up of a wick porous with dyed fatty acid ester, which goes to different layers and colors with the increase in temperature. The change in color is irreversible even after the product is returned to a more appropriate storage temperature. Many types of TTIs, also known as diffusion-based TTIs, are commercially available, such as 3 M MonitorMark indicators (9860C model), and have been applied to monitor the quality of nonpasteurized fresh juices. The indicator displays a change of color when the juice is stored above 10°C (Soltani Firouz et al., 2021).

Another type of indicator is a microbial TTI that is formed from a label containing lactic acid bacteria placed inside the packaging. The growth of bacteria in food with the increase in temperature leads to the release of more lactic acid compounds hence lower pH, which is proportional to the bacterial growth. A study to monitor food quality based on lactic acid vapor using bromothymol blue and methyl red-based indicator showed promising results as the indicator color changed from bright light green to bright red when exposed to 0.01 M lactic acid (Wanihsuksombat et al., 2010). The application of freshness indicators on meat or fish products shows interesting results in which a number of marker metabolites associated with muscle food products have been developed. Changes in the concentration of organic acid compounds such as butyrate, lactic acid, lactate, and acetic acid during storage can be used as an indicator of meat quality. Another compound such as ethanol has been reported to increase as a function of the storage time of anaerobically packaged marinated chicken hence a potential indicator for fermentative metabolism of lactic acid bacteria (Kerry et al., 2006).

Indicators of meat product decomposition can also be determined from the presence of biogenic amines such as histamine, putrescine, tyramine, and cadaverine. Natural pH dye such as anthocyanin has been successfully applied to monitor the freshness of chicken patties stored at 4°C for 21 days. The indicator stripe that is initially violet fades on the 18th day and turns to yellow on the 21st day. These changes in color correlate with the decrease in pH, the increase in TVB-N level, and the total viable count (Talukder et al., 2020).

Gas indicators can be used to monitor the change of metabolites such as CO_2, H_2S, NH_3, and dimethylamine ($(CH_3)_2NH$) and trimethylamine (C_3H_9N) as quality indices (Lee et al., 2019). A study on Kimchi using a BTB^-/tetra-butylammonium (TBA^+) ion-paired dye to monitor CO_2 level in the headspace during fermentation demonstrates an interesting result in which the indicator turns from blue to yellow with the decrease in pH (Lyu et al., 2019).

3.5 Artificial intelligence, machine learning, and Big Data

AI, machine learning, and BD are increasingly used in several applications in agriculture and the food industry.

3.5.1 Agriculture management

The agrifood supply chain is a complex set of activities that includes different stakeholders (e.g., farming, processing, storage, transportation, and distribution), enabling it to produce and provide quality foods to consumers. The agrifood supply chain needs to be constantly updated due to the increasing need for transparency, sustainability, and adapt to growing challenges, such as increased population, climate change, high production cost, reduced yield, soil degradation, and water quality (Papaioannou et al., 2020; Spanaki et al., 2021). AI, BD, machine learning, and IoT are examples of Industry 4.0 enablers that are proposed as solutions to these challenges for agricultural management (Ferrag et al., 2021; Kovács & Husti, 2018; Raj et al., 2021; Spanaki et al., 2021; Zhai et al., 2020). Although promising solutions for the application of AI in agriculture based on remote sensing, digital imaging for soil and crop tracking have emerged in the last 50 years, data gaps have hampered the efficient development and applications of such solutions (Zhang et al., 2021). However, BD, IoT, and machine learning have taken the role of AI in agriculture management to the next level. As the processing capacity of data increases and its beneficial uses is better understood, AI and machine learning are emerging in agricultural management as very useful tools for using IoT and BD leading to highly accurate predictions (Fig. 3.3). Bu and Wang (2019) have developed a smart agriculture IoT system based on deep reinforcement learning combined in the cloud layer to make immediate smart decisions such as determining the amount of water needed to be irrigated to improve crop growth environment.

It is very important to obtain data simultaneously on issues such as strengthening the decision-making mechanism, providing minimum input and maximum output, control, and follow-up in agricultural management. The simultaneous data provided by remote GPS-enabled farm equipment fitted with sensors about soil, weather, climate, and crops

FIGURE 3.3 Agriculture management 4.0.

constitute BD. BD can be used in agricultural management for simultaneous intervention to the target to strengthen raw material productivity and sustainability in agriculture to face climate change, market openings, and consumer expectations. Due to limited internet access and bandwidth, the use of BD in agricultural management has been delayed in some developing countries. However, the rapid spread of smart cell phones and mobile broadband connections will accelerate the use of BD in agricultural management in these countries (Belaud et al., 2019; Coble et al., 2018; Himesh et al., 2018; Jaipong et al., 2022). The FAO improved the Agricultural Metadata Element Set (AgMES) to process the enormous amount of agriculture data, including phenotypic and genomic data about plant growth, resistance capacity to diseases and pests, and high-yield characteristics, which are essential to improve high-yielding and resistant varieties. Companies, such as Deere, Dow, and Monsanto, have developed agricultural management using BD analytics (Sarkar et al., 2022). BD obtained by using environmental sensors, cameras, microphones, infrared cameras, wearable sensors, biosensors, and other sensor technologies for the monitoring of relative humidity levels, ammonia and carbon dioxide gas concentrations, food intake, animal weights, disease-causing viruses affecting the health of animals, and so on is a good resource for animal farming management. Thanks to BD for animal farming management, the need for manual observation and human decision-making has been reduced, and real-time monitoring and management of farm animals, transparency, and automated manipulation of certain inputs have been enhanced (Astill et al., 2020; Ellis et al., 2020; Neethirajan, 2020). In addition, there are studies reporting that production has increased using BD from sources such as soil moisture sensors, raindrop sensors, temperature sensors, and weather forecasts for automatic irrigation systems (Jha et al., 2019; Savitha, 2018; Talaviya et al., 2020).

There are reviews on the beneficial uses of machine learning in agricultural management in terms of product, livestock, water, and soil (Chlingaryan et al., 2018; Liakos et al., 2018). In agricultural management, machine learning has potential in areas such as disease detection, air temperature and humidity, carbon dioxide concentration detection, weed detection, yield estimation, crop quality, species identification, animal welfare, detection of water evaporation rates, and determination of soil moisture and temperature. Jin et al. (2016) reported that Particle Swarm Optimization algorithm assimilated field spectroscopic data into the AquaCrop model, which helped to improve the estimation accuracy of winter wheat yield under different planting dates and irrigation management strategies. Liang et al. (2019) demonstrated the successful validation of the SPACSYS model (mixed dimensional, multilayer, field scale, weather-driven, and daily-time-step dynamic simulation model) (Wu et al., 2007) for simulating crop yields and azote uptake in wheat–maize rotation in arid and semiarid regions and in maize monoculture system in semihumid regions of China. Xu et al. (2019) showed the usage of random forest and support vector machine algorithms for the prediction of metabolic status (concentrations of plasma β-hydroxybutyrate, free fatty acids, glucose, insulin, and insulin-like growth factor 1) of dairy cows in early lactation using on-farm cow data. Schmeling et al. (2021) developed a machine learning model (random forest models) for the sensor-based monitoring of lying behavior in dairy cows on a pasture and in the barn. Volkmann et al. (2021) proposed a method related to the detection of claw lesions based on the acoustic analysis of a cow's gait using a machine learning process with a random forest algorithm (sensitivity of 0.81 and specificity of 0.97). Goldstein et al. (2018) showed that field data, such as soil moisture, weather, irrigation characteristics, and resulting yield could be fused via machine learning algorithms to provide automated irrigation recommendations. Gumiere et al. (2020) compared the capabilities of a machine learning–based model and a physics-based model to predict soil matric potential in the root zone for real-time irrigation management. They concluded that the machine learning–based model could achieve better performance when compared to the physics-based model, but forecasting accuracy decreased rapidly toward the end of the 72-hour lead time. In addition, they emphasized that the machine learning model has the advantage of being easy to implement, as it needs a smaller computational effort for training. Studies on the use of random forest, instance-based models, regressions, artificial and deep neural networks, tree-based models, Bayesian models, K-nearest neighbor, and support vector machine learning models in soil management have been reported (Benos et al., 2021; Johann et al., 2016; Liakos et al., 2018; Motia & Reddy, 2021; Nahvi et al., 2016). Motia and Reddy (2021) emphasized that the use of machine learning algorithms is advantageous in determining physicochemical parameters such as temperature, pH, salinity, organic matter content, moisture/water content, bulk density of the soil, and soil classification. In addition, studies on the use of machine learning algorithms together with existing methods such as remote sensing used to monitor soil indicators in agriculture are important as they increase attention to the potential of machine learning in soil management (Diaz-Gonzalez et al., 2022).

3.5.2 Certification and traceability

Certification is very important in terms of quality, safety, and traceability of food, which could increase trust in food companies, promote sustainability, and satisfy customers in the supply process of foods from raw materials to consumers (Sun & Wang, 2019). Zhang et al. (2019) proposed a HACCP-based smart traceability strategy to ensure safety, quality control, and transparency during Chinese sturgeon transportation. For this, they combined a Quick Response code with the electronic product code of the RFID tag, providing users with wireless tracking. They proposed that potential risks can be reduced, quality control can be monitored with survival outcomes, and the volume of fish transport can be expanded at low cost thanks to intelligent monitoring. In the Food Industry 4.0 area, there are a few certificate-oriented research or review articles. Chkanikova and Sroufe (2021) conducted a study on three different certificates (unilateral Tesco Nurture, collaborative GlobalGAP, and multistakeholder UTZ certification) using a case study, collective action logic, and dynamic talent theory methods. They concluded in this study that retail efforts to develop certifications are driven by better alignment with the business's goal of improving the sustainability performance of supply chains. For example, the program, which uses the Ethereum blockchain of Gustav Gerig's Marine Stewardship Council-certified canned and bagged rose tuna series, sold under the Raimond Freres brand in Switzerland, was actively working. Khan et al. (2022) suggested that the distribution and accumulative score-based certification of shrimp exports from Bangladesh could strengthen food safety and traceability with a blockchain technology called ShrimChain. In addition, the researchers stated that involving shrimp farmers in the certification process can be an important parameter for high-quality shrimp production and consumer satisfaction in high-value markets.

Food traceability can be considered as a methodology allowing to examine the supply chain, collecting all basic data about products and processes, detailing the data, and making it understandable to ensure transparency and accountability from the raw material to the consumer (Hassoun, Alhaj Abdullah, et al., 2022; Liu, Wu, et al., 2022; Sinha et al., 2021). Food traceability 4.0 includes the use of Industry 4.0 technologies in this process (Hassoun, Alhaj Abdullah, et al., 2022). AI and BD, which are important enabling technologies of Industry 4.0, have the potential to take an active role in the traceability of foods. AI has found many applications in the fields of smart agriculture, food processing, food safety, food packaging, and marketing presenting various advantages for all collaborators of the food industry. Simultaneous observation and real-time monitoring, reduction of human error with digitalization, increase in efficiency, prevention of losses, estimation of quality and shelf life, and detection of fraud are some of these advantages (Hassoun, Alhaj Abdullah, et al., 2022; Marvin et al., 2022; Sharma et al., 2021). AI applications use machine learning and deep learning approaches to enhance these advantages in food traceability. Robots that can take human power in food processing processes, smart sensors that can predict the characteristics, and shelf life of foods that affect consumer acceptability can be combined with AI (Addanki et al., 2022; Kler et al., 2022; Kumar et al., 2021).

AI-related algorithms, such as machine learning, vector machines, and artificial neural networks, have been reported for various applications in food recognition, classification, adulteration and contamination detection, simultaneous food quality detection and analysis, and monitoring of foodborne diseases (Goyal et al., 2022; Lebelo et al., 2022; Wang et al., 2022). Marvin et al. (2016) demonstrated that Bayesian network modeling and various data sources were able to predict food fraud with 91.5% accuracy. Iymen et al. (2020) reported that an AI-related neural network algorithm was developed using sound vibrations to help consumers detect the presence of vegetable oils in butter. Moreover, the developed model can be embedded in a mobile phone application. Violino et al. (2020) reported that the geographical origin of extra virgin olive oil was predicted thanks to the AI model based on neural networks applied to spectral data with an accuracy of 94.6%. Laponogov et al. (2021) reported that a network-based machine learning algorithm comparing experimental anti-COVID-19 drugs at the genome level was used to identify potential anti-COVID-19 molecules in foods such as strawberries, cruciferous vegetables, citrus fruits, onions, and garlic.

Inaccurate data in BD have the potential to be misleading. Therefore cleaning and analysis of raw data are very important. With the analysis of BD, there is a high potential for users to improve their traceability processes and simultaneous decision-making mechanisms (Jin et al., 2020; Wolfert et al., 2017). Kler et al. (2022) reported the usability of BD provided by simultaneous IoT-based sensors to provide food quality assurance in foods where temperature control is important during the transportation and storage of food products, such as vegetables, fruits, milk, and ice cream.

3.5.3 Food supply chain management

Food supply chain refers to a series of activities involving companies implied in the procurement, transformation, product processing, packaging, storage, distribution, and marketing of raw materials and inputs. There are also collaborators operating in areas such as communication, labeling, analysis, financial activities, and legal regulations in the food supply chain. Each of the actors in the food supply chain is interconnected and has a responsibility for the end product. Moreover, the globalization of the food industry has made the food supply process longer and more complex and has also brought monitoring and security challenges. In addition, uncertainties such as counterfeit products and inefficient processes complicate the management of the food supply process (Kayikci et al., 2022). Therefore food supply chain management has to be in strong communication with each of these actors. All collaborators in the food supply chain must be controlled and monitored simultaneously to promote sustainability in food safety, reduce waste, food loss, economic burden, and cross-contamination, and obtain quality products. Cyber-physical systems, IoT, BD, automation and smart robotics, cloud computing, machine learning, and AI applications, which are the main enablers of Industry 4.0, have the potential to be used effectively to achieve this purpose (Barreto et al., 2017; Witkowski, 2017).

Studies on the potential of AI applications, which include algorithms, systems, and machines that symbolize intelligence in the digital future, are increasing in food supply chain management. Indeed, growing evidence shows the need for AI and its related applications for decreasing food wastage and providing food safety, cost savings, transparency, traceability, and better data management (Dora et al., 2022; Kollia et al., 2021). AI supports machine learning algorithms using BD to improve simultaneous tracking and decision-making in the food supply chain (Fig. 3.4) (Di Vaio et al., 2020; Olan et al., 2022).

Monitoring food storage can help minimize food losses. Szwedziak et al. (2020) proposed an artificial neural network model that can simultaneously monitor the moisture, protein, and gluten content of stored wheat using data from NIR spectroscopy combined with data obtained from the humidity and temperature sensors (AM2302). Osinenko et al. (2021) reported that BD from nondestructive sensors and vis/NIR spectroscopy was recognized as being able to detect fruit firmness and physiological storage disorders in Braeburn apples via autoregressive time series models after long-term controlled atmosphere storage. Lorenc et al. (2021) suggested an estimate method to prevent deterioration related to temperature changes in the cold chain supply of foods. BD obtained from wireless temperature sensors based on the Bluetooth low-energy technology during the entire operating cycle of the thermal container was analyzed using an artificial neural network. Yang, Liu, et al. (2021) reported a pathogen identification system using a paper chromogenic array enabled by a multilayer neural network algorithm to detect contaminated food. This system identified *Escherichia coli* O157:H7 and *Listeria monocytogenes* on lettuce with high accuracy (91%–95%). Xia et al. (2021) showed the usability of PLSR and back-propagation artificial neural network to determine simultaneously the moisture, fat content, and storage time of leisurely dried tofu thanks to data from a low-field nuclear magnetic resonance spectrometer. Gorzelany et al. (2022) modeled the mechanical properties of fresh and stored large cranberries using multiple linear regression and machine learning based on artificial neural networks and support vector algorithms. They emphasized that machine learning techniques were better than multiple linear regression.

It is hoped that estimation of process kinetics based on BD, AI, and machine learning will benefit the problems such as long processing time, high energy use, and decrease in product quality of food processing methods such as drying, frying, baking, heating, canning, extrusion, encapsulation, fermentation, and modified atmosphere packaging

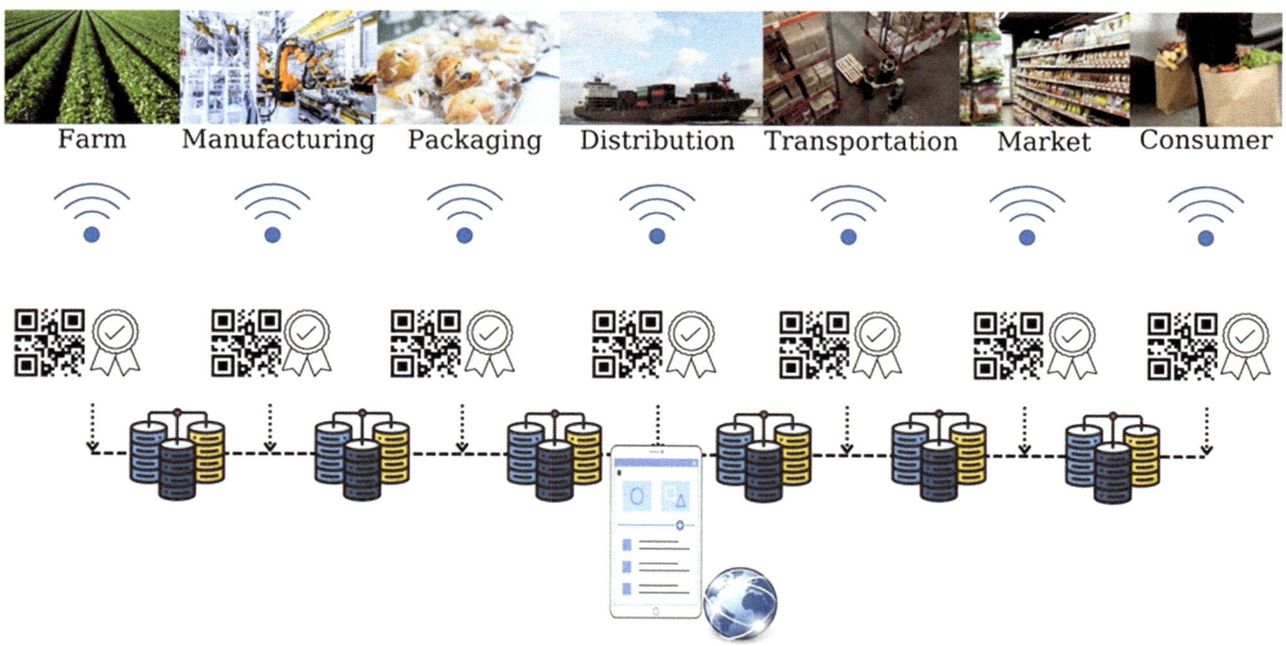

FIGURE 3.4 Food supply chain.

(Hürdoğan et al., 2022; Khan et al., 2022). Hürdoğan et al. (2022) reported that the kinetic data such as humidity rate, drying rate, and heat transfer coefficient can be used to predict the experimental performance of solar dryers over the drying of the shelled peanuts using multiple linear regression, decision tree, and support vector machine models. They stated that the humidity value could be estimated with the highest accuracy ($R^2 = 0.9972$) with the Random Tree Model. Gholamibozanjani et al. (2021) developed a mathematical model including the variable space network and the enthalpy method to explain the simultaneous transfer of heat and moisture during French fries.

In addition, there are potential applications of machine learning in energy consumption and calculation in food processing, waste management, and food sales forecasting (Garre et al., 2020; Milczarski et al., 2020; Tsoumakas, 2019).

3.5.4 Food safety and nutritional quality

The globalization of the food industry makes it difficult to ensure food safety and prevent foodborne diseases. For example, cases such as the *E. coli* contaminated lettuce, the Polish apple with a high chemical residual level exceeding eight times the permitted level, and the fipronil (insecticide) contaminated eggs caused the disposal of many food consignments including healthy products for protecting the public health and resulted in serious economic losses, market losses, and consumer concerns. Considering the increase in food demand and consumer expectations, it is understood that current practices may not be an effective option for ensuring future food safety (Walaszczyk et al., 2019). BD, AI, and machine learning strategies will be useful for the simultaneous monitoring and control of the safety of the nutritional quality of food from industry to marketing and consumer (Hassoun, Jagtap, Garcia-Garcia, Trollman, Pateiro, Lorenzo, Trif, Rusu, Aadil, Šimat, Cropotova, Câmara 2023; Savina et al., 2020).

BD is one of the most important forces that will guide future applications in food safety (Jin et al., 2020). Food industry collaborators can use online food safety databases such as The Rapid Alert System for Food and Feed or RASFF (EU key tool to ensure the flow of information to enable swift reaction when risks to public health are detected in the *food* chain), the Joint FAO/WHO Expert Committee on Food Additives Database, Food Contaminants database, the United States Import Refusal Report, WHO Collaborating Centers Database, Inspection Classification Database, and FAO/WHO Chronic individual food consumption database. There are reports on the use of digital images based on smartphones as a BD source for classification of milk samples, detection of protein content in milk, food contaminants, and pathogenic bacteria (*E. coli* O157:H7) in beef and spinach, fluoroquinolone antibiotics in food and simultaneous quality assessment (Jung et al., 2020; Liu et al., 2017; Tsakanikas et al., 2020; Ye et al., 2020). In addition, it was pointed out that BD from online surveys from social media platforms can be used to understand customers' perceptions of food safety and nutritional quality, to raise awareness, and to improve consumers' behavior (Soon, 2020). Moreover, sources including video monitoring, sensors and portable devices using IoT technology, Geographic Information System, and satellite imagery also helped to obtain data about food safety. For example, satellite images obtained with Sentinel-2, the FAO GeoNetwork, Landsat imagery database, and remote sensing technology have applications in the detection of plant diseases, crop growth, and contamination. Appropriate use of BD obtained from different sources in the monitoring of food safety and nutritional quality will improve the simultaneous access of food industry collaborators to information about the food product. It will also help optimize activity and logistics in this area, accelerate decision-making, reduce food spoilage and wastage, and prevent economic losses (Singh & Corradini, 2022).

Machine learning stands out as one of the analysis methods of BD obtained in food safety and nutritional quality. Batool et al. (2022) proposed a machine learning–based technique to improve food safety. The photos recorded with high-resolution cameras were sent to a cloud server using IoT. Attributes were determined using the principal component analysis technique from photographs separated by K-means clustering method, and finally, photographs were categorized by machine learning algorithms such as cuckoo optimization algorithm, k-nearest neighbors (KNN), and Naive Bayesian model. Na et al. (2020) analyzed 750 photographs of tomatoes with four diseases (bacterial cankers, late blight, leaf miners, and powdery mildew) using a deep learning algorithm combining Faster R-convolution neural network with ResNet-101 model, resulting in 90.8% accuracy.

Tan et al. (2022) developed a deep learning–assisted (convolutional neural network) visualized fluorometric array-based sensing method for the detection of biogenic amines in meat to monitor food quality and predict pathogenic microbial activities. Van De Looverbosch et al. (2021) used X-ray computed tomography scans along with the U-net-based model (image segmentation network) as a deep neural network to measure the internal defects of pear fruit caused by inadequate storage conditions. For pear fruit, excellent classification accuracies of 99.4% and 92.2% were obtained for the classification of "consumable" versus "nonconsumable" fruit on the one hand and "healthy" versus "defect but consumable" versus "nonconsumable" fruit on the other hand. The classification was done successfully thanks to the formation of internal voids in the fruit and browning data. Chakraborty et al. (2021) suggested the use of vis/NIR spectroscopy HSI and KNN (accuracy of 98.2%) and partial least-squares discriminant analysis (accuracy of 94.7%) algorithms to predict aflatoxin-B1 concentrations (25, 40, 70, 200, 300, and 500 ppb) in 240 maize kernels. Wang, Liang, et al. (2022) proposed a novel machine learning–based weight-of-evidence model for prioritizing food contact chemicals for carcinogenicity evaluation using structural alerts, toxicogenic scores, and quantitative structure–activity relationship models.

Studies have reported that machine learning can be used in genomic sequencing and metagenomic applications in foodborne pathogen detection (Kim & Kim, 2022; Lupolova et al., 2019). Bisgin et al. (2018) reported that an average accuracy of 85% was achieved in species identification with the support vector machine model using 625 features selected in 15 different insect wing photographs (6900 images), in the context of insect-based food contamination. Bağcıoğlu et al. (2019) demonstrated that machine learning (artificial neural network algorithm)–based FTIR can be effectively used to identify *Bacillus cereus, Bacillus cytotoxicus, Bacillus thuringiensis, Bacillus mycoides*, and *Bacillus weihenstephanensis* as the FTIR spectra exhibit different patterns for each microorganism (99.5% accuracy). Im et al. (2021) developed a machine learning model using the support vector machine algorithm to evaluate the pathogenic potential of the Shiga toxin–producing *E. coli* isolates using their whole-genome sequencing data.

Christodoulides et al. (2019) reported the opportunities of the electronic taste chip systems using BD and mimicking the perception of tastes, such as salty, sweet, sour, bitter, and glutamate, with the connection of AI and machine learning. In addition, Zhao et al. (2020) emphasized that AI can enhance sensors' ability to imitate human senses, that is, vision, hearing, smell, taste, and touch, providing opportunities for assessment of the sensory quality of food.

The data company Tellspec uses a cloud-based analysis engine and mobile application combined with food sensors based on NIR spectroscopy, bioinformatics methods, and machine learning algorithms to provide simultaneous molecular analysis of food. Tellspec has also been used in studies involving the simultaneous estimation of melamine and urea in wheat gluten samples, the differentiation of tenderloin and sirloin, and the aging of beef (Bazar et al., 2017; Kovacs et al., 2017).

3.6 Recommendations and future trends

Globalization has transformed the way we produce our food and has constrained the world food industry to rapidly shift toward high-quality, high-volume, and safe food production (Nakat & Bou-Mitri, 2021). To meet this purpose, several accurate, rapid, and nondestructive assessment tools and Industry-4.0 technologies have emerged over the past few decades to evaluate food quality and risk assessment. While traditional food quality assessment methods and technologies, such as sensory analysis and HPLC–based food testing, can produce meaningful data about food quality and safety; Industry 4.0 technologies, such as BD, deep learning, and AI technologies, can be used to evaluate data produced at all stages of the supply chain to make sustainable accurate food risk assessment and predictions. This proactive approach can help quality assurance and food safety professionals to handle and avoid situations such as major product recalls and prevent any damage to the brand's reputation and trust among consumers and the involved financial expenditures (Stoitsis et al., 2023). While there have been efforts from the stakeholders to safeguard food supply chains and to prepare for future crises, the COVID-19 pandemic has highlighted the importance of automation and digitalization of various operations in the food industry and the need to develop nondestructive, rapid, and accurate technologies to measure food quality to ensure food safety and uninterrupted food supply. The COVID-19 pandemic has given an impetus to the adoption and need for automation and robotics in the food industry during the last two years (Hassoun, Bekhit, et al., 2022). However, a limited number of studies have covered the application of Industry 4.0 technologies in the food industry compared to other industrial sectors. Initially, an appropriate combination of traditional and Industry 4.0 technologies would be an ideal strategy for quality production of diverse food products and to increase valorization of underutilized raw materials, by-products, and side-streams.

The introduction of Quality 4.0 concept in the food industry offers various advantages including automation and digitalization of food quality analysis. With an increase in industrial automation, AI and machine learning are finding more and more applications in a wide range of industries including food industry and agriculture, and digitalization is becoming crucial for business survival (Liu, Wang, et al., 2022). AI and BD are the two main enablers of Food Quality 4.0 and have the potential to revolutionize the food industry. The adoption of data analytical tools offers several benefits to the food industry, such as food safety, real-time decision-making, demand forecasting, and food waste management, whereas AI has a role to play in solving complex tasks, such as food quality and safety check, food sorting, and process optimization and has already transformed some areas of food processing and manufacturing environments. Barriers such as data security, limited shared data, lack of system standards, and legal issues need to be addressed before the full exploitation of BD in food production can be realized. Better-performing AI-based applications will be introduced to the food industry in the near future in several areas of food production increasing efficiency and accelerating the move toward intelligent food processing and quality management. Integration of smart sensors with other prediction technologies and AI and machine learning–based algorithms, such as artificial neural networks, are expected to be a source of greater resource efficiency and an enabler of massive environmental benefits. Moreover, these strategies are found successful in performing several food processing tasks, such as food grading, quality check, safety, traceability, packaging, food processing quality, and other stages in the supply and cold chains (Nayak et al., 2020). Adoption of Quality 4.0 will ensure the best quality management practices at all the levels of food supply chain. However, to optimally implement the Quality 4.0 system and to harness its full potential in food quality, close cooperation and collaboration are required between the actors involved. While a recent increase in interest has been seen in the research community and among the policymakers and managers regarding the adoption of Industry 4.0, there is a general lack of awareness about food quality aspects and uncertainties with respect to the wider adoption of these technologies in the food industry.

The implementation of Industry 4.0 technologies in medium and small-sized enterprises demands the identification of measures, such as roadmaps, incentives, and consultancy services, which will facilitate the successful transition to digital technologies (Matt et al., 2020). To facilitate the adoption process and accelerate the implementation of Food Quality 4.0, long-term policy-making strategies and extensive research are needed to expand our understanding of the challenges involved in the adoption of Quality 4.0 in agriculture and the food industry. Some of these challenges include the lack of availability of affordable technologies that are efficient under industrial environments and setups and the availability of skilled technical workers. The high capital and operational costs, inadequate infrastructure facilities especially in developing countries, and lack of adaptability to the existing industrial environments hinder the wider application of these technologies. These challenges need to be addressed for a successful transition from the traditional to Food Quality 4.0 system.

3.7 Conclusions

This chapter critically evaluated the potential and benefits of smart technologies and emerging analytical methods available for improving food quality in the context of the Quality 4.0 system. The role of various food quality sensors and the use of Industry 4.0 digital technologies, such as AI, BD, and machine learning, have been discussed in detail in areas such as agriculture and food supply chain management, certification and traceability, and food safety and nutritional quality. These technologies are revolutionizing the way food is produced and consumed worldwide, increasing quality through smart production and digitalization within agriculture and the food industry, and have led to the emergence of new food trends. The adoption of these technologies in the agrifood industry can offer valuable outcomes for public health and economic and environmental gains and can aid in achieving more sustainable food production and consumption systems. Extensive research is required to overcome the challenges involved in the adoption of the Quality 4.0 system and during the transition process. However, the adoption of Industry 4.0 and Quality 4.0 technologies alone will not be enough to transform the food industry, especially in the developing world, and efforts are required from all directions and all the stakeholders including change in international policies, especially climate and food-related strategies.

References

Addanki, M., Patra, P., & Kandra, P. (2022). Recent advances and applications of artificial intelligence and related technologies in the food industry. *Applied Food Research*, *2*(2), 100126. Available from https://doi.org/10.1016/j.afres.2022.100126.

Akyazi, T., Goti, A., Oyarbide, A., Alberdi, E., & Bayon, F. (2020). A guide for the food industry to meet the future skills requirements emerging with industry 4.0. *Foods*, *9*(4). Available from https://doi.org/10.3390/foods9040492, https://www.mdpi.com/2304-8158/9/4/492.

Alsaiari, M., Saleem, A., Alsaiari, R., Muhammad, N., Latif, U., Tariq, M., Almohana, A., & Rahim, A. (2022). $SiO_2/Al_2O_3/C$ grafted 3-n propylpyridinium silsesquioxane chloride-based non-enzdymatic electrochemical sensor for determination of carcinogenic nitrite in food products. *Food Chemistry*, *369*, 130970. Available from https://doi.org/10.1016/j.foodchem.2021.130970.

Anjos, O., Campos, M. G., Ruiz, P. C., & Antunes, P. (2015). Application of FTIR-ATR spectroscopy to the quantification of sugar in honey. *Food Chemistry*, *169*, 218–223. Available from https://doi.org/10.1016/j.foodchem.2014.07.138, http://www.elsevier.com/locate/foodchem.

AACC. (2000). *Approved Methods of the American Association of Cereal Chemists (AACC)*. (Volumes 1 á 2). From American Association of Cereal Chemists. Approved Methods Committee. 1200 pp.

AOAC. (1995). *Official methods of analysis of AOAC international*. Elsevier.

AOCS. (1996). Official Methods and Recommended Practices of the American Oil Chemist's Society, Physical and Chemical Characteristics of Oils, Fats and Waxes, Section I. Ed. The AOCS Methods Editor and the AOCS Technical Department. 54 pages. *Lipid/Fett, 99*(5). Available from https://doi.org/10.1002/lipi.19970990510.

Aprile, M. C., & Punzo, G. (2022). How environmental sustainability labels affect food choices: Assessing consumer preferences in southern Italy. *Journal of Cleaner Production*, *332*. Available from https://doi.org/10.1016/j.jclepro.2021.130046, https://www.journals.elsevier.com/journal-of-cleaner-production.

Aravind, S., Debabrata, B., Ragupathi, P., & Vignesh, G. (2021). Investigation on algae oil extraction from algae Spirogyra by Soxhlet extraction method. *Materials Today: Proceedings*, *43*, 308–313. Available from https://doi.org/10.1016/j.matpr.2020.11.668.

Astill, J., Dara, R. A., Fraser, E. D. G., Roberts, B., & Sharif, S. (2020). Smart poultry management: Smart sensors, big data, and the internet of things. *Computers and Electronics in Agriculture*, *170*. Available from https://doi.org/10.1016/j.compag.2020.105291, http://www.elsevier.com/inca/publications/store/5/0/3/3/0/4.

Ayankojo, A. G., Reut, J., Ciocan, V., Öpik, A., & Syritski, V. (2020). Molecularly imprinted polymer-based sensor for electrochemical detection of erythromycin. *Talanta*, *209*, 120502. Available from https://doi.org/10.1016/j.talanta.2019.120502.

Bağcıoğlu, M., Fricker, M., Johler, S., & Ehling-Schulz, M. (2019). Detection and identification of bacillus cereus, bacillus cytotoxicus, bacillus thuringiensis, bacillus mycoides and bacillus weihenstephanensis via machine learning based FTIR spectroscopy. *Frontiers in Microbiology*, *10*. Available from https://doi.org/10.3389/fmicb.2019.00902, https://www.frontiersin.org/journals/microbiology#.

Baranwal, J., Barse, B., Gatto, G., Broncova, G., & Kumar, A. (2022). Electrochemical sensors and their applications: A review. *Chemosensors*, *10*(9). Available from https://doi.org/10.3390/chemosensors10090363, http://www.mdpi.com/journal/chemosensors.

Barreto, L., Amaral, A., & Pereira, T. (2017). Industry 4.0 implications in logistics: An overview. *Procedia Manufacturing*, *13*, 1245–1252. Available from https://doi.org/10.1016/j.promfg.2017.09.045, http://www.journals.elsevier.com/procedia-manufacturing.

Bazar, G.K., Isabel,Z.H., Beyerer, J., Puente Leon, F., & Langle, T. (2017). 3 OCM: 3rd International conference on optical characterization of materials Karlsruhe, Germany. KIT Scientific Publishing Detection of beef aging combined with the differentiation of tenderloin and sirloin using a handheld NIR scanner http://file:///C:/Users/aaitkadd/Downloads/OCM_2017_ISBN_9783731506126-Proceedings.pdf.

Batool, A., Ganguli, S., Almashaqbeh, H. A., Shafiq, M., Vallikannu, A. L., Sankaran, K. S., Ray, S., & Sammy, F. (2022). An IoT and machine learning-based model to monitor perishable food towards improving food safety and quality. *Journal of Food Quality*, *2022*. Available from https://doi.org/10.1155/2022/6302331, https://www.hindawi.com/journals/jfq/.

Becerril, R., Nerín, C., & Silva, F. (2021). Bring some colour to your package: Freshness indicators based on anthocyanin extracts. *Trends in Food Science and Technology*, *111*, 495–505. Available from https://doi.org/10.1016/j.tifs.2021.02.042, http://www.elsevier.com/wps/find/journaldescription.cws_home/601278/description#description.

Beć, K. B., Grabska, J., & Huck, C. W. (2021). Principles and applications of miniaturized near-infrared (NIR) spectrometers. *Chemistry - A European Journal*, *27*(5), 1514–1532. Available from https://doi.org/10.1002/chem.202002838, http://onlinelibrary.wiley.com/journal/10.1002/(ISSN)1521-3765.

Belaud, J. P., Prioux, N., Vialle, C., & Sablayrolles, C. (2019). Big data for agri-food 4.0: Application to sustainability management for by-products supply chain. *Computers in Industry*, *111*, 41–50. Available from https://doi.org/10.1016/j.compind.2019.06.006, https://www.journals.elsevier.com/computers-in-industry.

BeMiller, J. N. (2017). *Carbohydrate analysis* (pp. 333–360). Springer Science and Business Media LLC. Available from http://doi.org/10.1007/978-3-319-45776-5_19.

Benos, L., Tagarakis, A. C., Dolias, G., Berruto, R., Kateris, D., & Bochtis, D. (2021). Machine learning in agriculture: A comprehensive updated review. *Sensors*, *21*(11). Available from https://doi.org/10.3390/s21113758, https://www.mdpi.com/1424-8220/21/11/3758/pdf.

Berton, M., Bovolenta, S., Corazzin, M., Gallo, L., Pinterits, S., Ramanzin, M., Ressi, W., Spigarelli, C., Zuliani, A., & Sturaro, E. (2021). Environmental impacts of milk production and processing in the Eastern Alps: A "cradle-to-dairy gate" LCA approach. *Journal of Cleaner Production*, *303*, 127056. Available from https://doi.org/10.1016/j.jclepro.2021.127056.

Bisgin, H., Bera, T., Ding, H., Semey, H. G., Wu, L., Liu, Z., Barnes, A. E., Langley, D. A., Pava-Ripoll, M., Vyas, H. J., Tong, W., & Xu, J. (2018). Comparing SVM and ANN based machine learning methods for species identification of food contaminating beetles. *Scientific Reports*, *8*(1). Available from https://doi.org/10.1038/s41598-018-24926-7, http://www.nature.com/srep/index.html.

Boudalia, S., Boudebbouz, A., Gueroui, Y., Bousbia, A., Benada, M., Leksir, C., Boukaabene, Z., Saihi, A., Touaimia, H., Aït-Kaddour, A., & Chemmam, M. (2020). Characterization of traditional Algerian cheese "Bouhezza" prepared with raw cow, goat and sheep milks. *Food Science and Technology*, *40*(suppl 2), 528–537. Available from https://doi.org/10.1590/fst.35919.

Boudebbouz, A., Boudalia, S., Bousbia, A., Gueroui, Y., Boussadia, M. I., Chelaghmia, M. L., Zebsa, R., Affoune, A. M., & Symeon, G. K. (2022). Determination of heavy metal levels and health risk assessment of raw cow milk in Guelma Region, Algeria. *Biological Trace Element Research*. Available from https://doi.org/10.1007/s12011-022-03308-1, http://www.springer.com/humana + press/journal/12011.

Boudebbouz, A., Boudalia, S., Bousbia, A., Habila, S., Boussadia, M. I., & Gueroui, Y. (2021). Heavy metals levels in raw cow milk and health risk assessment across the globe: A systematic review. *Science of the Total Environment*, *751*. Available from https://doi.org/10.1016/j.scitotenv.2020.141830, http://www.elsevier.com/locate/scitotenv.

Bouzembrak, Y., Klüche, M., Gavai, A., & Marvin, H. J. P. (2019). Internet of things in food safety: Literature review and a bibliometric analysis. *Trends in Food Science and Technology*, *94*, 54–64. Available from https://doi.org/10.1016/j.tifs.2019.11.002, http://www.elsevier.com/wps/find/journaldescription.cws_home/601278/description#description.

Brühl, L., & Unbehend, G. (2021). Precise color communication by determination of the color of vegetable oils and fats in the CIELAB 1976 (L*a*b*) color space. *European Journal of Lipid Science and Technology*, *123*(7). Available from https://doi.org/10.1002/ejlt.202000329, http://onlinelibrary.wiley.com/journal/10.1002/(ISSN)1438-9312.

Bu, F., & Wang, X. (2019). A smart agriculture IoT system based on deep reinforcement learning. *Future Generation Computer Systems*, *99*, 500–507. Available from https://doi.org/10.1016/j.future.2019.04.041.

Chakraborty, S. K., Mahanti, N. K., Mansuri, S. M., Tripathi, M. K., Kotwaliwale, N., & Jayas, D. S. (2021). Non-destructive classification and prediction of aflatoxin-B1 concentration in maize kernels using Vis–NIR (400–1000 nm) hyperspectral imaging. *Journal of Food Science and Technology*, *58*(2), 437–450. Available from https://doi.org/10.1007/s13197-020-04552-w, http://www.springerlink.com/content/121580/.

Chang, S. K. C., & Zhang, Y. (2017). *Protein analysis* (pp. 315–331). Springer Science and Business Media LLC. Available from https://doi.org/10.1007/978-3-319-45776-5_18.

Chen, L., & Opara, U. L. (2013). Approaches to analysis and modeling texture in fresh and processed foods—A review. *Journal of Food Engineering*, *119*(3), 497–507. Available from https://doi.org/10.1016/j.jfoodeng.2013.06.028.

Chen, Y. P., Feng, X., Blank, I., & Liu, Y. (2022). Strategies to improve meat-like properties of meat analogs meeting consumers' expectations. *Biomaterials*, *287*, 121648. Available from https://doi.org/10.1016/j.biomaterials.2022.121648.

Cheng, C., Peng, Y., Bai, J., Zhang, X., Liu, Y., Fan, X., Ning, B., & Gao, Z. (2014). Rapid detection of Listeria monocytogenes in milk by self-assembled electrochemical immunosensor. *Sensors and Actuators, B: Chemical*, *190*, 900–906. Available from https://doi.org/10.1016/j.snb.2013.09.041.

Chkanikova, O., & Sroufe, R. (2021). Third-party sustainability certifications in food retailing: Certification design from a sustainable supply chain management perspective. *Journal of Cleaner Production*, *282*. Available from https://doi.org/10.1016/j.jclepro.2020.124344, https://www.journals.elsevier.com/journal-of-cleaner-production.

Chlingaryan, A., Sukkarieh, S., & Whelan, B. (2018). Machine learning approaches for crop yield prediction and nitrogen status estimation in precision agriculture: A review. *Computers and Electronics in Agriculture*, *151*, 61–69. Available from https://doi.org/10.1016/j.compag.2018.05.012, http://www.elsevier.com/inca/publications/store/5/0/3/3/0/4.

Christodoulides, N., McRae, M. P., Simmons, G. W., Modak, S. S., & McDevitt, J. T. (2019). Sensors that learn: The evolution from taste fingerprints to patterns of early disease detection. *Micromachines*, *10*(4). Available from https://doi.org/10.3390/mi10040251, https://res.mdpi.com/micromachines/micromachines-10-00251/article_deploy/micromachines-10-00251.pdf?filename = &attachment = 1.

Coble, K. H., Mishra, A. K., Ferrell, S., & Griffin, T. (2018). Big data in agriculture: A challenge for the future. *Applied Economic Perspectives and Policy*, *40*(1), 79–96. Available from https://doi.org/10.1093/aepp/ppx056, http://aepp.oxfordjournals.org/.

Conzuelo, F., Gamella, M., Campuzano, S., Martínez-Ruiz, P., Esteban-Torres, M., De Las Rivas, B., Reviejo, A. J., Muñoz, R., & Pingarrón, J. M. (2013). Integrated amperometric affinity biosensors using Co^{2+}-tetradentate nitrilotriacetic acid modified disposable carbon electrodes: Application to the determination of β-lactam antibiotics. *Analytical Chemistry*, *85*(6), 3246–3254. Available from https://doi.org/10.1021/ac303604b.

Crocombe, R. A. (2018). Portable spectroscopy. *Applied Spectroscopy*, *72*(12), 1701–1751. Available from https://doi.org/10.1177/0003702818809719, https://journals.sagepub.com/loi/asp.

Daszykowski, M., Wrobel, M. S., Czarnik-Matusewicz, H., & Walczak, B. (2008). Near-infrared reflectance spectroscopy and multivariate calibration techniques applied to modelling the crude protein, fibre and fat content in rapeseed meal. *The Analyst*, *133*(11), 1523. Available from https://doi.org/10.1039/b803687j.

Del Borghi, A., Moreschi, L., & Gallo, M. (2020). *Life cycle assessment in the food industry. The interaction of food industry and environment* (pp. 63–118). Italy: Elsevier. Available from https://www.sciencedirect.com/book/9780128164495, 10.1016/B978-0-12-816449-5.00003-5.

Demeke, T., & Dobnik, D. (2018). Critical assessment of digital PCR for the detection and quantification of genetically modified organisms. *Analytical and Bioanalytical Chemistry*, *410*(17), 4039–4050. Available from https://doi.org/10.1007/s00216-018-1010-1, http://link.springer.de/link/service/journals/00216/index.htm.

De Marchi, M. (2013). On-line prediction of beef quality traits using near infrared spectroscopy. *Meat Science*, *94*(4), 455–460. Available from https://doi.org/10.1016/j.meatsci.2013.03.003.

Diaz-Gonzalez, F. A., Vuelvas, J., Correa, C. A., Vallejo, V. E., & Patino, D. (2022). Machine learning and remote sensing techniques applied to estimate soil indicators – Review. *Ecological Indicators*, *135*. Available from https://doi.org/10.1016/j.ecolind.2021.108517, http://www.elsevier.com/locate/ecolind.

Di Vaio, A., Boccia, F., Landriani, L., & Palladino, R. (2020). Artificial intelligence in the agri-food system: Rethinking sustainable business models in the COVID-19 scenario. *Sustainability (Switzerland)*, *12*(12). Available from https://doi.org/10.3390/SU12124851, https://www.mdpi.com/2071-1050/12/12/4851.

Dora, M., Kumar, A., Mangla, S. K., Pant, A., & Kamal, M. M. (2022). Critical success factors influencing artificial intelligence adoption in food supply chains. *International Journal of Production Research*, *60*(14), 4621–4640. Available from https://doi.org/10.1080/00207543.2021.1959665, http://www.tandfonline.com/toc/tprs20/current.

Ellis, J. L., Jacobs, M., Dijkstra, J., van Laar, H., Cant, J. P., Tulpan, D., & Ferguson, N. (2020). Review: Synergy between mechanistic modelling and data-driven models for modern animal production systems in the era of big data. *Animal*, *14*, s223–s237. Available from https://doi.org/10.1017/S1751731120000312, https://www.journals.elsevier.com/animal/.

El-Moghazy, A. Y., Zhao, C., Istamboulie, G., Amaly, N., Si, Y., Noguer, T., & Sun, G. (2018). Ultrasensitive label-free electrochemical immunosensor based on PVA-co-PE nanofibrous membrane for the detection of chloramphenicol residues in milk. *Biosensors and Bioelectronics*, *117*, 838–844. Available from https://doi.org/10.1016/j.bios.2018.07.025, http://www.elsevier.com/locate/bios.

ISO 14044:2006. (2006a). Environmental management, Life cycle assessment, Requirements and guidelines.

ISO 14040:2006. (2006b). Environmental management, Life cycle assessment, Principles and framework.

Esteki, M., Heydari, E., Simal-Gandara, J., Shahsavari, Z., & Mohammadlou, M. (2021). Discrimination of pistachio cultivars based on multi-elemental fingerprinting by pattern recognition methods. *Food Control*, *124*. Available from https://doi.org/10.1016/j.foodcont.2021.107889, https://www.journals.elsevier.com/food-control.

Famiglietti, J., Guerci, M., Proserpio, C., Ravaglia, P., & Motta, M. (2019). Development and testing of the product environmental footprint milk tool: A comprehensive LCA tool for dairy products. *Science of the Total Environment*, *648*, 1614–1626. Available from https://doi.org/10.1016/j.scitotenv.2018.08.142, http://www.elsevier.com/locate/scitotenv.

Fang, Z., Zhao, Y., Warner, R. D., & Johnson, S. K. (2017). Active and intelligent packaging in meat industry. *Trends in Food Science and Technology*, *61*, 60–71. Available from https://doi.org/10.1016/j.tifs.2017.01.002, http://www.elsevier.com/wps/find/journaldescription.cws_home/601278/description#description.

Ferrag, M. A., Shu, L., Djallel, H., & Choo, K. K. R. (2021). Deep learning-based intrusion detection for distributed denial of service attack in agriculture 4.0. *Electronics (Switzerland)*, *10*(11). Available from https://doi.org/10.3390/electronics10111257, https://www.mdpi.com/2079-9292/10/11/1257/pdf.

Ferronato, G., Corrado, S., De Laurentiis, V., & Sala, S. (2021). The Italian meat production and consumption system assessed combining material flow analysis and life cycle assessment. *Journal of Cleaner Production*, *321*, 128705. Available from https://doi.org/10.1016/j.jclepro.2021.128705.

Fu, X., & Chen, J. (2019). A review of hyperspectral imaging for chicken meat safety and quality evaluation: Application, hardware, and software. *Comprehensive Reviews in Food Science and Food Safety*, *18*(2), 535–547. Available from https://doi.org/10.1111/1541-4337.12428, http://onlinelibrary.wiley.com/journal/10.1111/(ISSN)1541-4337.

García-Oliveira, P., Fraga-Corral, M., Pereira, A. G., Prieto, M. A., & Simal-Gandara, J. (2022). Solutions for the sustainability of the food production and consumption system. *Critical Reviews in Food Science and Nutrition*, *62*(7), 1765–1781. Available from https://doi.org/10.1080/10408398.2020.1847028, https://www.tandfonline.com/loi/bfsn20.

Garre, A., Ruiz, M. C., & Hontoria, E. (2020). Application of machine learning to support production planning of a food industry in the context of waste generation under uncertainty. *Operations Research Perspectives*, *7*. Available from https://doi.org/10.1016/j.orp.2020.100147, http://www.journals.elsevier.com/operations-research-perspectives/.

Ghaani, M., Cozzolino, C. A., Castelli, G., & Farris, S. (2016). An overview of the intelligent packaging technologies in the food sector. *Trends in Food Science and Technology*, *51*, 1–11. Available from https://doi.org/10.1016/j.tifs.2016.02.008, http://www.elsevier.com/wps/find/journaldescription.cws_home/601278/description#description.

Gholamibozanjani, G., Leong, S. Y., Oey, I., Bremer, P., Silcock, P., & Farid, M. (2021). Heat and mass transfer modeling to predict temperature distribution during potato frying after pre-treatment with pulsed electric field. *Foods*, *10*(8). Available from https://doi.org/10.3390/foods10081679, https://www.mdpi.com/2304-8158/10/8/1679/pdf.

Gill, A., Zajda, J., & Meyerhoff, M. E. (2019). Comparison of electrochemical nitric oxide detection methods with chemiluminescence for measuring nitrite concentration in food samples. *Analytica Chimica Acta*, *1077*, 167–173. Available from https://doi.org/10.1016/j.aca.2019.05.065, http://www.journals.elsevier.com/analytica-chimica-acta/.

Giussani, B., Gorla, G., & Riu, J. (2022). Analytical chemistry strategies in the use of miniaturised NIR instruments: An overview. *Critical Reviews in Analytical Chemistry*, 1–33. Available from https://doi.org/10.1080/10408347.2022.2047607.

Goldstein, A., Fink, L., Meitin, A., Bohadana, S., Lutenberg, O., & Ravid, G. (2018). Applying machine learning on sensor data for irrigation recommendations: revealing the agronomist's tacit knowledge. *Precision Agriculture*, *19*(3), 421–444. Available from https://doi.org/10.1007/s11119-017-9527-4, http://www.wkap.nl/journalhome.htm/1385-2256.

Gorzelany, J., Belcar, J., Kuźniar, P., Niedbała, G., & Pentoś, K. (2022). Modelling of mechanical properties of fresh and stored fruit of large cranberry using multiple linear regression and machine learning. *Agriculture*, *12*(2), 200. Available from https://doi.org/10.3390/agriculture12020200.

Goyal, K., Kumar, P., & Verma, K. (2022). Food adulteration detection using artificial intelligence: A systematic review. *Archives of Computational Methods in Engineering*, *29*(1), 397–426. Available from https://doi.org/10.1007/s11831-021-09600-y, http://www.springerlink.com/content/1134-3060.

Guldiken, B., Karliga, S., Capanoglu, E., Yolci-Omeroglu, P., & Kamiloglu, S. (2020). *Food traceability innovative food analysis* (pp. 249–268). Canada: Elsevier. Available from https://www.elsevier.com/books/innovative-food-analysis/galanakis/978-0-12-819493-5, https://doi.org/10.1016/B978-0-12-819493-5.00009-1.

Gumiere, S. J., Camporese, M., Botto, A., Lafond, J. A., Paniconi, C., Gallichand, J., & Rousseau, A. N. (2020). Machine learning vs. physics-based modeling for real-time irrigation management. *Frontiers in Water*, *2*. Available from https://doi.org/10.3389/frwa.2020.00008, http://www.frontiersin.org/journals/water.

Guo, W., Li, X., & Xie, T. (2021). Method and system for nondestructive detection of freshness in Penaeus vannamei based on hyperspectral technology. *Aquaculture (Amsterdam, Netherlands)*, *538*, 736512. Available from https://doi.org/10.1016/j.aquaculture.2021.736512.

Guth, U., Vonau, W., & Zosel, J. (2009). Recent developments in electrochemical sensor application and technology—A review. *Measurement Science and Technology*, *20*(4), 042002. Available from https://doi.org/10.1088/0957-0233/20/4/042002.

Hassoun, A., Alhaj Abdullah, N., Aït-Kaddour, A., Ghellam, M., Beşir, A., Zannou, O., Önal, B., Aadil, R. M., Lorenzo, J. M., Mousavi Khaneghah, A., & Regenstein, J. M. (2022). Food traceability 4.0 as part of the fourth industrial revolution: key enabling technologies. *Critical Reviews in Food Science and Nutrition*. Available from https://doi.org/10.1080/10408398.2022.2110033, https://www.tandfonline.com/loi/bfsn20.

Hassoun, A., Aït-Kaddour, A., Abu-Mahfouz, A. M., Rathod, N. B., Bader, F., Barba, F. J., Biancolillo, A., Cropotova, J., Galanakis, C. M., Jambrak, A. R., Lorenzo, J. M., Måge, I., Ozogul, F., & Regenstein, J. (2022). The fourth industrial revolution in the food industry—Part I: Industry 4.0 technologies. *Critical Reviews in Food Science and Nutrition*. Available from https://doi.org/10.1080/10408398.2022.2034735, https://www.tandfonline.com/loi/bfsn20.

Hassoun, A., Bekhit, A. E. D., Jambrak, A. R., Regenstein, J. M., Chemat, F., Morton, J. D., Gudjónsdóttir, M., Carpena, M., Prieto, M. A., Varela, P., Arshad, R. N., Aadil, R. M., Bhat, Z., & Ueland, Ø. (2022). The fourth industrial revolution in the food industry—part II: Emerging food trends. *Critical Reviews in Food Science and Nutrition*. Available from https://doi.org/10.1080/10408398.2022.2106472, https://www.tandfonline.com/loi/bfsn20.

Hassoun, A., Cropotova, J., Rustad, T., Heia, K., Lindberg, S. K., & Nilsen, H. (2020). Use of spectroscopic techniques for a rapid and non-destructive monitoring of thermal treatments and storage time of sous-vide cooked cod fillets. *Sensors (Switzerland)*, *20*(8). Available from https://doi.org/10.3390/s20082410, https://www.mdpi.com/1424-8220/20/8/2410/pdf.

Hassoun, A., Jagtap, S., Garcia-Garcia, G., Trollman, H., Pateiro, M., Lorenzo, J. M., Trif, M., Rusu, A. V., Aadil, R. M., Šimat, V., Cropotova, J., & Câmara, J. S. (2023). Food quality 4.0: From traditional approaches to digitalized automated analysis. *Journal of Food Engineering*, *337*. Available from https://doi.org/10.1016/j.jfoodeng.2022.111216, http://www.sciencedirect.com/science/journal/02608774.

Hassoun, A., Jagtap, S., Trollman, H., Garcia-Garcia, G., Abdullah, N. A., Goksen, G., Bader, F., Ozogul, F., Barba, F. J., Cropotova, J., Munekata, P. E. S., & Lorenzo, J. M. (2023). Food processing 4.0: Current and future developments spurred by the fourth industrial revolution. *Food Control*, *145*. Available from https://doi.org/10.1016/j.foodcont.2022.109507, https://www.journals.elsevier.com/food-control.

Herreros-Chavez, L., Morales-Rubio, A., & Cervera, M. L. (2019). Green methodology for quality control of elemental content of infant milk powder. *LWT*, *111*, 484–489. Available from https://doi.org/10.1016/j.lwt.2019.05.055, https://www.journals.elsevier.com/lwt.

Himesh, S., Prakasa Rao, E. V. S., Gouda, K. C., Ramesh, K. V., Rakesh, V., Mohapatra, G. N., Kantha Rao, B., Sahoo, S. K., & Ajilesh, P. (2018). Digital revolution and big data: A new revolution in agriculture. *CAB Reviews: Perspectives in Agriculture, Veterinary Science, Nutrition and Natural Resources*, *13*. Available from https://doi.org/10.1079/PAVSNNR201813021, https://www.cabi.org/cabreviews/FullTextPDF/2018/20183235919.pdf.

Holotová, M., Horská, E., & Nagyová, Ĺ. (2021). Changing patterns of sustainable food consumption regarding environmental and social impact-insights from Slovakia. *Frontiers in Sustainable Food Systems*, *5*. Available from https://doi.org/10.3389/fsufs.2021.703827, http://www.frontiersin.org/journals/sustainable-food-systems#.

Hürdoğan, E., Çerçi, K. N., Saydam, D. B., & Ozalp, C. (2022). Experimental and modeling study of peanut drying in a solar dryer with a novel type of a drying chamber. *Energy Sources, Part A: Recovery, Utilization and Environmental Effects*, *44*(2), 5586–5609. Available from https://doi.org/10.1080/15567036.2021.1974126, http://www.tandf.co.uk/journals/titles/15567036.asp.

Im, H., Hwang, S. H., Kim, B. S., & Choi, S. H. (2021). Pathogenic potential assessment of the Shiga toxin-producing *Escherichia coli* by a source attribution-considered machine learning model. *Proceedings of the National Academy of Sciences of the United States of America*, *118*(20). Available from https://doi.org/10.1073/pnas.2018877118, https://doi.org/10.1073/pnas.2018877118.

Iymen, G., Tanriver, G., Hayirlioglu, Y. Z., & Ergen, O. (2020). Artificial intelligence-based identification of butter variations as a model study for detecting food adulteration. *Innovative Food Science and Emerging Technologies*, *66*, 102527. Available from https://doi.org/10.1016/j.ifset.2020.102527, http://www.elsevier.com/locate/ifset.

Jagtap, S., Saxena, P., & Salonitis, K. (2021). Food 4.0: Implementation of the augmented reality systems in the food industry. *Procedia CIRP*, *104*, 1137–1142. Available from https://doi.org/10.1016/j.procir.2021.11.191, 22128271; Elsevier B.V. United Kingdom, http://www.sciencedirect.com/science/journal/22128271.

Jaipong, P., Sriboonruang, P., Siripipattanakul, S., Sitthipon, T., Kaewpuang, P., & Auttawechasakoon, P. (2022). A review of intentions to use artificial intelligence in big data analytics for Thailand agriculture. *Review of advanced multidisciplinary Science, Engineering & Innovation (RAMSEY)*, *1*(2).

Jambrak, A. R., Nutrizio, M., Djekić, I., Pleslić, S., & Chemat, F. (2021). Internet of nonthermal food processing technologies (Iontp): Food industry 4.0 and sustainability. *Applied Sciences (Switzerland)*, *11*(2), 1–20. Available from https://doi.org/10.3390/app11020686, https://www.mdpi.com/2076-3417/11/2/686/pdf.

Jasson, V., Jacxsens, L., Luning, P., Rajkovic, A., & Uyttendaele, M. (2010). Alternative microbial methods: An overview and selection criteria. *Food Microbiology*, *27*(6), 710–730. Available from https://doi.org/10.1016/j.fm.2010.04.008.

Jeevanandam, J., Agyei, D., Danquah, M. K., & Udenigwe, C. (2021). *Food quality monitoring through bioinformatics and big data. Future foods: Global trends, opportunities, and sustainability challenges* (pp. 733–744). Portugal: Elsevier. Available from https://www.sciencedirect.com/book/9780323910019, https://doi.org/10.1016/B978-0-323-91001-9.00036-0.

Jespersen, B. M., & Munck, L. (2009). *Cereals and cereal products infrared spectroscopy for food quality analysis and control* (pp. 275–319). Denmark: Elsevier Inc. Available from http://www.sciencedirect.com/science/book/9780123741363, https://doi.org/10.1016/B978-0-12-374136-3.00011-0.

Jeswani, H. K., Burkinshaw, R., & Azapagic, A. (2015). Environmental sustainability issues in the food-energy-water nexus: Breakfast cereals and snacks. *Sustainable Production and Consumption*, *2*, 17–28. Available from https://doi.org/10.1016/j.spc.2015.08.001, http://www.journals.elsevier.com/sustainable-production-and-consumption/.

Jha, K., Doshi, A., Patel, P., & Shah, M. (2019). A comprehensive review on automation in agriculture using artificial intelligence. *Artificial Intelligence in Agriculture*, *2*, 1–12. Available from https://doi.org/10.1016/j.aiia.2019.05.004, http://www.keaipublishing.com/en/journals/artificial-intelligence-in-agriculture/.

Jin, C., Bouzembrak, Y., Zhou, J., Liang, Q., van den Bulk, L. M., Gavai, A., Liu, N., van den Heuvel, L. J., Hoenderdaal, W., & Marvin, H. J. P. (2020). Big data in food safety- A review. *Current Opinion in Food Science*, *36*, 24–32. Available from https://doi.org/10.1016/j.cofs.2020.11.006, http://www.journals.elsevier.com/current-opinion-in-food-science/.

Jin, X., Kumar, L., Li, Z., Xu, X., Yang, G., & Wang, J. (2016). Estimation of winter wheat biomass and yield by combining the aquacrop model and field hyperspectral data. *Remote Sensing*, *8*(12). Available from https://doi.org/10.3390/rs8120972, http://www.mdpi.com/2072-4292/8/12/972/pdf.

Johann, A. L., de Araújo, A. G., Delalibera, H. C., & Hirakawa, A. R. (2016). Soil moisture modeling based on stochastic behavior of forces on a no-till chisel opener. *Computers and Electronics in Agriculture*, *121*, 420–428. Available from https://doi.org/10.1016/j.compag.2015.12.020, http://www.elsevier.com/inca/publications/store/5/0/3/3/0/4.

Jung, Y., Heo, Y., Lee, J. J., Deering, A., & Bae, E. (2020). Smartphone-based lateral flow imaging system for detection of food-borne bacteria *E.coli* O157:H7. *Journal of Microbiological Methods*, *168*. Available from https://doi.org/10.1016/j.mimet.2019.105800, http://www.elsevier.com/locate/jmicmeth.

Kakani, V., Nguyen, V. H., Kumar, B. P., Kim, H., & Pasupuleti, V. R. (2020). A critical review on computer vision and artificial intelligence in food industry. *Journal of Agriculture and Food Research, 2*. Available from https://doi.org/10.1016/j.jafr.2020.100033, http://www.journals.elsevier.com/journal-of-agriculture-and-food-research.

Kamiloglu, S., Tomas, M., Ozdal, T., Yolci-Omeroglu, P., & Capanoglu, E. (2020). *Bioactive component analysis. Innovative food analysis* (pp. 41–65). Turkey: Elsevier. Available from https://www.elsevier.com/books/innovative-food-analysis/galanakis/978-0-12-819493-5, https://doi.org/10.1016/B978-0-12-819493-5.00002-9.

Kayikci, Y., Subramanian, N., Dora, M., & Bhatia, M. S. (2022). Food supply chain in the era of Industry 4.0: blockchain technology implementation opportunities and impediments from the perspective of people, process, performance, and technology. *Production Planning and Control, 33*(2-3), 301–321. Available from https://doi.org/10.1080/09537287.2020.1810757, http://www.tandf.co.uk/journals/titles/09537287.asp.

Kerry, J. P., O'Grady, M. N., & Hogan, S. A. (2006). Past, current and potential utilisation of active and intelligent packaging systems for meat and muscle-based products: A review. *Meat Science, 74*(1), 113–130. Available from https://doi.org/10.1016/j.meatsci.2006.04.024.

Khan, M. I. H., Sablani, S. S., Nayak, R., & Gu, Y. (2022). Machine learning-based modeling in food processing applications: State of the art. *Comprehensive Reviews in Food Science and Food Safety, 21*(2), 1409–1438. Available from https://doi.org/10.1111/1541-4337.12912, http://onlinelibrary.wiley.com/journal/10.1111/(ISSN)1541-4337.

Kim, S. S., & Kim, S. (2022). Impact and prospect of the fourth industrial revolution in food safety: Mini-review. *Food Science and Biotechnology, 31*(4), 399–406. Available from https://doi.org/10.1007/s10068-022-01047-6, http://www.springerlink.com/content/n3031612t262/.

Kler, R., Elkady, G., Rane, K., Singh, A., Hossain, M. S., Malhotra, D., Ray, S., & Bhatia, K. K. (2022). Machine learning and artificial intelligence in the food industry: A sustainable approach. *Journal of Food Quality, 2022*. Available from https://doi.org/10.1155/2022/8521236, https://www.hindawi.com/journals/jfq/.

Kollia, I., Stevenson, J., & Kollias, S. (2021). Ai-enabled efficient and safe food supply chain. *Electronics (Switzerland), 10*(11). Available from https://doi.org/10.3390/electronics10111223, https://www.mdpi.com/2079-9292/10/11/1223/pdf.

Kovács, I., & Husti, I. (2018). The role of digitalization in the agricultural 4.0 – how to connect the industry 4.0 to agriculture? *Hungarian Agricultural Engineering, 33*, 38–42. Available from https://doi.org/10.17676/hae.2018.33.38.

Kovacs, Z. B. G., Behafarid, D., Frederik, N., Isabel, H., Leon, B. J., & Langle, T. F. B. (2017). 3 OCM 2017: 3rd International Conference On Optical Characterization of Materials KARLSRUHE GERMANY KIT Scientific Publishing Simultaneous detection of melamine and urea in gluten with a handheld NIR scanner. http://file:///C:/Users/aaitkadd/Downloads/OCM_2017 ISBN_9783731506126-Proceedings.pdf.

Kumar, I., Rawat, J., Mohd, N., & Husain, S. (2021). Opportunities of artificial intelligence and machine learning in the food industry. *Journal of Food Quality, 2021*. Available from https://doi.org/10.1155/2021/4535567, https://www.hindawi.com/journals/jfq/.

Kuswandi, B., Maryska, C., Jayus, A. A., & Heng, L. Y. (2013). Real time on-package freshness indicator for guavas packaging. *Journal of Food Measurement and Characterization, 7*(1), 29–39. Available from https://doi.org/10.1007/s11694-013-9136-5.

Laponogov, I., Gonzalez, G., Shepherd, M., Qureshi, A., Veselkov, D., Charkoftaki, G., Vasiliou, V., Youssef, J., Mirnezami, R., Bronstein, M., & Veselkov, K. (2021). Network machine learning maps phytochemically rich "Hyperfoods" to fight COVID-19. *Human Genomics, 15*(1). Available from https://doi.org/10.1186/s40246-020-00297-x, https://rd.springer.com/journal/40246.

Lebelo, K., Masinde, M., Malebo, N., & Mochane, M. J. (2022). The surveillance and prediction of food contamination using intelligent systems: a bibliometric analysis. *British Food Journal, 124*(4), 1149–1169. Available from https://doi.org/10.1108/BFJ-04-2021-0366, https://www.emeraldinsight.com/loi/bfj.

Lee, K., Baek, S., Kim, D., & Seo, J. (2019). A freshness indicator for monitoring chicken-breast spoilage using a Tyvek® sheet and RGB color analysis. *Food Packaging and Shelf Life, 19*, 40–46. Available from https://doi.org/10.1016/j.fpsl.2018.11.016, http://www.journals.elsevier.com/food-packaging-and-shelf-life/.

Lee, S. J., & Rahman, A. T. M. M. (2013). *Intelligent packaging for food products innovations in food packaging* (2nd ed., pp. 171–209). South Korea: Elsevier Ltd. Available from http://www.sciencedirect.com/science/book/9780123946010, https://doi.org/10.1016/B978-0-12-394601-0.00008-4.

Li, J., Feng, H., Li, J., Feng, Y., Zhang, Y., Jiang, J., & Qian, D. (2015). Fabrication of gold nanoparticles-decorated reduced graphene oxide as a high performance electrochemical sensing platform for the detection of toxicant Sudan i. *Electrochimica Acta, 167*, 226–236. Available from https://doi.org/10.1016/j.electacta.2015.03.201, http://www.journals.elsevier.com/electrochimica-acta/.

Li, X., Zhang, L., Zhang, Y., Wang, D., Wang, X., Yu, L., Zhang, W., & Li, P. (2020). Review of NIR spectroscopy methods for nondestructive quality analysis of oilseeds and edible oils. *Trends in Food Science and Technology, 101*, 172–181. Available from https://doi.org/10.1016/j.tifs.2020.05.002, http://www.elsevier.com/wps/find/journaldescription.cws_home/601278/description#description.

Liakos, K. G., Busato, P., Moshou, D., Pearson, S., & Bochtis, D. (2018). Machine learning in agriculture: A review. *Sensors (Switzerland), 18*(8). Available from https://doi.org/10.3390/s18082674, http://www.mdpi.com/1424-8220/18/8/2674/pdf.

Liang, S., Zhang, X., Sun, N., Li, Y., Xu, M., & Wu, L. (2019). Modeling crop yield and nitrogen use efficiency in wheat and maize production systems under future climate change. *Nutrient Cycling in Agroecosystems, 115*(1), 117–136. Available from https://doi.org/10.1007/s10705-019-10013-4, http://www.wkap.nl/journalhome.htm/1385-1314.

Liu, H. Y., Wen, J. J., Huang, Z. H., Ma, H., Xu, H. X., Qiu, Y. B., Zhao, W. J., & Gu, C. C. (2019). Prussian blue analogue of copper-cobalt decorated with multi-walled carbon nanotubes based electrochemical sensor for sensitive determination of nitrite in food samples. *Chinese Journal of Analytical Chemistry, 47*(6), e19066–e19072. Available from https://doi.org/10.1016/S1872-2040(19)61168-0, https://www.journals.elsevier.com/chinese-journal-of-analytical-chemistry.

Liu, L., Wu, W., Zheng, L., Yu, J., Sun, P., & Shao, P. (2022). Intelligent packaging films incorporated with anthocyanins-loaded ovalbumin-carboxymethyl cellulose nanocomplexes for food freshness monitoring. *Food Chemistry, 387*. Available from https://doi.org/10.1016/j.foodchem.2022.132908, http://www.elsevier.com/locate/foodchem.

Liu, Y., Wang, J., Shi, Y., He, Z., Liu, F., Kong, W., & He, Y. (2022). Unmanned airboat technology and applications in environment and agriculture. *Computers and Electronics in Agriculture*, *197*, 106920. Available from https://doi.org/10.1016/j.compag.2022.106920.

Liu, Z., Zhang, Y., Xu, S., Zhang, H., Tan, Y., Ma, C., Song, R., Jiang, L., & Yi, C. (2017). A 3D printed smartphone optosensing platform for point-of-need food safety inspection. *Analytica Chimica Acta*, *966*, 81–89. Available from https://doi.org/10.1016/j.aca.2017.02.022, http://www.journals.elsevier.com/analytica-chimica-acta/.

Lorenc, A., Czuba, M., & Szarata, J. (2021). Big data analytics and anomaly prediction in the cold chain to supply chain resilience. *FME Transactions*, *49*(2), 315–326. Available from https://doi.org/10.5937/fme2102315L, http://www.mas.bg.ac.rs/transactions/index.html.

Lupolova, N., Lycett, S. J., & Gally, D. L. (2019). A guide to machine learning for bacterial host attribution using genome sequence data. *Microbial Genomics*, *5*(12). Available from https://doi.org/10.1099/mgen.0.000317, https://www.microbiologyresearch.org/content/journal/mgen/10.1099/mgen.0.000317.

Lyu, J. S., Choi, I., Hwang, K. S., Lee, J. Y., Seo, J., Kim, S. Y., & Han, J. (2019). Development of a BTB − /TBA + ion-paired dye-based CO2 indicator and its application in a multilayered intelligent packaging system. *Sensors and Actuators, B: Chemical*, *282*, 359–365. Available from https://doi.org/10.1016/j.snb.2018.11.073, https://www.journals.elsevier.com/sensors-and-actuators-b-chemical.

Magwaza, L. S., Opara, U. L., Nieuwoudt, H., Cronje, P. J. R., Saeys, W., Nicolaï, B., & Spectroscopy, N. I. R. (2012). Applications for internal and external quality analysis of citrus fruit-A review. *Food and Bioprocess Technology*, *5*(2), 425–444. Available from https://doi.org/10.1007/s11947-011-0697-1.

Marvin, H. J. P., Bouzembrak, Y., van der Fels-Klerx, H. J., Kempenaar, C., Veerkamp, R., Chauhan, A., Stroosnijder, S., Top, J., Simsek-Senel, G., Vrolijk, H., Knibbe, W. J., Zhang, L., Boom, R., & Tekinerdogan, B. (2022). Digitalisation and artificial intelligence for sustainable food systems. *Trends in Food Science and Technology*, *120*, 344–348. Available from https://doi.org/10.1016/j.tifs.2022.01.020, http://www.elsevier.com/wps/find/journaldescription.cws_home/601278/description#description.

Marvin, H. J. P., Bouzembrak, Y., Janssen, E. M., van der Fels- Klerx, H. J., van Asselt, E. D., & Kleter, G. A. (2016). A holistic approach to food safety risks: Food fraud as an example. *Food Research International*, *89*, 463–470. Available from https://doi.org/10.1016/j.foodres.2016.08.028, http://www.elsevier.com/inca/publications/store/4/2/2/9/7/0.

Matt, D. T., Modrák, V., & Zsifkovits, H. (2020). *Industry 4.0 for SMEs: Challenges, opportunities and requirements. Industry 4.0 for SMEs: Challenges, opportunities and requirements* (pp. 1–401). Italy: Palgrave Macmillan. Available from http://www.palgrave.com/us/book/9783030254247, 10.1007/978-3-030-25425-4.

McFarlane, D., & Sheffi, Y. (2003). The impact of automatic identification on supply chain operations. *The International Journal of Logistics Management*, *14*(1), 1–17. Available from https://doi.org/10.1108/09574090310806503.

Melini, V., & Melini, F. (2020). *Compositional and nutritional analysis. Innovative food analysis* (pp. 1–39). Italy: Elsevier. Available from https://www.elsevier.com/books/innovative-food-analysis/galanakis/978-0-12-819493-5, https://doi.org/10.1016/B978-0-12-819493-5.00001-7.

Meng, X., Kim, S., Puligundla, P., & Ko, S. (2014). Carbon dioxide and oxygen gas sensors-possible application for monitoring quality, freshness, and safety of agricultural and food products with emphasis on importance of analytical signals and their transformation. *Journal of the Korean Society for Applied Biological Chemistry*, *57*(6), 723–733. Available from https://doi.org/10.1007/s13765-014-4180-3, http://www.springer.com/chemistry/biotechnology/journal/13765.

Milczarski, P., Zieliński, B., Stawska, Z., Hłobaż, A., Maślanka, P., & Kosiński, P. (2020). *Lecture notes in computer science (including subseries lecture notes in artificial intelligence and lecture notes in bioinformatics). Machine learning application in energy consumption calculation and assessment in food processing industry* (pp. 369–379). Deutschland GmbH Poland: Springer Science and Business Media. Available from http://doi.org/10.1007/978-3-030-61534-5_33 16113349, https://www.springer.com/series/55812416.

Milovanovic, B., Djekic, I., Miocinovic, J., Djordjevic, V., Lorenzo, J. M., Barba, F. J., Mörlein, D., & Tomasevic, I. (2020). What is the color of milk and dairy products and how is it measured. *Foods*, *9*(11). Available from https://doi.org/10.3390/foods9111629, http://www.mdpi.com/journal/foods.

Min, David B., & Ellefson, W. C. (2010). *Fat analysis* (pp. 117–132). Springer Science and Business Media LLC. Available from 10.1007/978-1-4419-1478-1_8.

Mlalila, N., Kadam, D. M., Swai, H., & Hilonga, A. (2016). Transformation of food packaging from passive to innovative via nanotechnology: Concepts and critiques. *Journal of Food Science and Technology*, *53*(9), 3395–3407. Available from https://doi.org/10.1007/s13197-016-2325-6, http://www.springerlink.com/content/121580/.

Monago-Maraña, O., Durán-Merás, I., Muñoz de la Peña, A., & Galeano-Díaz, T. (2022). Analytical techniques and chemometrics approaches in authenticating and identifying adulteration of paprika powder using fingerprints: A review. *Microchemical Journal*, *178*, 107382. Available from https://doi.org/10.1016/j.microc.2022.107382.

Moradi, M., Molaei, R., & Guimarães, J. T. (2021). A review on preparation and chemical analysis of postbiotics from lactic acid bacteria. *Enzyme and Microbial Technology*, *143*, 109722. Available from https://doi.org/10.1016/j.enzmictec.2020.109722.

Motia, S., & Reddy, S. R. N. (2021). Exploration of machine learning methods for prediction and assessment of soil properties for agricultural soil management: A quantitative evaluation. *Journal of Physics: Conference Series*, *1950*, 1. Available from https://doi.org/10.1088/1742-6596/1950/1/012037, 17426596; IOP Publishing Ltd. India, http://iopscience.iop.org/journal/1742-6596.

Müller, P., & Schmid, M. (2019). Intelligent packaging in the food sector: A brief overview. *Foods*, *8*(1), 16. Available from https://doi.org/10.3390/foods8010016.

Na, M. H., Cho, W., & Kim, S. K. (2020). A construction of web application platform for detection and identification of various diseases in tomato plants using a deep learning algorithm. *Journal of Korean Society for Quality Management*, *48*(4), 581–596. Available from https://doi.org/10.7469/JKSQM.2020.48.4.581.

Nahvi, B., Habibi, J., Mohammadi, K., Shamshirband, S., & Razgan, O. S. Al (2016). Using self-adaptive evolutionary algorithm to improve the performance of an extreme learning machine for estimating soil temperature. *Computers and Electronics in Agriculture*, *124*, 150–160. Available from https://doi.org/10.1016/j.compag.2016.03.025, http://www.elsevier.com/inca/publications/store/5/0/3/3/0/4.

Nakat, Z., & Bou-Mitri, C. (2021). COVID-19 and the food industry: Readiness assessment. *Food Control*, *121*, 107661. Available from https://doi.org/10.1016/j.foodcont.2020.107661.

Nayak, J., Vakula, K., Dinesh, P., Naik, B., & Pelusi, D. (2020). Intelligent food processing: Journey from artificial neural network to deep learning. *Computer Science Review*, *38*, 100297. Available from https://doi.org/10.1016/j.cosrev.2020.100297.

Neethirajan, S. (2020). The role of sensors, big data and machine learning in modern animal farming. *Sensing and Bio-Sensing Research*, *29*. Available from https://doi.org/10.1016/j.sbsr.2020.100367, https://www.journals.elsevier.com/sensing-and-bio-sensing-research/.

Nielsen, S. S. (2006). *Proximate Assays in Food Analysis Encyclopedia of Analytical Chemistry*. Available from https://doi.org/10.1002/9780470027318.a1024.

Nopwinyuwong, A., Trevanich, S., & Suppakul, P. (2010). Development of a novel colorimetric indicator label for monitoring freshness of intermediate-moisture dessert spoilage. *Talanta*, *81*(3), 1126–1132. Available from https://doi.org/10.1016/j.talanta.2010.02.008.

Notarnicola, B., Sala, S., Anton, A., McLaren, S. J., Saouter, E., & Sonesson, U. (2017). The role of life cycle assessment in supporting sustainable agri-food systems: A review of the challenges. *Journal of Cleaner Production*, *140*, 399–409. Available from https://doi.org/10.1016/j.jclepro.2016.06.071.

Olan, F., Liu, S., Suklan, J., Jayawickrama, U., & Arakpogun, E. O. (2022). The role of artificial intelligence networks in sustainable supply chain finance for food and drink industry. *International Journal of Production Research*, *60*(14), 4418–4433. Available from https://doi.org/10.1080/00207543.2021.1915510, http://www.tandfonline.com/toc/tprs20/current.

Osborne, B. G. (2000). *Near-infrared spectroscopy in food analysis. Encyclopedia of analytical chemistry*. John Wiley & Sons. Available from 10.1002/9780470027318.a1018.

Osinenko, P., Biegert, K., McCormick, R. J., Göhrt, T., Devadze, G., Streif, J., & Streif, S. (2021). Application of non-destructive sensors and big data analysis to predict physiological storage disorders and fruit firmness in 'Braeburn' apples. *Computers and Electronics in Agriculture*, *183*. Available from https://doi.org/10.1016/j.compag.2021.106015, http://www.elsevier.com/inca/publications/store/5/0/3/3/0/4.

Özdoğan, G., Lin, X., & Sun, D. W. (2021). Rapid and noninvasive sensory analyses of food products by hyperspectral imaging: Recent application developments. *Trends in Food Science and Technology*, *111*, 151–165. Available from https://doi.org/10.1016/j.tifs.2021.02.044, http://www.elsevier.com/wps/find/journaldescription.cws_home/601278/description#description.

Pandey, S. K., & Kim, K. H. (2007). The relative performance of NDIR-based sensors in the near real-time analysis of CO_2 in air. *Sensors*, *7*(9), 1683–1696. Available from https://doi.org/10.3390/s7091683, http://www.mdpi.org/sensors/papers/s7091683.pdf.

Papaioannou, G., Mohammed, A. M., Despoudi, S., Saridakis, G., & Papadopoulos, T. (2020). The role of adverse economic environment and human capital on collaboration within agri-food supply chains. *International Journal of Information Management*, *52*. Available from https://doi.org/10.1016/j.ijinfomgt.2020.102077, https://www.journals.elsevier.com/international-journal-of-information-management.

Park, Y. W., Kim, S. M., Lee, J. Y., & Jang, W. (2015). Application of biosensors in smart packaging. *Molecular and Cellular Toxicology*, *11*(3), 277–285. Available from https://doi.org/10.1007/s13273-015-0027-1, http://www.springerlink.com/content/1738-642X.

Pasquini, C. (2018). Near infrared spectroscopy: A mature analytical technique with new perspectives – A review. *Analytica Chimica Acta*, *1026*, 8–36. Available from https://doi.org/10.1016/j.aca.2018.04.004, http://www.journals.elsevier.com/analytica-chimica-acta/.

Poyatos-Racionero, E., Ros-Lis, J. V., Vivancos, J. L., & Martínez-Máñez, R. (2018). Recent advances on intelligent packaging as tools to reduce food waste. *Journal of Cleaner Production*, *172*, 3398–3409. Available from https://doi.org/10.1016/j.jclepro.2017.11.075.

Qian, C., Wang, R., Wu, H., Ping, J., & Wu, J. (2018). Recent advances in emerging DNA-based methods for genetically modified organisms (GMOs) rapid detection. *TrAC - Trends in Analytical Chemistry*, *109*, 19–31. Available from https://doi.org/10.1016/j.trac.2018.09.021, http://www.elsevier.com/locate/trac.

Qiang, Y., Yun-min, W., Ming-ming, F., You-wen, Q., Hong-tao, W., Ming-hui, Z., Ying, L., & Jin-xia, A. (2016). Establishment of PCR-ELISA for detecting glyphosate resistant transgenic soybean. *Journal of Northeast Agricultural University (English Edition)*, *23*(2), 45–51. Available from https://doi.org/10.1016/s1006-8104(16)30046-0.

Qin, J., Vasefi, F., Hellberg, R. S., Akhbardeh, A., Isaacs, R. B., Yilmaz, A. G., Hwang, C., Baek, I., Schmidt, W. F., & Kim, M. S. (2020). Detection of fish fillet substitution and mislabeling using multimode hyperspectral imaging techniques. *Food Control*, *114*. Available from https://doi.org/10.1016/j.foodcont.2020.107234, https://www.journals.elsevier.com/food-control.

Qu, J. H., Liu, D., Cheng, J. H., Sun, D. W., Ma, J., Pu, H., & Zeng, X. A. (2015). Applications of near-infrared spectroscopy in food safety evaluation and control: A review of recent research advances. *Critical Reviews in Food Science and Nutrition*, *55*(13), 1939–1954. Available from https://doi.org/10.1080/10408398.2013.871693, http://www.tandf.co.uk/journals/titles/10408398.asp.

Raj, M., Gupta, S., Chamola, V., Elhence, A., Garg, T., Atiquzzaman, M., & Niyato, D. (2021). A survey on the role of internet of things for adopting and promoting Agriculture 4.0. *Journal of Network and Computer Applications*, *187*. Available from https://doi.org/10.1016/j.jnca.2021.103107, http://www.elsevier.com/inca/publications/store/6/2/2/8/9/3/index.htt.

Rao, T. N., Loo, B. H., Sarada, B. V., Terashima, C., & Fujishima, A. (2002). Electrochemical detection of carbamate pesticides at conductive diamond electrodes. *Analytical Chemistry*, *74*(7), 1578–1583. Available from https://doi.org/10.1021/ac010935d.

Romanello, R., & Veglio, V. (2022). Industry 4.0 in food processing: drivers, challenges and outcomes. *British Food Journal*, *124*(13), 375–390. Available from https://doi.org/10.1108/BFJ-09-2021-1056, https://www.emeraldinsight.com/loi/bfj.

Ruini, L., Marino, M., Pignatelli, S., Laio, F., & Ridolfi, L. (2013). Water footprint of a large-sized food company: The case of Barilla pasta production. *Water Resources and Industry*. *1-*, *2*, 7–24. Available from https://doi.org/10.1016/j.wri.2013.04.002.

Sadiku, M. N. O., Musa, S. M., & Ashaolu, T. J. (2019). Food industry: An introduction. *International Journal of Trend in Scientific Research and Development*, *3*(Issue-4), 128–130. Available from https://doi.org/10.31142/ijtsrd23638.

Sarkar, T., Salauddin, M., Kirtonia, K., Pati, S., Rebezov, M., Khayrullin, M., Panasenko, S., Tretyak, L., Temerbayeva, M., Kapustina, N., Azimova, S., Gruzdeva, L., Makhmudov, F., Nikitin, I., Kassenov, A., Shariati, M. A., & Lorenzo, J. M. (2022). A review on the commonly used methods for analysis of physical properties of food materials. *Applied Sciences (Switzerland)*, *12*(4). Available from https://doi.org/10.3390/app12042004, https://www.mdpi.com/2076-3417/12/4/2004/pdf.

Savina, A., Malyavkina, L., Baturina, N., Bolshakova, L., Zimina, L., & Vlasova, M. (2020). *Information and technological support of the system of food quality and safety management in the digital economical environment. ACM international conference proceeding series*. Association for Computing Machinery Russian Federation. Available from https://doi.org/10.1145/3446434.3446443, 9781450388900, http://portal.acm.org/.

Savitha, M. (2018). Smart crop field irrigation in iot architecture using sensors. *International Journal of Advanced Research in Computer Science*, *9* (1), 302–306. Available from https://doi.org/10.26483/ijarcs.v9i1.5348.

Schmeling, L., Elmamooz, G., Hoang, P. T., Kozar, A., Nicklas, D., Sünkel, M., Thurner, S., & Rauch, E. (2021). Training and validating a machine learning model for the sensor-based monitoring of lying behavior in dairy cows on pasture and in the barn. *Animals*, *11*(9). Available from https://doi.org/10.3390/ani11092660, https://www.mdpi.com/2076-2615/11/9/2660/pdf.

Serrano, N., Díaz-Cruz, J. M., Ariño, C., & Esteban, M. (2016). Antimony-based electrodes for analytical determinations. *TrAC – Trends in Analytical Chemistry*, *77*, 203–213. Available from https://doi.org/10.1016/j.trac.2016.01.011, http://www.elsevier.com/locate/trac.

Shalini Devi, K. S., Anusha, N., Raja, S., & Senthil Kumar, A. (2018). A new strategy for direct electrochemical sensing of a organophosphorus pesticide, triazophos, using a coomassie brilliant-blue dye surface-confined carbon-black-nanoparticle-modified electrode. *ACS Applied Nano Materials*, *1*(8), 4110–4119. Available from https://doi.org/10.1021/acsanm.8b00861, https://pubs.acs.org/journal/aanmf6.

Sharma, S., Gahlawat, V. K., Rahul, K., Mor, R. S., & Malik, M. (2021). Sustainable innovations in the food industry through artificial intelligence and big data analytics. *Logistics*, *5*(4), 66. Available from https://doi.org/10.3390/logistics5040066.

Silva, V. L., Sereno, A. M., & do Amaral Sobral, P. J. (2018). Food industry and processing technology: On time to harmonize technology and social drivers. *Food Engineering Reviews*, *10*(1). Available from https://doi.org/10.1007/s12393-017-9164-8, http://www.springer.com/life + sci/food + science/journal/12393.

Singh, M., & Corradini, M. G. (2022). *Big data and its role in mitigating food spoilage and quality deterioration along the supply chain* (pp. 93–112). Springer Science and Business Media LLC. Available from https://doi.org/10.1007/978-3-031-07179-9_5.

Sinha, A., Priyadarshi, P., Bhushan, M., & Debbarma, D. (2021). Worldwide trends in the scientific production of literature on traceability in food safety: A bibliometric analysis. *Artificial Intelligence in Agriculture*, *5*, 252–261. Available from https://doi.org/10.1016/j.aiia.2021.11.002, http://www.keaipublishing.com/en/journals/artificial-intelligence-in-agriculture/.

Soltani Firouz, M., Mohi-Alden, K., & Omid, M. (2021). A critical review on intelligent and active packaging in the food industry: Research and development. *Food Research International*, *141*, 110113. Available from https://doi.org/10.1016/j.foodres.2021.110113.

Soon, J. M. (2020). Consumers' awareness and trust toward food safety news on social media in Malaysia. *Journal of Food Protection*, *83*(3), 452–459. Available from https://watermark.silverchair.com/i0362-028x-83-3-452.pdf?token = AQECAHi208BE49Ooan9kkhW_Ercy7Dm3ZL_9Cf3qfKAc485ysgAAAnkwggJ1BgkqhkiG9w0BBwagggJmMIICYgIBADCCAlsGCSqGSIb3DQEHATAeBglghkgBZQMEAS4wEQQMtX239lXzWdD7WGs2AgEQgIICLHObwcUbKP, https://doi.org/10.4315/0362-028X.JFP-19-415.

Spanaki, K., Karafili, E., & Despoudi, S. (2021). AI applications of data sharing in agriculture 4.0: A framework for role-based data access control. *International Journal of Information Management*, *59*. Available from https://doi.org/10.1016/j.ijinfomgt.2021.102350, https://www.journals.elsevier.com/international-journal-of-information-management.

Squeo, G., De Angelis, D., Summo, C., Pasqualone, A., Caponio, F., & Manuel Amigo, J. (2022). Assessment of macronutrients and alpha-galactosides of texturized vegetable proteins by near infrared hyperspectral imaging. *Journal of Food Composition and Analysis*, *108*, 104459. Available from https://doi.org/10.1016/j.jfca.2022.104459.

Stoitsis, G., Papakonstantinou, M., Karvounis, M., & Manouselis, N. (2023). *The role of big data and artificial intelligence in food risk assessment and prediction* (pp. 1032–1044). Elsevier BV. Available from 10.1016/b978-0-12-819470-6.00041-x.

Stoytcheva, M., Gochev, V., & Velkova, Z. (2016). Electrochemical biosensors for direct determination of organophosphorus pesticides: A review. *Current Analytical Chemistry*, *12*(1), 37–42.

Sun, S., & Wang, X. (2019). Promoting traceability for food supply chain with certification. *Journal of Cleaner Production*, *217*, 658–665. Available from https://doi.org/10.1016/j.jclepro.2019.01.296, https://www.journals.elsevier.com/journal-of-cleaner-production.

Szwedziak, K., Polańczyk, E., Grzywacz, Z., Niedbała, G., & Wojtkiewicz, W. (2020). Neural modeling of the distribution of protein, water and gluten in wheat grains during storage. *Sustainability (Switzerland)*, *12*(12). Available from https://doi.org/10.3390/su12125050, https://res.mdpi.com/d_attachment/sustainability/sustainability-12-05050/article_deploy/sustainability-12-05050.pdf.

Talaviya, T., Shah, D., Patel, N., Yagnik, H., & Shah, M. (2020). Implementation of artificial intelligence in agriculture for optimisation of irrigation and application of pesticides and herbicides. *Artificial Intelligence in Agriculture*, *4*, 58–73. Available from https://doi.org/10.1016/j.aiia.2020.04.002, http://www.keaipublishing.com/en/journals/artificial-intelligence-in-agriculture/.

Talukder, S., Mendiratta, S. K., Kumar, R. R., Agrawal, R. K., Soni, A., Luke, A., & Chand, S. (2020). Jamun fruit (*Syzgium cumini*) skin extract based indicator for monitoring chicken patties quality during storage. *Journal of Food Science and Technology*, *57*(2), 537–548. Available from https://doi.org/10.1007/s13197-019-04084-y, http://www.springerlink.com/content/121580/.

Tan, X., Ye, Y., Liu, H., Meng, J., Yang, L. L., & Li, F. (2022). Deep learning-assisted visualized fluorometric sensor array for biogenic amines detection. *Chinese Journal of Chemistry*, *40*(5), 609–616. Available from https://doi.org/10.1002/cjoc.202100591, http://onlinelibrary.wiley.com/journal/10.1002/(ISSN)1614-7065.

Tomasevic, I., Djekic, I., Font-i-Furnols, M., Terjung, N., & Lorenzo, J. M. (2021). Recent advances in meat color research. *Current Opinion in Food Science*, *41*, 81–87. Available from https://doi.org/10.1016/j.cofs.2021.02.012, http://www.journals.elsevier.com/current-opinion-in-food-science/.

Tramuta, C., Decastelli, L., Barcucci, E., Ingravalle, F., Fragassi, S., Lupi, S., & Bianchi, D. M. (2022). Detection of peanut traces in food by an official food safety laboratory. *Foods*, *11*(5). Available from https://doi.org/10.3390/foods11050643, https://www.mdpi.com/2304-8158/11/5/643/pdf.

Tsakanikas, P., Karnavas, A., Panagou, E. Z., & Nychas, G. J. (2020). A machine learning workflow for raw food spectroscopic classification in a future industry. *Scientific Reports*, *10*(1). Available from https://doi.org/10.1038/s41598-020-68156-2, http://www.nature.com/srep/index.html.

Tsoumakas, G. (2019). A survey of machine learning techniques for food sales prediction. *Artificial Intelligence Review*, *52*(1), 441–447. Available from https://doi.org/10.1007/s10462-018-9637-z.

Valdés, A., Álvarez-Rivera, G., Socas-Rodríguez, B., Herrero, M., & Cifuentes, A. (2022). Capillary electromigration methods for food analysis and foodomics: Advances and applications in the period February 2019–February 2021. *Electrophoresis*, *43*(1–2), 37–56. Available from https://doi.org/10.1002/elps.202100201, http://onlinelibrary.wiley.com/journal/10.1002/(ISSN)1522-2683.

Van De Looverbosch, T., Raeymaekers, E., Verboven, P., Sijbers, J., & Nicolaï, B. (2021). Non-destructive internal disorder detection of Conference pears by semantic segmentation of X-ray CT scans using deep learning. *Expert Systems with Applications*, *176*. Available from https://doi.org/10.1016/j.eswa.2021.114925, 09574174 Elsevier Ltd. Belgium, https://www.journals.elsevier.com/expert-systems-with-applications.

Violino, S., Ortenzi, L., Antonucci, F., Pallottino, F., Benincasa, C., Figorilli, S., & Costa, C. (2020). An artificial intelligence approach for Italian EVOO origin traceability through an open source IoT spectrometer. *Foods*, *9*(6). Available from https://doi.org/10.3390/foods9060834, https://www.mdpi.com/2304-8158/9/6/834.

Volkmann, N., Kulig, B., Hoppe, S., Stracke, J., Hensel, O., & Kemper, N. (2021). On-farm detection of claw lesions in dairy cows based on acoustic analyses and machine learning. *Journal of Dairy Science*, *104*(5), 5921–5931. Available from https://doi.org/10.3168/jds.2020-19206, http://www.elsevier.com/wps/find/journaldescription.cws_home/721317/description#description.

Wang, C. C., Liang, Y. C., Wang, S. S., Lin, P., & Tung, C. W. (2022). A machine learning-driven approach for prioritizing food contact chemicals of carcinogenic concern based on complementary in silico methods. *Food and Chemical Toxicology*, *160*. Available from https://doi.org/10.1016/j.fct.2021.112802, http://www.elsevier.com/locate/foodchemtox.

Wang, Q., Shi, A., & Shah, F. (2019). *Rheology instruments for food quality evaluation. Evaluation technologies for food quality* (pp. 465–490). China: Elsevier. Available from https://www.sciencedirect.com/book/9780128142172, https://doi.org/10.1016/B978-0-12-814217-2.00018-4.

Walaszczyk, A., Jałmużna, I. B., & Lewandowski, J. (2019). *Production management and packaging: Food safety and Industry 4.0*. Wydawnictwo Politechniki Łódzkiej.

Wang, X., Bouzembrak, Y., Lansink, A. G. J. M. O., & van der Fels-Klerx, H. J. (2022). Application of machine learning to the monitoring and prediction of food safety: A review. *Comprehensive Reviews in Food Science and Food Safety*, *21*(1), 416–434. Available from https://doi.org/10.1111/1541-4337.12868, http://onlinelibrary.wiley.com/journal/10.1111/(ISSN)1541-4337.

Wanihsuksombat, C., Hongtrakul, V., & Suppakul, P. (2010). Development and characterization of a prototype of a lactic acid-based time-temperature indicator for monitoring food product quality. *Journal of Food Engineering*, *100*(3), 427–434. Available from https://doi.org/10.1016/j.jfoodeng.2010.04.027.

Witkowski, K. (2017). Internet of Things, Big Data, Industry 4.0—Innovative solutions in logistics and supply chains management. *Procedia Engineering*, *182*, 763–769. Available from https://doi.org/10.1016/j.proeng.2017.03.197, 18777058; Elsevier Ltd. Poland, http://www.sciencedirect.com/science/journal/18777058.

Wolfert, S., Ge, L., Verdouw, C., & Bogaardt, M. J. (2017). Big data in smart farming – A review. *Agricultural Systems*, *153*, 69–80. Available from https://doi.org/10.1016/j.agsy.2017.01.023, http://www.elsevier.com/inca/publications/store/4/0/5/8/5/1.

Wu, L., McGechan, M. B., McRoberts, N., Baddeley, J. A., & Watson, C. A. (2007). SPACSYS: Integration of a 3D root architecture component to carbon, nitrogen and water cycling-model description. *Ecological Modelling. 200 (3–4)*, 343–359. Available from https://doi.org/10.1016/j.ecolmodel.2006.08.010.

Wu, X., Liang, X., Wang, Y., Wu, B., & Sun, J. (2022). Non-destructive techniques for the analysis and evaluation of meat quality and safety: A review. *Foods*, *11*(22), 3713. Available from https://doi.org/10.3390/foods11223713.

Xia, A., Zhang, Y., Zhao, L., & Qin, P. (2021). Simultaneous, rapid and nondestructive determination of moisture, fat content and storage time in leisure dried tofu using LF-NMR. *Analytical Sciences*, *37*(2), 301–307. Available from https://doi.org/10.2116/ANALSCI.20P223, http://www.jstage.jst.go.jp/browse/analsci.

Xu, L., Yang, F., Dias, A. C. P., & Zhang, X. (2022). Development of quantum dot-linked immunosorbent assay (QLISA) and ELISA for the detection of sunset yellow in foods and beverages. *Food Chemistry*, *385*. Available from https://doi.org/10.1016/j.foodchem.2022.132648, http://www.elsevier.com/locate/foodchem.

Xu, W., van Knegsel, A. T. M., Vervoort, J. J. M., Bruckmaier, R. M., van Hoeij, R. J., Kemp, B., & Saccenti, E. (2019). Prediction of metabolic status of dairy cows in early lactation with on-farm cow data and machine learning algorithms. *Journal of Dairy Science*, *102*(11), 10186–10201. Available from https://doi.org/10.3168/jds.2018-15791, http://www.elsevier.com/wps/find/journaldescription.cws_home/721317/description#description.

Yang, M., Liu, X., Luo, Y., Pearlstein, A. J., Wang, S., Dillow, H., Reed, K., Jia, Z., Sharma, A., Zhou, B., Pearlstein, D., Yu, H., & Zhang, B. (2021). Machine learning-enabled non-destructive paper chromogenic array detection of multiplexed viable pathogens on food. *Nature Food*, *2*(2), 110–117. Available from https://doi.org/10.1038/s43016-021-00229-5, http://nature.com/natfood/.

Yang, Z., Albrow-Owen, T., Cai, W., & Hasan, T. (2021). Miniaturization of optical spectrometers. *Science (New York, N.Y.)*, *371*(6528). Available from https://doi.org/10.1126/science.abe0722, https://science.sciencemag.org/content/371/6528/eabe0722.full.pdf.

Yao, K., Sun, J., Chen, C., Xu, M., Zhou, X., Cao, Y., & Tian, Y. (2022). Non-destructive detection of egg qualities based on hyperspectral imaging. *Journal of Food Engineering*, *325*. Available from https://doi.org/10.1016/j.jfoodeng.2022.111024, http://www.sciencedirect.com/science/journal/02608774.

Ye, Y., Wu, T., Jiang, X., Cao, J., Ling, X., Mei, Q., Chen, H., Han, D., Xu, J. J., & Shen, Y. (2020). Portable smartphone-based QDs for the visual onsite monitoring of fluoroquinolone antibiotics in actual food and environmental samples. *ACS Applied Materials and Interfaces*, *12*(12), 14552–14562. Available from https://doi.org/10.1021/acsami.9b23167, http://pubs.acs.org/journal/aamick.

Yu, X., Wang, J., Wen, S., Yang, J., & Zhang, F. (2019). A deep learning based feature extraction method on hyperspectral images for nondestructive prediction of TVB-N content in Pacific white shrimp (Litopenaeus vannamei. *Biosystems Engineering*, *178*, 244–255. Available from https://doi.org/10.1016/j.biosystemseng.2018.11.018, http://www.elsevier.com/inca/publications/store/6/2/2/7/9/5/index.htt.

Zhai, Z., Martínez, J. F., Beltran, V., & Martínez, N. L. (2020). Decision support systems for agriculture 4.0: Survey and challenges. *Computers and Electronics in Agriculture*, *170*. Available from https://doi.org/10.1016/j.compag.2020.105256, http://www.elsevier.com/inca/publications/store/5/0/3/3/0/4.

Zhang, P., Guo, Z., Ullah, S., Melagraki, G., Afantitis, A., & Lynch, I. (2021). Nanotechnology and artificial intelligence to enable sustainable and precision agriculture. *Nature Plants*, *7*(7), 864–876. Available from https://doi.org/10.1038/s41477-021-00946-6, http://www.nature.com/nplants/.

Zhang, Y., Wang, W., Yan, L., Glamuzina, B., & Zhang, X. (2019). Development and evaluation of an intelligent traceability system for waterless live fish transportation. *Food Control*, *95*, 283–297. Available from https://doi.org/10.1016/j.foodcont.2018.08.018.

Zhang, Y., Zhao, M., Liu, C., Ma, K., Liu, T., Chen, F., Wu, L., Hu, D., & Lv, G. (2023). Comparison of two commercial methods with a UHPLC–MS/MS method for the determination of multiple mycotoxins in cereals. *Food Chemistry*, *406*, 135056. Available from https://doi.org/10.1016/j.foodchem.2022.135056.

Zhao, T., Feng, A., Jin, S., Shi, Y., Hou, B., & Yan, Y. (2020). Research progress on artificial intelligence human sensor. *Lecture Notes in Electrical Engineering*, *567*, 64–271. Available from https://doi.org/10.1007/978-981-13-8595-7_35, 18761119; Springer Verlag China, http://www.springer.com/series/7818.

Zhou, C., Zou, H., Sun, C., & Li, Y. (2021). Recent advances in biosensors for antibiotic detection: Selectivity and signal amplification with nanomaterials. *Food Chemistry*, *361*, 130109. Available from https://doi.org/10.1016/j.foodchem.2021.130109.

Zidarič, T., Jovanovski, V., Menart, E., Zorko, M., Kolar, M., Veber, M., & Hočevar, S. B. (2017). Multi-pulse galvanostatic preparation of nanostructured bismuth film electrode for trace metal detection. *Sensors and Actuators, B: Chemical*, *245*, 720–725. Available from https://doi.org/10.1016/j.snb.2017.01.162.

Chapter 4

Food Safety 4.0

Hana Trollman[1], Farah Bader[2], Yang Luo[3], Guillermo Garcia-Garcia[4], Sandeep Jagtap[5], Carlos Parra-López[4], James Colwill[6] and Frank Trollman[7]

[1]Department of Work, Employment, Management and Organisations, School of Business, University of Leicester, Brookfield, Leicester, United Kingdom, [2]Goody Products Marketing Company, Jeddah, Saudi Arabia, [3]School of Intelligent Manufacturing Ecosystem, Xi'an Jiaotong-Liverpool University, Suzhou, P.R. China, [4]Department of Agrifood System Economics, Institute of Agricultural and Fisheries Research & Training (IFAPA), Granada, Spain, [5]Sustainable Manufacturing Systems Centre, School of Aerospace, Transport and Manufacturing, Cranfield University, Cranfield, United Kingdom, [6]Wolfson School, Loughborough University, Loughborough, Leicestershire, United Kingdom, [7]Trauma and Orthopaedics Department, University Hospitals of Derby and Burton, Derby, United Kingdom

4.1 Introduction

Unsafe food affects an estimated 600 million people annually, causing acute or chronic illnesses such as diarrhea, cancers, permanent disability, and even death ("WHO Estimates of the Global Burden of Foodborne Diseases: Foodborne Disease Burden Epidemiology Reference Group," 2007; Rahimifard et al., 2018). The use of Industry 4.0 technologies has attracted increased attention from academics and practitioners to help address the frequent outbreaks of food safety problems (Zhou, Zhu, et al., 2022). Accordingly, this chapter aims to summarize how key Industry 4.0 technologies may be deployed in the food industry to benefit food safety by considering the use of food-grade robotics, the Internet of Things (IoT), and digital tools such as blockchain and artificial intelligence (AI). Each of these technologies has specific roles in predicting food safety risk, detecting foodborne pathogens and food adulteration, and managing food safety.

Safe food is not contaminated with potentially harmful levels of bacteria, parasites, viruses, toxins, chemicals, and/or radionuclides. Food can become unsafe at any point in the supply chain. Current food production and consumption are resource-intensive, yet about one-third of the food produced for human consumption is wasted (Zamuz et al., 2021). Unsafe food undermines food and nutritional security, burdens the food economy and public health system, and disrupts international trade. Foodborne diseases disproportionately affect populations in low- and middle-income countries in Africa and Asia, with an aggregate estimate of productivity loss of US $95 billion per year due to unsafe food (Jaffee et al., 2019).

Food safety is also an engineering challenge critical to the achievement of many of the sustainable development goals, including ending poverty, eliminating hunger, and promoting good health and well-being (Rahimifard & Trollman, 2018). The health and economic burdens of unsafe food may be largely avoided with appropriate preventive measures, investments, and behavioral changes from farm to fork. Recent advances in Industry 4.0 technologies mean that novel, low-cost solutions are becoming available to empower sustainability and ensure food safety (Jagtap et al., 2021; Yadav et al., 2022).

Industry 4.0 is being deployed to address food safety challenges across both global and local food supply chains. Blockchain is seen as a game changer for ensuring food safety and security by providing data integrity, immutability, trust, transparency, and full traceability of stored transaction records to all participants in the agrifood value chain (Casino et al., 2019; Menon & Jain, 2021; Trollman et al., 2022; Zhao et al., 2019). AI is invaluable for processing the vast amounts of data (Big Data) continuously generated by sensors, machines, systems, intelligent devices, and humans within food networks (Ramirez-Asis et al., 2022). The IoT facilitates the use of sensors that obtain data for agricultural products [e.g., radio-frequency identification (RFID) and GPS] that are uploaded to servers in real time (Khan et al., 2020). Robots are being introduced in every part of the food chain including farming, transport, food production, packaging, and delivery (Grobbelaar et al., 2021). Recent examples include the introduction of collaborative robots in the

Food Industry 4.0. DOI: https://doi.org/10.1016/B978-0-443-15516-1.00004-9

meat industry to promote higher standards of food safety (Romanov et al., 2022) and robots that may improve the safety of off-premise restaurant service (Byrd et al., 2021).

The potential of Industry 4.0 technologies to transform food safety is significant; however, a strategic approach to their deployment and implementation has not been the focus of recent literature (Rahimifard & Trollman, 2017; Zhou, Zhu, et al., 2022). This chapter aims to fill that gap by triangulating recent trends, approaches to food safety assessment, and technological developments to report on viable food safety and sustainable food production strategies.

4.2 Emerging trends in sustainable food production and consumption: One Health concept

Scientific advice on food safety needs to keep pace with our rapidly changing world. Food quality and safety may be directly affected by global pandemics, wars and other conflicts, and climate change (Jagtap, Trollman, et al., 2022; Trollman et al., 2021). Current food safety research has identified seven common health risks associated with food in the marketplace: microbial contamination of foods, chemical contamination of foods, food adulteration, misuse of food additives, mislabeling, genetically modified (GM) foods, and foods past their use by dates (Gizaw, 2019). These health risks are driving trends among researchers, food producers, and consumers.

One Health is trending worldwide with the Global Health Summit in May 2021 stating the need for this approach in the Rome Declaration, thus recognizing that the health of humans, domestic and wild animals, plants, and environmental systems are closely linked and interdependent (Mazzeo et al., 2022). The One Health conceptual framework has also been proposed as a bridge among food safety, nutrition, and sustainability, as shown in Fig. 4.1 (Devos et al., 2022). The importance of an integrated quality control system for food production in complex socioecological systems has been emphasized in the literature to assess food safety from farm to public health (Abukhattab et al., 2022; Ajose et al., 2022; Viegas et al., 2022). Integrated approaches such as One Health add value by optimizing human, animal, and environmental health compared to frameworks focused on individual elements of food safety (Gordon et al., 2021). Although the scientific literature on One Health is increasing globally, most research has focused on the human-animal health interface, neglecting environmental health (Miao et al., 2022).

Food traceability systems are key to implementing One Health as part of an integrated approach. The latest food traceability systems are mainly based on the IoT and supported by Industry 4.0 technologies (Hassoun, Abdullah, et al., 2022; Jagtap, Duong, et al., 2021). Current technologies in food traceability include RFID, IoT sensors, QR codes, cloud, and blockchain whereas future technology trends for upgrading IoT-based food traceability systems involve 5G and Distributed Ledger Technology not based on blockchain such as DAGs, Holochain, Hashgraph, and Tempo (Mehannaoui et al., 2023). However, much of the research on supply chain management investigates simplified (not real-world) problems based on artificial data, which means that industrial scale remains a future challenge (Rolf et al., 2022).

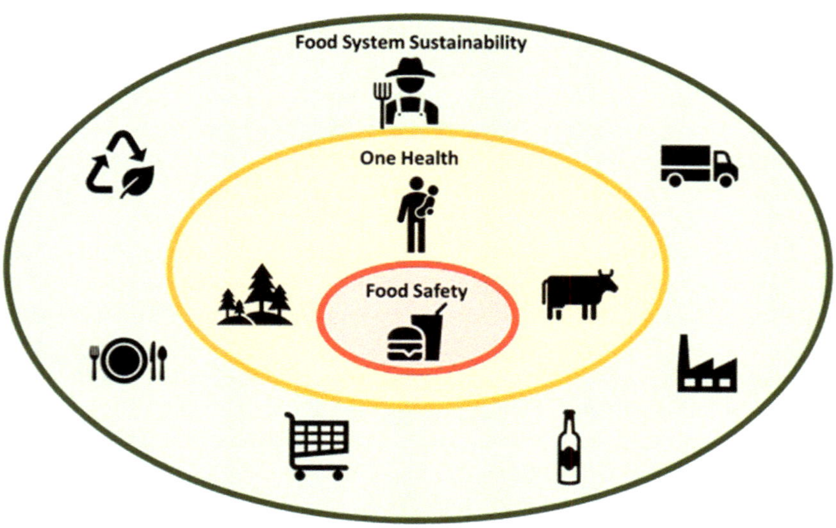

FIGURE 4.1 **One health.** The One Health conceptual framework connecting food safety to food system sustainability. *Adapted from Devos, Y., Arena, M., Ashe, S., Blanck, M., Bray, E., Broglia, A., Bronzwaer, S., Cafaro, A., Corsini, E., Dujardin, B., Dumont, A. F., Garcia, M. G., Gardi, C., Guerra, B., Kass, G. E. N., Maggiore, A., Martino, L., Merten, C., Percivaldi, C., … Hugas, M. (2022). Addressing the need for safe, nutritious and sustainable food: Outcomes of the "ONE – Health, Environment & Society – Conference 2022".* Trends in Food Science and Technology, 129, *164–178. http://www.elsevier.com/wps/find/journaldescription.cws_home/601278/description#description. https://doi.org/10.1016/j.tifs.2022.09.014.*

Microbial metabolomics is a trending new field for the qualitative and quantitative analysis of all metabolites in an organism or biological system. Since pathogens, toxins, and by-products produced by microbial degradation of food are closely related to food safety (Putri et al., 2022), monitoring these metabolites supports food safety and yields a new strategy for food safety assessment (Meng et al., 2023). In terms of One Health, microbial metabolomics has also been used to assess health, predict disease and diagnosis, and further the understanding of the interaction between body metabolism and nutrition quickly and effectively (Meng et al., 2023). Data processing and analysis are key processes in metabolomics research; however, signal processing and data analysis pose significant challenges requiring the development of new methods. Consequently, some researchers are turning to parallel computing and supervised machine learning to enable rapid optimization (Fu et al., 2022).

A trend related to One Health that is emerging among consumers is based on the idea that natural products are better due to perceived health benefits (Hassoun, Harastani, et al., 2022). This consumer behavior may lead to food safety issues and increased healthcare costs. For example, research has demonstrated a risk of developing campylobacteriosis from ingestion of raw milk or raw milk-products (Barata et al., 2022). Ensuring the safety and quality of both raw and processed food reflects a significant and trending challenge given increasing population and consumer demand for greater variety in food options. Conventional pathogen detection methods are time-consuming, unable to detect dormant forms of bacteria, consume large amounts of reagents, and have difficulty isolating target pathogens. Novel research in the area of microfluidics aims to integrate analysis procedures onto a miniaturized chip for rapid and portable detection of bacterial foodborne pathogens (Lonchamps et al., 2022). Another promising area for food hazard detection is nanozyme-activated food immunosensors (NAFIs), which significantly reduce food safety monitoring problems such as enzyme inhibition caused by complex food matrices (Shu et al., 2022). Increasing attention is also being focused on hydrogel-based materials for sensing due to their highly porous structure, excellent flexibility, self-cleaning capability, biocompatibility, and mechanical stability (Cheng et al., 2022; Ehtesabi et al., 2019; Jin et al., 2019).

The shift to alternative proteins and plant-based diets is a consumer-driven trend arising from concern for the natural environment in support of One Health. This trend is leading to the development of new food products with associated food safety and food quality concerns. Due to the novelty of this trend, little is known about the composition and dynamics of the microbiota present on such products (Duthoo et al., 2022; Wu et al., 2022). Examples of new food products include plant-based protein alternatives; animal-based protein alternatives (e.g., safe edible insects, Kopko et al., 2022); marine-based food alternatives (e.g., jellyfish, seaweed); cell-based food products (e.g., cultured meat, seafood, which are animal-derived meat products made by culturing animal cells in vitro); and products obtained by synthetic biology or new processing technologies. Food allergy is a major global human health problem with no effective cure. The consumption of these new products may have serious consequences for people with allergies and intolerances without appropriate assessment and accurate labeling. The risk lies in the inadvertent introduction of allergens into the diet through new exposures, new or altered proteins, or allergen cross-reactivity (Kopko et al., 2022).

Food quality may be improved with Industry 4.0 technologies (Hassoun, Jagtap, Garcia-Garcia, et al., 2023). A future trend that may support the precise monitoring of both raw and processed food products at home involves nanosensors associated with portable mobile devices (colorimetric, SERS, photothermal analysis) (Shu et al., 2022; Yang et al., 2022). In this respect, nanoenabled approaches to food safety hold considerable promise for the food industry. However, the potential risks to humans, animals, and the environment are not well understood, risk assessment methodologies for nanotoxicity need to be developed and validated, as well as analytical methods for the detection and characterization of engineered nanomaterials in complex food matrices and new regulatory environments need to be created (Eleftheriadou et al., 2017).

4.3 Food safety management

Consideration of nutritional and sustainability outcomes needs to be embedded in food safety assessments to enable a response to current global challenges (Devos et al., 2022). The growing global population must meet its needs without further depleting natural resources (Rahimifard et al., 2017). Consequently, alternative food sources are being increasingly investigated (Frezal et al., 2022; Ververis et al., 2020; Thinking about the future of food safety, 2022). These new foods may pose challenges for food safety management, requiring novel or revised methodologies.

The US FDA has released a food safety blueprint centered around four core elements: tech-enabled traceability, smarter tools and approaches for outbreak prevention and response, new business models and retail modernization, and food safety culture (U.S. Food & Drug Administration, 2021). However, a globally agreed framework for Open Science would make scientific outputs, including food safety assessments, more accessible, transparent, collaborative, and inclusive, particularly given the global nature of many food supply chains (Devos et al., 2022).

4.3.1 Food spoilage

Food spoilage occurs when microbiological, chemical, or physical changes occur that make food products unacceptable to consumers. Microorganisms cause microbiological food spoilage when they produce enzymes that generate objectionable by-products in food. Chemical food spoilage occurs when food components react with each other or with an added component, altering the sensory characteristics of the food, for example, oxidation, enzymatic and nonenzymatic browning. When moist foods become excessively dehydrated or dried foods absorb too much moisture, physical food spoilage results. From a microbiological-ecological perspective, failure of food safety and food spoilage are indistinguishable (Petruzzi et al., 2017).

To manage food safety, it is important to understand how different properties of a food, its environment, and event timeline can influence the development of spoilage. Classical microbiological methods of detecting food spoilage require multiple culturing steps and may not provide results in time for appropriate interventions. An additional complication is that universality has yet to be achieved for most microbial detection methods, leading to multiple analyses yielding enormous quantities of data. Consequently, the application of advanced AI-based statistical methods and approaches is warranted (Petruzzi et al., 2017).

In addition to traditional approaches to preventing microbiological spoilage such as heating, dehydration, fermentation, refrigeration, freezing, and preservatives, alternative approaches include nonthermal processing technologies, biopreservation, modified atmosphere packaging (MAP) (enclosing food in gas-barrier materials with a controlled gaseous environment), antimicrobial packaging, and hurdle technology (combining nonthermal food processing with traditional or emerging technologies) (Iordache et al., 2017). The utility of predictive microbiology for specific food spoilage flora has been recognized as a complementary approach to ensure safe food products.

4.3.2 Food contamination

Food contamination is a major public health concern with negative impacts on food quality and food safety (Gizaw, 2019). Food contaminants are substances that have not been intentionally added to food. A food contaminant is defined in the EU by Council Regulation (EEC) No. 315/93 (last consolidated version, 2009) as "any substance not intentionally added to food which is present in such food as a result of the production (including operations carried out in crop husbandry, animal husbandry and veterinary medicine), manufacture, processing, preparation, treatment, packing, packaging, transport or holding of such food, or as a result of environmental contamination" and excluding extraneous matter, such as, for example, insect fragments and animal hair.

There are three main categories of food contaminants: biological, chemical, and physical, which include sources that are from microorganisms, fungi, toxins, and chemicals—environmental pollutants, pesticides (environmental contamination), industrial cleaning processes (e.g., disinfection, cleaning, and sterilization), food preparation processes (process contaminants: produced during food processing, contact materials contamination, or cross-contamination), storage, or food contact materials (Sorbo et al., 2022). In addition, there exist contaminants of emerging concern that are currently unregulated but may be subject to future regulations, such as perchlorate, organophosphoric flame retardants, polybrominated flame retardants, perfluoroalkyl substances, microplastics, nanomaterials, and other toxins such as cyanogenic glucosides and pyrrolizidine alkaloids (Campo & Picó, 2020).

The physicochemical characteristics of some contaminants render many current analytical methodologies unsuitable, particularly for microplastics and nanomaterials. In addition, the recognition and control of emerging contaminants will be a key step in ensuring food safety from farm to fork and fundamental to achieving good levels of health.

4.3.3 Food fraud and food defense

Food fraud and food defense are part of a good food safety management system. Both pose risks to human health alongside legal and economic implications (Grace, 2018). Food fraud may be defined as any intentional action of businesses or individuals to deceive others regarding the integrity of food that is economically motivated, whereas food defense threats are behaviorally or ideologically motivated. Food fraud may include adulteration, substitution, dilution, tampering, simulation, counterfeiting, and misrepresentation. The global food supply chain needs improved food safety technological solutions and management systems to address, control, and prevent food fraud (Alrobaish et al., 2022).

Food integrity, going beyond traditional food safety, has been defined as "a multidimensional concept concerning the integrity of product, process, people, and data, implying the controlled status of a food product to be intact, safe, of quality, and authentic in its claims, as well as sourced, processed and distributed ethically throughout a food supply chain" as illustrated in Fig. 4.2 (Alrobaish et al., 2021; Wahab & Ling, 2019). However, the analysis of food fraud vulnerability factors based on real data is only beginning to be explored (Rezazade et al., 2021).

The adoption and conduct of food fraud vulnerability assessment is also still in its infancy (Soon et al., 2019). Many food businesses rely on in-house tools and internal experts for food fraud assessment (Soon et al., 2019), indicating a lack of networking and information sharing. While Vulnerability Assessment and Critical Control Point and Threat Assessment and Critical Control Point risk management approaches have been adopted by the Global Food Safety Initiative in Europe, neither have been formally accepted by the US Food and Drug Administration (US FDA) (Grace, 2018). Relatively new tools such as CARVER + Shock, Hazard Analysis Critical Control and Defense Points, and Hazard Analysis and Risk-Based Preventive Controls have yet to become widespread (Wiśniewska et al., 2021).

Traceability is important for the prevention of food fraud-related consequences. Traceability should include product traceability (physical location), process traceability (activity or sequence of activities applied to the product), genetic traceability, input traceability (raw materials, additives), disease and pest traceability, and measurement traceability (as related to accepted reference standards) (Opara, 2003). Four types of traceability exist as part of food integrity: track and trace, ensuring that certified products delivered to the end-user are uniquely identifiable and relatable to the producer; segregation, assuring the end-user that a product has met all compliance measures from production to retail and was kept separate from noncertified products; mass balance method, monitoring the volume of products downstream for equality with volume at origin; and the book and claim model based on certification at point of origin only (Mol & Oosterveer, 2015). Control measures supporting traceability that are currently incorporated into products and packaging include encrypted tags or machine-readable devices, such as QR or data matrix codes, RFID, tamper evident seals, holograms, and special inks (Manning et al., 2016; Wiśniewska et al., 2021).

4.3.4 Food packaging

In addition to protecting food products, packaging can convey nutritional information and potential safety issues. Active packaging and intelligent packaging are two emerging concepts designed as a response to global distribution chains, increased awareness of food waste, and changing consumer preferences. Fig. 4.3 classifies active and intelligent packaging according to their main characteristics.

Active packaging extends the shelf-life of food products through the addition of various components to the packaging material such as oxygen and ethylene scavengers, moisture regulators, and controlled-release antioxidant and antimicrobial agents that respond to environmental changes both inside and outside the packaging.

Intelligent packaging can monitor the condition of packaged food and the environment inside the package to alert manufacturers, retailers, and/or consumers of when a product could be spoiled or contaminated. Smart labels may be used to track products throughout the supply chain, confirm tampering, and enable rapid identification in the event of contamination. Smart labels can also share information about the place of origin and allergens.

Microfluidic devices show promising applications in monitoring the safety and quality of food packages (Pou et al., 2022; Sridhar et al., 2022). The microfluidic method has the potential to develop intelligent packaging systems, such as

FIGURE 4.2 Food integrity and food fraud. Food integrity and food fraud as a holistic approach that goes beyond food safety and food quality. *Adapted from Alrobaish, W. S., Jacxsens, L., Luning, P. A., & Vlerick, P. (2021). Food integrity climate in food businesses: Conceptualization, development, and validation of a self-assessment tool.* Foods, 10*(6). https://www.mdpi.com/2304-8158/10/6/1302/pdf. https://doi.org/10.3390/foods10061302 and Wahab, S. N., & Ling, E. K. (2019). Integrity of food supply chain: Going beyond food safety and food quality.* International Journal of Productivity and Quality Management, 1*(1), 1. https://doi.org/10.1504/ijpqm.2019.10019297.*

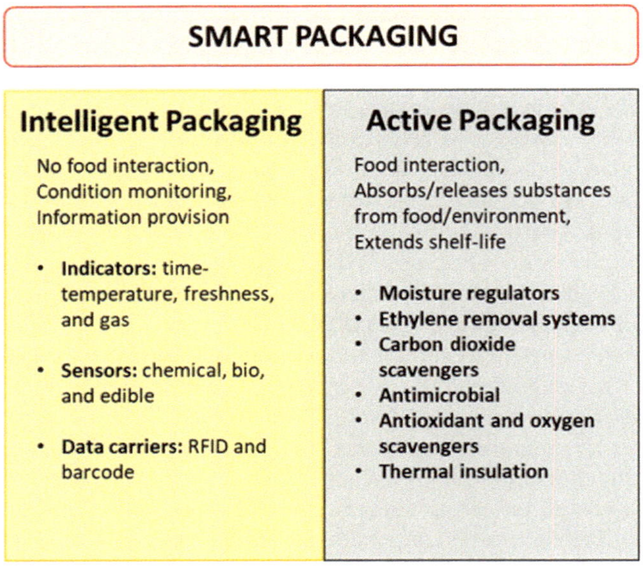

FIGURE 4.3 SMART packaging. Main characteristics of active and intelligent food packaging (Beshai et al., 2020; Drago et al., 2020; Pou et al., 2022).

time–temperature indicators, critical temperature indicators, microbial sensors, allergen detectors, heavy metal detectors, food additive detectors, food adulterant indicators, pesticide and insecticide residue detectors, nutritional content sensors, leakage indicators, gas indicators, pH indicators, biotoxin sensors, and other biochemical substance detectors in packaged foods (Pou et al., 2022).

Coating horticultural produce with edible materials to control interaction with the environment and prevent microbial contamination, while enhancing organoleptic and sensory attributes, has recently advanced preservation techniques (Santhosh et al., 2020). These biopolymers are a promising solution to postharvest losses of perishables, food insecurity, and environmental concerns. Intelligent and active food packaging made from silk has been proposed with the added benefit of intrinsic antimicrobial properties (Tracey et al., 2023). Exploiting such properties for food quality and food safety control promises to reduce food waste and provide high-quality, safe, and nutritious foods to consumers.

Researchers have generally focused on the development of individual active or intelligent packaging systems. However, combined intelligent and active packaging methods would particularly benefit perishable foods, with possible synergistic effects on food safety and quality. Commercial viability will be largely determined by cost.

4.4 Emerging technologies in food safety

This section describes some of the main emerging technologies connected with food safety. These include food-grade robotics, IoT, blockchain, AI, and Big Data.

4.4.1 Food-grade robotics

The introduction of automation in the food industry has been underway for many years, with the implementation of various levels of automation. Currently, the highest levels of automation are required to achieve the ambitions of Industry 4.0; the designed systems need to be interconnected, smart, flexible, and autonomous. For production, industrial robotics offer these capabilities. Moreover, they increase productivity, efficiency, and hygiene, enabling better legislation compliance. The food produced using industrial robots is of higher quality compared to manual handling or rigid automation and provides a better working environment in the facility (Iqbal et al., 2017). These industrial robots can be applied to a variety of tasks in food manufacturing. They are most commonly found in the following operations: pick and place, loading and unloading, decorating, packaging, and palletizing.

An industrial robot (IR) consists of a base, an arm, and a wrist, all connected by joints. Attaching these parts in different arrangements makes up the configuration of the robot, which is determined by the task(s) at hand that the IR is

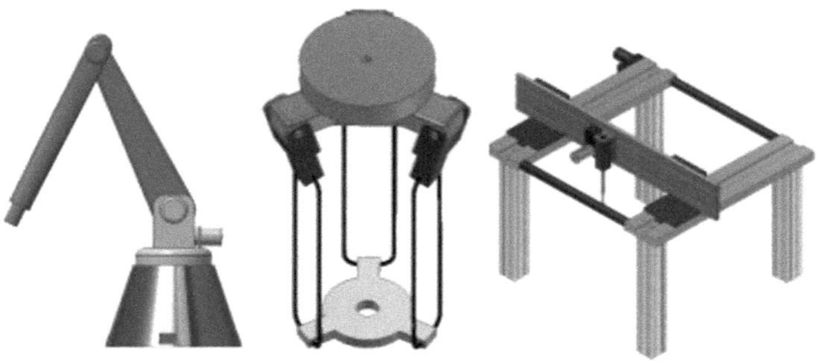

FIGURE 4.4 Industrial robots. Types of industrial robots used in food manufacturing.

required to perform (Bader & Rahimifard, 2018; Duong et al., 2020). Within the operations used by robotics in the food industry, there are three main configurations as described below and depicted in Fig. 4.4.

1. Articulated: these IRs resemble the human arm, comprising of a long shaft that rotates around a base. The robotic arms have the freedom to move flexibly in 360 degrees and to various distances away from the base in a spherical manner. These IRs are highly accurate and can manipulate a variety of payloads. Approximately 25% of all IRs found in manufacturing industries are articulated.
2. Parallel: these IRs comprise 3–6 arms linked to a top and bottom base that control a tool at the base in a hemispherical workspace. These are mainly implemented in assembly or pick-and-place tasks. These IRs are very accurate, flexible, and quick.
3. Cartesian: these IRs are triple-jointed with the ability to move vertically and horizontally in a quick manner, ensuring accuracy and high repeatability in a rectangular workspace. Typically, they are applied in assembly tasks, as well as pick-and-place operations due to their accuracy and high payload.

Attached to the end of the IR, otherwise referred to as the wrist of the IR, is the end effector or end-of-arm tool. This end tool carries out the required tasks and is found in various shapes and forms, all using different mechanisms to operate. Foods are challenging objects to manipulate with IRs due to their nature. Foods are often fragile and irregularly shaped items that can be slippery or sticky, and therefore, they need to be handled with the utmost care (Bader & Rahimifard, 2018; Hassoun, Jagtap, Trollman, et al., 2023). Consequently, specialized end tools are needed to manipulate the food.

Bader and Rahimifard (2020) detailed multiple end tools for the different IR applications in food manufacturing, highlighting four unique mechanisms that are most suitable for handling food. The first mechanism includes the use of air to create a vacuum through suction cups that connect with the food and seal it in place to perform assigned tasks. The high-speed air seal is tight enough to ensure a lifting force is created. Although this mechanism has great applications for foods with irregular surfaces, food residue may collect in the suction cups and ultimately pose a hygiene risk if not cleaned regularly. The second mechanism is direct contact between the arm tool and the food. Here, contact refers to the physical grip that the tool can use around the food to grasp and encompass it to perform the assigned task. Similar to what happens with the suction cups, residue may also build up in any end tool used for gripping, and there may also be damage to the food if it is easily deformable. The third mechanism of end tools is ingressive, including a stack of needles that can be retracted. These end tools create a grip by entering foods and manipulating them to carry out assigned tasks. Foods with flat surfaces are most suitable for this mechanism, however, it may result in the food bruising or breaking, depending on its fragility. The fourth mechanism operates by forming ice using the moisture content within the barrier of the foodstuff to create a temporary grip to carry out the task assigned. The grip achieved with the foodstuff is fragile and has a short holding time, nor can it handle heavy items. Therefore the required task should be quick. Based on the aforementioned mechanisms, there are seven common end tools that can be used to directly connect with food products and that are currently used in the food industry, as detailed in Fig. 4.5.

4.4.2 Internet of Things

Food supply chains consist of multiple actors from producing, processing, and distributing food until it reaches the consumer's plate. Each actor should be responsible for supplying safe food that is free from contamination and physical (including temperature) abuse while providing real time and accurate information about the food (Zhang et al., 2021).

End of Arm Tool	Mechanism	Advantages	Disadvantages	Example	Visualization
Pinching: two or more 'fingers' establishing a grip on the food, can be made of metal, plastic, or rubber	Contact	Simple, robust, and efficient. Suitable for small to medium foods that may be slippery	May cause bruising and possible contamination with food residue buildup	Pick and place of fruits and vegetables	
Enclosing: encompassing the food partially or wholly to achieve a grip	Contact			Pick and place of meat, poultry, and fish	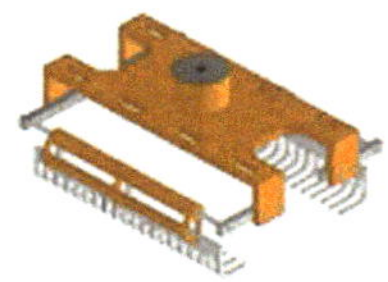
Pinning: short or long pin(s) achieve a grip by penetrating the food, typically made of metal	Ingressive	Strong grip, quick release, and low risk of contamination	Visual quality of food may be degraded as penetration may cause tissue tearing and bruising	Handling of fruits, vegetables, raw meat, poultry, and fish	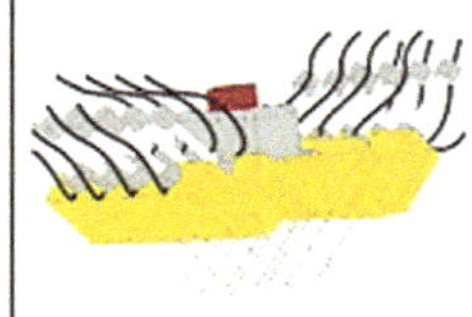
Pneumatic: a vacuum is established using pressurized air or gas to grip the food via suction, typically suction cups are used	Contact	Design can be compact; grip is reliable and non-intrusive	Possible to lose grip if hold time is long or food is heavy, sticky, or slippery; food residue build-up in suction cups probable	Pick and place of eggs, fruits, vegetables, and various packaging	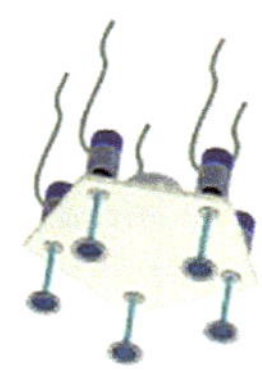
Freezing: grip established by forming a layer of ice between the end tool and the food's surface	Fluid	Highly hygienic as freeze grip is released via small injection of steam, grip is instant and non-intrusive	Grip time is very short, and release may be unpredictable; cannot handle foods that are sticky or slippery	Any frozen food, confectionary, and pastries	
Levitating: grip established by air pressure between end tool and food to manipulate light weight foods with flat surfaces	Air Pressure	Extremely hygienic as there is no direct contact with the food and it is non-intrusive	Release may be unpredictable as the holding time is limited	Lifting of lasagne sheets, pastry items, and baked goods	
Scooping: achieving manipulation by picking up foods from one place to another, typically the tool is flat or in a semi-spherical shape, can be made of metal or plastic	Contact	Fast and very simple tool that is non-intrusive and therefore has less of a contamination risk	Loss of control of material during transport may cause spills; if foods are soft, they can be deformed or bruised and food residue is possible in crevices	Transport of cut-up foods, powders, and dry ingredients	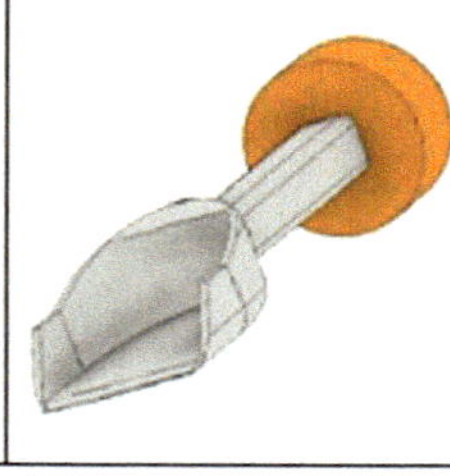

FIGURE 4.5 Table of industrial robot end tools. Types of Industrial Robot End Tools for use in food industry.

The food industry is facing significant disruption (Afrianto et al., 2020). New competitors are adopting digital technologies and challenging established market actors, competing with unprecedented innovation and agility while consumers demand more (Ragazou et al., 2022). To rise to the challenge, food companies need to increase flexibility while maintaining high productivity and compliance to ensure consumer confidence (Dutta et al., 2021). The digital revolution has begun, and food companies must leverage digital tools, including the IoT, digital twin, cloud, and analytics to drive agility and consumer responsiveness (Elahi & Tokaldany, 2020).

Leading companies go beyond just internal agility and responsiveness; they use the IoT to gain advanced insights to optimize recipes and production schedules (Hagen et al., 2019). Using the IoT, digital food companies tap into data from various sources and use analytics to gain deep consumer insights (Moradi & Dass, 2022). This may enable the reformulation of digital recipes for a quick response to trends (Jagtap & Duong, 2019). Plant automation may be leveraged to capitalize on these changing preferences. Digital plants can access information within their supply chains to rapidly adjust to changing supply and demand. Digitalization then allows food companies to get the product that consumers want to the market faster, safer, and of better quality than the competition.

Digital food companies also turn to digitalization to drive higher productivity (Jagtap & Rahimifard, 2019; Kittipanya-ngam & Tan, 2020; Martindale et al., 2022). They reduce downtime by monitoring equipment health and servicing equipment before it fails using Industrial IoT (IIoT) and predictive maintenance (Jagtap, Garcia-Garcia, et al., 2021; Jagtap, Rahimifard, et al., 2022; Jagtap, Skouteris, et al., 2021). They improve yields by identifying and correcting production issues in real time (Resende et al., 2021). They use the IoT and analytics to benchmark performance to identify improvement opportunities. They use the digital twin of their plant to simulate and validate the impact of changing equipment and production designs. They quickly deploy new recipes, packaging, and equipment using virtual commissioning. Finally, they identify opportunities to more accurately simulate products based on the IoT data and closed-loop digital twin to reflect reality better and improve optimization using real-world information (Lo et al., 2021). Digitalization helps food companies to improve throughput, decrease downtime, decrease cost, improve quality, and support a high level of traceability and compliance (Hassoun, Jagtap, Garcia-Garcia, et al., 2023; Hassoun, Jagtap, Trollman, et al., 2023; Hassoun, Kamiloglu, et al., 2023). This enables companies to raise the bar on innovation and agility without compromising production, cost, quality, or safety (Sgarbossa et al., 2022).

IoT has also impacted the food packaging market. Packaging has become interactive, aware, active, smart, and intelligent (Ahvenainen & Hurme, 1997; Chen, Brahma, et al., 2020). Food packaging interacts with its surroundings and its food to provide protection and reduce food waste (Jagtap, Duong, et al., 2021). Chemical or bio-sensors are an integral part of smart packaging to monitor the quality and safety of food throughout the supply chain, that is, from farm to fork (Griesche & Baeumner, 2020). This sensor technology enables the monitoring of freshness, presence of pathogens, leakage, emissions of gases, pH, time, and temperature in real time (Alam et al., 2021).

In summary, IoT is being utilized in the food industry to:

- Decrease food wastage via predictive analysis
- Extend the shelf-life of food products without relying on “expiry” or “use by” dates
- Ensure microbiological safety via remote monitoring and management
- Manage food quality via real-time monitoring during distribution and warehousing
- Predict better food hygiene via historical analysis of the data

Fig. 4.6 lists IoT sensors that are widely used in the food sector to monitor various parameters.

Big data application in the food supply chain is closely related to IoT deployment and occurs at every stage from farms to retailers’ shelves. For example, specific sensor units enable remote image processing of crops to assess pest issues or plant diseases in real time and relate this data to similar historical instances to suggest effective preventive measures (Huang et al., 2018). Big Data may also be deployed for forecasting weather changes or natural disasters and their impact on crops (Kamilaris et al., 2017). Researchers have demonstrated a case study wherein they applied Big Data for supplier selection through monitoring and scoring environmental performance (Belhadi et al., 2020), while others used Big Data for developing a sustainability assessment decision-support tool for by-product valorization (Belaud et al., 2019). Big Data may be used to understand consumer trends, design marketing, and perform demand forecasting based on historical data, footfall, and market data (Margaritis et al., 2022). New food product development may also employ Big Data (Jagtap & Duong, 2019) or be used to improve processes by deciphering market-related data. Big Data has been utilized in cold chains to predict and address food product deterioration during distribution as well as tracking food product quality (Rejeb et al., 2022). Researchers successfully harnessed social media data to reflect consumer opinion toward food quality and safety (Misra et al., 2022). Researchers have also used Big Data to develop a simulated pricing model to predict the real-time selling price of the food product based on its quality deterioration (Banerjee et al., 2019).

4.4.3 Other digital technologies

In the food industry, other digital technologies such as blockchain, AI, and Big Data can significantly improve food safety. Blockchain is a distributed digital ledger technology that records and stores data in an immutable and transparent

Sensor Type	Sensor Type	Detection	Food Application
TGS 822		Ethanol, Carbon Monoxide, Ammonia, Sulfur Dioxide, Alcohol, Gasoline, Methane, Acetone, etc.	Suitable for smoke detectors and alcohol detectors
SHT85	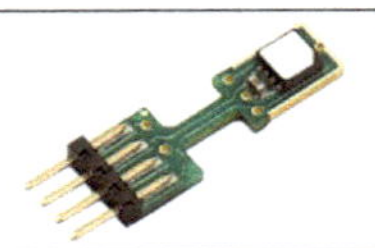	Temperature and Humidity	Suitable for measuring humidity and temperature
MH-Z14		Carbon Dioxide	Carbon-dioxide detection
MQ-131		Ozone	Monitoring air quality
C2H4		Ethylene	Monitoring ripening of fruits
4NO-250		Nitrogen Oxide	Monitoring food quality
SO2		Sulfur Dioxide	Monitor pungent odorous gas
MQ-7		Carbon Monoxide	Air quality monitoring

FIGURE 4.6 Table of IoT sensors. IoT sensors widely used in the food sector.

way. AI is a technology that allows machines to simulate human intelligence and behavior to solve complex problems, while Big Data is a term used to describe large amounts of data that can be analyzed to discover patterns, trends, and insights. These technologies applied to the food industry can enable early identification and resolution of food safety issues, ensure the quality and safety of food products, improve traceability and transparency of the food supply chain, and help detect fraud, as discussed in the following sections.

4.4.3.1 Predictive food safety

Predictive food safety refers to the use of data and advanced analytics to prevent and anticipate food safety problems before they occur, by identifying potential risks and taking proactive measures. It aims to improve food safety, reduce foodborne diseases, and make production more efficient and cost-effective. Zhou, Zhang, et al. (2022) review and describe the basic concepts of Big Data, AI, and blockchain, and their applications in the field of food safety from a data perspective, including data collection, storage and transfer, analysis, visualization, and security. The article also discusses the challenges and future developments of using these technologies in food safety.

Big data analytics can be used to analyze large amounts of food safety data to predict the likelihood of food safety incidents and to identify patterns in food safety data. Thus Sun et al. (2022) highlight the importance of rapid and accurate detection of foodborne pathogens to protect human health and the use of Big Data and omics-based technologies in combination with CRISPR-based technologies for more efficient and convenient food detection in the future. Likewise, Farber et al. (2021) review and suggest alternative approaches for *Listeria monocytogenes* risk management in low-risk foods, including the use of Big Data to inform microbial risk assessments instead of the traditional "zero tolerance" approach. In the same vein, Imanian et al. (2022) review how high-throughput sequencing technology, which has been integrated with Big Data analysis, could be incorporated into food safety systems to provide a better understanding of microorganisms in food and the environment and improve control measures. In addition, Soon and Saguy (2017) review how crowdsourcing (a new tool offered primarily via the internet for ideas, content, and funding) is being used in the food field and argue that it can be a valuable tool for Big Data, specifically in the use of crowdsourced data for shelf-life tracking, inventory control, foodborne disease surveillance, contaminated product identification, improved hygiene, food safety, communication, allergen management, and risk reduction for food businesses.

In addition, AI algorithms can be trained on food safety data to predict the likelihood of food safety incidents and identify patterns in food safety data. Thus Nychas et al. (2016) review how new technologies, such as AI, could help improve food safety management by providing tools to monitor, record, and control critical parameters throughout the entire life cycle of a product. In the same vein, Kim and Kim (2022) review how the integration of Big Data and AI in the food industry has had and will have a substantial impact, especially in the area of food safety, facilitating the prediction of food safety hazards, the detection of foodborne pathogens, and the effective management of food safety.

Blockchain also has great potential for food safety. Thus Xu et al. (2022) review the feasibility and application prospects of blockchain technology in plant food safety, animal food safety, and processed food safety. They found that biological contamination is a common food safety concern and that lack of transparency in the food supply chain is the main cause of these problems. They argued that blockchain technology has the potential to improve food safety by making the food supply chain more transparent, decentralized, and secure.

4.4.3.2 Quality control

Quality control in the food industry is the process of ensuring that food products meet or exceed established standards for safety, quality, and consistency. On this topic, Chen, Zhang, et al. (2020) review and discuss the use of AI technology in combination with physical field technologies, such as microwave, radiofrequency, infrared radiation, and ultrasonic fields, to dry fruits and vegetables in an energy-efficient and high-quality manner. Along the same lines, Wang et al. (2022) review the application of advanced sensing technologies and AI-based methods for detection and prediction during the sorting, drying, disinfection, sterilization, and freezing of berries. Similarly, Chen and Yu (2022) review the latest advances in noncontact techniques for food quality assessment and find that AI and image processing systems, as well as sensor technology, are increasingly used to improve the efficiency and accuracy of food safety inspections and meet consumer expectations for safe, high-quality food.

Otherwise, Chen et al. (2022) and Han et al. (2021) review and explore future trends in food cold chain logistics (FCCLs), highlighting the use of technologies such as IoT, AI, and Big Data to balance resource costs, food quality, safety, and environmental impact for a more sustainable FCCL. Lastly, Kittichotsatsawat et al. (2021) review previous research and find that the coffee supply chain leads in terms of blockchain implementation. They also conclude that Big Data and digital agriculture have the potential to bring benefits to the coffee sector by raising quality standards and improving production efficiency, and increasing its reliability for customers, with a focus on international acceptance and sustainability.

4.4.3.3 Traceability

Traceability in the food industry refers to the ability to trace the history, location, or movement of a food product through all stages of production, processing, and distribution. Traceability is important because it enables effective product recall in the event of contamination or other food safety problems. In this regard, Kaur et al. (2022) review the use of blockchain technology and the IoT in food supply chain management. They find that blockchain technology can improve food traceability and enhance business activities in the supply chain through collaborative relationships and operational efficiencies. However, they also identify several challenges, such as scalability, security, and privacy issues. In the same vein, Kampan et al. (2022) conduct a review on the use of blockchain technology for livestock product traceability, focusing specifically on a region in Thailand. The review concludes that the use of blockchain technology in the food industry has great potential, however, its adoption rate has been slow, especially in the livestock sector. The study identified several challenges impeding the adoption of blockchain technology, such as low awareness among stakeholders, poor coordination among industry players, and lack of resources and capacity among small-scale livestock farmers and processors. On the other hand, Yadav et al. (2022) address, in a systematic literature review, the need for Industry 4.0 technologies, including the IoT, blockchain, Big Data, Information and Communication Technologies (ICTs), Cloud Computing and Cyber-Physical Systems (CPSs), in the agricultural food supply chain. The authors suggest that the integration of these technologies can provide low-cost solutions and improve sustainability in the agricultural food supply chain, with blockchain being seen as a potential game changer for ensuring food safety and security.

4.4.3.4 Supply chain transparency

Transparency is a concept closely related to traceability, but they have distinct differences. While traceability is the ability to follow the movement of food products through the supply chain, transparency is the availability and accessibility of information about the food supply chain for use by consumers, regulators, and other stakeholders. Transparency in the supply chain helps to identify potential risks and issues, such as those related to food safety, and to take appropriate actions to address them. Blockchain excels in creating transparency throughout the food supply chain. In this context, Li et al. (2021) review the main blockchain platforms currently used in food supply chains and conduct a synthesis analysis to explore the benefits and challenges of blockchain technology in the food industry. The authors found that blockchain technology can improve traceability, increase transparency and security, reduce transaction costs and time, and improve operational efficiency in food supply chains. However, they also identified challenges to adoption, such as technological limitations, cost, governance, regulation, awareness, and education. Similarly, Keogh et al. (2023) suggested that blockchain technology can help improve food supply chain efficiency and reduce risk, but its adoption in the food industry is hampered by technical, organizational, and regulatory issues. They propose that blockchain technology could help the food industry meet the goals of improving traceability, transparency, and trust. In addition, Sharma and Singh (2022) discuss the potential of IoT, blockchain, and Industry 4.0 to revolutionize the food industry by improving supply chain transparency and efficiency through real-time data collection and analysis to optimize decision-making and reduce costs and inefficiencies in the supply chain.

4.4.3.5 Detection of food fraud

Food fraud detection in the food industry refers to the process of identifying and preventing intentional mislabeling, substitution, dilution, or addition of ingredients in food products for financial gain. Detecting food fraud is important because it helps to ensure that food products are safe for consumption and that consumers get what they have paid for. In this area, Echegaray et al. (2022) argue that Industry 4.0 technologies, such as robotics, IoT, Big Data, augmented reality, cybersecurity, and blockchain, have the potential to improve meat processing, preservation, and analysis, reduce food waste and loss, and prevent meat fraud. However, implementing these technologies can be difficult due to the complexity of the technologies themselves and other factors. In addition, Gopi et al. (2019) summarize current methodologies for determining the provenance and authenticity of seafood products, focusing on methods that could be used by regulators and the seafood industry to enforce regulations and protect against food fraud. These methods include DNA profiling, fatty acid profiling, and blockchain. The authors suggest that a combination of methods would be most effective in determining the seafood provenance, given the complex supply chain. Finally, Oliveira et al. (2019) review the use of analytical technologies such as near-infrared spectroscopy and hyperspectral imaging to monitor spice fraud. The review suggests that data fusion and Big Data analysis are promising tools that can improve the analysis of data obtained from these techniques and can be used for the control of spice fraud.

4.5 Food safety strategies for the use of digital tools

With the increasing amount and complexity of data generated in food supply chains, there is a need for practical methods to manage, share, and store data in a safe and efficient way. This is particularly important for data regarding food safety, which is arguably the most critical aspect of food production and distribution. Digital tools are increasingly being used to collect more detailed and larger amounts of food safety data in real time, as well as store and share it with interested stakeholders.

In this regard, the World Health Organization (WHO Global Strategy for Food Safety 2022–2030: Towards Stronger Food Safety Systems and Global Cooperation, 2022) recently set a global strategy for food safety with the objective of addressing current and emerging challenges that threaten the consumption of safe and healthy food by people from all over the world. They set five general strategic priorities, of which Strategy Priority 3 "Improving the use of food chain information, scientific evidence, and risk assessment in making risk management decisions" can be partially addressed with the use of digital tools, as can be seen in Fig. 4.7. Specifically, digital tools aid in gathering food safety

FIGURE 4.7 Digital tools. Digital tools that can support the achievement of Strategic Priority 3 of the World Health Organization (2022).

information to promote the use of scientific evidence and strengthen risk-management decisions for food control, therefore supporting the achievement of strategic objectives 3.1–3.4. Digital tools can also be used with the assessment tool designed by FAO/WHO (Food and Agriculture Organization of the United Nations & World Health Organization, 2019) to identify areas of potential improvement and plan sequential and coordinated activities in food control systems, particularly in its Dimension B, which covers the main control functions to ensure food safety along food supply chains.

Consequently, food businesses have been further motivated to start implementing digital tools to support their operations. However, despite the clear advantages digital tools provide, such implementation is challenging, due to the potential need for significant economic investment as well as the lack of modernization in the food sector. Morella et al. (2021) further investigated these challenges and classified them into technical, educational, and policy challenges. They proposed using key performance indicators to demonstrate the advantages of implementing digital tools, showcasing examples of their successful implementation to encourage wider acceptance, standardizing the implementation process, and integrating sustainability considerations into the implementation of digital tools. Other challenges that have been identified for the implementation of augmented reality in the food industry, which is common to other digital tools, are issues regarding privacy and security, absence of robust business models, risk of damage or harm, lack of social acceptance, insufficient quality of the digital tools, and technological issues (Jagtap, Saxena, et al., 2021). On the other hand, Ali and Aboelmaged (2022) identified lack of collaboration, organizational inertia, and lack of awareness as the critical barriers to implementing digital tools, while reduction in supply-demand misalignment, rapidly changing consumer needs, threat of legal penalties, and cost optimization are key drivers for such implementation.

Hassoun, Jagtap, Trollman, et al. (2023) assessed the implementation of a number of digital tools in food processing and identified several with the potential to improve food safety. Specifically, digital options such as IoT, smart sensors (during production and distribution as well as in intelligent packaging), AI, and Big Data have been shown to enable the monitoring, collection, and sharing of data for important indicators of food safety, such as humidity, temperature, variations in gas concentrations (such as oxygen and carbon dioxide), and changes in pH. Additional areas where food safety can be enhanced using digital tools include optimized hygiene, such as in live monitoring of cleaning-in-place operations (Escrig et al., 2020; Simeone et al., 2020), quality control systems that measure compliance to food safety regulations such as good manufacturing practices, and the use of blockchain to record transactions more safely, and share and validate data, improving traceability. Novel technologies that can also play an important role in optimizing food traceability as well as enabling faster testing include DNA sequencing technology (Tetra Pak, 2020). Furthermore, digital tools enable reactions to safety issues before contamination spreads, reducing business economic losses, maximizing operational performance, and enhancing overall food safety.

Incorporating digital tools in food production needs strategic planning as well as human, economic, and time resources. (Konur et al., 2021) presented an example in which they successfully designed, developed, and implemented digital tools, including IoT, Big Data analytics, machine learning, and CPS, in a traditional food manufacturing company. A novel data collection mechanism and an intelligent decision-support system were integrated into the company's existing equipment and tested to predict the performance of their operations. In another example, (Romanello & Veglio, 2022) described the process of implementing digital tools in a food-processing company, and highlighted the importance of early communication with employees and strategies to manage human resources to overcome resistance to change and to foster the necessary skills to implement digitalization successfully. They also noted that digital tools can improve the safety of the food product and also the safety of the employees, for example, by using robots and automation in the processes with the largest risk.

Akyazi et al. (2020) presented a strategy for digital transformation in the food industry based on the following six steps:

1. Evaluate the current use of digital tools and analyze novel key trends
2. Identify skills and professional job profiles needed to implement digital tools
3. Create training programs
4. Match skill requirements of the food industry and skills provided by training
5. Attract talented people
6. Monitor

Although this strategy is aimed at the food sector as a whole, individual companies may also benefit from applying some or all of the steps of this strategy in their business operations.

In conclusion, digital tools have demonstrated significant safety-enhancing advantages in the food sector. As the implementation of digital tools in the food industry is challenging, a step-wise implementation strategy is recommended, starting with a review of current operational activities in the business and the identification of digital tools with the potential to optimize performance. Any digital transformation in the food industry should be undertaken along with adequate staff training and monitoring of the results achieved by such change. Based on emerging trends, it is expected that food businesses will need to continue implementing digital tools to improve their operations as well as overall food safety.

4.6 Discussion

This chapter has revealed that some recently developed technologies may be promising options for food safety assessment. Specifically, the focus has been put on the implementation of the Food Safety 4.0 principle, based on advanced technologies such as food-grade robotics, IoT, and digital tools (e.g., blockchain and AI) for traceability and other food safety-related issues. The implementation of technology-based approaches to food safety is not without cost and may generate large amounts of data requiring potentially expensive storage and analysis. As most businesses have limited resources, to optimize the impact of investment in such technological solutions to food safety issues, a strategic approach that considers food safety holistically in terms of interconnected systems as part of sustainable food production and consumption strategy is warranted.

This chapter has highlighted the importance of robotics, IoT, and other digital tools in achieving Food Safety 4.0. The use of these digital enablers of food safety throughout the food supply chain has been demonstrated with specific examples. However, as illustrated by existing research, there is a significant gap between theory and application within the food industry. Hence, this chapter addresses the scant research investigating the application of Industry 4.0 technologies in the food industry. Food Safety 4.0 also identifies promising directions for future research in digitalization and automation technologies. This chapter is among the first to raise awareness of the importance of multidisciplinary and holistic approaches to food safety that simultaneously consider a broad range of emerging technologies and trends. In practice, this chapter may facilitate an understanding of the various challenges and opportunities offered by the adoption of Food Safety 4.0, supporting best practices in the management of production, processing, distribution, retail, and consumption. The research in this chapter may be used as a basis for understanding how to develop a sustainable food production strategy in support of food safety.

4.7 Conclusion

The main aim of this chapter was to discuss and describe the concept of Food Safety 4.0, highlighting the potential of Industry 4.0 technologies in advancing food safety. Industry 4.0 technologies have a significant role to play in sustainable social, environmental, and economic development. This chapter reports on applications of Industry 4.0 technologies, providing an up-to-date source of information about the latest developments and advances, as well as future directions in relation to emerging trends, for Food Safety 4.0. Consequently, this chapter may be used by policy-makers to advance and support transdisciplinary collaboration embracing the discussed technological innovations. A holistic and long-term policy-making approach is needed to facilitate an impactful adoption of Food Safety 4.0 through increased awareness of the applications of Industry 4.0 technologies.

The effectiveness of existing food safety assessment methods is questioned with each new food crisis, significantly undermining consumer confidence. However, food safety assessment approaches based on Industry 4.0 technologies aim to increase consumer confidence by providing accurate, real-time information, thereby minimizing food waste and loss, decreasing the use of energy and resources, and ensuring food quality. Traditional approaches to food safety focus on intrinsic attributes such as physical, chemical, microbial, and technical (processing) parameters through the application of time-consuming, laborious, and destructive methods. In contrast, Industry 4.0 technologies have the potential to overcome such limitations in a way that integrates physical and digital elements along the entire food value chain, improving the efficiency and reliability of food safety assessment techniques. The challenges facing a successful transition to Food Safety 4.0 include affordability, scalability, and commercial viability. Inadequate infrastructure is a critical constraint that needs to be addressed. Increased investment, pressure from consumers and policy-makers, and continued innovation are expected to drive the elimination of these barriers to adopting Food Safety 4.0. Optimal food safety assessment (achieved via the implementation of Food Safety 4.0) will help assure smart food safety controls and high-quality assessment of food products, facilitating the transition to more sustainable food systems.

Acknowledgments

Guillermo Garcia-Garcia acknowledges the Grant "Juan de la Cierva Incorporación," funded by MCIN/AEI/10.13039/501100011033 and "ESF Investing in your future."

References

Abukhattab, S., Kull, M., Abu-Rmeileh, N. M. E., Cissé, G., Crump, L., Hattendorf, J., & Zinsstag, J. (2022). Towards a one health food safety strategy for palestine: A mixed-method study. *Antibiotics*, *11*(10), 1359. Available from https://doi.org/10.3390/antibiotics11101359.

Afrianto, I., Djatna, T., Arkeman, Y., Sukaesih Sitanggang, I., Hermadi, I. (2020). Disrupting agro-industry supply chain in Indonesia with blockchain technology: Current and future challenges. In *2020 8th International conference on cyber and IT service management*, CITSM 2020. Institute of Electrical and Electronics Engineers Inc, Indonesia. http://ieeexplore.ieee.org/xpl/mostRecentIssue.jsp?punumber = 9268688, https://doi.org/10.1109/CITSM50537.2020.9268872.

Ahvenainen, R., & Hurme, E. (1997). Active and smart packaging for meeting consumer demands for quality and safety. *Food Additives and Contaminants*, *14*(6–7), 753–763. Available from https://doi.org/10.1080/02652039709374586.

Ajose, D. J., Abolarinwa, T. O., Oluwarinde, B. O., Montso, P. K., Fayemi, O. E., Aremu, A. O., & Ateba, C. N. (2022). Application of plant-derived nanoparticles (PDNP) in food-producing animals as a bio-control agent against antimicrobial-resistant pathogens. *Biomedicines*, *10*(10), 2426. Available from https://doi.org/10.3390/biomedicines10102426.

Akyazi, T., Goti, A., Oyarbide, A., Alberdi, E., & Bayon, F. (2020). A guide for the food industry to meet the future skills requirements emerging with industry 4.0. *Foods*, *9*(4). Available from https://doi.org/10.3390/foods9040492, https://www.mdpi.com/2304-8158/9/4/492.

Alam, A. U., Rathi, P., Beshai, H., Sarabha, G. K., & Deen, M. J. (2021). Fruit quality monitoring with smart packaging. *Sensors*, *21*(4), 1509. Available from https://doi.org/10.3390/s21041509.

Ali, I., & Aboelmaged, M. G. S. (2022). Implementation of supply chain 4.0 in the food and beverage industry: perceived drivers and barriers. *International Journal of Productivity and Performance Management*, *71*(4), 1426–1443. Available from https://doi.org/10.1108/IJPPM-07-2020-0393, http://www.emeraldinsight.com/info/journals/ijppm/ijppm.jsp.

Alrobaish, W. S., Jacxsens, L., Luning, P. A., & Vlerick, P. (2021). Food integrity climate in food businesses: Conceptualization, development, and validation of a self-assessment tool. *Foods*, *10*(6). Available from https://doi.org/10.3390/foods10061302, https://www.mdpi.com/2304-8158/10/6/1302/pdf.

Alrobaish, W. S., Jacxsens, L., Spagnoli, P., & Vlerick, P. (2022). Assessment of food integrity culture in food businesses through method triangulation. *Food Control*, *141*109168. Available from https://doi.org/10.1016/j.foodcont.2022.109168.

Bader, F., Rahimifard, S. (2018). Challenges for industrial robot applications in food manufacturing. In *ACM international conference proceeding series*. Association for Computing Machinery, United Kingdom. http://portal.acm.org/0.10.1145/3284557.3284723.

Bader, F., & Rahimifard, S. (2020). A methodology for the selection of industrial robots in food handling. *Innovative Food Science & Emerging Technologies*, *64*102379. Available from https://doi.org/10.1016/j.ifset.2020.102379.

Banerjee, T., Mishra, M., Debnath, N. C., & Choudhury, P. (2019). Implementing E-commerce model for agricultural produce: A research roadmap. *Periodicals of Engineering and Natural Sciences (PEN)*, *7*(1), 302. Available from https://doi.org/10.21533/pen.v7i1.353.

Barata, A. R., Nunes, B., Oliveira, R., Guedes, H., Almeida, C., Saavedra, M. J., da Silva, G. J., & Almeida, G. (2022). Occurrence and seasonality of Campylobacter spp. in Portuguese dairy farms. *International Journal of Food Microbiology, 383*. Available from https://doi.org/10.1016/j.ijfoodmicro.2022.109961, http://www.elsevier.com/locate/ijfoodmicro.

Belaud, J. P., Prioux, N., Vialle, C., & Sablayrolles, C. (2019). Big data for agri-food 4.0: Application to sustainability management for by-products supply chain. *Computers in Industry, 111*, 41–50. Available from https://doi.org/10.1016/j.compind.2019.06.006, https://www.journals.elsevier.com/computers-in-industry.

Belhadi, A., Kamble, S. S., Zkik, K., Cherrafi, A., & Touriki, F. E. (2020). The integrated effect of big data analytics, lean six sigma and green manufacturing on the environmental performance of manufacturing companies: The case of North Africa. *Journal of Cleaner Production, 252*. Available from https://doi.org/10.1016/j.jclepro.2019.119903, https://www.journals.elsevier.com/journal-of-cleaner-production.

Beshai, H., Sarabha, G, Rathi, P., Alam, A., & Deen, M. (2020). Freshness monitoring of packaged vegetables. *Applied Sciences, 10*(21), 7937. Available from https://doi.org/10.3390/app10217937.

Byrd, K., Fan, A., Her, E. S., Liu, Y., Almanza, B., & Leitch, S. (2021). Robot vs human: expectations, performances and gaps in off-premise restaurant service modes. *International Journal of Contemporary Hospitality Management, 33*(11), 3996–4016. Available from https://doi.org/10.1108/IJCHM-07-2020-0721, http://www.emeraldinsight.com/info/journals/ijchm/ijchm.jsp.

Campo, J., & Picó, Y. (2020). *Emerging contaminants and toxins* (pp. 729–758). Elsevier BV. Available from https://doi.org/10.1016/b978-0-12–813266-1.00017-6.

Casino, F., Kanakaris, V., Dasaklis, T. K., Moschuris, S., & Rachaniotis, N. P. (2019). Modeling food supply chain traceability based on blockchain technology. *IFAC-PapersOnLine, 52*(13), 2728–2733. Available from https://doi.org/10.1016/j.ifacol.2019.11.620, http://www.journals.elsevier.com/ifac-papersonline/, ISSN: 2405-8963.

Chen, J., Zhang, M., Xu, B., Sun, J., & Mujumdar, A. S. (2020). Artificial intelligence assisted technologies for controlling the drying of fruits and vegetables using physical fields: A review. *Trends in Food Science and Technology, 105*, 251–260. Available from https://doi.org/10.1016/j.tifs.2020.08.015, http://www.elsevier.com/wps/find/journaldescription.cws_home/601278/description#description.

Chen, Q., Qian, J., Yang, H., & Wu, W. (2022). Sustainable food cold chain logistics: From microenvironmental monitoring to global impact. *Comprehensive Reviews in Food Science and Food Safety, 21*(5), 4189–4209. Available from https://doi.org/10.1111/1541-4337.13014, http://onlinelibrary.wiley.com/journal/10.1111/(ISSN)1541-4337.

Chen, S., Brahma, S., Mackay, J., Cao, C., & Aliakbarian, B. (2020). The role of smart packaging system in food supply chain. *Journal of Food Science, 85*(3), 517–525. Available from https://doi.org/10.1111/1750-3841.15046, http://onlinelibrary.wiley.com/journal/10.1111/(ISSN)1750-3841.

Chen, T. C., & Yu, S. Y. (2022). The review of food safety inspection system based on artificial intelligence, image processing, and robotic. *Food Science and Technology (Brazil), 42*. Available from https://doi.org/10.1590/fst.35421, https://www.scielo.br/j/cta/a/cGFtGt7WstZDc7vY9LD3RGf/?format = pdf&lang = en.

Cheng, W., Wu, X., Zhang, Y., Wu, D., Meng, L., Chen, Y., & Tang, X. (2022). Recent applications of hydrogels in food safety sensing: Role of hydrogels. *Trends in Food Science and Technology, 129*, 244–257. Available from https://doi.org/10.1016/j.tifs.2022.10.004, http://www.elsevier.com/wps/find/journaldescription.cws_home/601278/description#description.

Devos, Y., Arena, M., Ashe, S., Blanck, M., Bray, E., Broglia, A., Bronzwaer, S., Cafaro, A., Corsini, E., Dujardin, B., Dumont, A. F., Garcia, M. G., Gardi, C., Guerra, B., Kass, G. E. N., Maggiore, A., Martino, L., Merten, C., Percivaldi, C., ... Hugas, M. (2022). Addressing the need for safe, nutritious and sustainable food: Outcomes of the "ONE – Health, Environment & Society – Conference 2022". *Trends in Food Science and Technology, 129*, 164–178. Available from https://doi.org/10.1016/j.tifs.2022.09.014, http://www.elsevier.com/wps/find/journaldescription.cws_home/601278/description#description.

Drago, E., Campardelli, R., Pettinato, M., & Pereo, P. (2020). Innovations in smart packaging concepts for food: An extensive review. *Foods, 9*(11), 1628. Available from https://doi.org/10.3390/foods9111628.

Duong, L. N. K., Al-Fadhli, M., Jagtap, S., Bader, F., Martindale, W., Swainson, M., & Paoli, A. (2020). A review of robotics and autonomous systems in the food industry: From the supply chains perspective. *Trends in Food Science and Technology, 106*, 355–364. Available from https://doi.org/10.1016/j.tifs.2020.10.028, http://www.elsevier.com/wps/find/journaldescription.cws_home/601278/description#description.

Duthoo, E., De Reu, K., Leroy, F., Weckx, S., Heyndrick + x, M., & Rasschaert, G. (2022). To culture or not to culture: careful assessment of metabarcoding data is necessary when evaluating the microbiota of a modified-atmosphere-packaged vegetarian meat alternative throughout its shelf-life period. *BMC Microbiology, 22*(1). Available from https://doi.org/10.1186/s12866-022-02446-9, http://www.biomedcentral.com/bmcmicrobiol/.

Dutta, G., Kumar, R., Sindhwani, R., & Singh, R. K. (2021). Digitalization priorities of quality control processes for SMEs: A conceptual study in perspective of Industry 4.0 adoption. *Journal of Intelligent Manufacturing, 32*(6), 1679–1698. Available from https://doi.org/10.1007/s10845-021-01783-2, http://www.kluweronline.com/issn/0956-5515/.

Echegaray, N., Hassoun, A., Jagtap, S., Tetteh-Caesar, M., Kumar, M., Tomasevic, I., Goksen, G., & Lorenzo, J. M. (2022). Meat industry. *Applied Sciences, 12*(14). Available from https://doi.org/10.3390/app12146986, 6986.

Ehtesabi, H., Roshani, S., Bagheri, Z., & Yaghoubi-Avini, M. (2019). Carbon dots-sodium alginate hydrogel: A novel tetracycline fluorescent sensor and adsorber. *Journal of Environmental Chemical Engineering, 7*(5). Available from https://doi.org/10.1016/j.jece.2019.103419, http://www.journals.elsevier.com/journal-of-environmental-chemical-engineering/.

Elahi, B., & Tokaldany, S. A. (2020). Application of internet of things-aided simulation and digital twin technology in smart manufacturing. *Advances in mathematics for Industry 4.0*, 335–359. Available from https://doi.org/10.1016/B978-0-12-818906-1.00015-2, Elsevier, United States, https://www.elsevier.com/books/advances-in-mathematics-for-industry-40/ram/978-0-12-818906-1.

Eleftheriadou, M., Pyrgiotakis, G., & Demokritou, P. (2017). Nanotechnology to the rescue: using nano-enabled approaches in microbiological food safety and quality. *Current Opinion in Biotechnology*, *44*, 87–93. Available from https://doi.org/10.1016/j.copbio.2016.11.012, http://www.elsevier.com/locate/copbio.

Escrig, J. E., Simeone, A., Woolley, E., Rangappa, S., Rady, A., & Watson, N. J. (2020). Ultrasonic measurements and machine learning for monitoring the removal of surface fouling during clean-in-place processes. *Food and Bioproducts Processing*, *123*, 1–13. Available from https://doi.org/10.1016/j.fbp.2020.05.003, http://www.elsevier.com/wps/find/journaldescription.cws_home/713883/description#description.

Farber, J. M., Zwietering, M., Wiedmann, M., Schaffner, D., Hedberg, C. W., Harrison, M. A., Hartnett, E., Chapman, B., Donnelly, C. W., Goodburn, K. E., & Gummalla, S. (2021). Alternative approaches to the risk management of *Listeria monocytogenes* in low risk foods. *Food Control*, *123*. Available from https://doi.org/10.1016/j.foodcont.2020.107601, https://www.journals.elsevier.com/food-control.

Food and Agriculture Organization of the United Nations & World Health Organization. (2019). Food control system assessment tool: Introduction and glossary https://apps.who.int/iris/handle/10665/329866.

Frezal, C., Nenert, C., & Gay, H. (2022). *Meat protein alternatives: Opportunities and challenges for food systems' transformation. OECD Food, Agriculture and Fisheries Papers, No* (182). Paris: OECD Publishing. Available from https://doi.org/10.1787/387d30cf-en.

Fu, J., Zhang, Y., Wang, Y., Zhang, H., Liu, J., Tang, J., Yang, Q., Sun, H., Qiu, W., Ma, Y., Li, Z., Zheng, M., & Zhu, F. (2022). Optimization of metabolomic data processing using NOREVA. *Nature Protocols*, *17*(1), 129–151. Available from https://doi.org/10.1038/s41596-021-00636-9, http://www.natureprotocols.com/.

Gizaw, Z. (2019). Public health risks related to food safety issues in the food market: A systematic literature review. *Environmental Health and Preventive Medicine*, *24*(1). Available from https://doi.org/10.1186/s12199-019-0825-5, http://www.springerlink.com/content/120858/?Content + Status = Accepted.

Gopi, K., Mazumder, D., Sammut, J., & Saintilan, N. (2019). Determining the provenance and authenticity of seafood: A review of current methodologies. *Trends in Food Science and Technology*, *91*, 294–304. Available from https://doi.org/10.1016/j.tifs.2019.07.010, http://www.elsevier.com/wps/find/journaldescription.cws_home/601278/description#description.

Gordon, I. J., Alonso, S., Crump, L., Dominguez-Salas, P., & de Garine-Wichatitsky, M. (2021). *Food Security and Nutrition* (pp. 327–343). CABI Publishing. Available from https://doi.org/10.1079/9781789242577.0327.

Grace, D. (2018). *Food fraud encyclopedia of food security and sustainability* (pp. 238–248). Kenya: Elsevier. Available from http://doi.org/10.1016/B978-0-08-100596-5.21577-1, https://doi.org/10.1016/B978-0-08-100596-5.21577-1.

Griesche, C., & Baeumner, A. J. (2020). Biosensors to support sustainable agriculture and food safety. *TrAC - Trends in Analytical Chemistry*, *128*. Available from https://doi.org/10.1016/j.trac.2020.115906, http://www.elsevier.com/locate/trac.

Grobbelaar, W., Verma, A., & Shukla, V. K. (2021). Analyzing human robotic interaction in the food industry. *Journal of Physics: Conference Series*, *1714*(1)012032. Available from https://doi.org/10.1088/1742-6596/1714/1/012032.

Hagen, B., Zucchella, A., & Ghauri, P. N. (2019). From fragile to agile: marketing as a key driver of entrepreneurial internationalization. *International Marketing Review*, *36*(2), 260–288. Available from https://doi.org/10.1108/IMR-01-2018-0023.

Han, J. W., Zuo, M., Zhu, W. Y., Zuo, J. H., Lü, E. L., & Yang, X. T. (2021). A comprehensive review of cold chain logistics for fresh agricultural products: Current status, challenges, and future trends. *Trends in Food Science and Technology*, *109*, 536–551. Available from https://doi.org/10.1016/j.tifs.2021.01.066, http://www.elsevier.com/wps/find/journaldescription.cws_home/601278/description#description.

Hassoun, A., Abdullah, N., Aït-Kaddour, A., Ghellam, M., Beşir, A., Zannou, O., Önal, B., Aadil, R. M., Lorenzo, J. M., Khaneghah., & Regenstein, J. M. (2022). Food traceability 4.0 as part of the fourth industrial revolution: Key enabling technologies. *Critical Reviews in Food Science and Nutrition*. Available from https://doi.org/10.1080/10408398.2022.2110033.

Hassoun, A., Harastani, R., Jagtap, S., Trollman, H., Garcia-Garcia, G., Awad, N. M. H., Zannou, O., Galanakis, C. M., Goksen, G., Nayik, G. A., Riaz, A., & Maqsood, S. (2022). Truths and myths about superfoods in the era of the COVID-19 pandemic. *Critical Reviews in Food Science and Nutrition*. Available from https://doi.org/10.1080/10408398.2022.2106939, https://www.tandfonline.com/loi/bfsn20.

Hassoun, A., Jagtap, S., Garcia-Garcia, G., Trollman, H., Pateiro, M., Lorenzo, J. M., Trif, M., Rusu, A. V., Aadil, R. M., Šimat, V., Cropotova, J., & Câmara, J. S. (2023). Food quality 4.0: From traditional approaches to digitalized automated analysis. *Journal of Food Engineering*, *337*. Available from https://doi.org/10.1016/j.jfoodeng.2022.111216, http://www.sciencedirect.com/science/journal/02608774.

Hassoun, A., Jagtap, S., Trollman, H., Garcia-Garcia, G., Abdullah, N. A., Goksen, G., Bader, F., Ozogul, F., Barba, F. J., Cropotova, J., Munekata, P. E. S., & Lorenzo, J. M. (2023). Food processing 4.0: Current and future developments spurred by the fourth industrial revolution. *Food Control*, *145*. Available from https://doi.org/10.1016/j.foodcont.2022.109507, https://www.journals.elsevier.com/food-control.

Hassoun, A., Kamiloglu, S., Garcia-Garcia, G., Parra-López, C., Trollman, H., Jagtap, S., Aadil, R. M., & Esatbeyoglu, T. (2023). Implementation of relevant fourth industrial revolution innovations across the supply chain of fruits and vegetables: A short update on Traceability 4.0. *Food Chemistry*, *409*. Available from https://doi.org/10.1016/j.foodchem.2022.135303, http://www.elsevier.com/locate/foodchem.

Huang, Y., Chen, Zx, Yu, T., Huang, Xz, & Gu, Xf (2018). Agricultural remote sensing big data: Management and applications. *Journal of Integrative Agriculture*, *17*(9), 1915–1931. Available from https://doi.org/10.1016/S2095-3119(17)61859-8, http://www.elsevier.com/journals/journal-of-integrative-agriculture/2095-3119.

Imanian, B., Donaghy, J., Jackson, T., Gummalla, S., Ganesan, B., Baker, R. C., Henderson, M., Butler, E. K., Hong, Y., Ring, B., Thorp, C., Khaksar, R., Samadpour, M., Lawless, K. A., MacLaren-Lee, I., Carleton, H. A., Tian, R., Zhang, W., & Wan, J. (2022). The power, potential, benefits, and challenges of implementing high-throughput sequencing in food safety systems. *npj Science of Food*, *6*(1). Available from https://doi.org/10.1038/s41538-022-00150-6, https://www.nature.com/npjscifood/about.

Iordache, F., Gheorghe, I., Lazar, V., Curutiu, C., Ditu, L. M., Grumezescu, A. M., & Holban, A. M. (2017). *Nanostructurated Materials for Prolonged and Safe Food Preservation* (pp. 305–335). Elsevier BV. Available from https://doi.org/10.1016/b978-0-12-804303-5.00009-2.

Iqbal, J., Khan, Z. H., & Khalid, A. (2017). Prospects of robotics in food industry. *Food Science and Technology*, *37*(2), 159–165. Available from https://doi.org/10.1590/1678-457x.14616.

Jaffee, S., Henson, S., Unnevehr, L., Grace, D., & Cassou, E. (2019). *The safe food imperative: Accelerating progress in low- and middle-income countries*. World Bank.

Jagtap, S., Bader, F., Garcia-Garcia, G., Trollman, H., Fadiji, T., & Salonitis, K. (2021). Food logistics 4.0: Opportunities and challenges. *Logistics*, *5* (1), 2. Available from https://doi.org/10.3390/logistics5010002.

Jagtap, S., & Duong, L. N. K. (2019). Improving the new product development using big data: a case study of a food company. *British Food Journal*, *121*(11). Available from https://doi.org/10.1108/BFJ-02-2019-0097, 2835-2848.

Jagtap, S., Duong, L., Trollman, H., Bader, F., Garcia-Garcia, G., Skouteris, G., Li, J., Pathare, P., Martindale, W., Swainson, M., & Rahimifard, S. (2021). *IoT technologies in the food supply chain food technology disruptions* (pp. 175–211). United Kingdom: Elsevier. Available from https://www.sciencedirect.com/book/9780128214701, https://doi.org/10.1016/B978-0-12-821470-1.00009-4.

Jagtap, S., Garcia-Garcia, G., & Rahimifard, S. (2021). Optimisation of the resource efficiency of food manufacturing via the Internet of Things. *Computers in Industry*, *127*. Available from https://doi.org/10.1016/j.compind.2021.103397, https://www.journals.elsevier.com/computers-in-industry.

Jagtap, S., Rahimifard, S., & Duong, L. N. K. (2022). Real-time data collection to improve energy efficiency: A case study of food manufacturer. *Journal of Food Processing and Preservation*, *46*(8). Available from https://doi.org/10.1111/jfpp.14338, http://onlinelibrary.wiley.com/journal/10.1111/(ISSN)1745-4549.

Jagtap, S., & Rahimifard, S. (2019). The digitisation of food manufacturing to reduce waste – Case study of a ready meal factory. *Waste Management*, *87*, 387–397. Available from https://doi.org/10.1016/j.wasman.2019.02.017, http://www.elsevier.com/locate/wasman.

Jagtap, S., Saxena, P., & Salonitis, K. (2021). *Food 4.0: Implementation of the augmented reality systems in the food industry*, . *Procedia CIRP* (104, pp. 1137–1142). United Kingdom: Elsevier B.V. Available from http://www.sciencedirect.com/science/journal/22128271, https://doi.org/10.1016/j.procir.2021.11.191, 22128271.

Jagtap, S., Skouteris, G., Choudhari, V., Rahimifard, S., & Duong, L. N. K. (2021). An internet of things approach for water efficiency: A case study of the beverage factory. *Sustainability*, *13*(6), 3343. Available from https://doi.org/10.3390/su13063343.

Jagtap, S., Trollman, H., Trollman, F., Garcia-Garcia, G., Parra-López, C., Duong, L., Martindale, W., Munekata, P. E. S., Lorenzo, J. M., Hdaifeh, A., Hassoun, A., Salonitis, K., & Afy-Shararah, M. (2022). The Russia-Ukraine conflict: Its implications for the global food supply chains. *Foods*, *11*(14), 2098. Available from https://doi.org/10.3390/foods11142098.

Jin, R., Kong, D., Yan, X., Zhao, X., Li, H., Liu, F., Sun, P., Lin, Y., & Lu, G. (2019). Integrating target-responsive hydrogels with smartphone for on-site ppb-level quantitation of organophosphate pesticides. *ACS Applied Materials and Interfaces*, *11*(31), 27605–27614. Available from https://doi.org/10.1021/acsami.9b09849, http://pubs.acs.org/journal/aamick.

Kamilaris, A., Kartakoullis, A., & Prenafeta-Boldú, F. X. (2017). A review on the practice of big data analysis in agriculture. *Computers and Electronics in Agriculture*, *143*, 23–37. Available from https://doi.org/10.1016/j.compag.2017.09.037, http://www.elsevier.com/inca/publications/store/5/0/3/3/0/4.

Kampan, K., Tsusaka, T. W., & Anal, A. K. (2022). Adoption of blockchain technology for enhanced traceability of livestock-based products. *Sustainability*, *14*(20), 13148. Available from https://doi.org/10.3390/su142013148.

Kaur, A., Singh, G., Kukreja, V., Sharma, S., Singh, S., & Yoon, B. (2022). Adaptation of IoT with blockchain in food supply chain management: An analysis-based review in development, benefits and potential applications. *Sensors*, *22*(21), 8174. Available from https://doi.org/10.3390/s22218174.

Keogh, J. G., Rejeb, A., Khan, N., & Zaid-Kaylani, K. (2023). Chapter 68 - blockchain: an enabler for safe food in global supply networks. *Present Knowledge in Food Safety*. Available from https://doi.org/10.1016/B978-0-12-819470-6.00008-1, 1045–1066.

Khan, P. W., Byun, Y.-C., & Park, N. (2020). IoT-blockchain enabled optimized provenance system for Food Industry 4.0 using advanced deep learning. *Sensors*, *20*(10), 2990. Available from https://doi.org/10.3390/s20102990.

Kim, S. S., & Kim, S. (2022). Impact and prospect of the fourth industrial revolution in food safety: Mini-review. *Food Science and Biotechnology*, *31*(4), 399–406. Available from https://doi.org/10.1007/s10068-022-01047-6, http://www.springerlink.com/content/n3031612t262/.

Kittichotsatsawat, Y., Jangkrajarng, V., & Tippayawong, K. Y. (2021). Enhancing coffee supply chain towards sustainable growth with big data and modern agricultural technologies. *Sustainability*, *13*(8), 4593. Available from https://doi.org/10.3390/su13084593.

Kittipanya-ngam, P., & Tan, K. H. (2020). A framework for food supply chain digitalization: lessons from Thailand. *Production Planning and Control*, *31*(2-3), 158–172. Available from https://doi.org/10.1080/09537287.2019.1631462, http://www.tandf.co.uk/journals/titles/09537287.asp.

Konur, S., Lan, Y., Thakker, D., Morkyani, G., Polovina, N., & Sharp, J. (2021). Towards design and implementation of Industry 4.0 for food manufacturing. *Neural Computing and Applications*. Available from https://doi.org/10.1007/s00521-021-05726-z, http://link.springer.com/journal/521.

Kopko, C., Garthoff, J. A., Zhou, K., Meunier, L., O'Sullivan, A. J., & Fattori, V. (2022). Are alternative proteins increasing food allergies? Trends, drivers and future perspectives. *Trends in Food Science and Technology*, *129*, 126–133. Available from https://doi.org/10.1016/j.tifs.2022.09.008, http://www.elsevier.com/wps/find/journaldescription.cws_home/601278/description#description.

Li, K., Lee, J.-Y., & Gharehgozli, A. (2021). Blockchain in food supply chains: A literature review and synthesis analysis of platforms, benefits and challenges. *International Journal of Production Research*, 1–20. Available from https://doi.org/10.1080/00207543.2021.1970849.

Lo, C. K., Chen, C. H., & Zhong, R. Y. (2021). A review of digital twin in product design and development. *Advanced Engineering Informatics, 48*. Available from https://doi.org/10.1016/j.aei.2021.101297, https://www.journals.elsevier.com/advanced-engineering-informatics.

Lonchamps, P.-L., He, Y., Wang, K., & Lu, X. (2022). Detection of pathogens in foods using microfluidic "lab-on-chip": A mini review. *Journal of Agriculture and Food Research, 10*100430. Available from https://doi.org/10.1016/j.jafr.2022.100430.

Manning, L., Smith, R., & Soon, J. M. (2016). Developing an organizational typology of criminals in the meat supply chain. *Food Policy, 59*, 44–54. Available from https://doi.org/10.1016/j.foodpol.2015.12.003, http://www.elsevier.com/inca/publications/store/3/0/4/1/9/index.htt.

Margaritis, I., Madas, M., & Vlachopoulou, M. (2022). Big data applications in food supply chain management: A conceptual framework. *Sustainability, 14*(7), 4035. Available from https://doi.org/10.3390/su14074035.

Martindale, W., Duong, L., & Jagtap, S. (2022). Food Industry 4.0: Unlocking advancement opportunities in the food manufacturing sector. *CABI*. Available from https://doi.org/10.1079/9781789248593.0000.

Mazzeo, A., Tremonte, P., Lombardi, S. J., Caturano, C., Correra, A., & Sorrentino, E. (2022). From the intersection of food-borne zoonoses and EU green policies to an in-embryo one health financial model. *Foods, 11*(18), 2736. Available from https://doi.org/10.3390/foods11182736.

Mehannaoui, R., Mouss, K. N., & Aksa, K. (2023). IoT-based food traceability system: Architecture, technologies, applications, and future trends. *Food Control, 145*. Available from https://doi.org/10.1016/j.foodcont.2022.109409, https://www.journals.elsevier.com/food-control.

Meng, X., Li, X., Yang, L., Yin, R., Qi, L., & Guo, Q. (2023). Review on microbial metabolomics of probiotics and pathogens: Methodologies and applications. *Biocell, 47*(1), 91–107. Available from https://doi.org/10.32604/biocell.2023.024310, https://www.techscience.com/biocell/v47n1/49938.

Menon, S., & Jain, K. (2021). Blockchain technology for transparency in agri-food supply chain: Use cases, limitations, and future directions. *IEEE Transactions on Engineering Management*. Available from https://doi.org/10.1109/TEM.2021.3110903, https://ieeexplore.ieee.org/servlet/opac?punumber = 17.

Miao, L., Li, H., Ding, W., Lu, S., Pan, S., Guo, X., Zhou, X. N., & Wang, D. (2022). Research priorities on one health: A bibliometric analysis. *Frontiers in Public Health, 10*. Available from https://doi.org/10.3389/fpubh.2022.889854, http://journal.frontiersin.org/journal/public-health/section/public-health-education-and-promotion#archive.

Misra, N. N., Dixit, Y., Al-Mallahi, A., Bhullar, M. S., Upadhyay, R., & Martynenko, A. (2022). IoT, big data, and artificial intelligence in agriculture and food industry. *IEEE Internet of Things Journal, 9*(9), 6305–6324. Available from https://doi.org/10.1109/JIOT.2020.2998584, http://ieeexplore.ieee.org/servlet/opac?punumber = 6488907.

Mol, A., & Oosterveer, P. (2015). Certification of markets, markets of certificates: Tracing sustainability in global agro-food value chains. *Sustainability, 7*(9), 12258–12278. Available from https://doi.org/10.3390/su70912258.

Moradi, M., & Dass, M. (2022). Applications of artificial intelligence in B2B marketing: Challenges and future directions. *Industrial Marketing Management, 107*, 300–314. Available from https://doi.org/10.1016/j.indmarman.2022.10.016, http://www.elsevier.com/locate/indmarman.

Morella, P., Lambán, M. P., Royo, J., & Sánchez, J. C. (2021). Study and analysis of the implementation of 4.0 technologies in the agri-food supply chain: A state of the art. *Agronomy, 11*(12). Available from https://doi.org/10.3390/agronomy11122526, https://www.mdpi.com/2073-4395/11/12/2526/pdf.

Nychas, G. J. E., Panagou, E. Z., & Mohareb, F. (2016). Novel approaches for food safety management and communication. *Current Opinion in Food Science, 12*, 13–20. Available from https://doi.org/10.1016/j.cofs.2016.06.005, http://www.journals.elsevier.com/current-opinion-in-food-science/.

Oliveira, M. M., Cruz-Tirado, J. P., & Barbin, D. F. (2019). Nontargeted analytical methods as a powerful tool for the authentication of spices and herbs: A review. *Comprehensive Reviews in Food Science and Food Safety, 18*(3), 670–689. Available from https://doi.org/10.1111/1541-4337.12436, http://onlinelibrary.wiley.com/journal/10.1111/(ISSN)1541-4337.

Opara, L. U. (2003). Traceability in agriculture and food supply chain: A review of basic concepts, technological implications and future prospects. *Food. Agriculture and Environment, 1*, 101–106.

Petruzzi, L., Corbo, M. R., Sinigaglia, M., & Bevilacqua, A. (2017). Microbial spoilage of foods: Fundamentals. *The microbiological quality of food: Foodborne spoilers*, 1–21. Available from https://doi.org/10.1016/B978-0-08-100502-6.00002-9, Elsevier Inc., Italy, http://www.sciencedirect.com/science/book/9780081005026.

Pou, K. R. J., Raghavan, V., & Packirisamy, M. (2022). Microfluidics in smart packaging of foods. *Food Research International, 161*. Available from https://doi.org/10.1016/j.foodres.2022.111873, http://www.elsevier.com/inca/publications/store/4/2/2/9/7/0.

Putri, S. P., Ikram, M. M. M., Sato, A., Dahlan, H. A., Rahmawati, D., Ohto, Y., & Fukusaki, E. (2022). Application of gas chromatography-mass spectrometry-based metabolomics in food science and technology. *Journal of Bioscience and Bioengineering, 133*(5), 425–435. Available from https://doi.org/10.1016/j.jbiosc.2022.01.011, http://www.elsevier.com/wps/find/journaldescription.cws_home/505516/description#description.

Ragazou, K., Passas, I., Garefalakis, A., & Dimou, I. (2022). Investigating the research trends on strategic ambidexterity, agility, and open innovation in SMEs: Perceptions from bibliometric analysis. *Journal of Open Innovation: Technology, Market, and Complexity, 8*(3). Available from https://doi.org/10.3390/joitmc8030118, http://www.mdpi.com/journal/JOItmC.

Rahimifard, S., Stone, J., & Trollman, H. (2018). Global food security: the engineering challenges. *International Journal of Sustainable Engineering, 11*(2), 77–78. Available from https://doi.org/10.1080/19397038.2018.1475091, http://www.tandf.co.uk/journals/titles/19397038.asp.

Rahimifard, S., & Trollman, H. (2017). Surpassing sustainability: making a 'net-positive' impact. *International Journal of Sustainable Engineering, 10*(6), 299–301. Available from https://doi.org/10.1080/19397038.2017.1401037, http://www.tandf.co.uk/journals/titles/19397038.asp.

Rahimifard, S., & Trollman, H. (2018). UN sustainable development goals: An engineering perspective. *International Journal of Sustainable Engineering, 11*(1), 1–3. Available from https://doi.org/10.1080/19397038.2018.1434985, http://www.tandf.co.uk/journals/titles/19397038.asp.

Rahimifard, S., Woolley, E., Webb, D. P., Garcia-Garcia, G., Stone, J., Jellil, A., Gimenez-Escalante, P., Jagtap, S., & Trollman, H. (2017). Forging new frontiers in sustainable food manufacturing. *Smart innovation, systems and technologies, 68*, 13–24. Available from https://doi.org/10.1007/978-3-319-57078-5_2, Springer Science and Business Media Deutschland GmbH United Kingdom, http://www.springer.com/series/8767.

Ramirez-Asis, E., Vilchez-Carcamo, J., Thakar, C. M., Phasinam, K., Kassanuk, T., & Naved, M. (2022). A review on role of artificial intelligence in food processing and manufacturing industry. *Materials Today: Proceedings*, *51*, 2462–2465. Available from https://doi.org/10.1016/j.matpr.2021.11.616, https://www.sciencedirect.com/journal/materials-today-proceedings.

Rejeb, A., Keogh, J. G., & Rejeb, K. (2022). Big data in the food supply chain: a literature review. *Journal of Data, Information and Management*, *4* (1), 33–47. Available from https://doi.org/10.1007/s42488-021-00064-0.

Resende, C., Folgado, D., Oliveira, J., Franco, B., Moreira, W., Oliveira-Jr, A., Cavaleiro, A., & Carvalho, R. (2021). TIP4.0: Industrial internet of things platform for predictive maintenance. *Sensors*, *21*(14), 4676. Available from https://doi.org/10.3390/s21144676.

Rezazade, F., Summers, J., & Lai Teik, D. O. (2021). Identifying food fraud vulnerability factors in food fraud incidents – a review of global incidents 2000–2018. *British Food Journal*, *124*(11), 4122–4142. Available from https://doi.org/10.1108/bfj-02-2021-0140.

Rolf, B., Jackson, I., Müller, M., Lang, S., Reggelin, T., & Ivanov, D. (2022). A review on reinforcement learning algorithms and applications in supply chain management. *International Journal of Production Research*, 1–29. Available from https://doi.org/10.1080/00207543.2022.2140221.

Romanello, R., & Veglio, V. (2022). Industry 4.0 in food processing: Drivers, challenges and outcomes. *British Food Journal.*, *124*(13). Available from https://doi.org/10.1108/BFJ-09-2021-1056, 375-390.

Romanov, D., Korostynska, O., Lekang, O. I., & Mason, A. (2022). Towards human-robot collaboration in meat processing: Challenges and possibilities. *Journal of Food Engineering*, *331*. Available from https://doi.org/10.1016/j.jfoodeng.2022.111117, http://www.sciencedirect.com/science/journal/02608774.

Santhosh, R., Tiwari, A., & Rawson, A. (2020). Biopolymers as edible coating for food. *Polymer Coatings.*, 233–274. Available from https://doi.org/10.1002/9781119655145.ch11.

Sgarbossa, F., Romsdal, A., Oluyisola, O. E., & Strandhagen, J. O. (2022). Digitalization in production and warehousing in food supply chains. *The digital supply chain*, 273–287. Available from https://doi.org/10.1016/B978-0-323-91614-1.00016-2, Elsevier, Norway, https://www.sciencedirect.com/book/9780323916141.

Sharma, S. K., & Singh, V. (2022). *Digitization of the Food Industry Enabled by Internet of Things, Blockchain, and Artificial Intelligence* (pp. 421–445). Elsevier BV. Available from https://doi.org/10.1016/b978-0-323-91158-0.00013-2.

Shu, R., Liu, S., Huang, L., Li, Y., Sun, J., Zhang, D., Zhu, M.-Q., & Wang, J. (2022). Enzyme-mimetic nano-immunosensors for amplified detection of food hazards: Recent advances and future trends. *Biosensors and Bioelectronics*, *217*114577. Available from https://doi.org/10.1016/j.bios.2022.114577.

Simeone, A., Woolley, E., Escrig, J., & Watson, N. J. (2020). Intelligent industrial cleaning: A multi-sensor approach utilising machine learning-based regression. *Sensors*, *20*(13), 3642. Available from https://doi.org/10.3390/s20133642.

Soon, J. M., Krzyzaniak, S. C., Shuttlewood, Z., Smith, M., & Jack, L. (2019). Food fraud vulnerability assessment tools used in food industry. *Food Control*, *101*, 225–232. Available from https://doi.org/10.1016/j.foodcont.2019.03.002.

Soon, J. M., & Saguy, I. S. (2017). Crowdsourcing: A new conceptual view for food safety and quality. *Trends in Food Science and Technology*, *66*, 63–72. Available from https://doi.org/10.1016/j.tifs.2017.05.013, http://www.elsevier.com/wps/find/journaldescription.cws_home/601278/description#description.

Sorbo, A., Pucci, E., Nobili, C., Taglieri, I., Passeri, D., & Zoani, C. (2022). Food safety assessment: Overview of metrological issues and regulatory aspects in the European Union. *Separations*, *9*(2), 53. Available from https://doi.org/10.3390/separations9020053.

Sridhar, A., Kapoor, A., Kumar, P. S., Ponnuchamy, M., Sivasamy, B., & Vo, D. V. N. (2022). Lab-on-a-chip technologies for food safety, processing, and packaging applications: a review. *Environmental Chemistry Letters*, *20*(1), 901–927. Available from https://doi.org/10.1007/s10311-021-01342-4, http://springerlink.metapress.com/app/home/journal.asp?wasp = d86tgdwvtg0yvw9gvkwp&referrer = parent&backto = browsepublicationsresults,140,541.

Sun, Y., Li, J., Zhu, L., & Jiang, L. (2022). Cooperation and competition between CRISPR- and omics-based technologies in foodborne pathogens detection: A state of the art review. *Current Opinion in Food Science*, *44*100813. Available from https://doi.org/10.1016/j.cofs.2022.100813.

Tetra Pak. (2020). A spotlight on food safety and Industry 4.0. https://www.tetrapak.com/content/dam/tetrapak/media-box/global/en/documents/industry4-food-safety-wp.pdf.

Thinking about the future of food safety. FAO. (2022). https://doi.org/10.4060/cb8667en.

Tracey, C. T., Kryuchkova, A. V., Bhatt, T. K., Krivoshapkin, P. V., & Krivoshapkina, E. F. (2023). Silk for post-harvest horticultural produce safety and quality control. *Journal of Food Engineering*, *337*. Available from https://doi.org/10.1016/j.jfoodeng.2022.111231, http://www.sciencedirect.com/science/journal/02608774.

Trollman, H., Garcia-Garcia, G., Jagtap, S., & Trollman, F. (2022). Blockchain for ecologically embedded coffee supply chains. *Logistics*, *6*(3), 43. Available from https://doi.org/10.3390/logistics6030043.

Trollman, H., Jagtap, S., Garcia-Garcia, G., Harastani, R., Colwill, J., & Trollman, F. (2021). COVID-19 demand-induced scarcity effects on nutrition and environment: investigating mitigation strategies for eggs and wheat flour in the United Kingdom. *Sustainable Production and Consumption*, *27*, 1255–1272. Available from https://doi.org/10.1016/j.spc.2021.03.001, http://www.journals.elsevier.com/sustainable-production-and-consumption/.

U.S. Food and Drug Administration (2021). New era of smarter food safety blueprint. https://www.fda.gov/food/new-era-smarter-food-safety/new-era-smarter-food-safety-blueprint.

Ververis, E., Ackerl, R., Azzollini, D., Colombo, P. A., de Sesmaisons, A., Dumas, C., Fernandez-Dumont, A., Ferreira da Costa, L., Germini, A., Goumperis, T., Kouloura, E., Matijevic, L., Precup, G., Roldan-Torres, R., Rossi, A., Svejstil, R., Turla, E., & Gelbmann, W. (2020). Novel foods in the European Union: Scientific requirements and challenges of the risk assessment process by the European Food Safety Authority. *Food Research International*, *137*109515. Available from https://doi.org/10.1016/j.foodres.2020.109515.

Viegas, C., Gomes, B., Oliveira, F., Dias, M., Cervantes, R., Pena, P., Gomes, A. Q., Caetano, L. A., Carolino, E., de Andrade, E. T., & Viegas, S. (2022). Microbial contamination in the coffee industry: An occupational menace besides a food safety concern? *International Journal of Environmental Research and Public Health*, *19*(20), 13488. Available from https://doi.org/10.3390/ijerph192013488.

Wahab, S. N., & Ling, E. K. (2019). Integrity of food supply chain: Going beyond food safety and food quality. *International Journal of Productivity and Quality Management*, *1*(1), 1. Available from https://doi.org/10.1504/ijpqm.2019.10019297.

Wang, D., Zhang, M., Mujumdar, A. S., & Yu, D. (2022). Advanced detection techniques using artificial intelligence in processing of berries. *Food Engineering Reviews*, *14*(1), 176−199. Available from https://doi.org/10.1007/s12393-021-09298-5, http://www.springer.com/life + sci/food + science/journal/12393.

WHO estimates of the global burden of foodborne diseases: Foodborne disease burden epidemiology reference group. World Health Organization. (2007).

WHO global strategy for food safety 2022-2030: Towards stronger food safety systems and global cooperation. World Health Organization, Geneva. (2022).

Wiśniewska, M. Z., Tarczyńska, A. S., & Ranadheera, C. S. (2021). Novel quality assurance systems against intentional contamination in dairy factories. *Dairy foods: Processing, quality, and analytical techniques*, 281−310. Available from https://doi.org/10.1016/B978-0-12-820478-8.00006-7, Elsevier, Poland, https://www.sciencedirect.com/book/9780128204788.

Wu, G., Zhuang, D., Chew, K. W., Ling, T. C., Khoo, K. S., Van Quyen, D., Feng, S., & Show, P. L. (2022). Current status and future trends in removal, control, and mitigation of algae food safety risks for human consumption. *Molecules (Basel, Switzerland)*, *27*(19), 6633. Available from https://doi.org/10.3390/molecules27196633.

Xu, Y., Li, X., Zeng, X., Cao, J., & Jiang, W. (2022). Application of blockchain technology in food safety control: current trends and future prospects. *Critical Reviews in Food Science and Nutrition*, *62*(10), 2800−2819. Available from https://doi.org/10.1080/10408398.2020.1858752, https://www.tandfonline.com/loi/bfsn20.

Yadav, V. S., Singh, A. R., Raut, R. D., Mangla, S. K., Luthra, S., & Kumar, A. (2022). Exploring the application of Industry 4.0 technologies in the agricultural food supply chain: A systematic literature review. *Computers and Industrial Engineering*, *169*. Available from https://doi.org/10.1016/j.cie.2022.108304, https://www.journals.elsevier.com/computers-and-industrial-engineering.

Yang, T., Luo, Z., Bewal, T., Li, L., Xu, Y., Mahdi Jafari, S., & Lin, X. (2022). When smartphone enters food safety: A review in on-site analysis for foodborne pathogens using smartphone-assisted biosensors. *Food Chemistry*, *394*. Available from https://doi.org/10.1016/j.foodchem.2022.133534, http://www.elsevier.com/locate/foodchem.

Zamuz, S., Munekata, P. E. S., Meiselman, H. L., Zhang, W., Xing, L., & Lorenzo, J. M. (2021). Consumer and market demand for sustainable food products. *Sustainable production technology in food*, 23−35. Available from https://doi.org/10.1016/B978-0-12-821233-2.00008-3, Elsevier, Spain, https://www.elsevier.com/books/sustainable-production-technology-in-food/lorenzo/978-0-12-821233-2.

Zhang, X., Guo, M., Ismail, B. B., He, Q., Jin, T. Z., & Liu, D. (2021). Informative and corrective responsive packaging: Advances in farm-to-fork monitoring and remediation of food quality and safety. *Comprehensive Reviews in Food Science and Food Safety*, *20*(5), 5258−5282. Available from https://doi.org/10.1111/1541-4337.12807, http://onlinelibrary.wiley.com/journal/10.1111/(ISSN)1541-4337.

Zhao, G., Liu, S., Lopez, C., Lu, H., Elgueta, S., Chen, H., & Boshkoska, B. M. (2019). Blockchain technology in agri-food value chain management: A synthesis of applications, challenges and future research directions. *Computers in Industry*, *109*, 83−99. Available from https://doi.org/10.1016/j.compind.2019.04.002.

Zhou, Q., Zhang, H., & Wang, S. (2022). Artificial intelligence, big data, and blockchain in food safety. *International Journal of Food Engineering*, *18*(1), 1−14. Available from https://doi.org/10.1515/ijfe-2021-0299, http://www.degruyter.com/view/j/ijfe.

Zhou, X., Zhu, Q., & Xu, Z. (2022). The mediating role of supply chain quality management for traceability and performance improvement: Evidence among Chinese food firms. *International Journal of Production Economics*, *254*108630. Available from https://doi.org/10.1016/j.ijpe.2022.108630.

Chapter 5

Food Traceability 4.0

Rizwan Matloob[1], Giulio Maria Bianco[2], Gaetano Marrocco[2], Cecilia Occhiuzzi[2], Lincoln C. Wood[3,4] and Alaa El-Din A. Bekhit[1]

[1]*Department of Food Science, The University of Otago, Dunedin, New Zealand,* [2]*Department of Civil Engineering and Computer Science Engineering, The University of Rome Tor Vergata, Rome, Italy,* [3]*Management Department, The University of Otago, Dunedin, New Zealand,* [4]*School of Management, Curtin University, Perth, Australia*

5.1 Introduction

Food traceability is the ability of concerned entities to monitor, track, and document the movement of food products, their ingredients, processing conditions, handling, and storage throughout the supply chain. In the context of foodstuff, traceability can be regarded as the ability to trace and track food throughout the various stages of the supply chain (Fanelli et al., 2021; Qian et al., 2020).

Food traceability is often linked with the management of logistics and is considered the ability to access information through timely reported identifications across logistics (Olsen & Borit, 2013). Indeed, Garaus and Treiblmaier (2021) argued that better traceability is directly linked with an increased flow of information among supply chain intermediaries and is subject to the performance of the intermediaries in managing specified responsibilities.

It is worth noting that the term "traceability" is subjective and there is no universal definition for it as it varies from location to location and is based on multiple jurisdictions (Golan et al., 2004; Olsen & Borit, 2013). The Codex Alimentarius Commission's (2005) definition of food traceability is probably the most widely accepted, and it states that "traceability is the ability to track the progression of a product over the food supply chain comprising stages of production, processing, logistics, and distribution." The European Commission (2008) definition declares the materials that will be used for food and the various stages that will go through and defines food traceability as "the ability to track any food, feed, food-producing animal, or substance that will be used for consumption, through all stages of production, processing, and distribution." The International Organization for Standardization provided ISO-2205 standards that define "the ability to follow the movement of a feed or food through specified stage(s) of production, processing, and distribution" (Olsen & Borit, 2013), which is adapted by the US Food and Drug Administration. The ability to implement food traceability in modern food production and trade is very important from a safety point of view (tracing foodborne outbreaks and management of recalls), and quality management (identification and control of hazards and defects, adulteration, fraud, and authentication).

5.1.1 Importance and need for food traceability

Food safety is a matter of global concern that is regulated by legislation and enforcement agencies. Several technologies and tools have been developed to detect and ensure food safety and integrity. Conventional anticounterfeit food traceability systems based on centralized networks are facing numerous issues related to data tempering (Lu et al., 2022). Additionally, the complexity of the modern food supply chain has escalated more than ever, making it harder to track and identify the flow of ingredients, products, and processes throughout the food supply chain (Casino et al., 2019). Several food safety incidents raised awareness of the need for strict and effective new technologies to prevent, trace, and control such incidences. For example, diluting milk with contaminated water has been reported in developing countries (Yang et al., 2019), peanut butter contamination by Salmonella and Big Olaf ice cream contamination by listeria in the United States, *Escherichia coli* outbreak through the consumption of cucumber in Germany, and the outbreak of mad cow disease in Britain are a few of many food safety incidents that had severe consequences and resulted in fatalities (Garaus & Treiblmaier, 2021). Due to the serious impact of such incidents, food traceability has become an

Food Industry 4.0. DOI: https://doi.org/10.1016/B978-0-443-15516-1.00005-0

important system for the management of such incidents (Liu et al., 2019; Robson et al., 2020). Therefore food traceability emerged as a critical aspect of legislation and its compliance is highly significant throughout the food supply chain (Casino et al., 2019).

An increase in food traceability requirements is also affecting the market competitiveness of an organization's supply chain. Modern consumers are concerned about the cleanliness, purity, authenticity, and overall quality of the food they consume (George et al., 2019). Consumers' purchases are highly motivated by ethical and environment-friendly production of food where the origin of the food and transparent food supply chain is communicated and readily accessible. Therefore it has now become increasingly important for organizations to adopt advanced traceability systems (Islam & Cullen, 2021). To provide maximum value to consumers, food supply chains are required to be equipped with the latest technologies to ensure transparency and integrity of the information, which enables trust and satisfaction across the supply chain as well as identifies and resolves food traceability and safety issues in a more productive manner (Casino et al., 2019; George et al., 2019).

5.1.2 Framework for food traceability

Considering the importance of traceability in the food supply chain, a conceptual framework was proposed by Aung and Chang (2014) that integrated key players in the food production chain and technologies that can be adapted for internal and external traceability (Fig. 5.1). The framework depicts different actors in the food supply chain such as farmers, processors, distributors, retailers, and consumers. Moreover, the framework covers the various processes involved in the food supply chain from the source of the materials until consumption. An integral part of the framework is that all the actors in the framework share information through a Food Safety Information System (FSIS). The framework for food traceability connects quality assurance and safety systems to ensure compliance with the food supply chain as well as links technologies with overall operations throughout the food supply chain (Aung & Chang, 2014).

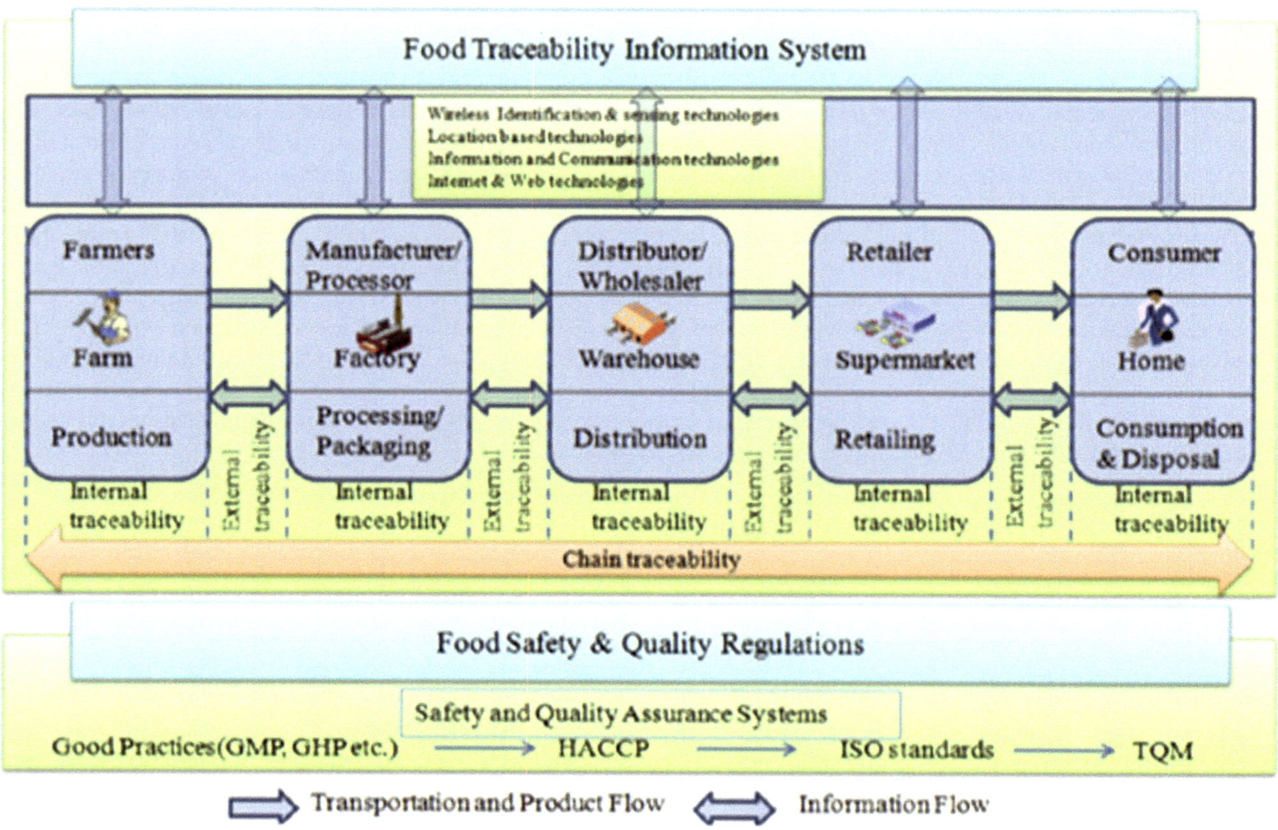

FIGURE 5.1 Conceptual framework for food traceability system by Aung and Chang (2014). *The figure is published with kind permission from the Elsevier (License Number 5426291171226).*

The conceptual framework in Fig. 5.1 refers to and distinguishes between the internal and external traceability of goods and processes. The difference between internal and external traceability was first identified by Moe (1998). According to Moe (1998), chain traceability, also known as external traceability, refers to the ability to ensure batch or product traceability throughout the production chain inclusive of harvesting, transportation, warehousing, production/processing, and sales. As a general rule, external traceability is widely considered as an ability to trace the overall supply chain and intermediaries in between the supply chain. On the other hand, internal traceability refers to the ability to ensure traceability within each step of the chain (Moe, 1998). Information on the movement and conditions employed in a specific process or by a given actor within the supply chain is documented (Behnke & Janssen, 2020).

As per Aung and Chang (2014), several identifiers are required for grouping products into traceable units, and based on the defined characteristics stated below, products can be grouped into traceable resource units. These traceable units can be classified into three categories:

- Batch unit products: These traceable resource units consist of products that encompass the same production process, such as batches of powder milk cans that have the same production date and shelf life.
- Trade units: These consist of products that are shifted simultaneously throughout the food supply chain by actors such as farmers, processors, distributors, retailers, and customers.
- Logistics units: These are the units that can be used to identify the products per their transport and storage mediums, for example, multiple pallets of powder milk that can have different batch numbers.

Batch, trade, and logistics units are strictly dependent on the type of data that is stored by the traceability systems (Behnke & Janssen, 2020). Considering these definitions, it is likely that trade units and batch units can overlap, such cases are more common when the products are subjected to the same production process. The framework incorporates a Food Safety Information System enabling all the actors and intermediaries in the food supply chain to access and share information through the use of multiple technologies.

5.1.3 Food Traceability 4.0 using digital technologies

A number of technologies, such as bar codes, radiofrequency identification (RFID), wireless sensor networks (WSNs), and DNA barcoding, have been developed over time to ensure the traceability of products across the food supply chain. However, due to the increase in population and changes over time in customers' preferences, substantial increases in varieties, volume, and marketing strategies of food products have been developed to meet these issues. Subsequently, this created the challenge of traceability of the products in the food supply chain, and more sophisticated systems are required. Manual record-keeping systems are not agile, real-time, and prone to errors. Thus, new more sophisticated systems for food traceability are needed (George et al., 2019; Yu et al., 2022).

The historical evolution and development of technologies can be classified into four distinct stages based on principles and technologies that led the progress. Generally, the four stages have been coined Industrial Revolution 1.0, 2.0, 3.0. and 4.0. Industry 1.0 refers to the mechanization of production systems that was afforded by the invention of the steam engine and subsequent improvement of the technology. Industry 2.0 refers to the wide adaptation of electricity in production systems after the discovery of electricity in the 19th century. The third industrial revolution resulted in the automation of processes through programmable computers and controls. The fourth industrial revolution is heavily dependent on the integration of cyber-physical systems, cloud computing, and industrial Internet of Things (IoT) manufacturing and production processes (Sadeghi et al., 2022)

The connectivity between the digital environment and the physical world, through the help of upcoming technologies, is known as Industrial Revolution 4.0 (Lee et al., 2015). The amalgamation of the digital and physical worlds implemented in the Industrial Revolution 4.0 helps industries to connect with machines and analyze data in more efficient and effective ways. Some of the important technologies that are used in the Industrial Revolution 4.0 are big data analytics, robotics, artificial intelligence, blockchain, and the IoT (Bai et al., 2020; Hassoun, Siddiqui, et al., 2022; Hassoun, Ait-Kaddour, et al., 2022).

The role of digital technologies as key enablers of traceability in the food supply chains is undeniably growing (Luque et al., 2017). Hassoun, Siddiqui, et al. (2022), Hassoun, Ait-Kaddour, et al. (2022) suggested that the inclusion of Industry 4.0 technologies and principles in traceability could maximize food safety and quality, preventing fraud, saving time and effort, as well as reducing food waste and food loss. In particular, blockchain and the IoT appear to be two of the most promising technologies contributing to Food Traceability 4.0 by enabling real-time digital traceability. These technologies will be discussed in the next sections of this chapter.

5.2 Radiofrequency identification technology for Food Traceability 4.0

5.2.1 Introduction to radiofrequency identification

At its core, an RFID system is composed of a reader and a tag that can also be named interrogator and transponder, respectively. The reader generates an electromagnetic (EM) field that scan tags. Then, the queried tags modulate and, at higher frequencies, backscatter the field back to the reader responding to its demands by letting it also know the unique identifier (ID) of the responding tag. The IDs allow for tracing and managing the object to whom the tags are attached. The reader-tag link is defined as forward, and the tag-reader link is backward or reverse. As a reader can interrogate multiple tags at once, it is considered the most complex and expensive component of the system. The tags consist of an antenna and an integrated circuit (IC), that is, a microchip, which elaborates the queries and their responses. Although chip-less transponders are currently a hot research topic, the technology is still in its infancy. A detailed review of the progress of chip-less RFIDs and possible future applications to Food Traceability 4.0 has been recently published (Fathi et al., 2020) and readers interested in more information are invited to consult that review.

In addition to the microchip and the antenna, a transponder can exploit a battery as a primary or secondary power source. The use of batteries can enable advanced sensing features and longer communication distances. Nevertheless, as a tag per object to be identified is needed, using batteries is highly expensive, and problems with battery life arise. A responder using a battery to communicate with the reader is categorized as active and passive. Active tags can even send signals in the absence of any readers' query. Passive transponders do not require any battery, and semipassive (also known as BAP; battery-assisted passive) transponders use the battery to help power the IC when an interrogation is received.

The maximum tag-reader distance depends on multiple parameters, such as the IC power sensitivity (viz., the minimum power needed to turn on the IC), the reader's and tag's antennas and their relative orientation and positioning, the polarization of the EM field generated by the interrogator and backscattered by the transponder, and the presence of metal or moisture. Deploying an RFID system is hence a complex task that requires substantial expertise. The most critical variable for the RFID performance is the operating frequency which can be low frequency (LF, in the range of 125–134 kHz), high frequency (HF, 13.56 MHz), ultra-high frequency (UHF, in the range of 860–960 MHz) and microwave that can be either UHF (2.45 GHz) or super high frequency (SHF; 5.8 GHz) (IEEE, 2019). Each type of RFID has different characteristics, typical uses, and adoption in Food Traceability 4.0, as shown in Fig. 5.2. Each type is more or less prone to some challenges such as implementation costs, the presence of metal and moisture that can hamper the tag's functioning, and even polarization of the EM waves used for the reader-tag link in the UHF and SHF RFID systems, which can hinder communications.

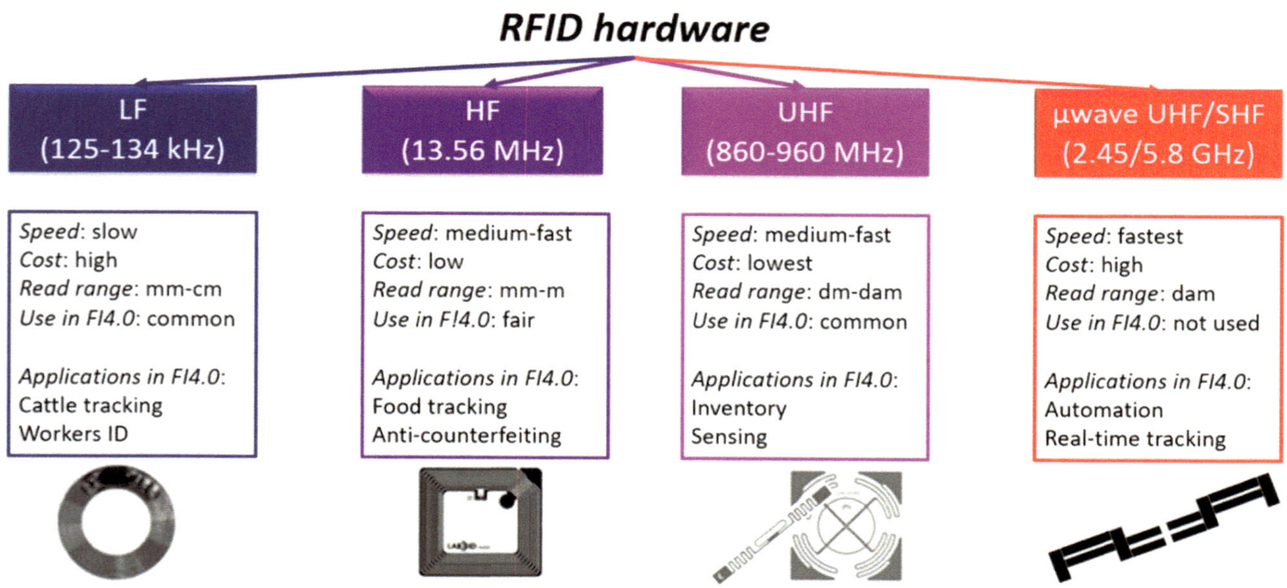

FIGURE 5.2 Types of existing RFID hardware, their characteristics, and some tags suitable for the different categories alongside some examples of possible applications in food traceability. The tags for the LF, HF, and UHF systems are adapted from Dabbene et al. (2016) and the microwave tags are from Jeon et al. (2006).

5.2.2 Radiofrequency identification for supply chain traceability

Historically, the most widespread utilization of RFID has been the traceability of goods thanks to their ability to track the movement of the goods. As both single and multiple items can be identified using one tag, the system will consider each tag assigned to a "traceable resource unit" (TRU) (Olsen & Borit, 2018). Fig. 5.3 depicts the most general architecture for tracking objects with RFID labels. The readers, which can be both handheld or fixed infrastructures, interrogate confined areas assigned to them, recognize any responding tag, and store the information on a register. The essential information that must be recorded by an ID, a location, and a time of reading. However, the tracking history by itself can support only small-scale traceability for local inventories like confined storage areas. Therefore supplementary information is usually inserted in the database when registering a tag into the traceability system by describing the corresponding TRU. Another field is typically left empty for the operators to record any important notes concerning the object, for example, if the TRU is damaged or has been sold and must be shipped.

Following the positive example provided by Wal-Mart (then Walmart) in 2003, other retail giants such as Tesco, Marks & Spencer, and Metro Group adopted the use of the UHF RFID for item-level traceability along the supply chain (Tao et al., 2022; Zhang & Yang, 2019). Indeed, the average inventory accuracy in retail is estimated to be about 65%, corresponding to a profit reduction of 10%, and RFID proved to be a good solution for increasing the accordance between the records of the inventory management systems and the stored material (Tao et al., 2022).

Traceability with HF RFID has gained much interest thanks to the diffusion of near field communication (NFC); a subset of HF RFID readers, and nowadays such a system is embedded in about three billion mobile phones (NFC Forum, 2022). The presence of the NFC interrogator in standard smartphones avoids the need to buy dedicated readers, hence reducing production costs. The NFC labels are particularly convenient for customer engagement since the interrogator retrieves the tracking history of the specific item purchased by the buyer. This feature can be helpful both for customer retention and anticounterfeiting (Khalil et al., 2020).

The LF RFID is primarily used to track livestock since its magnetic coupling guarantees reliable communications even after implantation. For other applications, generally, the cost per tag is too high and the reading distances are too short (Costa & Arnaud, 2011). Due to the need for batteries for proper functioning, microwave RFID has never been commercially applied for tracking goods as its use for traceability is inconvenient.

5.2.3 Radiofrequency identification for food traceability

By virtue of the benefits that RFID proved to provide in logistics, several radiofrequency identification devices and systems for food were investigated, developed, and even commercialized in the last decade. The first use in the food production chain was for tracing TRUs and livestock. In particular, livestock tracking is pursued with implanted tags (LF hardware) or ear tags (LF or UHF hardware). More advanced features for the authentication of products from the anticounterfeit standpoint and for checking the expiry date are commercialized through HF NFC tags that can interact with the customers' smartphones. Industrial RFID components for monitoring food production and storage chains are also available and exploited worldwide. Such systems are providing uninterrupted effectiveness and are appreciated by the food industry where more complex RFID environments providing multiple features are used to conduct multiple functions such as tracking the movement of products, providing information on product shelf life/quality, and management of expiry data.

Although commercial RFID devices for food traceability have become more employed and common in the last 10 years, academic research began both quantifying the benefits of RFID usage in food production systems and investigating more futuristic approaches. Examples of academic research aiming at industrial deployment in the near future are

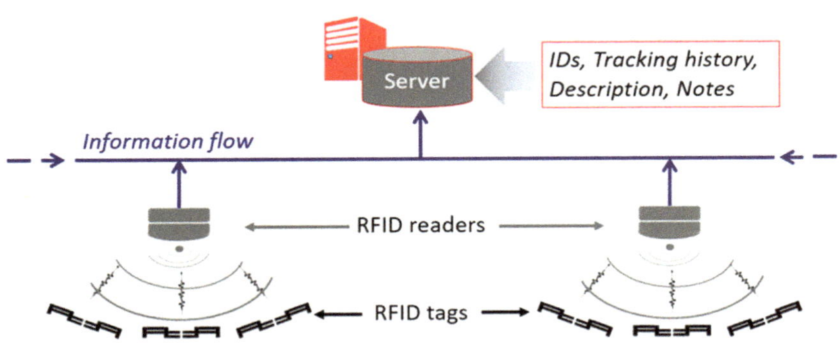

FIGURE 5.3 The general architecture of a typical traceability system through RFID. The readers collect information on the tags' locations and transmit it to a server that memorizes it in a database.

the creation of low-cost readers and the integration of RFID and barcode for interoperability with optical technology such as barcode or quick-response (QR) codes. The research aims at a higher impact in a more distant future and focuses on supporting the whole life of food products from their genesis to their disposal by exploiting a full array of RFID technologies, eventually empowered by sensing capabilities or machine learning, as shown by the case studies in the following section.

Table 5.1 provides examples of available current commercial-off-the-shelves components and some recent academic investigations on the use of RFID in the food production chain. It is worth noting that microwave RFID is absent as its use in the food industry is hindered by the high cost of its tags.

5.2.3.1 Radiofrequency identification in action—case studies in Food Traceability 4.0

Hereafter, three case studies will be discussed to show cases of research under industrial settings to delineate the future of radiofrequency use in the traceability of food. For the sake of simplicity, all the studies that will be considered fall under the UHF RFID category.

5.2.3.2 Optimization of plastic packaging recycling—RICIRCOLA project

Reduction of the environmental footprint is one of the main goals of Food Traceability 4.0. In particular, single-use, food packaging made of polyethylene terephthalate plastic is among the primary pollutants and is also relatively easy to recycle. In an investigation of the ViVACE managing framework for the circular economy (Bianchini & Rossi, 2021), the Radio6ense, Mecstar, and ILIP companies developed an RFID-tagged smart packaging system. The packaging had an RFID tag inserted in the typical label reporting the expiry date, the serial number, and all other mandatory information on the packaged food. The transponder allowed the implementation of a classical track and trace system for each smart package.

The RFID-enabled smart packaging was tested in October and November 2020 in Italy under commercial conditions in a project that was named the "RICIRCOLA project" (Bianchini & Rossi, 2021). A small fee was given to the customers (0.20 euros per tray) returning to the supermarket the purchased package for recycling after food consumption. Thanks to RFID traceability, compliance statistics were obtained, foreseeing a possible increase in plastic waste recovery efficiency higher than 120% after just one year of continuous utilization of the RFID packaging.

5.2.3.3 Traceability of perishable food in cold storage through radiofrequency identification and machine learning

Basic RFID track and trace approaches can be scaled up to integrate environmental sensing and more advanced functionalities through the use of artificial intelligence and machine learning. Alfian et al. (2020) developed a multifaceted infrastructure (Fig. 5.4) that exploited a classic RFID gate and an environmental sensor to measure the temperature and

TABLE 5.1 Recent examples of commercial components and academic research on radiofrequency identification (RFID) for food production applications.

Available commercial components		Current academic research	
Component	**Description**	**References**	**Description**
GaoRFID, 2022	LF (134 kHz) RFID ear-tag for livestock identification	Sanchez et al. (2017)	Design of a low-cost LF RFID reader for animal tracking
Identiv, 2022	NFC-tagged bottles of tequila for authentication	Conti (2022)	NFC-based traceability system for olive oil production and sale
Serva, 2022	NFC tags that track food items and their expiry date	Fan et al., (2019)	QR-code-to-UHF-RFID identifiers conversion and vice versa
Bionix, 2022	UHF RFIDs for controlling food TRUs during production	Barge, et al. (2020)	Integration of LF, HF, and UHF RFID for meat traceability
Impinj, 2022	UHF RFID system for monitoring the whole meat supply chain	Masudin et al. (2021)	Quantifications of the benefits of RFID usage for cold chain

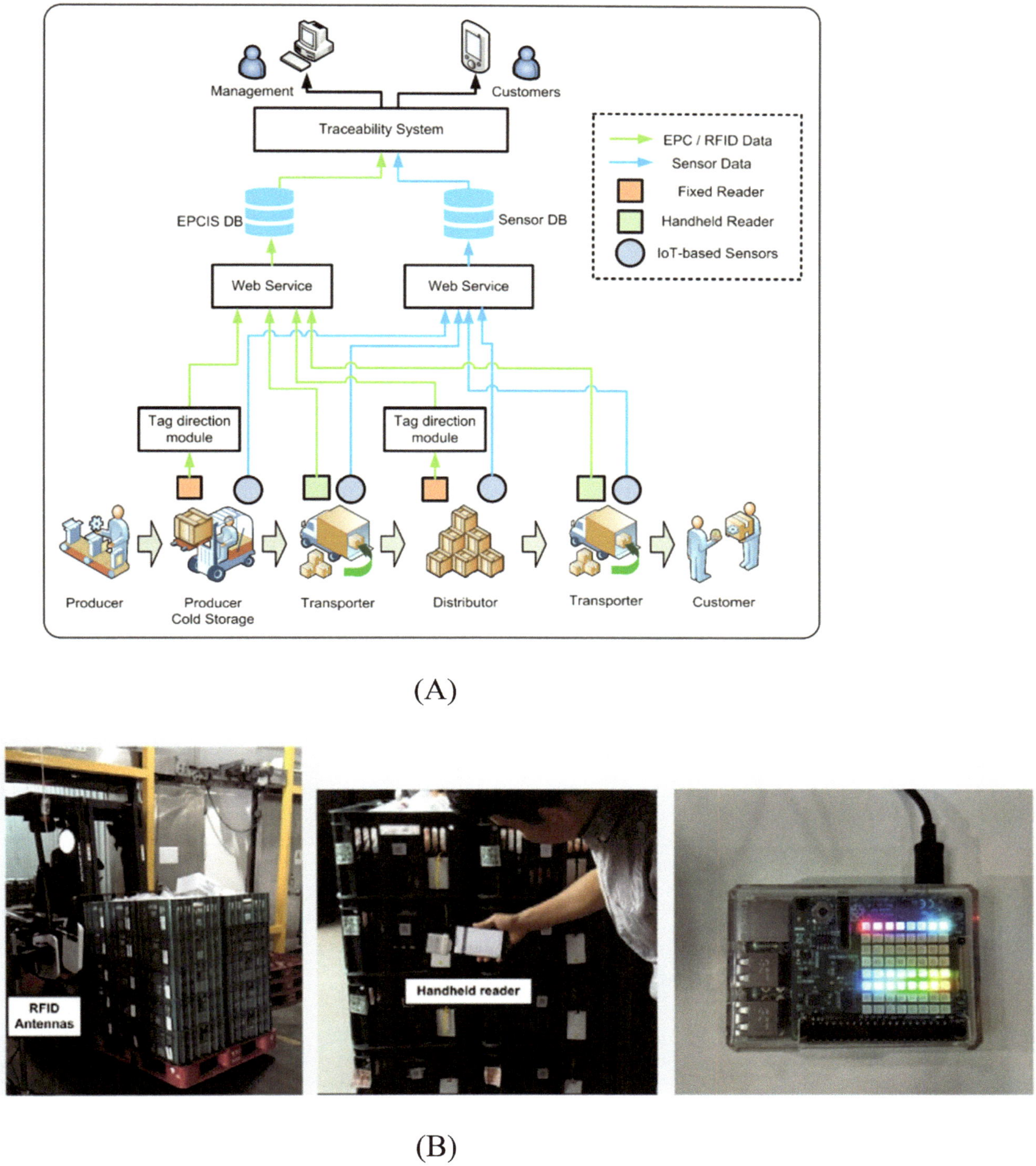

FIGURE 5.4 RFID system for food traceability and control of moving kimchi boxes. (A) High-level architecture of the whole system. (B) From left to right: a forklift moving through an RFID gate; handheld RFID reader interrogating tags; environmental sensor installed in the cold storage. *From Alfian, G., Syafrudin, M., Farooq, U., Ma'arif, M. R., Syaekhoni, M. A., Fitriyani, N. L., & Rhee, J. (2020). Improving the efficiency of RFID-based traceability system for perishable food by utilizing IoT sensors and machine learning model.* Food Control, 110, *107016.*

relative humidity of food products. The RFID fixed reading antennas interrogate each tag entering and exiting a cold storage unit, where food products (kimchi; Korean fermented spicy cabbage) are stored. Each box was tagged, and an environmental sensor monitor was used to detect the conditions that are optimal for food preservation to ensure the quality and safety of the kimchi (Fig. 5.4). A complete log of the boxes entering or exiting the storage facility was obtained thanks to a classification algorithm estimating the direction of movement from the electromagnetic features returned by the tags to the RFID gate.

5.2.3.4 Ripening of climacteric fruit estimations

Another example of machine learning algorithms empowering classical RFID infrastructures for food traceability is the device that is used for the estimation of the ripening status of avocado fruits as detailed by (Mostaccio et al., 2022). An RFID totem for checking the ripening of avocados in consumer distribution and a smart trolley monitoring ripening rooms are presented in Fig. 5.5. Due to the age variability of the fruits at the different checking stages, two different algorithms were trained by the system. The smart trolley in the ripening rooms has RFID tags and readers fixed to its skeleton. The monitored avocado fruits are maintained in contact with the tags and, during ripening, they change their electromagnetic response (Occhiuzzi et al., 2022). The same technique with a different hardware selection was employed for checking the avocado's ripening at the consumer distribution level (Mostaccio et al., 2022). This technique is currently under commercial investigation (Fig. 5.5), and it can be integrated into a classical RFID infrastructure for food traceability in warehouses.

5.3 Blockchain for Food Traceability 4.0

5.3.1 Introduction to blockchain

Treiblmaier (2018) defined blockchain as a distributed and decentralized digital ledger in which data can be recorded chronologically in a tamperproof manner. As the data cannot be altered, blockchain acts as a reliable source for recording data (Kamble et al., 2019). As the blockchain enhances trust and removes intermediaries, this allows users to directly connect and exchange goods and services (Alammary et al., 2019). The blockchain network is created collaboratively and is shared securely, thus making it easier for parties involved in the exchange of goods to trust each other without any need for a third party (Calatayud et al., 2019). Data on the blockchain are recorded on a peer-to-peer network, and thus it is accessible to all the computers in the network (Pence, 2019). Blockchain enhances visibility across the board by connecting every party in a secured network and multiple items such as agreements, milestones, and regulatory checks can easily be traced (Dolgui et al., 2020).

Smart contracts based on sensors and artificial intelligence can be connected with the blockchain and these smart contracts will be able to record and share data on the blockchain. Several rules can be applied to a blockchain network and all the parties involved can ensure compliance with rules and standard procedures. Additionally, smart contracts when combined with artificial intelligence and IoT will enable the prevention of errors at multiple stages of the supply chain (Andoni, et al., 2019; Feng et al., 2020).

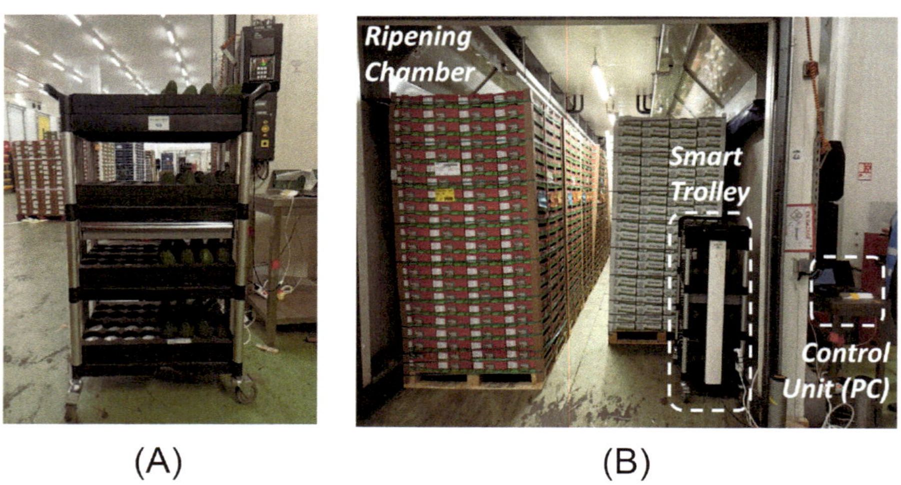

FIGURE 5.5 Smart trolley system used for monitoring the ripening of avocados. (A) Half-loaded trolley in the warehouse and (B) full-loaded trolley inside the ripening chamber and connected to the external control unit.

5.3.2 Blockchain for supply chain traceability

The food supply chain comprises all the activities involved in the production of food until the consumption and disposal of food (Dossa et al., 2022). Food products are subject to several variables such as shelf life, seasonality, and dependence on multiple factors such as climate and temperature. Several issues such as price fluctuation, food safety, food security, pest infestation, and disease breakout are common challenges in the food supply chain (Barbosa, 2021). Collectively, these issues create problems at the time of production, during storage, transport, and waste management (Barbosa, 2021; Fredriksson & Liljestrand, 2015).

As discussed earlier, due to the reliability of blockchain, the food supply chain can be improved by the reliable and accurate information presented in a transparent manner by the blockchain. As the data is shared and accessible to all the intermediaries, it will result in visibility across the food supply chain (Tan et al., 2022). Blockchain technology will enable real-time sharing of food information such as the origin of the product, batch code, and date of manufacturing as well as the movement of the product through the supply chain. Modern consumers are eager to learn about the food production environment and food compliance certifications that are required to be more transparent to enhance consumers' trust (Galvez et al., 2018). Undeniably, blockchain can enhance the traceability and management of the food supply chain (Kamilaris et al., 2019).

Blockchain will enable the various parties involved in the food supply chain to share assertions, and thus the quality of food can be validated across the food supply chain as each participant will be able to share evidence and evaluate each other's assertions, leading to transparency across the food supply chain. An efficient food supply chain will enable all participants in the food supply chain to verify multiple assertions such as the quality, timing, and quantity of food items throughout the food supply chain. Due to enhanced auditing and the fact that buyers and sellers would not have to spend much time to ensure the traceability of data, this thus results in more efficiency in the food supply chain (Galvez et al., 2018). Furthermore, the journey of a food item can be captured in an object of blockchain as a food bundle and this bundle will enable transparent reflection of information shared by food supply chain actors throughout the journey of a food item (Galvez et al., 2018). As blockchain simplifies data integration among multiple actors in the food supply chain (farmers, distributors, packers, processors, grocers, restaurants, and traders), more transparency, efficiency, security, and safety can be established in the food supply chain.

5.3.3 Blockchain for food traceability

Food traceability can vary depending on the food type, that is, plant-based foods or animal-based foods.

5.3.3.1 Blockchain for plant foods

Multiple processes are involved in the acquisition of plant food that can be summarized as follows (Xu et al., 2022):

- Purchase of seeds
- Planting/farming
- Harvesting
- Transportation and storage
- Processing
- Consumers and retail

    Fig. 5.6 depicts the architecture of a blockchain-based food traceability system. The architecture consists of four layers; a business, IoT traceability, blockchain, and application layer.

- *Business layer:* that layer consists of all business activities from farm to consumption (Feng et al., 2020).
- *IoT traceability layer:* The IoT layer records traceability information such as temperature, O_2 content, humidity, and CO_2 through devices such as RFIDs and Wireless Network Systems. Blockchain then acts as a ledger to record data from these devices (Imeri & Khadraoui, 2018).
- *Blockchain layer:* The blockchain layer enhances the transparency of information between different actors of the food supply chain and thus enables real-time sharing of data, resulting in better monitoring and control over food supply chain operations (Kshetri, 2018).
- *Application layer:* The application layer acts as an intermediary to connect actors in the food supply chain with the information platform. The actors in the food supply chain can easily view the whole process of the flow of food, information, and capital through this layer (Feng et al., 2020).

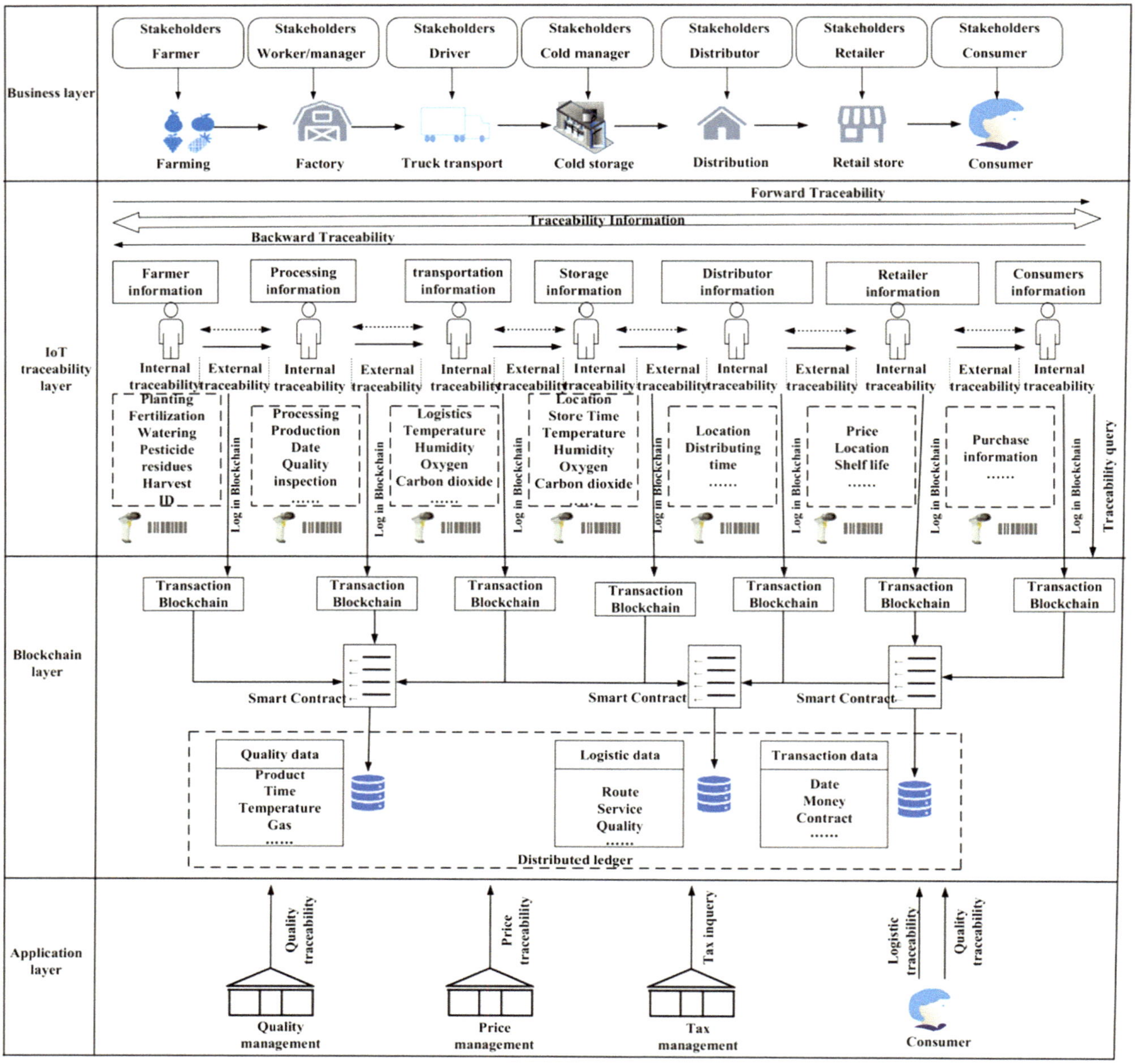

FIGURE 5.6 Architecture of blockchain-based food traceability system (Feng et al., 2020). *The figure is published with kind permission from the Elsevier (License Number 5446820599888).*

5.3.3.1.1 Purchase of seeds

The process starts with the acquisition of seeds. There are multiple varieties of seeds and there are chances that seeds might be genetically modified or of low quality. Blockchain technology will enable the traceability of the food quality of plants from the source by recording the origin of seeds until the end product (Xu et al., 2022). This will enable better control over the food quality and ensure farm-to-fork traceability.

5.3.3.1.2 Planting/farming

Due to rapid industrialization and intensive farming and food production, several ecological issues have occurred due to the intensive use of chemical fertilizers and pesticides that led to polluted soil and water becoming a major risk to food safety (Saeed et al., 2010). Another issue that compromises food safety is the presence of natural pollutants (e.g., heavy and toxic metals) and pesticide residues (Ramón & Lull, 2019). Due to the stated reasons, it is very important to monitor and document the quality of water, air, and soil as part of quality assurance programs (Xu et al., 2022). While the use of pesticides enhances the quality and yield of plant foods, many farmers violate the use of pesticides and do not allow sufficient breakdown periods. Excessive pesticide residues in plant foods are harmful to health (Strecker et al., 2013). A variety of traceability information such as quality, application, variety, and origin of soil, water, drugs, pesticides, and level of humidity and temperatures can be recorded through a blockchain-based traceability system on a date and time basis (Bastas & Liyanage, 2018; Caro et al., 2018).

5.3.3.1.3 Harvesting

The growth cycle of a crop varies and plays a major role in product quality, availability, and marketability. Hedonic attributes of the product such as the sugar-acid ratio and freshness of foods are dependent on harvest timings, and therefore, harvesting at the right time is an important factor for the best quality (Perboli et al., 2018). Traceability information such as date, time, yield, and weight can be effectively recorded, analyzed, and monitored through a blockchain-based traceability system (Feng et al., 2020). This will enable capturing the best quality time to direct products to markets as well as determining losses due to storage and product handling. Overall, such information can enable the optimization of revenue and financial operations.

5.3.3.1.4 Transportation and storage

The next step after harvesting is to transport the produce to processing plants. Due to the respiration process of plants after harvest, several environmental factors such as humidity, temperature, and oxygen concentrations will play an important role in dictating the metabolism of the plants and impact their quality and stability during storage and subsequent processing and marketing. Blockchain-based traceability systems equipped with artificially intelligent IoT and GPS sensors would enable real-time monitoring of cold chain temperature, storage, and transportation time and location of vehicles (Galvez et al., 2018).

5.3.3.1.5 Processing

Blockchain-based traceability system will enable the storage of information at various processing stages by tagging finished products on blockchain to store information regarding various processing steps employed, type of equipment used, disinfection techniques, name of operators, and details of batches and product tags (Feng et al., 2019). This will provide a permanent record of safe processing and parameters used for standard processing.

5.3.3.1.6 Consumers and retail

As the information about the product would be recorded on the blockchain, it would be available to all the intermediaries including end consumers. The origin of products plays an important role in ensuring food safety. There are numerous studies on the safety of products produced in polluted environments and their effect on human health. A recent study by He et al. (2019) on the soil-rice grain system in Wenling (China) concluded that metal-contaminated farmland soil is extremely damaging to the environment and human health. A blockchain-based traceability system will give a complete picture of the product's journey to retailers and consumers and information such as origin, sale price, shelf life, and production date would easily be verified (Garaus & Treiblmaier, 2021).

5.3.3.2 Blockchain for animal-based food

Several stages are involved in an animal-based food supply chain that involves processes such as feeding, medical care, transportation, slaughter, processing, storage, and retail (Xu et al., 2022). It is very important to observe the feed quality as it directly affects animals' health. The presence of contaminants such as metals and pesticides, or chemicals (e.g., used as additives or cleaning) in feed can compromise the health and performance of the animal as well as the safety of the products obtained from the animal. Moreover, it is equally important to ensure the quality of drinking water accessible to the animals. Sick animals are normally treated with various drugs and the correct application of drugs is also an important aspect to monitor (Fernández-Cisnal et al., 2018; Xu et al., 2022). Blockchain-based traceability systems will enable a transparent flow of information and will enable records of animal species, growth rates and animal performance, veterinary treatments, diet quality, and environmental conditions that the animal is subjected to.

Temperature plays a very important role at all stages of the food supply chain. Temperature not only affects the keeping quality of animal products but also impacts the health of the animal and its performance (i.e., growth rate), as well as the stress level of the animal and causes subsequent quality defects (e.g., formation of undesirable dark red color, commonly termed as "dark-cut") that reduce the product value. Dark-cut meat is the result of the depletion of muscle glycogen that is caused by several factors related to animal management (poor nutrition, age of the animal, subclinical diseases, hormone implant use, animal temperament, and aggressiveness) (Ponnampalam et al., 2017). Animal transportation to slaughterhouses is a critical process that if mishandled can lead to stress and dark meat cut occurrence. Exposure to high temperatures or cold during transportation can trigger stress. Similarly mixing animals from different mobs can cause stress and poor meat quality. Further fabrication of carcasses (breakdown of the carcass into various meat cuts distant to different processes and markets) is normally carried out at low temperatures to control microorganisms' growth, ensure safety, and maintain fresh meat color and quality (Xu et al., 2022). Smart contracts on the blockchain will be able to record traceability events at the logistics, distribution, and storage phase and information such as modes of transportation, transportation routes, duration, time and temperatures will easily be accessible (Feng et al., 2020). Information about animals' source and mob composition can also be stored. Collectively, the information can provide an extensive database for the optimization of meat quality and troubleshooting of any quality issues. As information stored in the blockchain once entered cannot be changed and it becomes accessible to interested parties, more accurate and transparent information becomes available. Through sensors and WSN, blockchain technology enables real-time monitoring and recording of temperature throughout the animal food supply chain. Temperatures can be recorded in a blockchain to document the temperature history from breeding to retail through a WSN (Xu et al., 2022).

Blockchain-based smart contracts play an important role in ensuring food safety where these contracts can monitor temperature and match it to regulatory requirements, and subsequently, the product is only released if all the underlying requirements are fulfilled. In case of a breach of any requirement, the parties involved will get the deviation notice and the contract will be automatically terminated. Blockchain technology will terminate the transaction of illegal food products and thus guarantee the microbiological safety of animal-based food products (Andoni et al., 2019).

As the variety and species can be detected through DNA analysis (genomics) and near-infrared spectroscopy imaging, such information can be recorded in QR Codes or RFID tags backed with blockchain networks to ensure the authenticity of meat (Schelm et al., 2017). A smart contract will also ensure the proper use of feed and drugs via the registration of the sale to farmers, through a smart contract sale, and purchase of feed/drugs can be mutually verified by a blockchain system, thus preventing meat adulteration from farm to fork (Kshetri, 2018).

5.3.4 Blockchain in action

The use of blockchain technology will enable the tracking of food from farm to fork in a transparent manner and enable stakeholders to have real-time visibility across the supply chain. Therefore, a number of well-reputed international companies have started to adopt blockchain technology. For example, Walmart has achieved unprecedented food supply chain transparency through the application of Hyperledger Fabric (a blockchain-based platform). Walmart has successfully finished two blockchain tests using IBM's blockchain technology based on Hyperledger Fabric; these are Pork in China and Mangoes in the Americas (IBM, 2021). Walmart's blockchain technology, which takes a farm-to-table approach, enhanced the process of tracking mango origins from seven days to 2.2 seconds and thus enabled increased transparency throughout Walmart's entire food supply chain (Kamath, 2018).

A notable example of blockchain is the application of the food traceability framework "IBM Food Trust Platform." IBM Food Trust Platform creates a collaborative network of multiple intermediaries in the food supply chain and connects parties such as producers/farmers, food processors, wholesalers, distributors, food manufacturers, retailers, and other related parties, and strives to enhance transparency and traceability and accountability across the food supply chain (IBM, 2021). Another application is the TE-FOOD platform. This platform combines blockchain and other technologies such as API and IoT to provide traceability solutions for several food industries and is currently being deployed by more than 6000 businesses (Te-food, 2020). Overall, the application of blockchain-based traceability systems will enhance food traceability and will contribute toward the efficiency and transparency of the food supply chain, thus ensuring food quality and the safety of consumers.

5.4 Conclusion

Traceability has been a crucial part of the food industry and traditional record-keeping practices no longer meet the consumers' demands and regulatory agencies of food safety and crisis management. Also, due to the perishable nature of foods as well as the high value of products, such as those carrying the protected designation of origin status, the industry is obligated to implement strategies that reduce waste and maintain the integrity of products. Blockchain and RFID technologies can help in addressing food safety and fraud challenges by connecting food supply chain actors and providing a tamper-proof record of events throughout the supply chain. These technologies will enable transparency and accountability throughout the food supply chain and thus reduce the risk of counterfeiting and fraud. Through real-time connectivity, these technologies would enable an agile response to any food-related hazards. A number of challenges such as the cost of implementation, the need for standardization, and effective interoperability of systems are still hurdles toward adopting these technologies. Future research needs to address these challenges to enhance the efficiency and reliability of food traceability.

Acknowledgment

This work was partly supported by Project ECS 0000024 Rome Technopole, – CUP B83C22002820006, NRP Mission 4 Component 2 Investment 1.5, Funded by the European Union – NextGenerationEU. Spoke: 2. Project: "Eco-friendly Electronic Labels for Food and Plastic Waste" (Flagship Project 3).

References

Alammary, A., Alhazmi, S., Almasri, M., & Gillani, S. (2019). Blockchain-based applications in education: A systematic review. *Applied Sciences*, *9* (12), 2400.

Alfian, G., Syafrudin, M., Farooq, U., Ma'arif, M. R., Syaekhoni, M. A., Fitriyani, N. L., & Rhee, J. (2020). Improving the efficiency of RFID-based traceability system for perishable food by utilizing IoT sensors and machine learning model. *Food Control*, *110*107016.

Andoni, M., Robu, V., Flynn, D., Abram, S., Geach, D., Jenkins, D., & Peacock, A. (2019). Blockchain technology in the energy sector: A systematic review of challenges and opportunities. *Renewable and Sustainable Energy Reviews*, *100*, 143–174.

Aung, M. M., & Chang, Y. S. (2014). Traceability in a food supply chain: Safety and quality perspectives. *Food Control*, *39*, 172–184.

Bai, C., Dallasega, P., Orzes, G., & Sarkis, J. (2020). Industry 4.0 technologies assessment: A sustainability perspective. *International Journal of Production Economics*, *229*107776.

Barbosa, M. W. (2021). Uncovering research streams on agri-food supply chain management: A bibliometric study. *Global Food Security*, *28*100517.

Barge, P., Biglia, A., Comba, L., Ricauda Aimonino, D., Tortia, C., & Gay, P. (2020). Radio frequency identification for meat supply-chain digitalisation. *Sensors*, *20*(17), 4957.

Bastas, A., & Liyanage, K. (2018). Sustainable supply chain quality management: A systematic review. *Journal of Cleaner Production*, *181*, 726–744.

Behnke, K., & Janssen, M. F. W. H. A. (2020). Boundary conditions for traceability in food supply chains using blockchain technology. *International Journal of Information Management*, *52*101969.

Bianchini, A., & Rossi, J. (2021). An innovative visualization tool to boost and monitor circular economy: An overview of its applications at different industrial sectors. In *Product life cycle - opportunities for digital and sustainable transformation*. doi:10.5772/intechopen.98761.

Bionix. (2022). RFID solutions in food [Online]. Retrieved October 25, 2022, from https://bionixtechnologies.com/en/rfid-food/.

Calatayud, A., Mangan, J., & Christopher, M. (2019). The self-thinking supply chain. *Supply Chain Management: An International Journal*, *24*, 22–38. Available from https://doi.org/10.1108/SCM-03-2018-0136.

Caro, M.P., Ali, M.S., Vecchio, M., & Giaffreda, R. (2018). Blockchain-based traceability in agri-food supply chain management: A practical implementation. In *2018 IoT vertical and topical summit on agriculture-Tuscany* (IOT Tuscany), 08–09 May 2018, Tuscany, Italy (pp. 1–4). IEEE. doi:10.1109/IOT-TUSCANY.2018.8373021.

Casino, F., Kanakaris, V., Dasaklis, T. K., Moschuris, S., & Rachaniotis, N. P. (2019). Modeling food supply chain traceability based on blockchain technology. *Ifac-Papers Online*, *52*, 2728–2733.

Codex Alimentarius Commission (2005). Principles and guidelines for monitoring the performance of national food control systems. https://www.fao.org/fao-who-codexalimentarius/sh-proxy/en/?lnk = 1&url = https%253A%252F%252Fworkspace.fao.org%252Fsites%252Fcodex%252FStandards%252FCXG%2B91-2017%252FCXG_091e.pdf.

Conti, M. (2022). EVO-NFC: Extra virgin olive oil traceability using NFC suitable for small-medium farms. *IEEE Access*, *10*, 20345–20356.

Costa, G.N., & Arnaud, A. (2011). A low frequency RFID temperature data logger. In *2011 Argentine School of Micro-Nanoelectronics, Technology and Applications* (pp. 1–4). IEEE.

Dabbene, F., Gay, P., & Tortia, C. (2016). *Radio-frequency identification usage in food traceability. Advances in food traceability techniques and technologies* (pp. 67–89). Woodhead Publishing.

Dolgui, A., Ivanov, D., Potryasaev, S., Sokolov, B., Ivanova, M., & Werner, F. (2020). Blockchain-oriented dynamic modelling of smart contract design and execution in the supply chain. *International Journal of Production Research*, *58*(7), 2184–2199.

Dossa, A. A., Gough, A., Batista, L., & Mortimer, K. (2022). Diffusion of circular economy practices in the UK wheat food supply chain. *International Journal of Logistics Research and Applications*, *25*(3), 328–347.

Fan, B., Qian, J., Wu, X., Du, X., Li, W., Ji, Z., & Xin, X. (2019). Improving continuous traceability of foodstuff by using barcode-RFID bidirectional transformation equipment: Two field experiments. *Food Control*, *98*, 449–456.

Fanelli, V., Mascio, I., Miazzi, M. M., Savoia, M. A., De Giovanni, C., & Montemurro, C. (2021). Molecular approaches to agri-food traceability and authentication: An updated review. *Foods*, *10*(07), 1644.

Fathi, P., Karmakar, N. C., Bhattacharya, M., & Bhattacharya, S. (2020). Potential chipless RFID sensors for food packaging applications: A review. *IEEE Sensors Journal*, *20*(17), 9618–9636.

Feng, H., Chen, J., Zhou, W., Rungsardthong, V., & Zhang, X. (2019). Modeling and evaluation on WSN-enabled and knowledge-based HACCP quality control for frozen shellfish cold chain. *Food Control*, *98*, 348–358.

Feng, H., Wang, X., Duan, Y., Zhang, J., & Zhang, X. (2020). Applying blockchain technology to improve agri-food traceability: A review of development methods, benefits and challenges. *Journal of Cleaner Production*, *260*121031.

Fernández-Cisnal, R., García-Sevillano, M. A., García-Barrera, T., Gómez-Ariza, J. L., & Abril, N. (2018). Metabolomic alterations and oxidative stress are associated with environmental pollution in *Procambarus clarkii*. *Aquatic Toxicology*, *205*, 76–88.

Fredriksson, A., & Liljestrand, K. (2015). Capturing food logistics: a literature review and research agenda. *International Journal of Logistics Research and Applications*, *18*(1), 16–34.

Galvez, J. F., Mejuto, J. C., & Simal-Gandara, J. (2018). Future challenges on the use of blockchain for food traceability analysis. *TrAC Trends in Analytical Chemistry*, *107*, 222–232.

GaoRFID (2022). Gao RFID Inc. 134 kHz LF livestock Ear RFID Tag. Manhattan, NY, USA, 2022 [Online]. Retrieved October 30, 2022, from https://gaorfid.com/product/tag-animal-cattle-sheep-pig-track-ear-lf-134-khz-rfid/.

Garaus, M., & Treiblmaier, H. (2021). The influence of blockchain-based food traceability on retailer choice: The mediating role of trust. *Food Control*, *129*108082.

George, R. V., Harsh, H. O., Ray, P., & Babu, A. K. (2019). Food quality traceability prototype for restaurants using blockchain and food quality data index. *Journal of Cleaner Production*, *240*118021.

Golan, E.H., Krissoff, B., Kuchler, F., Calvin, L., Nelson, K.E., & Price, G.K. (2004). *Traceability in the US food supply: Economic theory and industry studies* (No. 1473-2016-120760). https://www.ers.usda.gov/webdocs/publications/41623/28673_aer830_1_.pdf?v = 0.

Hassoun, A., Siddiqui, S. A., Smaoui, S., Ucak, İ., Arshad, R. N., Garcia-Oliveira, P., & Bono, G. (2022). Seafood processing, preservation, and analytical techniques in the age of industry 4.0. *Applied Sciences*, *12*(3), 1703.

Hassoun, A., Aït-Kaddour, A., Abu-Mahfouz, A. M., Rathod, N. B., Bader, F., Barba, F. J., & Regenstein, J. (2022). The fourth industrial revolution in the food industry—Part I: Industry 4.0 technologies. *Critical Reviews in Food Science and Nutrition*, 1–17. Available from https://doi.org/10.1080/10408398.2022.2034735.

He, M., Shen, H., Li, Z., Wang, L., Wang, F., Zhao, K., & Xu, J. (2019). Ten-year regional monitoring of soil-rice grain contamination by heavy metals with implications for target remediation and food safety. *Environmental Pollution*, *244*, 431–439.

IBM. (2021). https://www.ibm.com/ IBM Food Trust - Blockchain for the world's food supply. https://www.ibm.com/dk-en/products/supply-chain-intelligence-suite/food-trust 2021.

Identiv. (2022). Identiv NFC-enabled smart packaging drives outstanding results for OTACA tequila through consumer engagement [Online]. Retrieved October 21, 2022, from https://www.identiv.com/company/press-releases/identiv-nfc-enabled-smart-packaging-drives-outstanding-results-for-otaca-tequila-through-consumer-engagement.

IEEE. (2019). IEEE standard definitions of terms for radio wave propagation. IEEE Std 211–2018 (revision of IEEE Std 211–1997) (pp. 1–57). doi:10.1109/IEEESTD0.2019.8657413.

Imeri, A., & Khadraoui, D. (2018). The security and traceability of shared information in the process of transportation of dangerous goods. In *2018 9th IFIP international conference on new technologies, mobility and security (NTMS)* (pp. 1–5). IEEE.

Impinj (2022). *Chainway commits to consumer safety with RAIN RFID food traceability* [Online]. Retrieved October 30, 2022, from https://www.impinj.com/library/partner-solutions/chainway-rain-rfid-food-traceability-for-consumer-safety.

Islam, S., & Cullen, J. M. (2021). Food traceability: A generic theoretical framework. *Food Control*, *123*107848.

Jeon, S., Yu, Y., & Choi, J. (2006). Dual-band slot-coupled dipole antenna for 900 MHz and 2.45 GHz RFID tag application. *Electronics Letters*, *42*(22), 1.

Kamath, R. (2018). Food traceability on blockchain: Walmart's pork and mango pilots with IBM. *The Journal of the British Blockchain Association*, *1* (1), 3712.

Kamble, S., Gunasekaran, A., & Arha, H. (2019). Understanding the Blockchain technology adoption in supply chains-Indian context. *International Journal of Production Research*, *57*(7), 2009–2033.

Kamilaris, A., Fonts, A., & Prenafeta-Boldú, F. X. (2019). The rise of blockchain technology in agriculture and food supply chains. *Trends in Food Science & Technology*, *91*, 640–652.

Khalil, G., Doss, R., & Chowdhury, M. (2020). A novel RFID-based anti-counterfeiting scheme for retail environments. *IEEE Access*, *8*, 47952–47962.

Kshetri, N. (2018). 1 Blockchain's roles in meeting key supply chain management objectives. *International Journal of Information Management*, *39*, 80–89.

Lee, J., Bagheri, B., & Kao, H. A. (2015). A cyber-physical systems architecture for industry 4.0-based manufacturing systems. *Manufacturing Letters*, *3*, 18–23.

Liu, R., Gao, Z., Nayga, R. M., Jr, Snell, H. A., & Ma, H. (2019). Consumers' valuation for food traceability in China: Does trust matter? *Food Policy*, *88*101768.

Lu, Y., Li, P., & Xu, H. (2022). A Food anti-counterfeiting traceability system based on Blockchain and Internet of Things. *Procedia Computer Science*, *199*, 629–636.

Luque, A., Peralta, M. E., De Las Heras, A., & Córdoba, A. (2017). State of the Industry 4.0 in the Andalusian food sector. *Procedia Manufacturing*, *13*, 1199–1205.

Masudin, I., Ramadhani, A., Restuputri, D. P., & Amallynda, I. (2021). The effect of traceability system and managerial initiative on Indonesian food cold chain performance: A Covid-19 pandemic perspective. *Global Journal of Flexible Systems Management*, *22*(4), 331–356.

Moe, T. (1998). Perspectives on traceability in food manufacture. *Trends in Food Science & Technology*, *9*(5), 211–214.

Mostaccio, A., Bianco, G.M., Amendola, S., Marrocco, G., & Occhiuzzi, C. (2022). RFID for Food Industry 4.0—current trends and monitoring of fruit ripening. In 2022 *IEEE 12th international conference on RFID technology and applications (RFID-TA)* (pp. 109–112). IEEE.

NFC Forum (2022), Worldwide NFC technology use surge over last 24 months [Online]. Retrieved October 21, 2022, from https://nfc-forum.org/news/2022-07-worldwide-nfc-technology-use-surges-over-last-24-months/

Occhiuzzi, C., Camera, F., D'Orazio, M., D'Uva, N., Amendola, S., Bianco, G. M., & Marrocco, G. (2022). Automatic monitoring of fruit ripening rooms by UHF RFID sensor network and machine learning. *IEEE Journal of Radio Frequency Identification*, *6*, 649–659. Available from https://doi.org/10.1109/JRFID.2022.3174272.

Olsen, P., & Borit, M. (2013). How to define traceability. *Trends in Food Science & Technology*, *29*(2), 142–150.

Olsen, P., & Borit, M. (2018). The components of a food traceability system. *Trends in Food Science & Technology*, *77*, 143–149.

Pence, H. E. (2019). Blockchain: will better data security change chemical education? *Journal of Chemical Education*, *97*(7), 1815–1818.

Perboli, G., Musso, S., & Rosano, M. (2018). Blockchain in logistics and supply chain: A lean approach for designing real-world use cases. *Ieee Access*, *6*, 62018–62028.

Ponnampalam, E. N., Hopkins, D. L., Bruce, H., Li, D., Baldi, G., & Bekhit, A. E. D. (2017). Causes and contributing factors to "dark cutting" meat: Current trends and future directions: A review. *Comprehensive Reviews in Food Science and Food Safety*, *16*(3), 400–430.

Qian, J., Ruiz-Garcia, L., Fan, B., Villalba, J. I. R., McCarthy, U., Zhang, B., & Wu, W. (2020). Food traceability system from governmental, corporate, and consumer perspectives in the European Union and China: A comparative review. *Trends in Food Science & Technology*, *99*, 402–412.

Ramón, F., & Lull, C. (2019). Legal measures to prevent and manage soil contamination and to increase food safety for consumer health: *The case of Spain. Environmental Pollution*, *250*, 883–891.

Robson, K., Dean, M., Brooks, S., Haughey, S., & Elliott, C. (2020). A 20-year analysis of reported food fraud in the global beef supply chain. *Food Control*, *116*107310.

Sadeghi, K., Kim, J., & Seo, J. (2022). Packaging 4.0: The threshold of an intelligent approach. *Comprehensive Reviews in Food Science and Food Safety*, *21*, 2615–2638.

Saeed, A., Safina, N., Sultan, M. T., Seema, M., Muhammad, N., & Anwaar, A. (2010). Physico-chemical attributes and heavy metal content of mangoes (*Mangifera indica* L.) cultivated in different regions of Pakistan. *Pakistan Journal of Botany*, *42*(4), 2691–2702.

Sanchez, J., Garcia, M., & Moncada, M. V. (2017). *Design and Implementation of a Low Cost RFID ISO 11784/11785 Reader for the Automatic Livestock Management in Nicaragua. In Advances in Automation and Robotics Research in Latin America* (pp. 173–185). Cham: Springer.

Schelm, S., Haase, I., Fischer, C., & Fischer, M. (2017). Development of a multiplex real-time PCR for determination of apricot in marzipan using the plexor system. *Journal of Agricultural and Food Chemistry*, *65*(2), 516–522.

Serva (2022). Best Buy. "SERVA NFC food tracker 5-pack," Vancouver, Canada, 2022 [Online]. Retrieved October 30, 2022, from https://www.bestbuy.ca/en-ca/product/serva-nfc-food-tracker-5-pack/14682870.

Strecker, R., Weigt, S., & Braunbeck, T. (2013). Cartilage and bone malformations in the head of zebrafish (*Danio rerio*) embryos following exposure to disulfiram and acetic acid hydrazide. *Toxicology and Applied Pharmacology*, *268*(2), 221–231.

Tan, A., Gligor, D., & Ngah, A. (2022). Applying blockchain for halal food traceability. *International Journal of Logistics Research and Applications*, *25*(6), 947–964.

Tao, F., Wang, L., Fan, T., & Yu, H. (2022). RFID adoption strategy in a retailer-dominant supply chain with competing suppliers. *European Journal of Operational Research*, *302*(1), 117–129.

Te-food. (2020). The #1 end-to-end. https://te-food.com/: https://te-food.com/solution/blockchain/

Treiblmaier, H. (2018). The impact of the blockchain on the supply chain: a theory-based research framework and a call for action. *Supply Chain Management: an International Journal*, *23*, 545–559.

Xu, Y., Li, X., Zeng, X., Cao, J., & Jiang, W. (2022). Application of blockchain technology in food safety control: current trends and future prospects. *Critical Reviews in Food Science and Nutrition*, *62*(10), 2800–2819.

Yang, Y., Huisman, W., Hettinga, K. A., Liu, N., Heck, J., Schrijver, G. H., & Van Ruth, S. M. (2019). Fraud vulnerability in the Dutch milk supply chain: Assessments of farmers, processors and retailers. *Food Control*, *95*, 308–317.

Yu, Z., Jung, D., Park, S., Hu, Y., Huang, K., Rasco, B. A., & Chen, J. (2022). Smart traceability for food safety. *Critical Reviews in Food Science and Nutrition*, *62*(4), 905–916.

Zhang, L. H., & Yang, H. (2019). Incentives for RFID adoption with imperfect read rates: Wholesale price premium versus cost sharing. *Journal of the Operational Research Society*, *70*(9), 1440–1456.

Chapter 6

Food processing and preservation in the Food Industry 4.0 era

Gulay Ozkan[1], Sebahattin Serhat Turgut[2,3], Busra Gultekin Subasi[4], Esra Capanoglu[1], Tuba Esatbeyoglu[5] and Aberham Hailu Feyissa[3]

[1]Department of Food Engineering, Faculty of Chemical and Metallurgical Engineering, Istanbul Technical University, Maslak, Istanbul, Turkey, [2]Department of Food Engineering, Faculty of Engineering and Natural Sciences, Suleyman Demirel University, Cunur, Isparta, Turkey, [3]Research Group for Food Production Engineering, National Food Institute, Technical University of Denmark, Kgs Lyngby, Denmark, [4]Center for Innovative Food (CiFOOD), Department of Food Science, Aarhus University, Agro Food Park 48, Aarhus N 8200, Denmark, [5]Department of Molecular Food Chemistry and Food Development, Institute of Food and One Health, Gottfried Wilhelm Leibniz University Hannover, Hannover, Germany

6.1 Introduction

The Food Industry 4.0 era frames the rise in digital technologies that are being used throughout the whole food production chain, from farm to fork. This trend is drawing the attention hence becoming quite popular and significant in the food industry every other day. The most frequent examples of digital transformation in the food industry are e-commerce and delivery (such as the use of mobile platforms for orders and delivery tracking), customer engagement (use of social platforms, mobile apps, etc., to collect information about consumer preferences), supply chain transparency (use of sensors, blockchain, etc., to track, manage and improve the supply chain), and food safety and quality control (such as the use of sensors, cameras, digital twins, etc., to ensure food quality and reduce the risk of foodborne illness) (van Soest, 2021).

The market for food and beverage processing equipment was valued at USD 58.3 billion in 2021, and according to a MarketsandMarkets report, it will grow to USD 76.0 billion by 2026. In the same report, the increasing significance of automation and autonomy in the food production and processing sector due to advancements in computerized food processing technologies is emphasized. It was noted that automation contributes to a reduction in manufacturing time and an increase in output as demand continues to soar along with costs. The report stressed that food processors are recognizing the capacity of data-driven insights to maximize their use of raw resources, improve the quality and safety of their products, ensure traceability, and enable continuous innovation (MarketsandMarkets, 2022a). According to a different MarketsandMarkets analysis, the global market for digital twins is predicted to increase from USD 6.9 billion in 2021 to USD 73.5 billion by 2025, with a compound yearly growth rate of 60.6% of the projected period. The report also highlights that the food and beverage industry is one of the top industries expected to adopt digital twin technology (MarketsandMarkets, 2022b). Moreover, according to a survey made with 121 decision-makers from 29 countries, 86% of respondents believe that digital transformation will have a major impact on the food manufacturing industry in the next 20 years (Schaffrath, 2018).

In accordance with the given information, there is no doubt that the digital transformation of the food sector with the Industry 4.0 era will have crucial importance in the future of food processing. The digital transformation of food manufacturing also involves the application of sophisticated modeling and simulation techniques and the development of digital twins in food processing and preservation. They have the potential to completely change the industry by raising productivity, lowering waste, and improving the quality and safety of the final product. So, in the Food Industry 4.0 era, modeling (the use of mathematical and computer-based simulations to forecast the behavior of food processing and preservation systems) and digital twins (virtual replicas of the food processes to improve production, safety, and quality by integrating data and simulating processes) have become a crucial tool in the food sector

Food Industry 4.0. DOI: https://doi.org/10.1016/B978-0-443-15516-1.00006-2

(Hassoun et al., 2023). Therefore in the rest of the section, we will take a look at the role that modeling and digitalization (focusing on digital twins) play in food processing and preservation, as well as how they work together and where they are headed in the future.

6.2 Emerging food processing

Over the past years, novel processing technologies have emerged to improve the nutritional, functional, microbial, and organoleptic properties of food products. Pulsed electric field (PEF), ultrasound (US), high hydrostatic pressure (HHP) processing, and cold plasma (CP) are nonthermal food processing techniques that have been researched in the last decades by applying alone or combined with other tools to develop food products. The use of these technologies will be discussed in this section by providing the purpose of the study, food sample, operating conditions, and outcomes (Table 6.1). The vast amount of the provided research here focused on not only the induced effects of emerging technologies for the targeted properties but also on determining the optimum processing conditions. Mathematical optimization approaches were used occasionally to determine the best operational parameters for these emerging applications. On the other hand, screening out a range of a chosen independent parameter (pressure, voltage, power, time, temperature etc.) while keeping the rest of the variables constant is the most common approach. As discussed in the following sections, modeling and digitalization have a magnificent potential for process optimization, and Table 6.1 aims to present briefly the most used parameters for novel technologies that might be used as data for digitalization as input and/or output, as well as the results of these treatments.

6.2.1 Pulsed electric field

PEF processing depends on applying high high-intensity electric field in the form of pulses for a few μs on a sample that is placed within two electrodes. PEF treatment results in electroporation by increasing the transmembrane potential of the cell membrane and thus formation of pores after reaching the critical value. The level of the electric field intensity determines whether the effect of the electroporation is temporary or permanent (Toepfl et al., 2006). In the most recent studies, PEF has been applied for increasing the shelf life of food products, protection during supercooling, and extraction of bioactive compounds, improving the bioaccessibility and bioavailability of phenolic substances, modifying the characteristics of the fermented product, and developing a sodium reduction strategy in processed meat products.

Regarding the effect of PEF application on the shelf life, fruits, fruit juices, chicken, meat, shrimp, and fish products are commonly subjected to electric fields. For instance, Younis et al. (2023) investigated the impact of PEF application by varying the intensity (10, 20, 30, and 40 kV/cm), application time (40, 80, 120, and 160 ms), and pulse numbers (50, 100, 150, and 200 pulses) on the nutritional, microbiological, and physical quality of fresh Barhi dates during the storage period. According to the outcomes, optimum conditions were determined as 10.3 kV/cm intensity, 46.73 ms treatment time, and 169.9 pulses for PEF treatment and 18.7°C temperature and 21 days time for storage conditions. At these conditions, the total viable count was detected as 0.098 log CFU/g without alteration on the physicochemical and nutritional characteristics of the fresh date fruits. In addition to this study, Li et al. (2021) evaluated the PEF treatment conditions combined with temperature as well as tea polyphenols and natamycin as natural preservatives considering the shelf life of cantaloupe juice. Application of PEF (20 kV/cm for 200 μs) combined with mild thermal treatment at 55°C for 5 minutes or use of preservatives demonstrated a comparable effect on the inhibition of *S. cerevisiae* as thermal pasteurization. While 5 log reduction was achieved by combined PEF-mild thermal treatment, 3.4 log reduction was obtained with combined application of tea polyphenols and then PEF treatment. In another research, PEF treatment with extremely low frequency (ELF) was analyzed in terms of the influence on the shelf life of tilapia (Wang, Wang, et al., 2022). ELF-PEF treatment was identified as a promising processing technique due to enhancing the shelf life by 1–2 days, decreasing the softening of texture, lowering the protein oxidation and denaturation as well as reducing the endogenous enzyme activity, compared to the control. Moreover, the hurdle effect of PEF (15 kV/cm and 800 pulses), Chamuang leaf extract impregnation (1–2%), and high-voltage cold atmosphere plasma (HVCAP) (16 kV (rms) at atmospheric pressure) on the shelf life of Pacific white shrimp was studied by Shiekh et al. (2021). Exposure to PEF pretreatment and use of 2% plant extract during vacuum impregnation as well as HVCAP application led to the lowest microbial load (5.12 log CFU/g), which is lower than the limit (6 log CFU/g) and control (8.90 log CFU/g) after storage at 4°C for 18 days. Besides, Kang et al. (2022) applied PEF and oscillating magnetic field (OMF) to beef samples with different fat amounts to prevent freezing damage during supercooling. Fat-rich beef was treated by both PEF-OMF, whereas lean-rich samples were subjected to PEF alone. Due to the fact that

TABLE 6.1 Effects of novel processing systems on the shelf life and other related topics of food products.

Processing system	Food sample	Research area	Processing conditions	Results	References
PEF combined with temperature and natural preservatives	Cantaloupe juice	Shelf life	5, 20, 25, and 30 kV/cm electric field intensity for 400 μs; 200, 400, 600, and 800 μs treatment times at 20 kV/cm	5 log reduction in *S. cerevisiae* by combined PEF (20 kV/cm for 200 μs PEF) and mild heat treatment (55°C for 5 min) 3.4 log reduction in *S. cerevisiae* by combined tea polyphenols application and then PEF (20 kV/cm for 200 μs) treatment	Li et al. (2021)
MI PEF	Chicken breast meat	Shelf life	2.50, 4.67, and 7 kV/cm electric field intensity, 0.2 ms pulse width, and 11.6 ms pulse duration	Increase in shelf life more than 2 days after PEF treatment at 4.67 and 7 kV/cm electric field intensity	Aşık-Canbaz et al. (2022)
ELF PEF	Tilapia	Shelf life	500 kV/m electric field strength, 6.5 s pulse width, and 0.1 Hz pulse frequency	Enhanced shelf life by 1–2 days with high quality during storage at 4°C for 6 days	Wang, Wang, et al. (2022)
Combined PEF, CO_2, and HHP application	Salmon	Shelf life	1 or 2 kV/cm electrical field, 20 μs pulse width, 20 Hz frequency, 100 pulses, and 15 s total treatment time CO_2 for 10 s at a 70% concentration 150 MPa for 5 min	6.08–6.91 days shelf life for control No change in the shelf life after PEF treatment alone 7–11 days shelf life of prerigor processed salmon after PEF at 1 kV/cm + CO_2 6–11 days shelf life of postrigor processed salmon after PEF at 1 kV/cm + CO_2 8 days shelf life of postrigor processed salmon after PEF at 1 kV/cm + CO_2 + HHP	Pérez-Won et al. (2021)
HVCAP and combined PEF with vacuum impregnation	Pacific white shrimp	Shelf life	697 kJ/kg specific energy, 15 kV/cm field strengths, 800 pulses for PEF 16 kV (rms) at atmospheric pressure using Ar/air (80:20) atmosphere for 10 min for HVCAP	5.12 and 8.90 log CFU/g microbial load after HVCAP and combined PEF with vacuum impregnation of 2% Chamuang leaf extract and control, respectively	Shiekh et al. (2021)

(*Continued*)

TABLE 6.1 (Continued)

Processing system	Food sample	Research area	Processing conditions	Results	References
PEF	Fresh Barhi date	Shelf life	10, 20, 30, and 40 kV/cm for 40, 80, 120, and 160 μs PEF exposure time, and 50, 100, 150, and 200 pulses.	Optimum conditions: 10.3 kV/cm intensity; 46.73 ms treatment time and 169.9 pulses, 18.7°C storage temperature and 21 days storage time; 0.098 log CFU/g total viable count No change in the physicochemical and nutritional characteristics	Younis et al. (2023)
PEF and OMF	Beef	Supercooling preservation	PEF (100 V_p/m at 20 kHz) OMF (10 mT at 10 Hz)	Inhibitory effect on ice nucleation during freezing conditions (−18°C) by combined PEF-OMF application	Kang et al. (2022)
PEF	*Acheta domesticus*	Extraction	1.5 kV/cm electric field strength, 15 μs pulse width, 20 Hz pulse frequency, 100−1000 pulses	Increase in protein and fat yields by 18% and 40% after PEF at 4.90 kJ/kg, respectively Increase in oil-binding capacity, emulsifying capacity, and antioxidant activity of the cricket flour by 28.10%, 64.88%, and 58.20% after PEF at 24.53 kJ/kg, respectively	Psarianos et al. (2022)
PEF	Olive	Extraction	2 kV/cm electric field, 30 μs pulse width, 90 Hz frequency, 5 kJ/kg specific energy	Increase in oil yield by up to 25%	Navarro et al. (2022)
PEF	Rosehip infusion	Bioavailability	1 kV/cm electric field strength, 15 kJ/kg specific energy, 135 pulse numbers	4.62% and 3.42% bioavailability of catechin in PEF- and thermal-treated samples, respectively	Ozkan et al. (2022a)
PEF	Cranberrybush puree	Bioaccessibility	20 μs pulse width, 3 kV/cm electric field strength, 2 Hz pulse frequency, 15 kJ/kg specific energy	Increase in the bioaccessibility of chlorogenic acid by 30%	Ozkan et al. (2022b)
PEF	Yogurt	Fermentation	4−20 kV voltage, 20−80 pulse number, 1−21 Hz frequency, 5−8 μs pulse width	Accelerated fermentation, enhanced stiffness, and diminished syneresis in the yogurt	Stühmeier-Niehe et al. (2023)

(Continued)

TABLE 6.1 (Continued)

Processing system	Food sample	Research area	Processing conditions	Results	References
PEF	Meat	Sodium reduction	20 Hz frequency and 10 kV voltage	Selection of PEF-treated sample by most of the panelists (84%) in terms of saltiness	Bhat et al. (2020)
Microfiltration combined with US	Skim milk	Shelf life	720 W, 5 s pulse time, 5 s intermittent time, 3, 9, and 15 min total ultrasonication time	Complete removal of bacteria in skim milk after combined microfiltration and US treatment at 1296 J/mL Enhanced shelf life up to 40 days at 4°C	Zhang et al. (2021)
HI-US	Beef	Shelf life	40 kHz, 11 W/cm^2, 30 min	Lower bacterial counts after HI-US treatment rather than that of control	Valenzuela et al. (2021)
Combined US with microwave	Sugarcane juice	Shelf life	24 kHz frequency 70% power set 20°C temperature 5 s on and 5 s off pulse for 20 min	Enhanced shelf life after combined US with microwave treatment rather than US alone or control	Zia et al. (2019)
US with pH shifting	Amaranth protein	Modification	50% amplitude for 10-min pH shifting at pH 2 and 12	1167, 438, and 2268.7 nm particle size for native, US + pH12, and US + pH2, respectively Increased protein solubility from ~70% to 97.1% by US + pH12 85.4%, 88.9%, 89.1%, and 90.2% in vitro protein digestibility for native, US, US + pH12, and pH2, respectively	Figueroa-González et al. (2022)
US	Sago starch films	Modification	600 W at 100% amplitude and 20 kHz for 2.5, 5, or 10 min	More compact structure Decreased moisture vapor permeability Increased transparency and thermal resistance	Abral et al. (2019)
US	Curcumin nanoemulsion	Drug delivery	10 min ultrasonication time, 40% ultrasonication intensity, and 50°C temperature	Higher wound healing and antiinflammatory properties	Ahmad et al. (2019)

(*Continued*)

TABLE 6.1 (Continued)

Processing system	Food sample	Research area	Processing conditions	Results	References
US combined with vacuum cold plasma	Starch nanoparticles	Drug delivery	5 or 10 min ultrasonication time, 40% ultrasonication intensity	High colloidal stability	Chang et al. (2019)
US	Rosehip nectar	Sedimentation	50% and 100% amplitude for 5, 10, 15, and 30 min	No phase separation after 15 days of storage	Atalar et al. (2020)
Combined US-Alcalase	Rice bran proteins	Hydrolysis treatment	50, 100, and 150 W power 10, 25, and 40 min treatment time	The highest degree of hydrolysis 27.35% by combined US (150 W, 40 min) with Alcalase	Fathi et al. (2021)
US	Fermented goat milk	Bioactivity	100 W, 30 kHz at 0%, 30%, 60%, and 90% amplitudes for 15 min	18.09-, 12.79-, 12.44-, 9.4-, and 1.92-fold increase in the exopolysaccharide content, a-amylase inhibition, a-glucosidase inhibition, antioxidant activity, and anticancer activity after US treatment at 60% amplitude, respectively	Hashemi and Gholamhosseinpour (2020)
HHP	Carrot–orange juice blend	Shelf life	Scanning between 200 and 400 MPa up to 5 min	More than 6-log CFU/mL reduction was obtained with 300 MPa^{-2} min and 400 MPa^{-1} min applications after HHP treatment it was below 2-log CFU/mL after 28 days of storage with a stable color	Pokhrel et al. (2022)
HHP	Oat starch	Modification and digestibility	500 MPa for 5–30 min treatment	15 min treatment was proposed as optimum for better gelation, crystalline disintegration, and in vitro digestibility	Zhang et al. (2022)
HHP	Tomato juice	Metabolic profile	A comparative study of 550 MPa 10 min at room temperature HHP treatment against high-temperature short-time thermal treatment at 110°C for 8.6 s	HHP provided significant protection for 425 metabolites and increased total aldehydes, lycopene, and carotenoids out of 14 carotenoids and 56 volatile markers rather than thermal treatment	Wang, Chen, et al. (2022)

(Continued)

TABLE 6.1 (Continued)

Processing system	Food sample	Research area	Processing conditions	Results	References
HHP with controlled storage	Oyster	Shelf life	200, 400, and 600 MPa HHP treatment for 3 min at 20°C, and a storage was conducted either at 4 or −20°C	*Vibrio vulnificus* and *Vibrio parahemolyticus* colonies were undetectable when treated with >400 MPa and stayed lower under −20°C storage. Free amino acid contents of day 0 and day 15 were almost the same for 400 and 600 MPa	Liu et al. (2022)
HHP with controlled pH	Goji berry and raspberry	Extraction	Acidic (pH 1.2) and alkaline (pH 12.5) polysaccharide extractions were executed with 500 MPa pressure for 10 min at room temperature	HHP-assisted alkaline extraction yielded the lowest pancreatic lipase activity for the extracts as well as higher rhamnogalacturonan-I regions (both for acid and alkaline extractions)	Ahmadi et al. (2022)
HHP-assisted thermal processing	Cream cheese	Processing	HHP treatment at 450 and 550 MPa for 20 min and pasteurization at 65°C for 30 min	In the presence of germinant (L-alanine), HHP-assisted pasteurization, produced cream cheese had 63% and 81% lower Maillard byproducts and furosine content fluorescence intensity compared to ultra-high sterilization (137°C for 4 s) and high-temperature sterilization (121°C for 10 min), respectively	Song et al. (2023)
HHP-assisted vacuum impregnation	Seabream fillets	Marination/ shelf life	HHP treatment as 250 MPa for 6 min and vacuum impregnation as 50 mbar for 5 min, at 10 and 4°C, respectively	Lemon aroma perception increased by 77% with HHP-assisted marination, and total viable bacteria was lower than 4-log CFU/mL while it was 6-log for conventional marination at the end of 16 days storage at below 2°C	Bou et al. (2023)

(*Continued*)

TABLE 6.1 (Continued)

Processing system	Food sample	Research area	Processing conditions	Results	References
HHP-assisted germination	Soybean	Antinutrient elimination and bioavailability	300, 450, and 600 MPa for 5 and 10 min HHP treatment	Bioavailabilities of Ca, Fe, and Zn increased by 4.0, 2.3, and 2.7 times, respectively; however, germination alone lowered the phytic acid and tannin content more than HHP-assisted processing	Wu et al. (2023)
HHP	Human milk	Shelf life	400 MPa for 5 min at 20°C	Based on the proteomics data, fat globule distributions and microstructure as well as the protein profile of HHP-treated sample were less affected and almost the same with the raw control	Zhang et al. (2023)
CP	Okra pods	Extraction	3 L/min plasma flow for 20 kHz with 750 W power consumption, for 5, 15, and 30 s. Gap between the electrodes was 6 cm	Extraction yields of water-soluble and chelator-soluble pectins increased by 60% with 15 s CP and 126% with 30 s CP treatments, respectively. However, CP treatments decreased the gel network formation of extracted pectins	Zielinska et al. (2022)
In-package CP	Beef steaks	Prevention	85 kV for 180 s. Gap between the electrodes was 4.8 cm	After 1–2 days of storage of in-package CP processed raw beef steaks, polycyclic aromatic hydrocarbon formation following the charcoal grilling inhibited by 35%–96%	Hu et al. (2022)
CP-assisted drying	Apple slices	Drying	Laminar flow (0.5 m/s) for 60 kHz with 100 W at 60°C. Gap between the electrodes was 4.8 cm	CP alone has no affect on drying. In combination with conventional drying (0.5 m/s air flow at 60°C) requiring time for complete drying reduced from 9600 s to 3600 s with a x85 fold lower energy consumption	Ranjbar Nedamani and Hashemi (2022)

(*Continued*)

TABLE 6.1 (Continued)

Processing system	Food sample	Research area	Processing conditions	Results	References
CP-assisted US	Spearmint	Essential oil extraction	CP as 15–24 kV, 50 Hz for 5 min and US as 220–240 V bath at 30°C–80°C for 1–30 min	Central composite design (CCD) and response surface methodology (RSM) were used to find optimum processing parameters. Oil extraction yield increased by 156% with determined optimum CP-US parameters (22 kV for CP with 40°C for 24 min for US)	Moradi-Sadr et al. (2023)
CP	Shrimp tropomyosin	Modification	50 kV for 5, 10, 15, and 20 min	IgE binding capacity of shrimp tropomyosin decreased by 96%, while free amino acids and surface hydrophobicity increased by 74% and 58%, respectively, with a 20-min CP treatment	Cheng et al. (2023)
CP	Pea protein isolate	Modification	100 W for 2.5, 5, 7.5, 10, 12.5, 15, and 20 min	Conductivity, light transmittance, foamability, and foam stability of pea protein isolate were increased up to 17%, 87%, 112%, and 77%, respectively	Qu et al. (2023)
CP	Zein	Controlled release/ encapsulation	40, 55, 70, 85, and 100 W for 2 min 85 W for 0.5, 1, 1.5, 2, 2.5, and 3 min	85 W for 2-min treatment decreased the particle size by 30%, providing the best wall material properties as the largest surface area and highest zeta potential. Burst release property increased from 53% to 76%	Zhou et al. (2023)

CO_2, Carbon dioxide; *CP*: cold plasma; *ELF*, extremely low frequency; *HHP*, high hydrostatic pressure; *HI*, high intensity; *HVCAP*, high-voltage cold atmosphere plasma; *MI*, moderate intensity; *OMF*, oscillating magnetic field; *US*, ultrasound; *S. cerevisiae*, *Saccharomyces cerevisiae*.

increasing the level of fat causes low electrical conductivity, combined PEF-OMF application exhibited an inhibitory effect on ice nucleation during freezing conditions (−18°C).

Apart from these studies, PEF treatment was also utilized for the extraction of bioactive compounds, for example, protein, fat, and chitin extraction from cricket (*Acheta domesticus*) flour (Psarianos et al., 2022) and olive oil extraction (Navarro et al., 2022) with high extraction yields and improved quality of the final products. In addition to these, rosehip infusions (Ozkan et al., 2022a) and cranberrybush puree (Ozkan et al., 2022b) were subjected to PEF treatment at

15 kJ/kg energy input to enhance the recovery of phenolic compounds after in vitro digestion. Results obtained from these investigations showed that the bioaccessibility of chlorogenic acid in cranberrybush puree was increased by 30% after PEF treatment rather than the control sample. Moreover, catechin in PEF-treated rosehip infusion (4.62%) had higher transepithelial transportation in comparison to the thermal-treated sample (3.42%).

PEF treatment was also applied for some other purposes that PEF treatment could alter the performance of starter cultures, including *Lactobacillus delbrueckii* subsp. *bulgaricus* and *Streptococcus thermophilus*, and resulted in accelerated fermentation, enhanced stiffness, and diminished syneresis in the final yogurt product (Stühmeier-Niehe et al., 2023). In another study, the potential of PEF application as a sodium-reducing strategy for meat products was assessed by Bhat et al. (2020). It was stated that although the PEF-applied sample (1.2%) had a lower amount of sodium rather than the control (2%), most of the panelists (84%) selected the PEF-treated sample for saltiness.

6.2.2 Ultrasound treatment

Ultrasonication, being an environmentally friendly technology, is identified by the application of sonic waves with a frequency in the range of 20–100 kHz for a few seconds to several minutes (Chemat et al., 2017; Deng et al., 2018). The formation of cavitation bubbles by the action of these waves contributes to the rupture of cell walls and membranes, release of intracellular components (Vinatoru, 2001), reduction in the size of fat globules, and providing homogeneity of liquids (Jabbar et al., 2014). Recently, there have been various investigations about the effect of sonication on the shelf life (Valenzuela et al., 2021; Zhang et al., 2021), modification of proteins (Figueroa-González et al., 2022) or starch films (Abral et al., 2019), preparation of nanoemulsion (Ahmad et al., 2019) or nanoparticles (Chang et al., 2019), and so on.

In case of increasing the shelf life, different food products such as skim milk (Zhang et al., 2021), beef (Valenzuela et al., 2021), and sugarcane juice (Zia et al., 2019) were subjected to US treatment alone or combined with other techniques at specific operating conditions. According to the results, shelf life of all these food products was found to be extended after exposure to US, depending on the initial microbial load, microbial diversity, process intensity, and storage conditions such as temperature and duration.

In the context of modifying the physicochemical, structural, and functional properties of the food constituents, amaranth protein was subjected to combined pH shifting (pH 2.0 and 12.0) and US treatment (Figueroa-González et al., 2022). It was obtained that the particle size of the native protein (1167 nm) was decreased by combined US with pH shifting to 12.0 (438 nm), while it was increased to 2268.7 nm by combined US with pH shifting to 2.0. Among other treatments, the solubility of the protein was also enhanced from ~70% to 97.1% by the application of US and pH shifting to 12.0 due to reduced particle size and increased contact area between water and protein as well as conformational changes at extreme alkaline conditions (Jiang et al., 2017). In addition to these, in vitro protein digestibility of the native protein (85.4%) improved by US treatment alone (88.9%), combined US and pH shifting to 12.0 (89.1%), and only pH shifting to 2.0 (90.2%). Results suggested that these increases are related to the ascended availability of digestive enzymes and protein by the action of US (Jin et al., 2021). On the other hand, the increase in the protein digestibility for the samples at acidic conditions may be linked to the susceptibility of more trypsin cleavage sites to proteolytic activity (He et al., 2020).

Besides, US treatment has been widely used for the development of drug delivery systems. For instance, Ahmad et al. (2019) prepared curcumin-loaded nanoemulsions by using US technique and revealed their wound healing and antiinflammatory properties in in vitro rat model. Similarly, Chang et al. (2019) could produce starch nanoparticles with high colloidal stability using combined vacuum CP with US treatment.

US treatment was also applied for some other purposes that US treatment decreased the particle size of rosehip nectar and allowed the particles to stay as a suspension without any phase separation after a 15-day storage period (Atalar et al., 2020). In another study, it was aimed to investigate the effects of US application on the bioactive properties of fermented goat milk. According to the outcomes, 18.09-, 12.79-, 12.44-, 9.4-, and 1.92-fold increases were determined in the exopolysaccharide content, α-amylase inhibition, α-glucosidase inhibition, antioxidant activity, and anticancer activity of US-treated samples at 60% amplitude in comparison to that of untreated sample, respectively (Hashemi & Gholamhosseinpour, 2020).

6.2.3 High hydrostatic pressure

HHP (or high-pressure processing) is one of the most popular and industrially scaled-up nonthermal processing methods that expose biosamples to define pressure in a range of 100–800 MPa, which is generated in a pressure pump and transferred to the sample through an aqueous media. HHP enables to protect and maintain the nutritional quality as well as sensorial attributions of the processing sample while ensuring microbial safety (Jayathunge et al., 2017; Oey et al., 2008). The processing parameters as independent variables are pressure, holding time, and temperature, and HHP technology can either be used alone for food processing or as a combination with other treatments, which is known as the "hurdle strategy" (Wu et al., 2022).

In a study, carrot–orange juice blends (pH 4.0, 5.0, 6.0) are processed with HHP treatment at 200, 300, and 400 MPa for 1–5 minutes at 20°C. The juice blends were initially inoculated with *Listeria innocua* to surrogate *Listeria monocytogenes* with a 7-log approximate load. Following the HHP treatment, a reduction of 6-log CFU/mL was obtained with 300 MPa for 2 minutes and 400 MPa for 1 minute applications. No significant decrease was observed following the HHP treatment of juice blends for total ascorbic acid, carotenoid, and phenolic contents compared with an untreated control sample. The count of total mesophilic aerobic bacteria was still below 2-log CFU/mL after 28 days of storage and color attributes as L^*, a, and b values were not changed. Milder pressure applications were identified as eligible to inactive the microbial load successfully (Pokhrel et al., 2022).

Aroma compounds in plant-based food products have been investigated to understand their degradation and/or formation mechanisms during processing or storage periods. Tomato juice was processed with HHP at 550 MPa for 10 minutes at room temperature. In this study, metabolic profiles of volatile and nonvolatile compounds in tomato juice were investigated for different processing conditions: HHP against high-temperature short-time thermal treatment at 110°C for 8.6 seconds. Following in-depth metabolic profiling for 425 metabolites, 14 carotenoids, and 56 volatile compounds, it was pointed out that HHP provided significant protection for defined constitutes of tomato juice particularly for ascorbic acid, quercetin, amino acid derivatives, and alcohols compared to the heat-treated sample (an untreated sample was the reference point). On the other hand, it was observed that some aldehydes, lycopene, and total carotenoid contents increased when treated with HHP. All outcomes verified that HHP is a better way of processing foods mainly for sensorial and chemical attributes (Wang, Wang, et al., 2022). There is an enormous gap in the literature for the processing effect on the volatile/nonvolatile constituents of plant-based foods and their formation/degradation mechanisms. Digitalization and biosensor usages might provide some promising solutions for this challenging zone, for a better understanding of the behavior of compounds during processing and might be helpful to determine tailored optimal processing conditions for varying food materials.

HHP treatment was used not only to ensure microbial safety for a muscle-based food (seabream fillet) but also for forced impregnation of the marination compound toward the tissue. The seabream fillets were exposed to subsequent vacuum impregnation as 50 mbar for 5 minutes at 4°C then treated with HHP as 250 MPa for 6 minutes at 10°C. The marinating solution consisted of 14% lemon concentrate, 0.6% white vinegar, and 0.08% lemon peel aroma, and the rest was water. Samples were stored below 2°C for 1, 5, 9, 12, and 16 days. When compared with the conventional marination, lemon aroma perception in the HHP-assisted marinated sample was higher by 77%; furthermore, total viable bacteria were lower than 4-log CFU/mL for HHP-assisted marinated sample, while it was 6-log CFU/mL for conventional marination at the end of 16 days (Bou et al., 2023). Microbial safety, aka microbial load decrease, and the alteration of chemical constituents within the biomass during the processing might be monitored, optimized, and digitalized however, mass transfer efficiency is another crucial data for food processing as explained in the previous study.

6.2.4 Cold plasma

A gaseous material is converted into ionized plasma matter using an adequate level of energy (dielectric barrier discharge, plasma jet, corona discharge, radio frequency, or microwave plasma systems), composed of electrons, neutral molecules and atoms, and charged reactive species. Application of the created plasma to the food samples mainly enables microbial inactivation, degradation of microbial spores and toxins, protection of nutritional and organoleptic attributes of foods, improved material extraction, and avoidance of biofilm formation under nonthermal conditions (Nwabor et al., 2022). Scaling up the potential and industrial applicability of CP processing is not common; nevertheless, it is one of the most promising emerging technologies and is being investigated/applied for distinctive purposes in food science.

In a simulation study, apple slices (6 cm diameter, 0.3 cm thickness) were numerically modeled using COMSOL Multiphysics, for their drying behaviors under conventional air drying (0.5 m/s air flow at 60°C), CP drying (laminar flow (0.5 m/s) for 60 kHz with 100 W at 60°C. Gap between the electrodes was 4.8 cm) and synergetic affect of their concurrent treatment. For that purpose, drying kinetics, shrinkage model, energy consumption, and related governing equations (continuity and momentum equations, water evaporation) were defined numerically. Experimental and simulated data were in good correlation ($R^2 = 0.95$). It was observed that CP alone has no affect on drying while in combination with conventional drying, requiring time for complete drying reduced from 9600 to 3600 seconds with an 85-fold lower energy consumption. The diameter of the dried sample was so close to the fresh apple slice with a very low bulk shrinkage of 0.008 (Ranjbar Nedamani & Hashemi, 2022). This study can be nominated as a successful example for modeling of food processing.

CP was used concurrently with the bath US process in another study to extract the essential oils from spearmint efficiently, in comparison with US treatment alone. Varying parameters for both CP, US as well as different water/material ratios, were employed to conduct the most efficient extraction from the biomass. Instead of pairing all the variables and analyzing their extraction efficiencies together with final product quality, some traditional and common statistical optimization methods, central composite design (CCD), and response surface methodology (RSM) were used to find the optimum coupling variables. Eventually, it was determined that oil extraction yield increased by 156% when treated concurrently with CP as 22 kV and US as 40°C for 24 minutes (Moradi-Sadr et al., 2023). Using CCD and RSM enabled the elimination of every possible/potential parameter for CP and US treatments.

Zein was used as the wall material for the encapsulation of curcumin, following the exposure to CP aiming the improved encapsulation efficiency, stability, and better controlled release performance. CP parameters were employed as two different series: power series (40, 55, 70, 85, and 100 W for 2 minutes) and time series (85 W for 0.5, 1, 1.5, 2, 2.5, and 3 minutes). Both series indicated the same parameter coupling, 85 W for 2 minutes, as the most efficient CP treatment for decreasing the particle size by 30%, providing the best wall material properties with the largest surface area and highest zeta potential. The burst release property of the CP-treated capsules increased from 53% to 76% (Zhou et al., 2023). Modified wall materials and their controlled release performances (mass transfer ratios) might be potentially monitored and simulated for a better understanding of the relation of applied treatment and its functional consequences.

6.3 Modeling in Food Industry 4.0

Mathematical modeling involves creating a mathematical representation of a real-world phenomenon to provide insight or generate predictions. In the context of food production, modeling refers to using mathematical expressions to understand and analyze physical and biochemical changes in food. This involves examining the impact of various input variables, which can range from process conditions to raw material properties and product recipes, and so on. In food processes, mathematical modeling is used for many different purposes (including enzyme kinetics, predictive microbiology, chemometrics, process control, simulation of food processes, optimization, etc.), and it can be classified under three main categories: (1) data-driven modeling, (2) physics-based modeling, and (3) hybrid applications (An et al., 2015; Erdoğdu, 2013; Liao & Köttig, 2014).

6.3.1 Data-driven modeling

Data-driven techniques determine the features of foods or food production processes using information from previously gathered data (training data), then use the trained model to forecast future variation. Neural networks, support vector machines, decision trees, partial least squares regression, and principal component analysis are a few examples of data-driven models used in food processing and safety. A mathematical model, on the other hand, that uses physical rules and principles to describe a system's or phenomenon's behavior is known as a physics-based model. These models rely on fundamental scientific principles like mechanics, thermodynamics, fluid dynamics, and reaction kinetics to forecast system behavior under various conditions. Models based on physics such as those for heat transmission, mass transfer, fluid flow, and structural mechanics are mostly utilized in the food process modeling. The main differences between these two different modeling approaches originate from the underlying principles and application methods. The data-driven models employ statistical analysis and machine learning approaches to find patterns and links in massive datasets. Then, the generated patterns and links are used to make future predictions.

These models require a large amount of data to train the model, and generally, there is no or limited knowledge about the underlying principles and mechanism of the system. Therefore when the system to be modeled is too complex or there is no need to relate the causes and effects of the system behaviors, data-driven models may be preferred.

6.3.1.1 Physics-based modeling

However, to achieve a physics-based model, a good understanding of the physics, chemistry, and biology of the system is required. A limited amount of data to validate and refine the developed model is sufficient most of the time. So, it can be summarized that the physics-based models are helpful when there is a good understanding of the physical processes involved and when accurate predictions with stronger cause—effect relationships of the system characteristics are required. Hybrid methods combine the given techniques, and they are used to enhance the prediction accuracy of the developed models. By combining these models with advances in sensor technology and data analytics, the food industry can continue to innovate and improve the safety, quality, and sustainability of food production.

Both main modeling categories given are extensively used in the food processes. Especially with the improvements in advanced technologies such as the Internet of Things (IoT), cloud computing, big data analytics, smart sensors, and artificial intelligence (AI) in the Industry 4.0 era, the collection and processing of data have become more economical and straightforward. Therefore modeling has become more widely used in the different fields food industry for different purposes, such as process control and optimization, quality assurance, safety evaluation, process design, and innovation. One of the areas where modeling is extensively used is the control and optimization of food processes. This is done to maximize process efficiency, product quality, and profits while minimizing waste, energy consumption, economic and quality losses, and so on. For example, the optimum parameters of HHP and sous-vide cooking were determined considering the physicochemical properties (minimization of hardness and cooking loss) of veal patties (Janardhanan et al., 2022). In another study, HHP was used for waste valorization purposes. In the study conducted by Agcam et al. (2021), black carrot pomace (waste of black carrot juicing process) was evaluated for the extraction of bioactive compounds and the HHP conditions were optimized to maximize the amount of extracted anthocyanins, phenolics, and antioxidant activity. PEF conditions were also optimized to maximize the extraction yield from orange peel (Athanasiadis et al., 2022) and biosurfactants from Chubak root (Khosrow Shahi et al., 2021). PEF was also used and its parameters (including pH, pulse width, and voltage) were optimized to reduce aflatoxin levels in foods (Vijayalakshmi et al., 2018). One of the other novel food processing methods, namely CP, is used as a pretreatment for drying corn kernels (Li et al., 2019). The effects of pretreatment time, power, and drying temperature on drying time and rate were investigated. The optimization targeted the minimization of these drying characteristics, which also helped to improve efficiency and potentially reduce energy consumption. Although it is a very traditional method, RSM, a parametric and experimental optimization method, has been used in all the studies mentioned earlier, indicating that it is still a popular method of optimization in the food sector. Moreover, the modeling of variables was performed using multivariate polynomial regression in the studies given earlier.

Numerical optimization methods, such as genetic algorithm (GA), one of the best known of the heuristic optimization methods, are also extensively used for the optimization of food systems. With the use of GA, the optimal parameters are found gradually by iteratively improving a population of candidate solutions through selection, mutation, and crossover. GA uses a process like natural selection to solve problems. It starts with a group of possible solutions and improves them over time by picking the best ones and creating new ones based on them (Deb et al., 2002). GA was recently used to find the optimal concentration of nanoparticles to achieve the best performance of nanocomposite films used for food packaging. The study focused on polyvinyl alcohol films that were prepared with different combinations of Montmorillonite and titanium oxide nanoparticles, and the support vector machine model was used to predict their performance (Zamanian et al., 2023). Ohale et al. (2022) aimed to optimize the extraction process of oil from mango kernels and characterize the physicochemical and instrumental properties of the extracted oil. RSM coupled with GA was used to find the optimal temperature, dosage, and extraction time for maximum oil yield. The pseudo-second-order model was found to be the most suitable for describing the extraction process. Another study that employed multiobjective GA, focused on the optimization of the nutritional quality of fried Catla fish using mustard oil. The researchers experimented with temperature, time, and oil amount to find the best combination. An artificial neural network was developed to find the nonlinear correlation between frying conditions and nutritional quality properties. Then various optimization methods, such as GA and metaheuristic optimization methodologies, were used to simultaneously optimize the indices for health benefits, reduce frying time, and minimize the wastage of culinary media (Sadhu et al., 2022). Moreover, Bozzato et al. (2021) examined the use of air impingement as an alternative thawing method for the service of frozen food products. In the study, frozen chicken fingers were thawed in an air impingement prototype with different combinations of process parameters identified by GA optimization. When compared to refrigerator thawing, air impingement reduced the thawing time by up to 85% while still preserving the food structure.

Apart from the GA, there are various other multiobjective optimization algorithms that do not require complicated mathematical equations available such as Pareto and scalarization methods like multiobjective evolutionary algorithms, particle swarm optimization, and ant colony optimization (Gunantara, 2018). Various optimization algorithms, including the given ones, can be implemented effortlessly using libraries available in different programming languages. For instance, in R, one can use rmoo (Benitez & Roa, 2022) and nsga2R (Tsou & Lee, 2022), while in

Python, libraries like PulP (Mitchell et al., 2009), pymoo (Blank & Deb, 2020), pyOtp (Perez & Jansen, 2008), and DEAP (Fortin et al., 2012) can be utilized. MATLAB's Global Optimization Toolbox is another useful option for multiobjective optimization.

One of the growing fields regarding the use of mathematical models is quality assurance and safety evaluation. Especially, the advancing sensor technology, the accessibility of these technologies, and the increase in the processability of the obtained data provide a basis for the control of food quality using mathematical modeling such that more than 89% of big data have been produced within the last 2 years (Michael, 2021). As the amount of data generated from food processing and preservation operations continues to increase, mathematical modeling is being used to generate more innovative solutions. On the other hand, when it comes to food safety, applications such as predictive microbiology and shelf life determination are areas where mathematical modeling is frequently used. For example, a recent study examined the relationship between egg shape and egg quality traits such as internal and eggshell quality using digital images. It hypothesizes that the shape index can be used to classify eggs and improve the grading process. Using machine learning models (k-nearest neighbors, decision trees, random forest, and neural network), the study found that rounder eggs had higher quality than elongated eggs, and the random forest had the best accuracy of 97.9% (de Oliveira-Boreli et al., 2023). In another study, nondestructive spectral sensors (NIR and FT-NIR) were used with data-driven modeling methods to estimate sensory quality attributes of black tea samples, including appearance, body, color, overall quality, bulk density, cellulose, water extract, and moisture. Partial least squares regression (PLSR) and principal components regression were used to develop the mathematical models suggesting that PLSR combined with FT-NIR technology may be useful for the rapid and economical evaluation of sensory scores and related properties in the tea industry (Turgut et al., 2022). Kowalik et al. (2018) investigated the behavior of *Escherichia coli* in mascarpone cheese during storage at different temperatures and developed predictive models using the Baranyi, Gompertz, and logistic approaches. The Gompertz model provided the most accurate fit and was used to generate growth rates for secondary modeling using polynomial regression to analyze the effect of temperature.

6.3.1.2 Hybrid applications

The hybrid modeling approach is also used for the prediction of the quality property change during food processing. Rabeler and Feyissa (2018) combined physics-based modeling with kinetic models to predict moisture and temperature changes in chicken breast meat, as well as the resulting textural change, during roasting. Moreover, the cooking loss and dimensional change of fish flesh during heating were successfully defined using a semiempirical modeling understanding (Blikra et al., 2020). Within a study investigating the microbial inactivation probability of the HHP treatment, a probabilistic model was developed based on logistic regression to determine the probabilities of achieving target log reductions of microorganisms. The study pointed out that the models developed could predict the target log reduction of *Salmonella*, *Typhimurium*, and *E. coli* in different food matrices at various pressure, temperature, and time levels (Buzrul, 2019). Therefore it can be stated that probabilistic modeling can be useful to forecast whether a food product is safe to consume after a particular treatment is applied. Another study targeting food safety during the ohmic heating of chicken meat assessed the effect of the product and ohmic heating properties on cold spot generation using the developed physics-based model. The study highlights the importance of special care during the ohmic heating of meat samples to avoid underprocessing on sample surfaces (Turgut et al., 2022).

Mathematical modeling is commonly used in food processing and safety to design processes and equipment, as well. The most widely used methods for designing equipment and processes involve physics-based modeling, which includes mechanistic modeling and the use of process simulators. While there are various process simulation software options available for the conceptual design of food systems, such as CHEMCAD, AspenHysys, AspenPlus, DWSIM, SuperPro Designer, and ProSimPlus, numerical simulation in two- or three-dimensional spaces is more commonly used for simulation of specific equipment within the processes. For example, Karaduman et al. (2012) explored the use of toroid cans as an innovative approach to increase heat transfer rates and decrease processing times in the canning process. The designed toroid can (filled with water, pineapple slices, and tomato paste for different experiments) was modeled and analyzed in-depth comparing with thermal processing experiments with simulation results.

In another study, researchers explored the potential of PEF technology as a nonthermal treatment method for liquid food products. The study focused on identifying the key parameter for PEF treatment, which is the uniform electric field in the treatment zone, and evaluating the impact of treatment chamber design on achieving homogenous treatment. To this end, the researchers used COMSOL Multiphysics to investigate the homogeneity of electric field strength in different treatment chambers and proposed a coaxial treatment chamber design for continuous liquid sample treatment (Arshad et al., 2020). Turgut et al. (2021) introduced a novel drying apparatus, called ohmic-assisted drying (OAD), to combine ohmic heating with convectional air drying to improve the drying characteristics of potato slices. The OAD system has been found to significantly reduce drying time by 20–60% compared to conventional air drying systems, and a 3D mechanistic model has been developed to better understand the mechanisms of OAD and predict the spatial distribution of temperature, moisture, and pressure in the drying material. Another study focuses on the food waste valorization in Jeju Island, Republic of Korea, and proposes a plasma reactor design as a solution to the problem of rotten citrus fruits. The reactor uses microwave plasma and liquefied petroleum gas as a heat source for drying and composting. A numerical study using computational fluid dynamics in ANSYS-fluent found that the temperature of the reactor decreased with an increase in the airflow rate of the centrifugal fan (Ko et al., 2021).

Overall, modeling is an important tool in the Food Industry 4.0 era because of the benefits it has for the food industry. They include improved efficiency, reduced waste, and enhanced product quality and safety. By using models to predict the behavior of food processing and preservation systems, food manufacturers can optimize their operations to reduce energy consumption, minimize waste, and improve product quality. Additionally, by using predictive modeling to identify potential food safety risks, manufacturers can implement preventive measures to ensure that their products are safe for consumption. Therefore the design and analysis of the food processes and equipment using mathematical modeling are very convenient ways due to reduced time and money requirements.

6.4 Digitalization in Food Industry 4.0

Digitalization is a critical component and an emerging topic of the Food Industry 4.0 era. Digitalization refers to the process of using digital technologies to transform physical data and processes into digital formats that can be easily accessed, analyzed, and manipulated. In many fields, including food manufacturing, digitalization enables the collection, processing, and analysis of large amounts of data to improve production operations. As a result, digitalization has enabled the creation of more sophisticated and accurate mathematical models that are more detailed and better able to capture the complexities of real-world systems.

Mathematical modeling is closely related to digitalization because it provides a way to analyze and interpret the vast amounts of data that are generated through digitalization. It was explained in the previous section that mathematical models, no matter the modeling type used, can be defined as the representation of a real-life phenomenon that is built using mathematical equations and algorithms. However, mathematical models and simulations are offline efforts to predict the system behaviors for the "future" (to be implemented after the simulation) and identify the "potential" problems (Verboven et al., 2020). Digitalization, on the other hand, has made mathematical models easier to apply to real-world problems by combining models with sensor data and other sources of information to solve problems "on time." In that way, engineers and scientists can create digital twins that provide real-time monitoring and analysis of physical systems.

6.4.1 Digital twins

Digital twins are virtual replicas of physical objects, processes, or systems that allow for "real-time" monitoring, analysis, and optimization (Singh et al., 2022; Söderberg et al., 2017). So it means that the digital twin incorporates the computer simulation into a system throughout its lifecycle by connecting it to the real world with the use of sensors (Verboven et al., 2020). To do it, five dimensions of a digital twin were given by Qi et al. (2021) as follows: (1) physical entities, (2) virtual models, (3) digital data, (4) services, and (5) connections. A physical entity refers to a real-world object or system that can be described and modeled using physical parameters and laws, such as devices, products, physical systems, activities, processes, or organizations. A virtual model is a digital representation of a physical entity that accurately reproduces the entity's geometry, properties, and behaviors. It typically consists of a 3D geometric model that describes the entity's shape and structure, a physics model that reflects the physical phenomena of the entity, a behavior model that describes the entity's responses to changes in the environment, and a rule model that enables logical abilities such as reasoning, judgment, evaluation, and autonomous decision-making (Qi et al., 2021). Both data-driven modeling approaches, especially deep learning methods, and physics-based modeling approaches can be employed to build the virtual model of a digital twin. However, compared to data-driven methods, mechanistic virtual models can predict process or product variables and impacts that are not directly measurable and incorporate them into data analysis. It can also function as a virtual sensor and a digital replica of the actual process, provided the model produces outputs timely and accounts for variability (Gwanpua et al., 2013; Hertog et al., 2007). Regarding digital twin data, it is a crucial aspect of the concept, and it deals with multitemporal scale, multidimension, multisource, and heterogeneous data. The data can be obtained from physical entities, virtual models, services, knowledge, and fusion data that results from combining all these types of data (Qi et al., 2018). Digital twins provide application services to users and require third-party and platform services for functioning. Lastly, the virtual models are linked with their corresponding physical entities dynamically, allowing for advanced simulation, operation, and analysis. Connections between physical entities, virtual models, services, and data facilitate the exchange of information and data (Qi et al., 2021).

6.4.2 Current usage of digital twins in the food industry

In the food industry, digital twins can be used to create virtual replicas of food processing and preservation systems, or food material enabling manufacturers to optimize their operations and improve product quality and safety. More specifically, Udugama et al. (2023) classified the digital food twins as (1) process historian digital twins, (2) digital twins that monitor the real-time behavior of entities, and (3) autonomous digital twins (using AI). There are, however, few ideas for digital twins that are particularly implemented in the automated food industry, but the available applications are rather done for future prediction. For example, in their follow-up studies, Schudel et al. (2022) and Shrivastava et al. (2023) developed physics-based digital twins to find the optimal packaging for ventilated strawberries. In the first study, they evaluated the performance of different packaging types (top-sealed paperboard, open and closed plastic clamshell) for strawberries during the supply chain. The study used laboratory experiments and physics-based simulations to measure fruit quality attributes, such as mass loss, total soluble solids, acidity content, firmness, color change, and incidence of decay. The results showed that the top-sealed paperboard packaging had the best performance in terms of respiration-dependent quality, mass loss, and time of wetness (Schudel et al., 2022). The latter study employed the digital twins to mimic how the strawberries evolved during 21 different types of packaging, and actionable metrics of fruit quality and shelf life were predicted. The study pinpointed the critical quality loss points in the supply chain for every package and identified the package that performs best in balancing the trade-off between the respiration-driven biochemical shelf life, transpiration-driven physical shelf life, and mold growth-driven microbial shelf life of fruit (Shrivastava et al., 2023). In another study, a physics-based digital twin was used to compare and quantify the impact of different measures (such as time length of shipment, shelf temperatures, relative humidity, and evaporative cooling) in preserving the quality of fresh cucumber, eggplant, strawberry, and raspberry in the cold chain. The study monitored 95 shipments in the cold chain for the fruits that were imported from Spain to Switzerland, and simulations were run on the digital twins to evaluate different optimization scenarios (Shoji et al., 2022). A similar study describing the development of a physics-based digital twin of citrus fruit is conducted to identify when, why, and to what extent fruit quality and marketability are reduced in the supply chain. Sensor data from 47 commercial shipments were used to create actionable metrics for stakeholders for real-time monitoring of food quality and marketability. Results showed a large spread in the shipments for different quality and marketability metrics, with half of the shipments lying outside the ideal range (Shrivastava et al., 2022).

In addition to examining food quality and safety throughout the food chain, digital twins are also used in the design and optimization process of food production equipment. A recent study investigated the effects of the spatial configuration of chicken meat on the heating pattern in a custom radio frequency-assisted heating oven. Both experiments and simulations revealed that the vertically oriented samples exhibited a larger heating efficiency than the horizontally oriented ones, regardless of the power. However, temperature uniformity in vertical configurations was poorer than in horizontal ones (Goñi et al., 2022). Lysova et al. (2022) simulated the behavior of a counter-current tube-in-tube heat exchanger for liquid foods using computational fluid dynamics. The study found that the impact of product velocity on pressure drop was positive, while higher inlet product and water temperatures resulted in lower pressure drop values. A response surface was generated and validated, which could be used to develop a digital twin of the device, leading to a deeper knowledge and finer control of the system. Another study explored the effects of the electric field on microwave heating in a microwave-assisted thermal sterilization system used for ready-to-eat meals. A digital twin simulation model was developed using the finite-difference time-domain method to be used to improve the heating uniformity and throughput of ready-to-eat meals in industrial systems. It was shown that the electric field intensity was greatest on the surface and side edge of the microwave cavity and of the food (Hong et al., 2021).

Scheduling the production of food ingredients and products is another potential application for digital twins in food processing. A recent study discussed the use of digital twins in the food processing industry for product safety, quality, cost reduction, and timely delivery. The authors proposed a recipe-based approach to create realistic and accurate digital models of production processes using SuperPro Designer v11 and SchedulePro v9.1 for model development and planning/scheduling, respectively. An industrial-scale beer production facility was considered as a case study to demonstrate the process of developing and using a model in the context of simulation and scheduling. The authors emphasized that the proposed methods and generated solutions are not necessarily optimal but could have a significant impact on improving the design and operation of food manufacturing facilities (Koulouris et al., 2021).

According to the literature research made among the up-to-date studies, it is seen that the current digital food twin approaches are mostly focused on monitoring, prediction, design, and optimization, but very few are focused on autonomous control. Moreover, implementing digital food twins requires "real-time" process assessment/control with multidisciplinary knowledge, particularly in modeling the biophysical processes that affect food properties due to the variability of raw materials. However, this aspect is not sufficiently integrated into the literature (Henrichs et al., 2022; Krupitzer et al., 2022). A potential obstacle to the implementation of digital twins in real time might be the long computational time required especially for physics-based models. Although physics-based digital twins generate more in-depth and explainable results about the processes they represent, the complex nature of these models sometimes requires long processing times (hours or even days) for analysis. This limitation restricts the simultaneous use of these digital twins in continuous production processes. To overcome this drawback, reduced-order models (ROMs) have become increasingly popular in recent years. The main idea behind ROMs is to reduce the complexity of physics-based models while retaining the model's fidelity including nonlinear behaviors. In that way, ROMs aim to offer precise and quantitative depictions of system dynamics while keeping the computational expense lower than that of the original numerical model. Moreover, ROMs provide an approach for interpreting system dynamics in a more accessible manner (Lucia et al., 2004). As an example, Kannapinn et al. (2022) proposed a physics-based, data-driven digital twin framework for autonomous food processing, which could gather real-time information on the product's current state to make thermal processing controllable. The study focused on a parsimonious experimental design for training nonintrusive ROMs of a thermal process, enabling efficient selection of training data. The proposed digital twin framework was designed to be applicable within the industry and allows on-device model predictive control. Another study about the use of ROMs in food processing and safety discussed the ROM techniques to address large-scale optimization problems in flow control involving time-dependent Navier–Stokes equations. In their scenario, the authors consider the sterilization of packaged foods in an autoclave to eliminate existing microorganisms. Trust-region proper orthogonal decomposition method was proposed to save computational work during the control problem solution while still providing accurate solutions (Fahl, 2001). Similarly, Xiao et al. (2015) presented two new nonintrusive ROMs for solving the Navier–Stokes equations. The models have been demonstrated to provide accurate solutions for modeling flow problems while colony-forming unit (CPU) times are reduced by several orders of magnitude in comparison to high-fidelity models. Different approaches for reducing the computational physics model order are extensively discussed in the literature (Lucia et al., 2004; Noack et al., 2011).

6.5 Challenges and future directions

While digitalization and modeling have a lot of potential for the food industry, there are also big obstacles to overcome before they can reach their full potential. One major challenge is the need to integrate data from multiple sources and systems into a unified platform for analysis and decision-making. This is a problem that might be solved with commercial services with a single platform providing full control of the digital twin and all the elements of the digital system together. Such systems are already available for other industries like construction and machinery. However, there are not enough advanced alternatives for food production, and significant expenditures in information technology infrastructure and data analytics tools are needed for this. Providing the technological infrastructure alone is not enough, as a workforce that can manage this technology to fully adapt the digitalization in the food system is also required. Rahimifard et al. (2022) have pointed out that there is a deficiency in technical proficiency and ambiguity regarding the methods of luring younger individuals with digital skills to the industry, as well as how to teach new skills to and improve the current workforce. Furthermore, considerable adjustments to current organizational structures and cultures are necessary to support Industry 4.0 technologies such as digital twins. Companies, especially small and medium-sized enterprises, must be willing to invest in new technologies and undergo significant training to ensure that employees have the necessary skills to operate and maintain them. The food industry must also address concerns about data security and privacy, particularly with regard to delicate data like product formulas and customer information.

Looking to the future, there are several promising areas for further research and development in the food industry. For instance, using blockchain technology to manage supply chains and track products has significant promise for enhancing food safety and quality assurance. There are already studies in the literature in this direction (Mondal et al., 2019; Tian, 2017). On the other hand, the impact of machine learning algorithms on big data analytics in the food industry is expected to keep growing and increase in the future, leading to improvements in product quality and reduction in waste. The development of more sophisticated digital twins that include capabilities for machine learning and predictive analytics is another potential field. As a result, manufacturers would be able to mimic and enhance their procedures in real time (which is limited for now), resulting in even greater gains in productivity, safety, and product quality.

The adoption of Industry 4.0 technologies, such as digitalization and modeling, has the potential to revolutionize the food industry, increasing productivity, lowering waste, and improving product quality and safety, even though there are many obstacles to overcome. It will be crucial for food manufacturers to stay flexible and adaptable as the sector develops and new technologies are introduced to take advantage of the opportunities provided by Industry 4.0.

References

Abral, H., Basri, A., Muhammad, F., Fernando, Y., Hafizulhaq, F., Mahardika, M., Sugiarti, E., Sapuan, S. M., Ilyas, R. A., & Stephane, I. (2019). A simple method for improving the properties of the sago starch films prepared by using ultrasonication treatment. *Food Hydrocolloids*, *93*, 276–283. Available from https://doi.org/10.1016/j.foodhyd.2019.02.012.

Agcam, E., Akyıldız, A., Kamat, S., & Balasubramaniam, V. M. (2021). Bioactive compounds extraction from the black carrot pomace with assistance of high pressure processing: An optimization study. *Waste Biomass Valor*, *12*, 5959–5977. Available from https://doi.org/10.1007/s12649-021-01431-z.

Ahmad, N., Ahmad, R., Al-Qudaihi, A., Alaseel, S. E., Fita, I. Z., Khalid, M. S., & Pottoo, F. H. (2019). Preparation of a novel curcumin nanoemulsion by ultrasonication and its comparative effects in wound healing and the treatment of inflammation. *RSC Advances*, *9*(35), 20192–20206. Available from https://doi.org/10.1039/c9ra03102b.

Ahmadi, S., Yu, C., Zaeim, D., Wu, D., Hu, X., Ye, X., & Chen, S. (2022). Increasing RG-I content and lipase inhibitory activity of pectic polysaccharides extracted from goji berry and raspberry by high-pressure processing. *Food Hydrocolloids*, *126*, 107477. Available from https://doi.org/10.1016/j.foodhyd.2021.107477.

An, D., Kim, N. H., & Choi, J. H. (2015). Practical options for selecting data-driven or physics-based prognostics algorithms with reviews. *Reliability Engineering & System Safety*, *133*, 223–236. Available from https://doi.org/10.1016/j.ress.2014.09.014.

Arshad, R. N., Malek, Z. A., Ahmad, M. H., Buntat, Z., Kumara, C. L. G. P., & Abdulameer, A. Z. (2020). Coaxial treatment chamber for liquid food treatment through pulsed electric field. *Indonesian Journal of Electrical Engineering and Computer Science*, *19*, 1169–1176. Available from https://doi.org/10.11591/ijeecs.v19.i3.pp1169-1176.

Aşık-Canbaz, E., Çömlekçi, S., & Can Seydim, A. (2022). Effect of moderate intensity pulsed electric field on shelf-life of chicken breast meat. *British Poultry Science*, *63*(5), 641–649. Available from https://doi.org/10.1080/00071668.2022.2051431.

Atalar, I., Saricaoglu, F. T., Odabas, H. I., Yilmaz, V. A., & Gul, O. (2020). Effect of ultrasonication treatment on structural, physicochemical and bioactive properties of pasteurized rosehip (Rosa canina L.) nectar. *Lwt*, *118*, 108850. Available from https://doi.org/10.1016/j.lwt.2019.108850.

Athanasiadis, V., Chatzimitakos, T., Kotsou, K., Palaiogiannis, D., Bozinou, E., & Lalas, S. I. (2022). Optimization of the extraction parameters for the isolation of bioactive compounds from orange peel waste. *Sustainability*, *14*(21), 13926. Available from https://doi.org/10.3390/su142113926.

Benitez, F., & Roa, D. P. (2022). *rmoo: Multi-objective optimization in R*. https://github.com/Evolutionary-Optimization-Laboratory/rmoo/

Bhat, Z. F., Morton, J. D., Mason, S. L., & Bekhit, A. E. D. A. (2020). The application of pulsed electric field as a sodium reducing strategy for meat products. *Food Chemistry*, *306*, 125622. Available from https://doi.org/10.1016/j.foodchem.2019.125622.

Blank, J., & Deb, K. (2020). Pymoo: Multi-objective optimization in python. *IEEE Access*, *8*, 89497–89509. Available from https://doi.org/10.1109/ACCESS.2020.2990567.

Blikra, M. J., Hodnefjell, Å. V., Feyissa, A. H., & Skipnes, D. (2020). Dimensional change and cook loss during heating of fish: Problem formulation and semi-empirical modeling approach. *Journal of Food Engineering*, *281*, 110004. Available from https://doi.org/10.1016/j.jfoodeng.2020.110004.

Bou, R., Guerrero, L., López, M., Claret, A., López-Mas, L., & Castellari, M. (2023). Effect of vacuum impregnation and high hydrostatic pressure treatments on shelf life, physicochemical, and sensory properties of seabream fillets. *Food and Bioprocess Technology*, *16*(5), 1089–1100. Available from https://doi.org/10.1007/s11947-022-02980-4.

Bozzato, A., Pippia, E., Tiberi, E., & Manzocco, L. (2021). Air impingement to reduce thawing time of chicken fingers for food service. *Journal of Food Processing and Preservation*, *45*, e15962. Available from https://doi.org/10.1111/jfpp.15962.

Buzrul, S. (2019). High hydrostatic pressure inactivation of microorganisms: A probabilistic model for target log-reductions. *International Journal of Food Microbiology*, *309*, 108330. Available from https://doi.org/10.1016/j.ijfoodmicro.2019.108330.

Chang, R., Ji, N., Li, M., Qiu, L., Sun, C., Bian, X., Qiu, H., Xiong, L., & Sun, Q. (2019). Green preparation and characterization of starch nanoparticles using a vacuum cold plasma process combined with ultrasonication treatment. *Ultrasonics Sonochemistry*, *58*(June). Available from https://doi.org/10.1016/j.ultsonch.2019.104660.

Chemat, F., Rombaut, N., Meullemiestre, A., Turk, M., Perino, S., Fabiano-Tixier, A. S., & Abert-Vian, M. (2017). Review of green food processing techniques. Preservation, transformation, and extraction. *Innovative Food Science and Emerging Technologies*, *41*(May), 357–377. Available from https://doi.org/10.1016/j.ifset.2017.04.016.

Cheng, J.-H., Li, J., & Sun, D.-W. (2023). Effects of dielectric barrier discharge cold plasma on structure, surface hydrophobicity and allergenic properties of shrimp tropomyosin. *Food Chemistry*, *409*, 135316. Available from https://doi.org/10.1016/j.foodchem.2022.135316.

Deb, K., Pratap, A., Agarwal, S., & Meyarivan, T. (2002). A fast and elitist multiobjective genetic algorithm: NSGA-II. *IEEE Transactions on Evolutionary Computation*, *6*, 182–197. Available from https://doi.org/10.1109/4235.996017.

Deng, Y., Bi, H., Yin, H., Yu, J., Dong, J., Yang, M., & Ma, Y. (2018). Influence of ultrasound assisted thermal processing on the physicochemical and sensorial properties of beer. *Ultrasonics Sonochemistry*, *40*, 166–173. Available from https://doi.org/10.1016/j.ultsonch.2017.07.017.

de Oliveira-Boreli, F. P., Pereira, D. F., Gonçalves, J. A., da Silva, V. Z., Nääs, I., & de, A. (2023). Non-destructive assessment of hens' eggs quality using image analysis and machine learning. *Smart Agricultural Technology*, *4*, 100161. Available from https://doi.org/10.1016/j.atech.2022.100161.

Erdoğdu, F. (2013). Mathematical modeling of transport phenomena for simulation and optimization of food processing operations. In N. G. Stoforos, P. Taoukis, S. Yanniotis, & V. T. Karathanos (Eds.), *Food engineering series* (pp. 473–487). Springer. Available from https://doi.org/10.1007/978-1-4614-7906-2_23.

Fahl, M. (2001). *Trust-region methods for flow control based on reduced order modelling*. Doctoral thesis. Fachbereich IV der Universitat Trier.

Fathi, P., Moosavi-Nasab, M., Mirzapour-Kouhdasht, A., & Khalesi, M. (2021). Generation of hydrolysates from rice bran proteins using a combined ultrasonication-Alcalase hydrolysis treatment. *Food Bioscience*, *42*, 101110. Available from https://doi.org/10.1016/j.fbio.2021.101110.

Figueroa-González, J. J., Lobato-Calleros, C., Vernon-Carter, E. J., Aguirre-Mandujano, E., Alvarez-Ramirez, J., & Martínez-Velasco, A. (2022). Modifying the structure, physicochemical properties, and foaming ability of amaranth protein by dual pH-shifting and ultrasound treatments. *Lwt*, 153. Available from https://doi.org/10.1016/j.lwt.2021.112561.

Fortin, F.-A., De Rainville, F.-M., Gardner, M.-A., Parizeau, M., & Gagne, C. (2012). DEAP: Evolutionary algorithms made easy. *Journal of Machine Learning Research*, *13*, 2171–2175.

Goñi, S. M., D'amore, M., Della Valle, M., Olivera, D. F., Salvadori, V. O., & Marra, F. (2022). Effect of load spatial configuration on the heating of chicken meat assisted by radio frequency at 40.68 MHz. *Foods*, *11*(8), 1096. Available from https://doi.org/10.3390/foods11081096.

Gunantara, N. (2018). A review of multi-objective optimization: Methods and its applications. *Cogent Engineering*, *5*, 1502242. Available from https://doi.org/10.1080/23311916.2018.1502242.

Gwanpua, S. G., Verlinden, B. E., Hertog, M. L. A. T. M., Van Impe, J., Nicolai, B. M., & Geeraerd, A. H. (2013). Towards flexible management of postharvest variation in fruit firmness of three apple cultivars. *Postharvest Biology and Technology*, *85*, 18–29. Available from https://doi.org/10.1016/j.postharvbio.2013.04.015.

Hashemi, S. M. B., & Gholamhosseinpour, A. (2020). Effect of ultrasonication treatment and fermentation by probiotic *Lactobacillus plantarum* strains on goat milk bioactivities. *International Journal of Food Science and Technology*, *55*(6), 2642–2649. Available from https://doi.org/10.1111/ijfs.14517.

Hassoun, A., Jagtap, S., Trollman, H., Garcia-Garcia, G., Abdullah, N. A., Goksen, G., Bader, F., Ozogul, F., Barba, F. J., Cropotova, J., Munekata, P. E. S., & Lorenzo, J. M. (2023). Food processing 4.0: Current and future developments spurred by the fourth industrial revolution. *Food Control*, *145*, 109507.

He, S., Zhao, J., Cao, X., Ye, Y., Wu, Z., Yue, J., Yang, L., Jin, R., & Sun, H. (2020). Low pH-shifting treatment would improve functional properties of black turtle bean (*Phaseolus vulgaris* L.) protein isolate with immunoreactivity reduction. *Food Chemistry*, *330*, 127217. Available from https://doi.org/10.1016/j.foodchem.2020.127217.

Henrichs, E., Noack, T., Pinzon Piedrahita, A. M., Salem, M. A., Stolz, J., & Krupitzer, C. (2022). Can a byte improve our bite? An analysis of digital twins in the food industry. *Sensors*, *22*, 115. Available from https://doi.org/10.3390/s22010115.

Hertog, M. L. A. T. M., Lammertyn, J., De Ketelaere, B., Scheerlinck, N., & Nicolaï, B. M. (2007). Managing quality variance in the postharvest food chain. *Trends in Food Science & Technology*, *18*, 320–332. Available from https://doi.org/10.1016/j.tifs.2007.02.007.

Hong, Y.-K., Stanley, R., Tang, J., Bui, L., & Ghandi, A. (2021). Effect of electric field distribution on the heating uniformity of a model ready-to-eat meal in microwave-assisted thermal sterilization using the FDTD method. *Foods*, *10*(2), 0311. Available from https://doi.org/10.3390/foods10020311.

Hu, Y., Tian, H., Hu, S., Dong, L., Zhang, J., Yu, X., Han, M., & Xu, X. (2022). The effect of in-package cold plasma on the formation of polycyclic aromatic hydrocarbons in charcoal-grilled beef steak with different oils or fats. *Food Chemistry*, *371*, 131384. Available from https://doi.org/10.1016/j.foodchem.2021.131384.

Jabbar, S., Abid, M., Hu, B., Muhammad Hashim, M., Saeeduddin, M., Lei, S., Wu, T., & Zeng, X. (2014). Influence of sonication and high hydrostatic pressure on the quality of carrot juice. *International Journal of Food Science and Technology*, *49*(11), 2449–2457. Available from https://doi.org/10.1111/ijfs.12567.

Janardhanan, R., Virseda, P., Huerta-Leidenz, N., & Beriain, M. J. (2022). Effect of high–hydrostatic pressure processing and sous-vide cooking on physicochemical traits of Biceps femoris veal patties. *Meat Science*, *188*, 108772. Available from https://doi.org/10.1016/j.meatsci.2022.108772.

Jayathunge, K. G. L. R., Stratakos, A. Ch, Cregenzán-Albertia, O., Grant, I. R., Lyng, J., & Koidis, A. (2017). Enhancing the lycopene in vitro bioaccessibility of tomato juice synergistically applying thermal and non-thermal processing technologies. *Food Chemistry*, *221*, 698–705. Available from https://doi.org/10.1016/j.foodchem.2016.11.117.

Jiang, S., Ding, J., Andrade, J., Rababah, T. M., Almajwal, A., Abulmeaty, M. M., & Feng, H. (2017). Modifying the physicochemical properties of pea protein by pH-shifting and ultrasound combined treatments. *Ultrasonics Sonochemistry*, *38*, 835–842. Available from https://doi.org/10.1016/j.ultsonch.2017.03.046.

Jin, J., Okagu, O. D., Yagoub, A. E. G. A., & Udenigwe, C. C. (2021). Effects of sonication on the in vitro digestibility and structural properties of buckwheat protein isolates. *Ultrasonics Sonochemistry*, *70*, 105348. Available from https://doi.org/10.1016/j.ultsonch.2020.105348.

Kang, T., Lee, D., Ko, Y., & Jun, S. (2022). Effects of pulsed electric field (PEF) and oscillating magnetic field (OMF) on supercooling preservation of beef at different fat levels. *International Journal of Refrigeration*, *136*, 36–45. Available from https://doi.org/10.1016/j.ijrefrig.2022.01.004.

Kannapinn, M., Pham, M. K., & Schäfer, M. (2022). Physics-based digital twins for autonomous thermal food processing: Efficient, non-intrusive reduced-order modeling. *Innovative Food Science & Emerging Technologies*, *81*, 103143. Available from https://doi.org/10.1016/j.ifset.2022.103143.

Karaduman, M., Uyar, R., & Erdogdu, F. (2012). Toroid cans – An experimental and computational study for process innovation. *Journal of Food Engineering*, *111*, 6–13. Available from https://doi.org/10.1016/j.jfoodeng.2012.02.009.

Khosrow Shahi, S., Didar, Z., Hesarinejad, M. A., & Vazifedoost, M. (2021). Optimized pulsed electric field-assisted extraction of biosurfactants from Chubak (*Acanthophyllum squarrosum*) root and application in ice cream. *Journal of the Science of Food and Agriculture*, *101*, 3693–3706. Available from https://doi.org/10.1002/jsfa.11000.

Ko, H. Y., Shin, D. H., Oh, J. H., Kang, H., Choi, D. E., & Choi, S. (2021). Numerical simulation of thermal flow characteristics in plasma reactor for rotten citrus fruits drying. *Applied Science and Convergence Technology*, *30*, 70–73. Available from https://doi.org/10.5757/ASCT.2021.30.3.70.

Koulouris, A., Misailidis, N., & Petrides, D. (2021). Applications of process and digital twin models for production simulation and scheduling in the manufacturing of food ingredients and products. *Food and Bioproducts Processing*, *126*, 317–333. Available from https://doi.org/10.1016/j.fbp.2021.01.016.

Kowalik, J., Lobacz, A., Zulewska, J., & Dec, B. (2018). Analysis and mathematical modelling of the behaviour of *Escherichia coli* in the mascarpone cheese during cold storage. *International Journal of Food Science & Technology*, *53*, 1541–1548. Available from https://doi.org/10.1111/ijfs.13736.

Krupitzer, C., Noack, T., & Borsum, C. (2022). Digital food twins combining data science and food science: System model, applications, and challenges. *Processes*, *10*, 1781. Available from https://doi.org/10.3390/pr10091781.

Li, L., Yang, R., & Zhao, W. (2021). The effect of pulsed electric fields (PEF) combined with temperature and natural preservatives on the quality and microbiological shelf-life of cantaloupe juice. *Foods*, *10*(11), 2606. Available from https://doi.org/10.3390/foods10112606.

Li, S., Chen, S., Han, F., Xv, Y., Sun, H., Ma, Z., Chen, J., & Wu, W. (2019). Development and optimization of cold plasma pretreatment for drying on corn kernels. *Journal of Food Science*, *84*, 2181–2189. Available from https://doi.org/10.1111/1750-3841.14708.

Liao, L., & Köttig, F. (2014). Review of hybrid prognostics approaches for remaining useful life prediction of engineered systems, and an application to battery life prediction. *IEEE Transactions on Reliability*, *63*, 191–207. Available from https://doi.org/10.1109/TR.2014.2299152.

Liu, C., Gu, Z., Lin, X., Wang, Y., Wang, A., Sun, Y., & Shi, Y. (2022). Effects of high hydrostatic pressure (HHP) and storage temperature on bacterial counts, color change, fatty acids and non-volatile taste active compounds of oysters (*Crassostrea ariakensis*). *Food Chemistry*, *372*, 131247. Available from https://doi.org/10.1016/j.foodchem.2021.131247.

Lucia, D. J., Beran, P. S., & Silva, W. A. (2004). Reduced-order modeling: new approaches for computational physics. *Progress in Aerospace Sciences*, *40*, 51–117. Available from https://doi.org/10.1016/j.paerosci.2003.12.001.

Lysova, N., Solari, F., & Vignali, G. (2022). Optimization of an indirect heating process for food fluids through the combined use of CFD and response surface methodology. *Food and Bioproducts Processing*, *131*, 60–76. Available from https://doi.org/10.1016/j.fbp.2021.10.010.

MarketsandMarkets. (2022a). *Food & beverage processing equipment market. Global outlook, trends, and forecast to 2026*. MarketsandMarkets. https://www.marketsandmarkets.com/Market-Reports/food-processing-equipment-market-121668697.html. Accessed 03.12.23.

MarketsandMarkets. (2022b). *Digital twin market size & forecast*. MarketsandMarkets. https://www.marketsandmarkets.com/Market-Reports/digital-twin-market-225269522.html. Accessed 03.12.23.

Michael, A. P. (2021). *Technology statistics: How fast is tech advancing?* https://mediapeanut.com/how-fast-is-technology-growing-statistics-facts/. Accessed 12.02.22.

Mitchell, S., Kean, A., Mason, A., O'Sullivan, M., Phillips, A., & Peschiera, F. (2009). *Optimization with PuLP — PuLP 2.7.0 documentation*. https://coin-or.github.io/pulp/. Accessed 03.20.23.

Mondal, S., Wijewardena, K. P., Karuppuswami, S., Kriti, N., Kumar, D., & Chahal, P. (2019). Blockchain inspired RFID-based information architecture for food supply chain. *IEEE Internet of Things Journal*, *6*, 5803–5813. Available from https://doi.org/10.1109/JIOT.2019.2907658.

Moradi-Sadr, J., Ebadi, M.-T., Ayyari, M., & Ghomi, H. (2023). Optimization of ultrasonic Bath and cold plasma pre-treatments in the spearmint essential oil isolation process. *Food Science & Nutrition*, *00*(1–12). Available from https://doi.org/10.1002/fsn3.3224.

Navarro, A., Ruiz-Méndez, M. V., Sanz, C., Martínez, M., Rego, D., & Pérez, A. G. (2022). Application of pulsed electric fields to pilot and industrial scale virgin olive oil extraction: Impact on organoleptic and functional quality. *Foods*, *11*(14). Available from https://doi.org/10.3390/foods11142022.

Noack, B. R., Morzynski, M., & Tadmor, G. (2011). *Reduced-order modelling for flow control*. Springer Science & Business Media.

Nwabor, O. F., Onyeaka, H., Miri, T., Obileke, K., Anumudu, C., & Hart, A. (2022). A cold plasma technology for ensuring the microbiological safety and quality of foods. *Food Engineering Reviews*, *14*(4), 535–554. Available from https://doi.org/10.1007/s12393-022-09316-0.

Oey, I., Van der Plancken, I., Van Loey, A., & Hendrickx, M. (2008). Does high pressure processing influence nutritional aspects of plant based food systems? *Trends in Food Science and Technology*, *19*(6), 300–308. Available from https://doi.org/10.1016/j.tifs.2007.09.002.

Ohale, P. E., Nwajiobi, O. J., Onu, C. E., Madiebo, E. M., & Ohale, N. J. (2022). Solvent extraction of oil from three cultivars of Nigerian mango seed kernel: Process modeling, GA - optimization, nonlinear kinetics and comparative characterization. *Applied Food Research*, *2*. Available from https://doi.org/10.1016/j.afres.2022.100227.

Ozkan, G., Esatbeyoglu, T., & Capanoglu, E. (2022a). Bioavailability of Rosehip (*Rosa canina* L.) infusion phenolics prepared by thermal, pulsed electric field and high pressure processing. *Foods*, *11*(13). Available from https://doi.org/10.3390/foods11131955.

Ozkan, G., Kostka, T., Dräger, G., Capanoglu, E., & Esatbeyoglu, T. (2022b). Bioaccessibility and transepithelial transportation of cranberrybush (Viburnum opulus) phenolics: Effects of non-thermal processing and food matrix. *Food Chemistry*, *380*, 132036. Available from https://doi.org/10.1016/j.foodchem.2021.132036.

Pérez-Won, M., Cepero-Betancourt, Y., Reyes-Parra, J. E., Palma-Acevedo, A., Tabilo-Munizaga, G., Roco, T., Aubourg, S. P., & Lemus-Mondaca, R. (2021). Combined PEF, CO_2 and HP application to chilled coho salmon and its effects on quality attributes under different rigor conditions. *Innovative Food Science and Emerging Technologies*, *74*, 102832. Available from https://doi.org/10.1016/j.ifset.2021.102832.

Perez, R.E., & Jansen, P.W. (2008). http://www.pyopt.org/. Accessed 03.20.23.

Pokhrel, P. R., Boulet, C., Yildiz, S., Sablani, S., Tang, J., & Barbosa-Cánovas, G. V. (2022). Effect of high hydrostatic pressure on microbial inactivation and quality changes in carrot-orange juice blends at varying pH. *LWT*, *159*, 113219. Available from https://doi.org/10.1016/j.lwt.2022.113219.

Psarianos, M., Dimopoulos, G., Ojha, S., Cavini, A. C. M., Bußler, S., Taoukis, P., & Schlüter, O. K. (2022). Effect of pulsed electric fields on cricket (*Acheta domesticus*) flour: Extraction yield (protein, fat and chitin) and techno-functional properties. *Innovative Food Science and Emerging Technologies*, *76*, 102908. Available from https://doi.org/10.1016/j.ifset.2021.102908.

Qi, Q., Tao, F., Hu, T., Anwer, N., Liu, A., Wei, Y., Wang, L., & Nee, A. Y. C. (2021). Enabling technologies and tools for digital twin. *Journal of Manufacturing Systems*, *58*, 3–21. Available from https://doi.org/10.1016/j.jmsy.2019.10.001.

Qi, Q., Tao, F., Zuo, Y., & Zhao, D. (2018). Digital twin service towards smart manufacturing. In *Procedia CIRP, 51st CIRP conference on manufacturing systems* (Vol. 72, pp. 237–242). https://doi.org/10.1016/j.procir.2018.03.103

Qu, Z., Chen, G., Wang, J., Xie, X., & Chen, Y. (2023). Preparation, structure evaluation, and improvement in foaming characteristics of fibrotic pea protein isolate by cold plasma synergistic organic acid treatment. *Food Hydrocolloids*, *134*, 108057. Available from https://doi.org/10.1016/j.foodhyd.2022.108057.

Rabeler, F., & Feyissa, A. H. (2018). Modelling the transport phenomena and texture changes of chicken breast meat during the roasting in a convective oven. *Journal of Food Engineering*, *237*, 60–68. Available from https://doi.org/10.1016/j.jfoodeng.2018.05.021.

Rahimifard, S., Brewer, S., Garcia-Garcia, G., & Jagtap, S. (2022). Digitalising food manufacturing. *Food Science and Technology*, *36*, 24–27. Available from https://doi.org/10.1002/fsat.3603_5.x.

Ranjbar Nedamani, A., & Hashemi, S. J. (2022). Energy consumption computing of cold plasma-assisted drying of apple slices (Yellow Delicious) by numerical simulation. *Journal of Food Process Engineering*, *45*(5), e14019. Available from https://doi.org/10.1111/jfpe.14019.

Sadhu, T., Lahiri, S. K., Bhattacharjee, A., & Chakrabarty, J. (2022). Application of artificial neural network with metaheuristic optimization for improving the nutritive value of fried fish. *Journal of Food Processing and Preservation*, *46*, e170190. Available from https://doi.org/10.1111/jfpp.17190.

Schaffrath, T. (2018). Digitization in the food and beverage industry: opportunity or challenge? https://food-blog.csb.com/us-en/digitization-in-the-food-and-beverage-industry-opportunity-or-challenge. Accessed 03.12.23.

Schudel, S., Shrivastava, C., Rebeaud, S. G., Karafka, L., Shoji, K., Onwude, D., & Defraeye, T. (2022). Combining experiments and mechanistic modeling to compare ventilated packaging types for strawberries from farm to retailer. *Food Packaging and Shelf Life*, *34*, 100944. Available from https://doi.org/10.1016/j.fpsl.2022.100944.

Shiekh, K. A., Benjakul, S., Qi, H., Zhang, B., & Deng, S. (2021). Combined hurdle effects of pulsed electric field and vacuum impregnation of Chamuang leaf extract on quality and shelf-life of Pacific white shrimp subjected to high voltage cold atmospheric plasma. *Food Packaging and Shelf Life*, *28*, 100660. Available from https://doi.org/10.1016/j.fpsl.2021.100660.

Shoji, K., Schudel, S., Shrivastava, C., Onwude, D., & Defraeye, T. (2022). Optimizing the postharvest supply chain of imported fresh produce with physics-based digital twins. *Journal of Food Engineering*, *329*, 111077. Available from https://doi.org/10.1016/j.jfoodeng.2022.111077.

Shrivastava, C., Berry, T., Cronje, P., Schudel, S., & Defraeye, T. (2022). Digital twins enable the quantification of the trade-offs in maintaining citrus quality and marketability in the refrigerated supply chain. *Nature Food*, *3*, 413–427. Available from https://doi.org/10.1038/s43016-022-00497-9.

Shrivastava, C., Schudel, S., Shoji, K., Onwude, D., da Silva, F. P., Turan, D., Paillart, M., & Defraeye, T. (2023). Digital twins for selecting the optimal ventilated strawberry packaging based on the unique hygrothermal conditions of a shipment from farm to retailer. *Postharvest Biology and Technology*, *199*, 112283. Available from https://doi.org/10.1016/j.postharvbio.2023.112283.

Singh, M., Srivastava, R., Fuenmayor, E., Kuts, V., Qiao, Y., Murray, N., & Devine, D. (2022). Applications of digital twin across industries: A review. *Applied Sciences*, *12*, 5727. Available from https://doi.org/10.3390/app12115727.

Söderberg, R., Wärmefjord, K., Carlson, J. S., & Lindkvist, L. (2017). Toward a digital twin for real-time geometry assurance in individualized production. *CIRP Annals*, *66*, 137–140. Available from https://doi.org/10.1016/j.cirp.2017.04.038.

Song, B., Zhu, P., Zhang, Y., Ju, N., Si, X., Pang, X., Lv, J., & Zhang, S. (2023). Preparation and quality assessment of processed cream cheese by high hydrostatic pressure combined thermal processing and spore-induced germination. *Journal of Food Engineering*, *341*, 111319. Available from https://doi.org/10.1016/j.jfoodeng.2022.111319.

Stühmeier-Niehe, C., Lass, L., Brocksieper, M., Chanos, P., & Hertel, C. (2023). Pre-treatment of starter cultures with mild pulsed electric fields influences the characteristics of set yogurt. *Foods*, *12*(3), 442. Available from https://doi.org/10.3390/foods12030442.

Tian, F. (2017). A supply chain traceability system for food safety based on HACCP, blockchain & Internet of things. In *2017 International conference on service systems and service management* (pp. 1–6). https://doi.org/10.1109/ICSSSM.2017.7996119.

Toepfl, S., Mathys, A., Heinz, V., & Knorr, D. (2006). Review: Potential of high hydrostatic pressure and pulsed electric fields for energy efficient and environmentally friendly food processing. *Food Reviews International*, *22*(4), 405–423. Available from https://doi.org/10.1080/87559120600865164.

Tsou, C.S. (Vince), Lee, M.C. (Alan). (2022). *nsga2R: Elitist non-dominated sorting genetic algorithm.* https://cran.r-project.org/web/packages/nsga2R/nsga2R.pdf.

Turgut, S. S., Entrenas, J. A., Taşkın, E., Garrido-Varo, A., & Pérez-Marín, D. (2022). Estimation of the sensory properties of black tea samples using non-destructive near-infrared spectroscopy sensors. *Food Control*, *142*, 109260. Available from https://doi.org/10.1016/j.foodcont.2022.109260.

Turgut, S. S., Küçüköner, E., Feyissa, A. H., & Karacabey, E. (2021). A novel drying system – simultaneous use of ohmic heating with convectional air drying: System design and detailed examination using CFD. *Innovative Food Science and Emerging Technologies*, *72*, 102727. Available from https://doi.org/10.1016/J.IFSET.2021.102727.

Turgut, S.S., Siamos, E., & Feyissa, A.H. (2022). Simulation and parametric sensitivity evaluation for ohmic heating of chicken breast. *Presented at the 8th international food operations and processing simulation workshop, FoodOPS 2022*. https://doi.org/10.46354/i3m.2022.foodops.009

Udugama, I. A., Kelton, W., & Bayer, C. (2023). Digital twins in food processing: A conceptual approach to developing multi-layer digital models. *Digital Chemical Engineering*, *7*, 100087. Available from https://doi.org/10.1016/j.dche.2023.100087.

Valenzuela, C., Garcia-Galicia, I. A., Paniwnyk, L., & Alarcon-Rojo, A. D. (2021). Physicochemical characteristics and shelf life of beef treated with high-intensity ultrasound. *Journal of Food Processing and Preservation*, *45*(4), 1–11. Available from https://doi.org/10.1111/jfpp.15350.

van Soest, T. (2021). Digital transformation in the food industry: Trends, examples, and benefits. *The Future of Commerce*. Available from https://www.the-future-of-commerce.com/2021/04/23/digital-transformation-in-food-industry/. (accessed 3.12.23).

Verboven, P., Defraeye, T., Datta, A. K., & Nicolai, B. (2020). Digital twins of food process operations: the next step for food process models? *Current Opinion in Food Science, Food Engineering & Processing*, *35*, 79–87. Available from https://doi.org/10.1016/j.cofs.2020.03.002.

Vijayalakshmi, S., Nadanasabhapathi, S., Kumar, R., & Sunny Kumar, S. (2018). Effect of pH and pulsed electric field process parameters on the aflatoxin reduction in model system using response surface methodology. *Journal of Food Science and Technology*, *55*, 868–878. Available from https://doi.org/10.1007/s13197-017-2939-3.

Vinatoru, M. (2001). An overview of the ultrasonically assisted extraction of bioactive principles from herbs. *Ultrasonics Sonochemistry*, *8*(3), 303–313. Available from https://doi.org/10.1016/S1350-4177(01)00071-2.

Wang, J., Wang, Q., Xu, L., & Sun, D. W. (2022). Effects of extremely low frequency pulsed electric field (ELF-PEF) on the quality and microstructure of tilapia during cold storage. *Lwt*, *169*, 113937. Available from https://doi.org/10.1016/j.lwt.2022.113937.

Wang, X., Chen, F., Ma, L., Liao, X., & Hu, X. (2022). Non-volatile and volatile metabolic profiling of tomato juice processed by high-hydrostatic-pressure and high-temperature short-time. *Food Chemistry*, *371*, 131161. Available from https://doi.org/10.1016/j.foodchem.2021.131161.

Wu, C.-P., Wu, S.-M., Lin, Y.-H., Wu, Y.-H., Huang, B.-C., Huang, H.-W., & Wang, C.-Y. (2022). High pressure processing-based hurdle strategy for microbial shelf life of packed food in the cold chain. *Food Packaging and Shelf Life*, *34*, 100983. Available from https://doi.org/10.1016/j.fpsl.2022.100983.

Wu, X., Tan, M., Zhu, Y., Duan, H., Ramaswamy, H. S., Bai, W., & Wang, C. (2023). The influence of high pressure processing and germination on anti-nutrients contents, in vitro amino acid release and mineral digestibility of soybeans. *Journal of Food Composition and Analysis*, *115*, 104953. Available from https://doi.org/10.1016/j.jfca.2022.104953.

Xiao, D., Fang, F., Buchan, A. G., Pain, C. C., Navon, I. M., & Muggeridge, A. (2015). Non-intrusive reduced order modelling of the Navier–Stokes equations. *Computer Methods in Applied Mechanics and Engineering*, *293*, 522–541. Available from https://doi.org/10.1016/j.cma.2015.05.015.

Younis, M., Mohamed Ahmed, I. A., Ahmed, K. A., Yehia, H. M., Abdelkarim, D. O., Fickak, A., El-Abedein, A. I. Z., Alhamdan, A., & Elfeky, A. (2023). Pulsed electric field as a novel technology for fresh barhi date shelf-life extension: Process optimization using response surface methodology. *Horticulturae*, *9*(2), 155. Available from https://doi.org/10.3390/horticulturae9020155.

Zamanian, M., Sadrnia, H., Khojastehpour, M., Rohani, A., Thibault, J., & Hosseini, F. (2023). Multi-objective optimization of PVA/TiO2/MMT mixed matrix membrane for food packaging. *Journal of Polymers and the Environment*, *31*, 90–101. Available from https://doi.org/10.1007/s10924-022-02632-w.

Zhang, J., Duley, J. A., Cowley, D. M., Shaw, P. N., Koorts, P., & Bansal, N. (2023). Comparative proteomic analysis of donor human milk pasteurized by hydrostatic high-pressure. *Food Chemistry*, *403*, 134264. Available from https://doi.org/10.1016/j.foodchem.2022.134264.

Zhang, J., Zhang, M., Bai, X., Zhang, Y., & Wang, C. (2022). The impact of high hydrostatic pressure treatment time on the structure, gelatinization and thermal properties and in vitro digestibility of oat starch. *Grain & Oil Science and Technology*, *5*(1), 1–12. Available from https://doi.org/10.1016/j.gaost.2022.01.002.

Zhang, W., Liu, Y., Li, Z., Xu, S., Zhang, J., Hettinga, K., & Zhou, P. (2021). Effects of microfiltration combined with ultrasonication on shelf life and bioactive protein of skim milk. *Ultrasonics Sonochemistry*, *77*, 105668. Available from https://doi.org/10.1016/j.ultsonch.2021.105668.

Zhou, J., Bian, Y., Yang, T., Wang, J., Kong, T., Zhang, Y., Chen, Y., & Li, S. (2023). A non-thermal modification method to enhance the encapsulation efficiency, stability, and slow-release performance of zein-based delivery systems – Cold plasma. *Journal of Food Engineering*, *345*, 111415. Available from https://doi.org/10.1016/j.jfoodeng.2023.111415.

Zia, S., Khan, M. R., Zeng, X. A., Sehrish., Shabbir, M. A., & Aadil, R. M. (2019). Combined effect of microwave and ultrasonication treatments on the quality and stability of sugarcane juice during cold storage. *International Journal of Food Science and Technology*, *54*(8), 2563–2569. Available from https://doi.org/10.1111/ijfs.14167.

Zielinska, S., Cybulska, J., Pieczywek, P., Zdunek, A., Kurzyna-Szklarek, M., Staniszewska, I., Liu, Z.-L., Pan, Z., Xiao, H.-W., & Zielinska, M. (2022). Structural morphology and rheological properties of pectin fractions extracted from okra pods subjected to cold plasma treatment. *Food and Bioprocess Technology*, *15*(5), 1168–1181. Available from https://doi.org/10.1007/s11947-022-02798-0.

Chapter 7

Robotics as key enabler technology in Food Industry 4.0 and beyond

Dele Raheem[1], Horst Treiblmaier[2], Wael M. Mohammed[3], Borja Ramis Ferrer[4] and Jose L. Martinez-Lastra[3]
[1]Artic Centre, University of Lapland, Rovaniemi, Finland, [2]School of International Management, Modul University Vienna, Vienna, Austria, [3]FAST-Lab, Faculty of Engineering and Natural Sciences, Tampere University, Tampere, Finland, [4]The Institute of Electrical and Electronics Engineers, Madrid, Spain

7.1 Introduction

The increasing challenges that are caused by population growth, climate change, environmental problems, and shortage of skilled human labor in many advanced economies are strong drivers for Food Industry 4.0 and beyond to adopt innovative robotic applications. To ensure sustainability in worldwide production and consumption practices, the food industry can benefit from the application of robotics and automation.

The methods of food production and consumption are constantly modified and must be tailored to meet customer demands in increasingly digitalized societies. As of 2023, the impact of the COVID-19 pandemic still makes a strong case for the integration of robotics and automation into all stages of food production systems with the goal of ensuring food safety and food supply (Henry, 2020). To realize this vision, machines, storage systems, and utilities must be able to share information across the value chain, as well as control each other autonomously (Thoben et al., 2017). Henceforth, this is expected to result in systems where all processes are seamlessly integrated through modern equipment and numerous control points (Demartini et al., 2018).

The need for databases to develop deep learning models was previously highlighted in the works of Wang et al. (2022), who advocated to maintain databases that contain food property information related to robotic handling, the properties of viscoelasticity or rheology, friction, geometry, and weight as well as the integration of new databases for bin-picking scenarios to ensure satisfactory performance (Wang et al., 2022). Currently still in its infancy, we expect that the data-driven agriculture and food industry will mature in the coming decades, shaping the food systems of the future (Marvin et al., 2022; Rejeb, Rejeb, et al., 2022). A further important driver toward sustainable food systems will be the increasing digitalization of farms, the food industry, and supply chains with robotic technology.

In this chapter, we summarize and discuss the transformations that are brought about by integrating robotics in agriculture and the food industry. In Section 7.2, several enablers of robotics and their impact on farming and manufacturing environments are presented. Section 7.3 gives a brief overview on the fourth industrial revolution and its implications in the food industry. This section also highlights the collaborative activities of humans and robots, as well as the interconnection of cobots, automation, and wireless networks with the goal of optimized food production. In Section 7.4, the emergence of the Internet of Things (IoT) in the food sector is discussed in some detail. Section 7.5 briefly outlines food processing and handling with the application of Industry 4.0 technologies. Section 7.6 discusses automation that engages robots and robotic systems in the food industry. Finally, Section 7.7 concludes the chapter with several practical examples.

7.2 Enablers of robotics

The food industry includes vast topics and activities such as collecting, processing, and packaging. In this regard, Fig. 7.1 depicts a generic food industry chain flow, which illustrates the focus of this chapter: the usage of robotics in food industry. From an automation perspective, two environments can be identified: farming and manufacturing.

Food Industry 4.0. DOI: https://doi.org/10.1016/B978-0-443-15516-1.00007-4

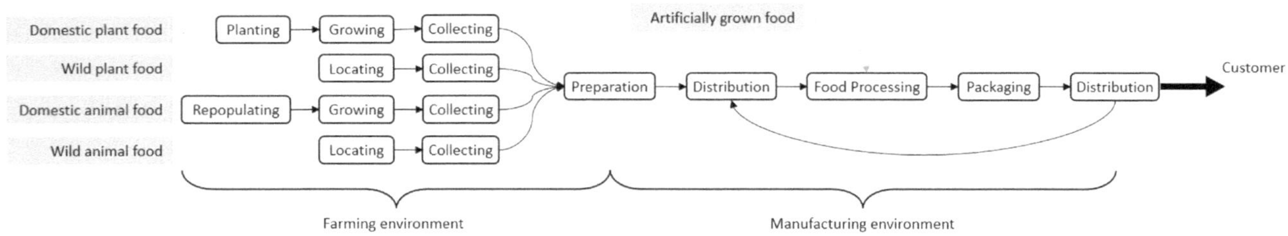

FIGURE 7.1 Generic food industry chain flow (Cimini & Moresi, 2019). *Based on Cimini, A., & Moresi, M. (2019). A progressive approach towards a more sustainable food industry.* Chemical Engineering Transactions, 75, *565–570.*

The farming environment includes the initial stages of the food industry chain and mostly depends on the type of origin of the food to be produced. The manufacturing environment is a controlled environment where sophisticated machinery is used to process the food with the goal of providing the final product.

As presented in Fig. 7.1, the farming environment is subdivided into four main categories. The first category is the domestic plant food, such as tomatoes, olives, and potatoes that grow in farms or greenhouses and depend on systematic activities and agriculture methods. Secondly, the wild plant food category that may include wild mushrooms and berries, which require only collection and sorting. Thirdly, domestic animal food, such as chicken, beef, honey, dairy, and eggs. This category also follows systematic methods for growing and preparing the food. Finally, the fourth category includes wild animal food such as deer meat and seafood. Due to its complexity, the farming environment frequently depends on humans to perform tasks that need a certain degree of dexterity. Still, this creates opportunities for robots to be involved in supporting human workers.

The second environment shown in Fig. 7.1 is the manufacturing environment. This environment is advanced in terms of the use of digital and industrial technologies. It follows well-established and defined processes and already employs robots to perform numerous activities like material handling. However, due to the increasing pressure to adopt more sustainable measures, this environment may also benefit from more robotic system involvement. As an example, the changes in packaging materials and waste generation already drive the implementation of new robotic technology in production processes.

It is important to highlight some of the special cases in the chain as presented in Fig. 7.1. The first special case addresses the production of artificially grown food such as cultured meat where the chain starts directly in the processing stage. Another special case includes the processing of frozen fish where all the production of the fish products is conducted on fishing ships. Lastly, some food products serve as ingredients for other food products. For example, the mayonnaise production, where oil and eggs are produced and packaged by a supplier and later are distributed to factories that will finally produce mayonnaise.

It is well established that robotics technology is a cornerstone of reaching high-performance production. Thanks to their reliability, versatility, dexterity, flexibility, accuracy, and repeatability, robots are increasingly employed in factories (Jung & Lim, 2020). The food industry is a prominent example where robots are heavily used in material handling and packaging among other applications. In 2020, 126 robots have been used for every 10,000 persons worldwide in the manufacturing sector according to the International Federation of Robotics. This use is reflected in the annual sales of industrial robots, which peaked in 2018 with 422,000 sold units and 384,000 in 2020 (Statista Research Department, Industrial robots—worldwide installations, Statista, 2023). According to McHugh (McHugh, 2017), in 2017 robot usage in the food industry increased by 26% in food packaging and 40% in palletizing. Due to a steadily increasing population, the food industry heavily depends on the utilization of enabling technologies to sufficiently satisfy the growing needs. In this section, the term *Enabler* refers to any technology, tool, method, technique, system, or concept that permits an improvement in the performance of the respective industry. As an example, one of the key enabling technologies of navigators is the IoT. In more detail, the IoT enables navigators to monitor roads, traffic, and transportation systems in a real-time manner using the smartphones of passengers and pedestrians. Thus the navigators can provide traffic estimations, accurate trip planning, and arrival time predictions. In summary, a technology can be considered to be an enabling technology if it can provide tangible and relevant improvements for the industry. To better understand the role of robotics as an enabling technology for the food industry, a brief description of robots will be given.

The term Robots, as it is known today, was popularized in 1942 by the science fiction story Runaround by Isaac Asimov (Jazar, 2010). Even earlier, in 1921, the famous Czech playwriter Karel Čapek introduced the word robot in his play Rossum's Universal Robots (Sciavicco & Siciliano, 2000). Currently, robots, as automation assets, are categorized by several standards and taxonomies. As an example, the Japanese Industrial Robots Association categorizes robots into six categories based on autonomy level, control type, physical features, and behavior. Other organizations like the Robotics Institute of America and the Association Francaise de Robotique use more compact categorizations in their classifications. Regardless of the classifications, in the literature, robots are categorized based on several aspects (Vermesan & Bacquet, 2017). For instance, based on their typology, robots can be categorized into parallel robots, spherical robots, and cylindrical robots. Another criterion is the driving power, as robots can be actuated using hydraulics, pneumatics, or electrical motors. Additionally, robots can be categorized based on their mobility into either mobile or stationary and also be divided into two basic categories: collaborative and noncollaborative. Collaborative robots (cobots) are robots that can share the workspace with humans and conduct work in a collaborative and synchronized manner. A more detailed description on the use of cobots in food industry is given in Section 7.3.1. In this chapter, we understand robots are programmable cyber-physical systems that are able to work autonomously. Their main objective is to improve the overall performance of the system.

In the food industry, the increasing demands pertaining to high quality, high quantity, and low-cost products, and the increase in competition at national and international levels drive the industry to exploit all possible enabling technologies. In addition, the new concepts and initiatives may drive changes in the food industry to adopt more process-sustainable solutions. For instance, the switch between packaging materials in some industries requires the investigation of enabling technologies, including robotics. In this regard, different types of robots and robotics systems are exploited in the food industry following the respective needs of each production phase. As an example, in a manufacturing environment, one of the main challenges is the material of food itself. It is deformable, irregular, and sometimes slippery. To address these challenges, the concepts of soft robotics and soft grippers were introduced (Dinakaran et al., 2023). These solutions allow the grasping and handling of food products without damaging them. Furthermore, these solutions are employed where the packaging has been regulated to use only soft wrapping biodegradable materials instead of hard plastic containers. Another example includes the transportation of food products within the production system. This transportation is traditionally solved with conveyor systems. Due to their low customizability, conveyor systems may not always be effective, especially if the product is personalized and customizable. Furthermore, traditional conveyor systems limit the change of factory layouts as they are fixed once the production system is installed. To overcome these challenges, robots and robotic systems can be used as an enabling technology for allowing the production system to be physically more flexible. As an example, the usage of mobile robots for transporting food increases the flexibility of production and enables the shopfloor to be more modular. Another example is the use of levitated transportation systems. These state-of-the-art systems use robotic coordination systems to transfer the food. The main advantage of using such systems is the zero-maintenance feature. Moreover, these systems are specifically targeting the food industry as they provide clean and safe food production environments.

In the farming environment, the potential of using robots as enabling technologies is higher than in the manufacturing environment. Due to the high dependency on human workers, several research projects have been conducted to assess the benefits of robots with a focus on mobile robots (Siegwart et al., 2011). As an example, the usage of robotic drones in farming and monitoring the fields has been increasing in recent years (Rejeb, Abdollahi, et al., 2022). These systems can also function as remote mobile sensing nodes that can be important to model and then estimate the yield of the crops. Additionally, robotic drones are recently employed to monitor farm animals that are allowed free movement and require continuous monitoring. The exploitation of such robots is also on the rise for off-road mobile duties. Such robots can carry heavy loads and work long hours, which makes them suitable for monitoring ripened products and harvesting them at the optimal moment. For example, these technologies have been tested and demonstrated in tomato and grape farms in Europe.

7.3 Fourth industrial revolution

Climate change, human activities, and environmental conditions of the food system are intricately linked. The production of food heavily depends on favorable climate conditions, which in turn are dependent on the way in which humans produce their food. Food production on a massive scale was initially fostered by the industrial revolution that, over time, led to the increasing use of fossil fuels. The current practices of food production and distribution contribute significantly to global warming (Adamo et al., 2021). As a response, the concept of the fourth industrial revolution (called Industry 4.0) carries the promise to mitigate the negative consequences of modern production systems in numerous ways.

The first industrial revolution used water and steam to mechanize production, the second applied electric energy to create mass production, and the third introduced electronics and information technology to automate production. Finally, the current fourth industrial revolution builds upon the third revolution and integrates the digital transformation that has been ongoing since approximately the middle of the last century. The emerging concept of Industry 4.0—often

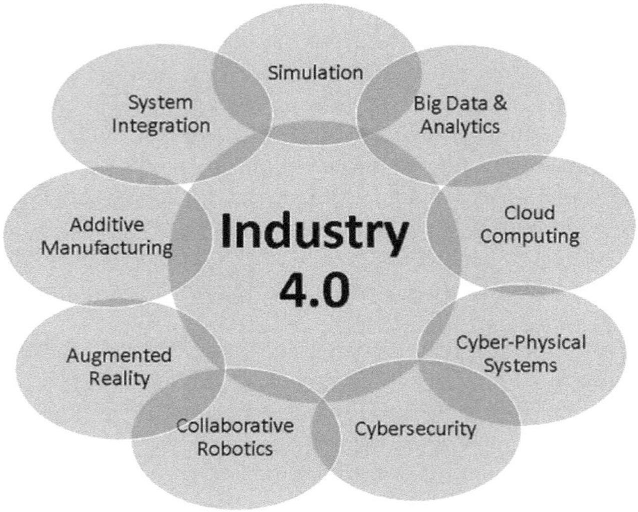

FIGURE 7.2 Technologies related to Industry 4.0. *Adapted from Bahrin, M. A. K., Othman, M. F., Azli, N. H. N., & Talib, M. F. (2016). Industry 4.0: A review on industrial automation and robotic.* Jurnal Teknologi, 78, *6–13.*

mixed up with related concepts such as "smart manufacturing" and "digital transformation"—has since gained significant attention among practitioners and academics alike. The term itself is defined in numerous ways and was comprehensively analyzed by Culot et al. (2020) who scrutinized 81 academic papers and differentiated between key enabling technologies, distinctive characteristics, and possible outcomes. Industry 4.0 connects IoT alongside cyber-physical systems in a way where the combination of software, sensors, actuators, and communication technology enables inanimate objects to exchange data. "Things" now have the potential to feed information into networks and thus contribute value to increasingly automated manufacturing processes (Bahrin et al., 2016).

Enabling technologies include physical–digital interface technologies, network technologies, data processing technologies, and physical–digital process technologies. Relevant characteristics of Industry 4.0 are process integration, real-time information transparency, virtual representation of the real world, autonomy, servitization of products and manufacturing capabilities, predictability, and, finally, modularity and reconfigurability. The outcomes include productivity, flexibility, mass customization, environmental sustainability, reduced time and cost to market, higher quality, improved lead time, and, ideally, economic growth. Given the manifold potential of Industry 4.0, it is no surprise that the expected impact on the food industry is substantial. Specifically, Industry 4.0 is predicted to fulfill new customer requirements pertaining to knowledge regarding product origins and the environmental impact of the consumed products. Furthermore, supply chains can be streamlined by integrating data along the chain and allowing for data availability in real time. Taken together, these measures will help to improve food security and increase trust in food safety (Oltra-Mestre et al., 2021). An overview of the technologies and concepts that are related to Industry 4.0 is shown in Fig. 7.2.

7.3.1 Human–robot collaboration

As previously presented in Section 7.2, robots may share the workspace with humans. This possibility has been always a concern, as it is critical to keep humans safe from any harm that can be caused by robots. This concern was regularly solved by the addition of safety fences to protect the human from the robot while it is working. Once the robot is in the manual or programming mode, these fences can be opened, which allows access to the robot for the robot specialist. This approach served very well for most industrial applications. However, some industries still rely on human craft and touch capabilities for tasks that require complex manipulations or require sophisticated skills as in woodworking, food manipulation, and fabric handling. Therefore a new concept has emerged to permit robots and humans to work simultaneously together. This concept is known as Human–Robot Collaboration (HRC). The HRC concept is a subtopic of Human–Robot Interaction (HRI) where humans and robots work together at the same time to accomplish certain tasks like assembling products. With a common goal, the interaction between the human and the robot is illustrated in Table 7.1. The interaction can be unnecessary or, as noted in the table as *Independence* (both robot and human perform independent activities) if the resources are enough and both robot and human are skilled to perform the task alone. Then, in a case where the resources are sufficient and either the robot or the human lacks some skills, the interaction can be noted as *Simple Collaboration*. As an example, a screwing task in assembly work can be done by a human but

TABLE 7.1 Taxonomy of the human–robot interactions.

Resources	Skills	Type of activity
Sufficient	Sufficient	Independence
Sufficient	Insufficient	Simple collaboration
Insufficient	Sufficient	Obstruction
Insufficient	Insufficient	Coordinated collaboration

Source: Based on the classification in O'Donovan, P., Leahy, K., Bruton, K., & O'Sullivan, D. T. J. (2015). Big data in manufacturing: A systematic mapping study. *Journal of Big Data, 2*, 20. https://doi.org/10.1186/s40537-015-0028-x.

not as fast as the robot, so the human can relay this task to the robot to improve process performance. Thirdly, if the robot and the human are skilled to perform the tasks but the resources are not enough, then the situation is noted as *Obstruction*. In other words, the robot or the human must wait for the other to be in an idle state or ready for collaboration. Lastly, if the resources are insufficient and the skills are also insufficient, then this situation is noted as *Coordinated Collaboration*.

HRI, which includes HRC, therefore represents all forms of activities between the human and the robot. Traditionally, the physical interaction between the human and the robot was only permitted while the robot was in the manual or programming mode. In this mode, the human can physically touch the robot and do all the needed work in the robot cell. Once the robot is put on automatic or operational mode, the human is required to close the fence and stop the interaction with the robot. This limitation has slowed the involvement of robotics in some cases, including all tasks that can be improved by collaboration with robots. Thus several robot producers introduced new robots known as collaborative robots or cobots. Cobots are special types of robots that are safe to be operated around humans without the need for safety distance or fences. They are equipped with additional sensing systems that allows detecting collisions and lead-through programming (i.e., a robot programming method where the operator physically guides the robot to the targets in the working space). As they are relatively new, cobots are still under development and will be further evolving in the near future. Moreover, they are a main topic in the research of HRC. As an example, Toichoa Eyam et al. (2021) present an innovative approach for robot adaptation to human worker emotions using electroencephalogram signals. Furthermore, Hormaza et al. (2019) present a novel approach for training the human worker to work with a collaborative robot using virtual reality technology. Another study that includes HRC is the Cobot self-adaptation using a deep learning vision system by De Miguel Lazaro et al. (2019).

In the food industries, collaboration between humans is a key requirement for numerous tasks, as this industry heavily relies on human workers, especially in the farming environment. Existing procedures can be especially enhanced in the food collection where robots can be used to take over dull and repetitive tasks. As an example, in some farms, workers are required to visit the farms on a daily basis to examine the fruits before harvesting. Such tasks can be done more efficiently using mobile robots that can operate with/around humans. Overall, as technology advances, more possibilities are expected to be available for robots to collaborate with human workers in the food industry.

7.3.2 Cobot, automation, wireless sensor networks

The fourth industrial revolution demands a high connectivity between all stakeholders in the value chain. Thanks to the advances in Information and Communication Technologies (ICTs), production can be distributed between several factories and facilities. The concept of connected factories enables modern production systems to be collaborative, agile, and resilient to disruptions. This connectivity is established at the information level where suppliers, manufacturers, and customers can exchange information. The aim is to optimize production planning and scheduling (Katasonov et al., 2017). To reach this level of automation, researchers and system developers must exploit technologies such as sensing networks, artificial intelligence (AI), and Big Data.

The amount of generated data in the manufacturing sector is exceptionally large and can be a challenge to analyze and process (O'Donovan et al., 2015). These data are mainly generated from the sensors and devices on the shopfloor. AI-based applications can model production based on the data to support companies in making decisions and estimating production outcomes (Borja et al., 2021). These models can be evolving and adapting to newly generated data. Another usage for this data is to create sophisticated digital twins that can simulate the manufacturing system for safer and better production scheduling as presented by Mohammed et al. (2022). In this regard, the concept of a smart connected factory depends on sharing this data among all the involved partners in the supply chain. As an example, a manufacturer can predict material shortage based on the data of the supplier, and the customer can expect a change in packaging due to production and delivery changes on the manufacturer's side. As another example, Borja et al. (2021) compare the use of databases in knowledge-based manufacturing. They develop a solution exploiting numerous synergies, which is a winning formula for Big Data scenarios.

This interconnected, yet distributed, production system depends on the collected data. In this regard, many published research projects adopted the use of wireless low-power sensors to provide more precise modeling capabilities. As an example, the use of wearable sensors on the shopfloor gives insights into the status of the human, which can be essential to avoid accidents as explained by Sedighi Maman et al. (2017). Moreover, and in connection with the HRC, modeling the human at work requires sensing human emotions, feelings, biosignals, and surroundings. This means that the developed solution will need haptic sensing to collect signals such as, for example, electroencephalogram, electrocardiogram, electromyogram, and blood oxygen level. Additionally, nonhaptic sensing can be used to perceive human body language. With these approaches, a human can function as a mobile wireless sensing station. The challenge is to develop comfortable, lightweight sensors and related devices that can be carried for several hours.

In relation to the food industry, and as described in Section 7.2, a substantial part of the tasks is performed in farming environments. This means that the need for wireless sensing is larger than it is in the manufacturing environment. Adding to that, and as farming environments might lack proper telecommunication infrastructure, developers are encouraged to provide solutions for remotely located devices. As an example, Koster et al. (2011) present a novel approach for the Mobile Internet Protocol Version 6. Besides, Aqeel-ur-Rehman et al. (2014) present a review on using wireless sensors in the agricultural sector and predict substantial improvements in crop yield. In the review, the authors list several reasons for using these sensors, including monitoring the weather, soil, and crops, monitoring the quality of the irrigation and fertilization of the soil, monitoring the product status and ripeness, and pesticides control.

7.4 Internet of Things

According to the Industrial Internet Consortium, the IoT can be defined as "a concept where components are connected via a computer network, and where one or more of those components interacts with the physical world" (Vavra, 2020). The IoT is an important driver to facilitate value networks in which sensors are applied on physical objects, which can include not only pallets in supply chains, industrial machinery, and household appliances but also consumer goods (Rejeb et al., 2019).

In IoT, the "thing" is defined as a physical object that can be assigned an internet protocol address, with the ability to send data over a network. This ability to suddenly gain access to transmittable data has translated into new ways to monitor machines during production (Dey et al., 2018). The opportunities for IoT in the food industry are almost limitless. Food supply chains are better managed with a wide range of sensors and the availability of software solutions to analyze data at all levels from raw materials within the factory to the delivery of products to consumers (Bigliardi et al., 2022). The main applications of IoT in the food sector are aimed at reducing production costs, improving productivity through shorter set-up times, reducing errors and downtime, increasing information related to the production process, improving the quality of the products offered and reducing waste, customization and greater involvement of the end customer, and flexibility in the production of small batches (Bigliardi et al., 2022).

The combination of advanced technologies and data processing capabilities, such as the IoT, machine learning, blockchain, and deep learning techniques in supply chain activities will help to build reliable and intelligent supply chains (Sadeghi Darvazeh et al., 2020). In this regard, blockchain (or, in a more general way: distributed ledgers) can help to connect numerous players along the supply chain and allow access to immutable data in basically real time. This not only helps to create trust in the data itself but also facilitates auditing procedures. By extending the information chain to final consumers, they can become part of the overall information flow and learn about the origin and distribution of their food products, which helps them to assess the quality of their purchases (Treiblmaier & Garaus, 2023).

Machine learning algorithms can be applied to accurately predict imminent machine failures. For instance, IoT can provide real-time telemetry data through the monitoring of machinery in supply chains to reveal the details of production processes (Sadeghi Darvazeh et al., 2020). In this respect, digital technologies such as the IoT can bridge gaps created by the sudden disruption of the food supply chain as a result of the lack of information regarding consumer preferences and attitudes in the postlockdown period due to the pandemic (Galanakis et al., 2021). Similarly, the IoT can also support the creation of sustainable food security systems and help to fight poverty and hunger. A crucial issue in this regard is the avoidance of food waste, which can be achieved by creating efficient distribution systems that minimize wastage through efficient monitoring and distribution systems (Kaur, 2019).

7.5 Food processing and handling

The application of Industry 4.0 innovations in food processing and handling systems enables the aggregation of vast amounts of data. These data provide the basis for further improvements in precision technology that can lead to solutions that can improve traditional agrifood systems (Spanaki et al., 2022; Wolfert et al., 2017). Both terrestrial and aquatic resources can be better valorized at production, consumption, and waste disposal stages with robotic Industry 4.0 innovations. Examples of this include the application of drones in digital aquaponics farms that involve remote monitoring and control of ecosystem parameters, a high degree of automation, and intelligent decision-making to ensure high crop yields and quality (Hemathilake & Gunathilake, 2022). In this example, the application of Industry 4.0 technologies is needed to enhance the production capabilities of the aquaponics system and maximize crop yield on a commercial level (Abbasi et al., 2022).

Hydroponics, aquaponics, and precision/smart agricultural technologies contribute to the enhancement of agricultural productivity (Hemathilake & Gunathilake, 2022). Current developments in nanotechnology also provide many promising applications in the food industry, such as the use of former food waste and by-products in sustainable food packaging strategies (Freitas et al., 2021). The trendy valorization of foods such as plant-based, insect-based, cell-based, 3D-printed, personalized, and gene-edited foods, as well as foods resulting from by-products and ugly produce (i.e., food that is deemed too "ugly" to consume), is recently reviewed as new production, processing, and handling systems that can benefit from smart and digital solutions (Hassoun et al., 2022).

Robotic machinery has been successfully deployed to harvest apples and other crops. A mobile robot examines the health of the crops and carries out simple operations such as pruning, observing, and controlling ripening profiles (Golombek, 2019). Robotic harvesting vehicles can replace farm workers and perform labor-intensive tasks by combining recent developments in visual sensor technologies, automation, and autonomous propulsion (Raheem, 2020).

Smart Agri-Food operations can also be achieved by combining the capabilities of AI and Agricultural Technology (Agri-Tech) drones. These unmanned drones mimic the ways in which bird swarms behave and open up new opportunities to operate in inaccessible and remote areas. The overall goal is to enhance productivity, which is especially important in areas that exhibit limited capacity (Spanaki et al., 2022). Among others, the applications operating in these drones can measure nitrogen deficiencies in the soil and determine the right amount of fertilizer. These drones can also identify the amount of insecticide and herbicide needed for specific areas and selectively apply it in specific regions, thereby minimizing the use of potentially harmful substances. These procedures can contribute to ensuring food security by maximizing the yield while optimizing the needed resources, which is especially important in light of the growing world population. Overall, due to an aging society and labor shortages in many countries, automation in the food industry is in high demand to maintain profit margins.

7.6 Robots and applications of robotics systems

As illustrated earlier, the food industry includes all the activities that are required to produce food products starting from the initial planting and growing processes, passing by the picking, cleaning, preparing, processing, and ending in packing and delivering the final food products. Thus automation, and especially robotics systems are intensively exploited technologies due to their versatility and dexterity. In fact, several types of robots and robotic systems are used in the food industry including single robotic arms, duel robotic arms, cobots, drones, off-road wheeled mobile robots, legged mobile robots, and parallel robots. The use of modern technologies has been driven by the evolvement of product specifications such as size and material. As an example, the change from plastic hard containers to sustainable soft wrapping plastic transformed the products that need to be handled into three-dimensional deformable objects. Therefore this section presents innovative applications where robots and robotic systems are deployed in the food industry. Moreover, the common topic in these applications is to highlight robots as an enabling technology for the food industry.

7.6.1 Food as a deformable material: handling and manipulations

As previously highlighted, the need for changing the packaging might introduce new challenges for industries where robots are used for handling the products. The food industry is currently subjected to such a change due to the amount of waste generated and the envisioned reduction of plastic material usage as it should be food-safe packaging.

Thus robotics specialists face the challenge to reprogram robots that handle soft deformable material rather than hard plastic containers. Robotics material handling is mainly divided into two phases: grasping and moving. In the case of nondeformable material, once the object is grasped by the robot, the object movement will be restricted to the motion of the robot. However, in a deformable materials case, the object will move and deform after grasping. This deformation increases the complexity of material handing.

A solution has been introduced in the European project titled as Smart Grading, Handling, and Packaging Solutions for Soft and Deformable Products in Agile and Reconfigurable Lines (AGILEHAND) (European Commission, 2023). The AGILEHAND project aims to provide solutions for handling and manipulating food-deformable products. The proposed solution will be demonstrated in four pilots, which include meat production, berries production, fish production, and vegetable production. In more detail, the solution is divided into three main topics: smart sensing, robotic material handling, and agile and flexible production systems. The smart sensing will provide information such as material specifications, size, and weight. The smartness of the sensing system is a result of creating a reasoning layer. This layer collects raw data from sensors and then provides food object information. As an example, detecting the malleability of a material cannot be achieved without physical contact; however, by knowing the material, its temperature, and its surface color, a reasoner may identify its rigidity. For instance, a frozen fish can be handled as a rigid body, whereas a shiny fish at room temperature is malleable and slippery. Once the material specification is identified, the information is transferred to the robotic handling system. This information will be needed for the robots to (1) know how to approach the object for grasping, (2) know the gasping location and grasping force, and (3) calculate the motion profile such that the material is not damaged. Moreover, the robotic handling system uses robots to transfer material. This activity requires bimanual robotic material handling technologies. In other words, the motion of the robots will depend on the forces at the griper. These forces must be maintained at a specific level to avoid damaging the material. As a result of such a robotic handling system, AGILEHAND will be able to enable flexible and agile production systems.

7.6.2 Drones as an enabling technology for harvesting wild berries

As discussed in Section 7.2, the farming environment seeks robotic systems to support humans at work, and food collection is one of the major activities that may benefit from robotics deployment. When it comes to wild food products such as berries and mushrooms, the majority of the food remains in the forest unpicked. According to Vaara et al. (2013), around 10% of the berries were collected from the forests in Finland in 2011. When transferring this percentage to weight, the remained food in the forest may equal 500 million kg. The cause of such problem includes a shortage of human workers to collect the products, short seasons (i.e., the time between food to be ready for picking and being inedible), difficulty to reach and navigate in the forests, and the limitation of what human can pick and carry per day. With all the mentioned factors, Finland and Nordic countries, in general, continuously look for solutions to increase the production of such "free food" (Ihalainen et al., 2002; Sainz et al., 2010). Actually, one of the suggested approaches is to ease the procedures for hiring foreign pickers. This practice has been done to help in collecting the berries for several years. Nonetheless, this approach creates many challenges such as language barriers, possible human abuse and trafficking, health issues, and long working hours (Steensig, 2021; Yle News, 2020).

With these challenges and opportunities for business prosperity, several technical solutions have been planned, designed, and already tested. One of these solutions is proposed by the newly awarded European project titled as *Fostering and Enabling AI, Data and Robotics Technologies for Supporting Human Workers in Harvesting Wild Food (FEROX)* (European Commission, 2022). The FEROX project aims to provide robotic solutions for supporting the workers in the forest while collecting the berries. The solution is planned to be demonstrated in selected areas in Finland with the collaboration of the Arctic Flavors Association. These demonstrations will employ foreign and local workers with the presence of Social Sciences and Humanities experts alongside the technical teams. This will help to assess the benefits of the project in supporting these workers.

The technical solution includes the application of self-navigated drones equipped with Light Detection and Ranging (LiDAR), cameras, and sensors. Some of these drones will be deployed to scan and identify the locations of the berries. Once a drone scans a certain area, a detailed 3D model will be created. These models will include information about the trees, bushes, and terrain. For an AI system to detect the locations of the berries, the generated 3D model will be used together with continuous weather and meteorological data such as temperature, humidity, rain, and snow levels. Another group of drones will be deployed while the workers collect the berries. These drones will be equipped with navigation systems and cameras to help allocate the workers in the forest. This service will also help in supporting the workers in case of injuries, dehydration, disorientation, or loneliness. Moreover, these drones can direct the first aid responders to the exact location of the workers in case of an accident. A third group of drones will be deployed to help the workers carrying the heavy buckets of collected berries. On average, a worker collects up to 150 kg of berries every day (Vaara et al., 2013). Due to the heavy load and the irregular terrain, injuries may occur. In this regard, these heavy-lifting drones will be coordinated by an AI system to run shifts for collecting berries from the workers.

7.7 Conclusions

With the rise in its deployments at factories, robotics technologies are forecasted to be a key enabling technology for the coming years. This is already evident in the momentum of research in related topics such as robots, AI, IoT, and Big Data. In this regard, this chapter presented robotics technology as an enabling technology for the food industry. In the wake of the fourth revolution in the food industries, robotics emerges as a key solution for complex applications like supporting workers in the field and in the factory. Moreover, robotics, and especially soft robotics, can be essential for elevating food handling and manipulation.

This involvement of robotics is highly important to provide sustainable solutions both for environmental and production problems. Thanks to recent research and development, robots can safely work and share the working space with humans. This human–robot collaboration can be critical for deploying robots to work in the farms for planting, irrigating, monitoring, and collecting food. In addition, with the advances in AI, Big Data, and IoT technologies, robots can be equipped with devices that enable cognition capabilities. These capabilities may empower robots to reach human-friendly behavior where trust and understanding are needed.

In the food industries, the overall research direction must focus on strengthening the collaboration between humans and robots. In this regard, robots can perform dull, dirty, and repetitive tasks, while humans can focus on intellectual tasks that require reasoning. The merge between these capabilities will elevate the food industry in general, and specifically help in farming environments where robots are currently less employed. This goal of finding synergies between humans and robots will require reskilling and training for humans to better understand how robots work and what they can achieve. Finally, this transition of seamlessly integrating robots in nontechnical industries will also require substantial efforts at a societal level.

References

Abbasi, R., Martinez, P., & Ahmad, R. (2022). An ontology model to represent aquaponics 4.0 system's knowledge. *Information Processing in Agriculture*, *9*, 514–532. Available from https://doi.org/10.1016/j.inpa.2021.120.001.

Adamo, N., Al-Ansari, N., & Sissakian, V. (2021). Review of Climate Change Impacts on Human Environment: Past, Present and Future Projections. *ENG*, *13*, 605–630. Available from https://doi.org/10.4236/eng.2021.1311044.

Aqeel-ur-Rehman., Abbasi, A. Z., Islam, N., & Shaikh, Z. A. (2014). A review of wireless sensors and networks' applications in agriculture. *Computer Standards & Interfaces*, *36*, 263–270. Available from https://doi.org/10.1016/j.csi.2011.030.004.

Bahrin, M. A. K., Othman, M. F., Azli, N. H. N., & Talib, M. F. (2016). Industry 4.0: A Review on Industrial Automation and Robotic. *Jurnal Teknologi*, *78*. Available from https://doi.org/10.11113/jt.v78.9285.

Bigliardi, B., Bottani, E., & Filippelli, S. (2022). A study on IoT application in the Food Industry using Keywords Analysis. *Procedia Computer Science*, *200*, 1826–1835. Available from https://doi.org/10.1016/j.procs.2022.010.383.

Borja, R. F., Wael, M., Mussawar, M. A., Sergii, I., Jiayi, Z., Robert, H., & Jose, L. M. L. (2021). Comparing ontologies and databases: a critical review of lifecycle engineering models in manufacturing. *Knowledge and Information Systems*.

Cimini, A., & Moresi, M. (2019). A progressive approach towards a more sustainable food industry. *Chemical Engineering Transactions*, *75*, 565–570. Available from https://doi.org/10.3303/CET1975095.

Culot, G., Nassimbeni, G., Orzes, G., & Sartor, M. (2020). Behind the definition of Industry 4.0: Analysis and open questions. *International Journal of Production Economics*, *226*, 107617. Available from https://doi.org/10.1016/j.ijpe.2020.107617.

De Miguel Lazaro, O., Mohammed, W. M., Ferrer, B. R., Bejarano, R., & Martinez Lastra, J. L. (2019). *An Approach for adapting a Cobot Workstation to Human Operator within a Deep Learning Camera. IEEE 17th International Conference on Industrial Informatics (INDIN)* (2019, pp. 789–794). Helsinki, Finland: IEEE. Available from https://doi.org/10.1109/INDIN41052.2019.8972238.

Demartini, M., Pinna, C., Tonelli, F., Terzi, S., Sansone, C., & Testa, C. (2018). Food industry digitalization: from challenges and trends to opportunities and solutions. *IFAC-PapersOnLine*, *51*, 1371–1378. Available from https://doi.org/10.1016/j.ifacol.2018.080.337.

Dey, N., Hassanien, A. E., Bhatt, C., Ashour, A. S., & Satapathy, S. C. (2018). *Internet of Things and Big Data Analytics Toward Next-Generation Intelligence*. Cham: Springer International Publishing. Available from https://doi.org/10.1007/978-3-319-60435-0.

Dinakaran, V. P., Balasubramaniyan, M. P., Le, Q. H., Alrubaie, A. J., Al-khaykan, A., Muthusamy, S., Panchal, H., Jaber, M. M., Dixit, A. K., & Prakash, C. (2023). A novel multi objective constraints based industrial gripper design with optimized stiffness for object grasping. *Robotics and Autonomous Systems*, *160*, 104303. Available from https://doi.org/10.1016/j.robot.2022.104303.

European Commission, FEROX Project, Fostering and Enabling AI, Data and Robotics Technologies for Supporting Human Workers in Harvesting Wild Food. (2022). https://cordis.europa.eu/project/id/101070440 (accessed January 28, 2023).

European Commission, AGILEHAND Project, Smart Grading, Handling and Packaging Solutions for Soft and Deformable Products in Agile and Reconfigurable Lines | (2023). https://doi.org/10.3030/101092043 (accessed January 28, 2023).

Freitas, L. C., Barbosa, J. R., da Costa, A. L. C., Bezerra, F. W. F., Pinto, R. H. H., & Carvalho Junior, R. N. de (2021). From waste to sustainable industry: How can agro-industrial wastes help in the development of new products? *Resources, Conservation and Recycling*, *169*, 105466. Available from https://doi.org/10.1016/j.resconrec.2021.105466.

Galanakis, C. M., Rizou, M., Aldawoud, T. M. S., Ucak, I., & Rowan, N. J. (2021). Innovations and technology disruptions in the food sector within the COVID-19 pandemic and post-lockdown era. *Trends in Food Science & Technology*, *110*, 193–200. Available from https://doi.org/10.1016/j.tifs.2021.020.002.

Golombek, A. (2019). *Robots are in the process of revolutionizing farming | RealClearMarkets*. <https://www.realclearmarkets.com/articles/2019/02/19/robots_are_in_the_process_of_revolutionizing_farming_103628.html/> Accessed 30.01.23.

Hassoun, A., Cropotova, J., Trif, M., Rusu, A. V., Bobiş, O., Nayik, G. A., Jagdale, Y. D., Saeed, F., Afzaal, M., Mostashari, P., Khaneghah, A. M., & Regenstein, J. M. (2022). Consumer acceptance of new food trends resulting from the fourth industrial revolution technologies: A narrative review of literature and future perspectives. *Frontiers in Nutrition*, *9*, 972154.

Hemathilake, D. M. K. S., & Gunathilake, D. M. C. C. (2022). High-productive agricultural technologies to fulfill future food demands: Hydroponics, aquaponics, and precision/smart agriculture. In R. Bhat (Ed.), *Future foods* (pp. 555–567). Academic Press. Available from https://doi.org/10.1016/B978-0-323-91001-9.00018-9.

Henry, R. (2020). Innovations in agriculture and food supply in response to the COVID-19 pandemic. *Molecular Plant*, *13*, 1095–1097. Available from https://doi.org/10.1016/j.molp.2020.070.011.

Hormaza, L. A., Mohammed, W. M., Ferrer, B. R., Bejarano, R., & Martinez Lastra, J. L. (2019). On-line training and monitoring of robot tasks through virtual reality. In *2019 IEEE 17th international conference on industrial informatics (INDIN)*, pp. 841–846. <https://doi.org/10.1109/INDIN41052.2019.8971967/>.

Ihalainen, M., Alho, J., Kolehmainen, O., & Pukkala, T. (2002). Expert models for bilberry and cowberry yields in Finnish forests. *Forest Ecology and Management*, *157*, 15–22. Available from https://doi.org/10.1016/S0378-1127(00)00653-8.

Jazar, R. N. (2010). Theory of applied robotics: Kinematics, dynamics, and control, *2nd ed*. New York: Springer.

Jung, J. H., & Lim, D.-G. (2020). Industrial robots, employment growth, and labor cost: A simultaneous equation analysis. *Technological Forecasting and Social Change*, *159*, 120202. Available from https://doi.org/10.1016/j.techfore.2020.120202.

Katasonov, A., Lastusilta, T., Korvola, T., Saari, L., Bendas, D., Mohammed, W. M., & Lee, A. N. (2017). *An approach to production scheduling optimization a case of an oil lubrication and hydraulic systems manufacturer. 2017 International conference on engineering, technology and innovation (ICE/ITMC)* (pp. 1123–1130). IEEE. Available from https://doi.org/10.1109/ICE.2017.8280007.

Kaur, H. (2019). Modelling internet of things driven sustainable food security system. *Benchmarking: An International Journal*, *28*, 1740–1760. Available from https://doi.org/10.1108/BIJ-12-2018-0431.

Koster, V., Dorn, D., Lewandowski, A., & Wietfeld, C. (2011). A novel approach for combining micro and macro mobility in 6LoWPAN enabled networks. In *2011 IEEE vehicular technology conference (VTC Fall)*, pp. 1–5. https://doi.org/10.1109/VETECF.2011.6093126.

Marvin, H. J. P., Bouzembrak, Y., van der Fels-Klerx, H. J., Kempenaar, C., Veerkamp, R., Chauhan, A., Stroosnijder, S., Top, J., Simsek-Senel, G., Vrolijk, H., Knibbe, W. J., Zhang, L., Boom, R., & Tekinerdogan, B. (2022). Digitalisation and artificial Intelligence for sustainable food systems. *Trends in Food Science and Technology*, *120*, 344–348. Available from https://doi.org/10.1016/j.tifs.2022.010.020.

McHugh, T. (2017). *Robotics on the rise in the food industry. Food technology magazine*. <https://www.ift.org/news-and-publications/food-technology-magazine/issues/2017/october/columns/processing-robotics-food-beverage-processing-foodservice/> Accessed 31.01.23.

Mohammed, W. M., Haber, R. E., & Martinez Lastra, J. L. (2022). Ontology-driven guidelines for architecting digital twins in factory automation applications. *Machines*, *10*, 861. Available from https://doi.org/10.3390/machines10100861.

Oltra-Mestre, M. J., Hargaden, V., Coughlan, P., & Segura-García del Río, B. (2021). Innovation in the agri-food sector: Exploiting opportunities for Industry 4.0. *Creativity and Innovation Management*, *30*, 198–210. Available from https://doi.org/10.1111/caim.12418.

O'Donovan, P., Leahy, K., Bruton, K., & O'Sullivan, D. T. J. (2015). Big data in manufacturing: A systematic mapping study. *Journal of Big Data*, *2*, 20. Available from https://doi.org/10.1186/s40537-015-0028-x.

Raheem, D. (2020). Digitalisation in a local food system: Emphasis on Finnish Lapland. *Open Agriculture*, *5*, 496–508. Available from https://doi.org/10.1515/opag-2020-0049.

Rejeb, A., Keogh, J. G., & Treiblmaier, H. (2019). Leveraging the internet of things and blockchain technology in supply chain management. *Future Internet*, *11*, 161. Available from https://doi.org/10.3390/fi11070161.

Rejeb, A., Abdollahi, A., Rejeb, K., & Treiblmaier, H. (2022). Drones in agriculture: A review and bibliometric analysis. *Computers and Electronics in Agriculture*, *198*, 107017. Available from https://doi.org/10.1016/j.compag.2022.107017.

Rejeb, A., Rejeb, K., Abdollahi, A., Al-Turjman, F., & Treiblmaier, H. (2022). The interplay between the internet of things and agriculture: A bibliometric analysis and research agenda. *Internet of Things*, *19*, 100580. Available from https://doi.org/10.1016/j.iot.2022.100580.

Sadeghi Darvazeh, S., Raeesi Vanani, I., & Mansouri Musolu, F. (2020). Big data analytics and its applications in supply chain management. In L. Romeral Martínez, R. A. Osornio Rios, & M. Delgado Prieto (Eds.), *New trends in the use of artificial intelligence for the Industry 4.0*. IntechOpen. Available from https://doi.org/10.5772/intechopen.89426.

Sainz, R. C., Tome, M., Sánchez-González, M., Miina, J., Spanos, K., & Palahi, M. (2010). Modelling non-wood forest products in Europe: A review. *Forest Systems*, *19*, 69–85. Available from https://doi.org/10.5424/fs/201019S-9324.

Sciavicco, L., & Siciliano, B. (2009). *Modelling and control of robot manipulators*, *2000*. <https://doi.org/10.1007/978-1-4471-0449-0/> Accessed 25.02.19.

Sedighi Maman, Z., Alamdar Yazdi, M. A., Cavuoto, L. A., & Megahed, F. M. (2017). A data-driven approach to modeling physical fatigue in the workplace using wearable sensors. *Applied Ergonomics*, *65*, 515–529. Available from https://doi.org/10.1016/j.apergo.2017.020.001.

Siegwart, R., Nourbakhsh, I. R., & Scaramuzza, D. (2011). *Introduction to autonomous mobile robots* (2nd ed.). MIT Press.

Spanaki, K., Karafili, E., Sivarajah, U., Despoudi, S., & Irani, Z. (2022). Artificial intelligence and food security: Swarm intelligence of AgriTech drones for smart AgriFood operations. *Production Planning & Control*, *33*, 1498–1516. Available from https://doi.org/10.1080/09537287.2021.1882688.

Statista Research Department, Industrial robots—worldwide installations, Statista. (2023). <https://www.statista.com/statistics/264084/worldwide-sales-of-industrial-robots/> Accessed 23.01.23.

Steensig, S.L. (2021). *Did Finland do enough to protect its foreign berry pickers from COVID? Euronews*. <https://www.euronews.com/2021/09/16/did-finland-do-enough-to-protect-its-foreign-berry-pickers-from-covid/> Accessed 02.02.23.

Thoben, K.-D., Wiesner, S., & Wuest, T. (2017). "Industrie 4.0" and smart manufacturing – A review of research issues and application examples. *International Journal of Automation Technology*, *11*, 4–16. Available from https://doi.org/10.20965/ijat.2017.p0004.

Toichoa Eyam, A., Mohammed, W. M., & Martinez Lastra, J. L. (2021). Emotion-driven analysis and control of human-robot interactions in collaborative applications. *Sensors*, *21*, 4626. Available from https://doi.org/10.3390/s21144626.

Treiblmaier, H., & Garaus, M. (2023). Using blockchain to signal quality in the food supply chain: The impact on consumer purchase intentions and the moderating effect of brand familiarity. *International Journal of Information Management*, *68*, 102514. Available from https://doi.org/10.1016/j.ijinfomgt.2022.102514.

Vaara, M., Saastamoinen, O., & Turtiainen, M. (2013). Changes in wild berry picking in Finland between 1997 and 2011. *Scandinavian Journal of Forest Research*, *28*, 586–595. Available from https://doi.org/10.1080/02827581.2013.786123.

Vavra, C. (2020). Industrial internet of things vocabulary terms updated: The industrial internet consortium (IIC) released updated definitions for terms such as the internet of things, cloud computing and more. *Control Engineering*, *67*, 45.

Vermesan, O., & Bacquet, J. (Eds.), (2017). *Cognitive hyperconnected digital transformation: Internet of things intelligence evolution* (1st ed.). River Publishers. Available from https://doi.org/10.1201/9781003337584.

Wang, Z., Hirai, S., & Kawamura, S. (2022). Challenges and opportunities in robotic food handling: A review. *Frontiers in Robotics and AI*, *8*. Available from https://doi.org/10.3389/frobt.2021.789107.

Wolfert, S., Ge, L., Verdouw, C., & Bogaardt, M.-J. (2017). Big data in smart farming – A review. *Agricultural Systems*, *153*, 69–80. Available from https://doi.org/10.1016/j.agsy.2017.010.023.

Yle News. (2020). *Lapland police report busy summer rescuing lost berry pickers. News*. <https://yle.fi/a/3-11458780/> Accessed 02.02.23.

Chapter 8

Significant roles of smart sensors in the modern agriculture and food industry

Sebahattin Serhat Turgut[1,2], Gonca Bilge[3], José Antonio Entrenas[4] and Dolores Pérez-Marín[4]

[1]*Research Group for Food Production Engineering, National Food Institute, Technical University of Denmark, Kgs Lyngby, Denmark,* [2]*Department of Food Engineering, Faculty of Engineering and Natural Sciences, Suleyman Demirel University, Cunur, Isparta, Turkey,* [3]*Department of Food Engineering, Faculty of Engineering, Yeditepe University, Istanbul, Türkiye,* [4]*Department of Animal Production, Faculty of Agriculture and Forestry Engineering, University of Cordoba, Campus Rabanales, Córdoba, Spain*

8.1 Introduction

The Fourth Industrial Revolution, often known as Industry 4.0, is a term used to characterize the current trend of integrating digital technologies into industrial production and supply chain management. One of the key areas in which these technologies are being applied is in the food and agriculture sector, where smart sensors are being used to improve quality, safety, efficiency, and sustainability. The use of smart systems is expected to increase by accelerating so that the Food and Agriculture Organization (FAO) estimates that present agricultural production must grow by 70% by 2050 to meet the rising demand for high-quality, environmentally friendly food (FAO, 2009). Similarly, the United Nations (UN) general assembly approved the agenda "Transforming our World: the 2030 Agenda for Sustainable Development" in 2015, which is an action plan for the prosperity of people and the planet. Under this agenda, 17 Sustainable Development Goals have been embodied with a timeframe of completion from 2015 to 2030 and two of these goals are "Zero Hunger" and "Industry, Innovation and Infrastructure" (UN, 2015).

Ensuring food quality, safety, and production sustainability is crucial not only to meet the growing food demand, but also for producers, legislators, and control mechanisms. Given the very complex nature of foods, a continuous effort is necessary to guarantee their safety and quality. Moreover, foods can be sensitive to process-induced impacts such as temperature, oxygen, light, pressure, and electricity. As a result, quality aspects are sacrificed to some extent to assure safety throughout the agrofood chain. Besides that, producers may resort to illegal imitation and fraud practices to maximize economic benefit, improve the visible value of the product, conceal food defects, reduce production costs, etc., which is another food safety and the quality threat that must be avoided. However, the risks in the agrofood industry are not limited to those. In addition, factors such as pollutants, pesticides/antibiotic residues, microbial spoilage, infestation, and insect residues that are transmitted to agricultural products from polluted/waste waters in the agricultural field or on farms and diversification of the given risk factors due to evolving technology and precautions implemented are among the factors to be prevented (FAO, 2022; UNCTAD, 2017). As a result, to ensure food safety and quality, existing measurement methods should be updated regularly, and/or new methods must be introduced. That is why maintenance and improvement of required safety and quality attributes of foods is one of the most prominent areas in food science and researchers are constantly exploring food analysis technologies to detect all possible sources of threats to food from the farm to the fork.

It is not possible to provide food safety, quality, and sustainability through the whole agrofood chain with a single action. It is therefore required to deal with it by breaking it into smaller subunits whenever feasible. Then, control measures should be taken for the steps such as the cultivation period (pre- and postharvesting) and food production (pre- and postproduction) including the proper storage and processing of raw materials, followed by packaging and storage. The postproduction phase can be extended to entail transportation and storage of the product in markets until consumers purchase it.

Food Industry 4.0. DOI: https://doi.org/10.1016/B978-0-443-15516-1.00008-6

There are many different methods available to take measurements at the determined control points, starting from the preharvest to the postprocessing steps. Among them, besides the traditional methods (such as titration and microbial plating), instrumental analytical methods (such as spectroscopic, chromatographic, and electrophoretic methods) have become increasingly routine in the last few decades (Otles et al., 2015). The most common instrumental analytical methods for food analysis are listed as liquid and gas chromatography, mass spectroscopy, capillary electrophoresis, infrared spectroscopy, and nuclear magnetic resonance spectroscopy (Tang et al., 2019). These approaches (together with traditional methods) have a wide range of applications, including the detection of food composition and contaminants, identification of pathogenic and spoilage microorganisms in agrifood items, and analysis of quality properties. However, these methods have some limitations and disadvantages such as (1) having an extremely low amount of target analyte and/or (2) a higher minimum detection limit than the concentration of the component to be analyzed; (3) being time-consuming and labor-intensive sample preparation methods; (4) being economically less efficient due to the lab space, trained personnel, and consumables, and (5) being destructive (Lu, 2017). It is crucial to develop nondestructive, active, and rapid testing methodologies to manage food quality and safety while also addressing green production and sustainability (El-Mesery et al., 2019; Hassoun et al., 2023). As a result, by the end of the 20th century, the food processing industry had gradually abandoned or transformed traditional inspection procedures in favor of automated process control. The new approach checks critical process control points to ensure that the end items fulfill standards. These advancements were made possible by the benefits of computers, microchips, and smart sensor technologies (Gao et al., 2020; Kondakci & Zhou, 2017; Sharma & Sharma, 2022).

Sensors are electrical devices or components of systems that identify events or changes, sense data, and send it to a processing system. They can be classified as active/passive sensors, industrial/nonindustrial sensors, and real-time/batch-processing sensors (Mukherjee et al., 2020). Parallel with the rise of the Industry 4.0 revolution in the 21st century, the growing technologies in computers, electronics, and data science such as arise of artificial intelligence, the Internet of Things (IoT), and cloud computing, the sensor understanding transformed into the *smart sensors* (Bouzembrak et al., 2019; Chowdhury et al., 2022; Galvan et al., 2022; Özdoğan et al., 2021). The basic difference between smart sensors and standard sensor technologies is the intelligence capacity of devices. Smart sensors are characterized as elements with sensing, signal processing, and understanding capabilities, with goals ranging from simple viewing to complex remote sensing, surveillance, search and track, robotics, perceptronics, and intelligence applications (Correa-Hernando et al., 2011). For a sensor to be called smart, it must have at least one of the following seven properties: ability to cooperate with other devices, adaptability (learning from the environment to meet users' wishes), autonomy (making decisions and taking actions alone), human-like interaction, multifunctionality, personality (being proactive), and reactivity (reacting to the changes in the environment) (Barreiro Elorza et al., 2011).

Several literature overviews are done so far considering the use of different smart sensing applications from different perspectives (Issad et al., 2019; Akhtar et al., 2021; Antonacci et al., 2018; Chen et al., 2020; Ha et al., 2020; Halonen et al., 2020; Hassan et al., 2021; Abdo Hassoun, Alhaj Abdullah, et al., 2022; Hassoun, Aït-Kaddour, et al., 2022; Hassoun et al., 2022; Hassoun et al., 2023; He et al., 2020; Kakani et al., 2020; Lakhwani et al., 2019; Mustafa & Andreescu, 2018; Režek Jambrak et al., 2021; Suma, 2021; Yousefi et al., 2019). However, Industry 4.0 understanding grows so rapidly that the technological development rate increases twofold every 18 months according to computing measures. Moreover, more than 89% of big data has been produced within the last 2 years (Michael, 2021). The consequences of this rapid development are undoubtedly affecting smart sensors, which are also developing at an unprecedented speed. For this reason, modern smart sensor applications in food and agriculture have been attempted to be described within this chapter from the perspective of Industry 4.0. To achieve this, the common sensor technologies (spectral and nonspectral sensors) used in food and agricultural systems are briefly introduced first for readers who lack fundamental knowledge. Next, the most recent advancements in smart sensor technology in food and agricultural applications are evaluated to show the potential of sensor technologies used along the entire agrifood chain from the perspective of smart processing. Detailed overviews are then provided for the cultivation period of agricultural products (preharvest and postharvest periods), the food production chain (packaging, quality and safety, process control, and optimization), and the use of joint technologies.

8.2 Sensor technologies for the food and agriculture

Sensor technologies have their roots in the 1800s. Current clamps and coils were used to measure current in the early 1800s. Instruments were installed in the 1890s to record atmospheric conditions that influenced the performance of grid components (Swain et al., 2022). The first aerial photography was taken in 1858 from a tethered balloon being the ancestor of remote sensors (Cracknell, 2018). With the improvement in electronics, the size of microprocessors, microcontrollers, and computers have shrunk in size and the sensors are affected by this evaluation. For example, image sensors for eye and face tracking in size of 0.575 mm × 0.575 mm can be produced today (Burgett, 2019), and many sensors, which were initially used with large devices and for specific purposes, have portable and smarter versions available in the market. In this context, brief information about the working principles of the sensors, which are mostly used within the agrofood sector within the scope of smart applications, are shared under two groups namely spectral and nonspectral sensors. Apart from them, there are different and/or traditional sensor technologies including fluorescence spectroscopy, X-ray spectroscopy, bio-speckling, etc., and temperature, pressure, level, flow sensors, etc. (which are mostly used as process control equipment). However, the sensors described in this chapter are only tools for smart sensor technologies. The real progress is driven by developments in big data, cloud computing, the use of graphical processing units, IoT, wireless technologies, and robotics.

A representative illustration showing the evaluation of the used technologies and sensors used in the agrofood chain is given in Fig. 8.1. So, while different sensors are still used only to collect data, calculations and decisions and actions are done by computer systems assisting the sensors such as remotely piloted aircraft systems, unmanned air vehicles (UAVs), global navigation satellite systems, geographic information systems, big data analysis, IoT, the Internet of Everything (IoE), cloud computing, wireless sensors technologies, decision support systems, and autonomous robots. However, the computational methods are outside the present chapter since numerous sources cover them in detail (Burns & Ciurczak, 2007; Ozaki et al., 2006; Williams & Norris, 1987; Williams et al., 2019). This chapter will focus on showing the potential of smart sensors through the entire agrifood chain from the perspective of smart processing. To illustrate it, the authors will mention first the fundamentals of some sensors very briefly, providing some examples of the most relevant research advances and studies published in recent years.

8.2.1 Spectral sensors

8.2.1.1 Infrared spectroscopy

Infrared (IR) spectroscopy is one of the nondestructive, noninvasive, rapid, and effective analysis techniques and is considered one of the ideal methods for monitoring the quality of food and agricultural products (Zareef et al., 2020). A schematic representation of IR reflectance spectroscopy is given in Fig. 8.2. IR (714–10^6 nm) is a vibrational spectroscopic method that investigates the relationship between the interaction of light and matter (solids, liquids, or gases) in the electromagnetic spectrum beyond the visible region (390–700 nm) (Bureau et al., 2019). The IR region is categorized as near-IR (714–2.5×10^2 nm, NIR), mid-IR (2.5×10^2–2.5×10^3 nm, MIR), and far-IR (2.5×10^3–10^6 nm, FIR) regions (Raja & Barron, 2016). However, NIR and MIR, primarily NIR, are preferred for food and agricultural product analyses (Bureau et al., 2019; O'Sullivan, 2020). Covalent bonds in organic molecules are not rig meaning that they are always in motion like vibration. When a molecule is irradiated with polychromatic IR electromagnetic radiation, which matches the frequency of its vibrational modes (symmetric stretching, antisymmetric stretching, rocking, wagging, twisting, scissoring), IR-active vibrations absorb the radiative energy. This increases the amplitude of the vibration, but the frequency remains the same. When the exciting and transmitted/reflected spectra are compared, it is seen that the intensity of certain wavelengths (or wavenumbers) has significantly dropped due to absorption by molecules. The information about the frequency of absorbed energy and density is used for qualitative and quantitative analyses (Schaller et al., 2015; Sun, 2009). IR spectroscopy is particularly sensitive to molecular overtone and combination vibrations of some hydrogen-based functional groups such as O–H, C–H, C–O, and N–H (Zareef et al., 2020), which makes IR spectroscopy suitable for detection of the four main food components (lipids, proteins, carbohydrates, and water) (Fellows, 2009).

In conventional IR spectroscopy, the light beams having varying wavelengths are generated mechanically using diffraction grating. However, today, Fourier Transform IR (FTIR) spectrometers offer significant advantages such as improved signal-to-noise ratio, faster IR spectra acquisition, and superior wavenumber accuracy. FTIR spectroscopy uses an interferometer to generate modulated light and measures all IR wavelengths simultaneously. The interferogram is then converted into a conventional spectrum using the Fourier transform algorithm and then the final spectrum is obtained by subtracting the blank spectrum (measured without sample) and the sample's spectrum (measured with sample) (O'Sullivan, 2020; Sun, 2009).

8.2.1.2 UV/Vis spectroscopy

The working principle of UV/Vis spectroscopy is a bit similar to that of IR spectroscopy except for some differences. The UV/Vis spectroscopy covers an energy region of 1.5–6.2 eV which relates to a wavelength range of 200–800 nm

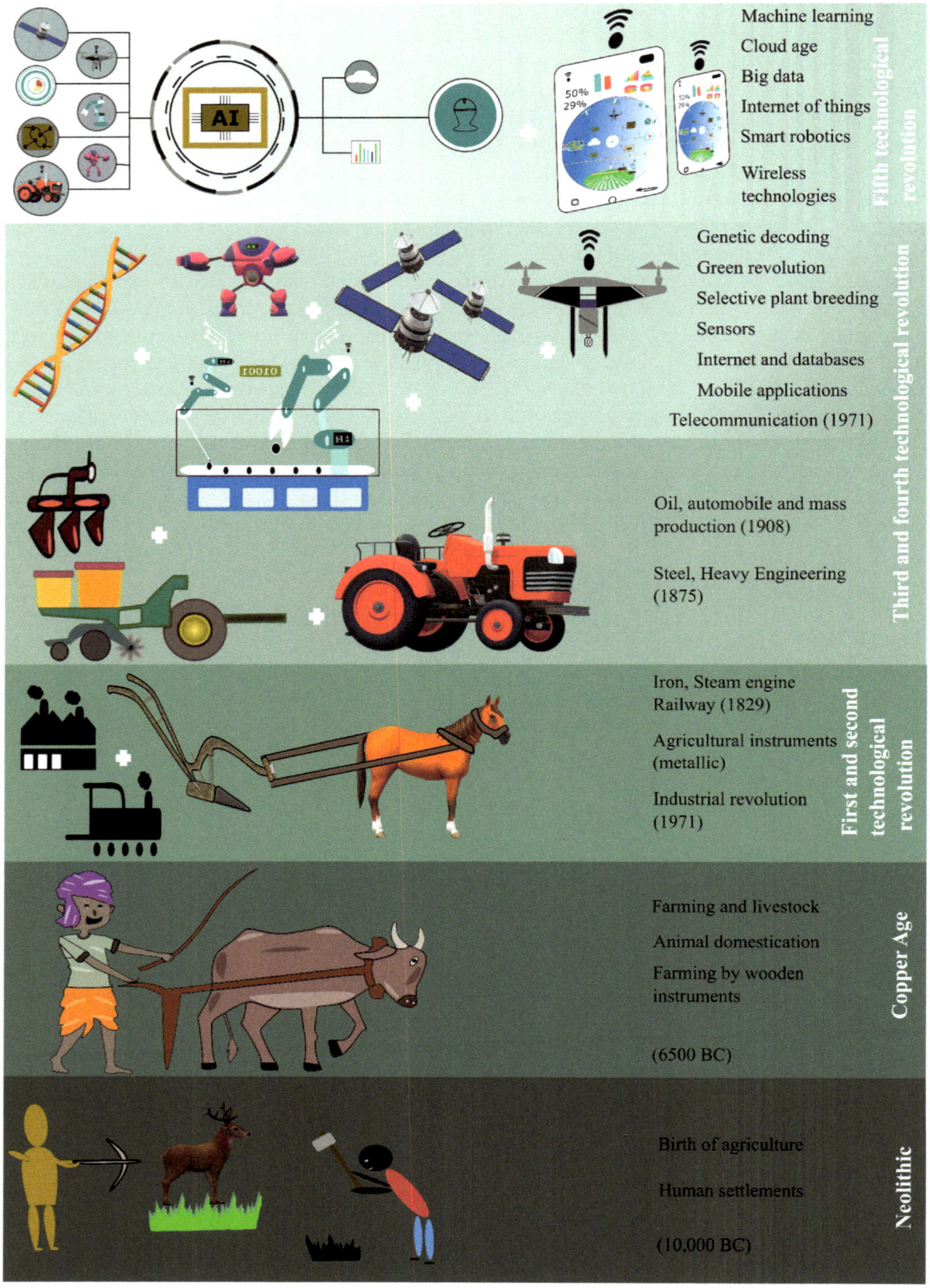

FIGURE 8.1 A representative illustration showing the evaluation of the used technologies and sensors used in the agrofood chain. *From Martos, V., Ahmad, A., Cartujo, P., & Ordoñez, J. (2021). Ensuring agricultural sustainability through remote sensing in the era of Agriculture 5.0.* Applied Sciences, 11*(13), 5911. https://doi.org/10.3390/app11135911.*

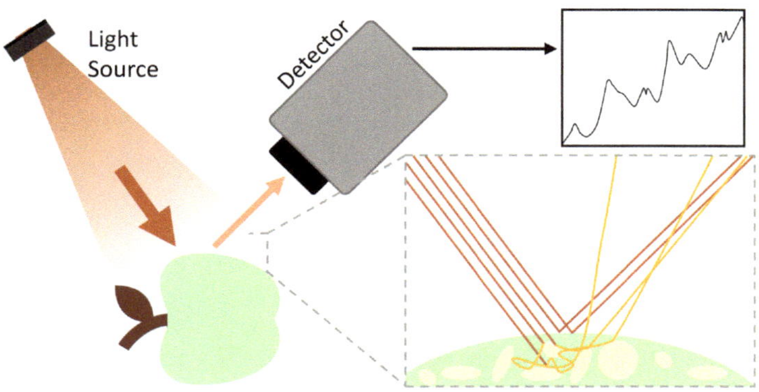

FIGURE 8.2 Schematic representation of the basic working principle of near-infrared diffuse reflectance spectroscopy.

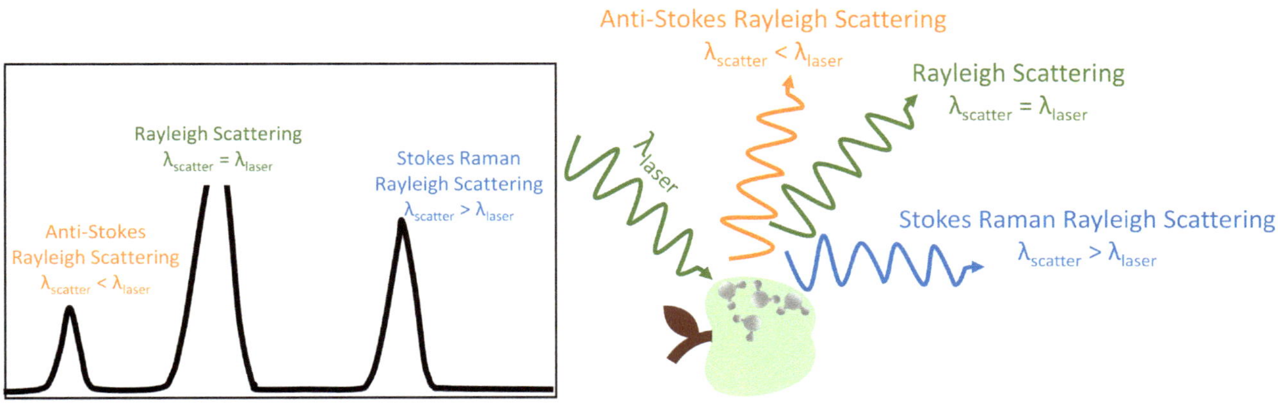

FIGURE 8.3 Schematic representation of the basic working principle of Raman spectroscopy.

(Raja & Barron, 2016). When sample molecules are exposed to light with the same energy as a potential electronic transition within the molecule, some of the light energy is absorbed when the electron is promoted to a higher energy orbital. However, UV/Vis spectroscopy provides much less information about molecular structure compared to IR and NMR spectroscopy, because the electronic excitation that may occur in organic molecules is limited by the energies available in the 200–800 nm range (Schaller et al., 2015). Apart from its common and simple use, it also has some restrictions. It does not perform well on liquids with suspended particles. In such a case, the sample will scatter the light more than it absorbs causing skewness of obtained data. Besides, it is possible to use UV/Vis instruments to analyze solid samples or suspensions with a diffraction apparatus; however, this is not common (Raja & Barron, 2016).

8.2.1.3 Raman spectroscopy

When a food sample is subjected to monochromatic light in the Vis region, the majority of light is transmitted and/or absorbed depending on the optical properties of the sample. However, some of the light is scattered in all directions. When an incident monochromatic light interacts with an electron in a sample, the electron rises to a virtual energy state and then falls back to the initial energy level by losing energy. In this case, the released photon has the same frequency as the incident photon. If the scattered light has the same frequency as the incident light, the scattering is known as *Rayleigh Scattering*, and it accounts for the majority of the total scattered light. However, sometimes only a tiny fraction of electrons can fall back to a different energy level than the incident photon while losing energy. This is possible when the frequency of the emitted photon is different from the incident photon and gives rise to *Raman Scattering* (Bumbrah & Sharma, 2016). Moreover, when the electron absorbs energy, the frequency of the scattered photon is less than the frequency of the incident photon and stokes lines are observed on the Raman spectrum (Bumbrah & Sharma, 2016; Lu, 2017). On the contrary, in the case of energy released by the electron, the frequency of the emitted photon is greater than the incident photon and antistokes lines form (Fig. 8.3). As a result, by examining the spectra and intensity of the Raman lines, a sample may be subjected to both quantitative and qualitative evaluation accurately, rapidly and nondestructively (Wang et al., 2018).

8.2.1.4 Nuclear magnetic resonance spectroscopy

The principle behind nuclear magnetic resonance (NMR) spectroscopy relies on the nuclei of an atom placed in a magnetic field and subjected to electromagnetic radiation resonating at a certain frequency (Fig. 8.4). NMR requires the nuclei to have a certain property which is called *spin*. Only *spin-active* or *NMR-active* nuclei are detectable by NMR spectroscopy, and it is a characteristic of nuclei that have an odd number of protons and/or neutrons (such as ^{1}H, ^{2}H, ^{11}B, ^{13}C, ^{31}P, ^{15}N, and ^{29}Si). Nuclei with an odd number of protons and an odd number of neutrons have integer spins, while nuclei with an odd number of protons or neutrons have half-integer spins. The most often utilized nuclei in NMR spectroscopy have half-integer spins (such as ^{1}H, ^{13}C, and ^{19}F). Because protons exist, nuclei are charged, and spin-active nuclei produce their own, small magnetic field as they move. Under normal circumstances, the spins of any given set of nuclei are disordered and their energy levels are equal. This situation is known as *degeneracy*. As the magnetic field is applied, some nuclei are oriented against the field (higher energy spin state) while others are aligned with it (lower energy spin state). The difference between these two energy states corresponds to a change in energy (ΔE). When variable frequency radio-frequency radiation is applied to this system, nuclei can only absorb energy that matches ΔE. The absorbed energy has a particular frequency that causes resonance, resulting in a peak in the NMR spectra. Most NMR spectra contain several signals at different frequencies because samples, in general, contain many nuclei with distinct ΔEs. This spectrum is then utilized for structural (quantitative) examination of the atoms in the sample (Lambert et al., 2019; Khaneja, 2020; Macomber, 1998).

8.2.1.5 Laser-induced breakdown spectroscopy

Laser-induced breakdown spectroscopy (LIBS) is an analytical atomic emission spectroscopic technique that especially has a key role in elemental analysis (Hussain Shah et al., 2020). For LIBS analysis, a high-energy laser (Neodymium-doped Yttrium Aluminum Garnet (Nd:YAG)) pulse is used to break down the sample's atoms and to form plasma (Yu et al., 2020). For the formation of plasma, the laser fluence should be greater than the latent heat of evaporation of the target (Jiang et al., 2017). The plasma emits radiations which are received by a spectrometer with the help of optical fiber to generate the spectrum. Afterward, the spectrum is used/processed for elemental detection. A schematic representation of LIBS is given in Fig. 8.5. Although the generation of the plasma destroys the product surface, its ablation is so small (in microns) that LIBS is considered as a nondestructive analysis method. LIBS can be used for both quantitative and qualitative analysis (Hussain Shah et al., 2020). While the presence of wavelengths, which is characteristic of each atom, is used to determine existing elements in the sample, spectral intensity is used to find the amount of these elements (Gaudiuso et al., 2010).

8.2.1.6 Computer vision and hyperspectral imaging

Computer vision is a branch of artificial intelligence that allows computers and systems to derive meaningful information from various visual inputs such as digital photos and videos. Therefore computer vision is the eyes of the intelligent system to think and decide (IBM, 2022). Computer vision technology mimics the impact of human vision in assessing the

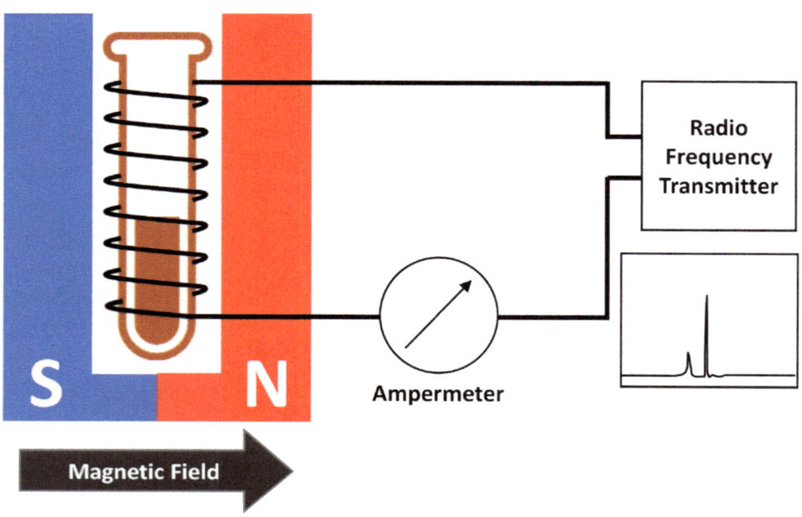

FIGURE 8.4 Schematic representation of the basic working principle of nuclear magnetic resonance spectroscopy.

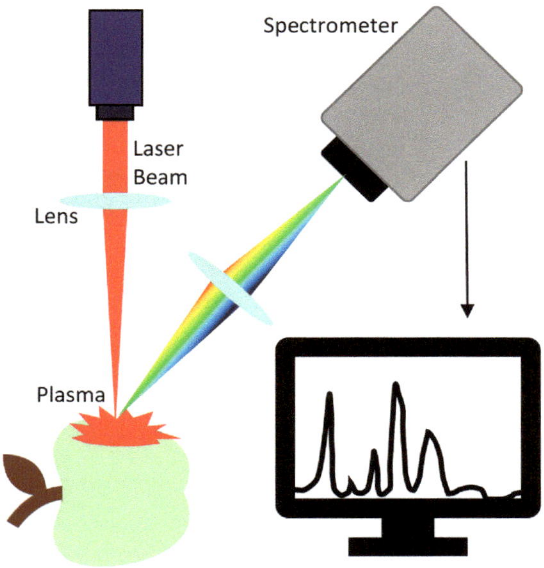

FIGURE 8.5 Schematic representation of the basic working principle of laser-induced breakdown spectroscopy.

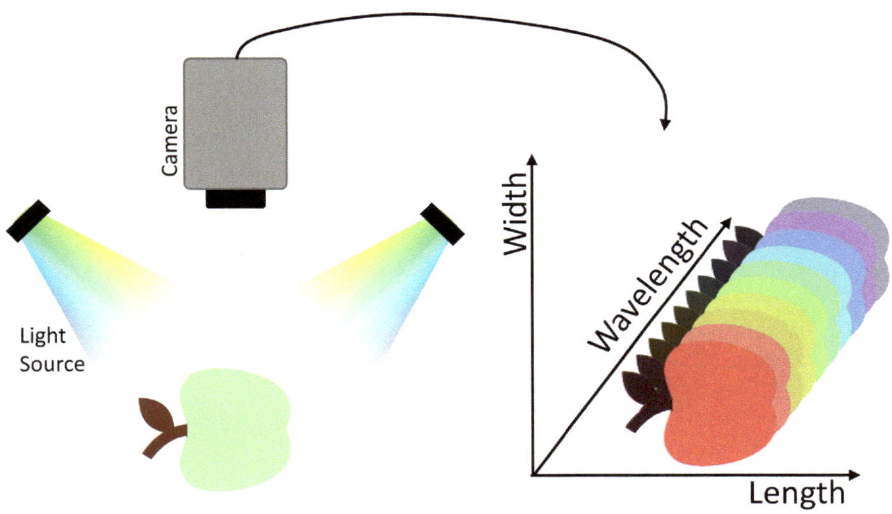

FIGURE 8.6 Schematic representation of the basic working principle of hyperspectral imaging.

quality of fruits and vegetables, for example, by electronically seeing a picture, interpreting and recognizing the characters, and providing information to the quality grading and sorting equipment (Bhargava & Bansal, 2021). For traditional computer vision systems, the digital information acquired with RGB (red, green, and blue segments of the images taken in the visible spectrum) cameras that grab the visible light (400–700 nm) can be used to obtain many characteristics such as texture, shape, color, size, and defects (Kempelis et al., 2022; Uğuz & Uysal, 2021). The benefits of spectroscopy and imaging are combined in hyperspectral imaging. It has been used effectively to quantify the internal and exterior properties of several food items. The reflectance and transmission spectra of samples are captured using hyperspectral detectors approximately in the range of 900–3360 nm (Fig. 8.6) (Saha & Manickavasagan, 2021). The obtained data by RGB and/or hyperspectral cameras are processed with five sequential steps, which are image acquisition, preprocessing, image segmentation, feature extraction, and classification with the use of classical machine learning (Bhargava & Bansal, 2021). However, deep learning is a more effective method for computer vision since the used neural networks can extract patterns for classification that are inspired by the human understanding of how brains function (Babich, 2020). Therefore the convolutional neural network and recurrent neural network are among the most used deep learning algorithms to teach computers how to look and see what images and videos tell, respectively (IBM, 2022).

8.2.1.7 Remote sensors

Remote sensing is the process of detecting and monitoring the physical characteristics of an area. This is achieved by analyzing the reflected/emitted radiation by the Earth's surface. The data are collected by special cameras (sensors) placed typically on satellites or aircrafts including drones (NOAA, 2021). The satellite cameras can capture radiation on various wavelengths that typically range from visible/optical spectrum (400–700 nm), infrared spectrum (NIR, MIR, and FIR that corresponds to 700–1300 nm, 1300–3000 nm, and 3000–14000 nm, respectively), and microwave spectrum (1 mm–1 m wavelength) (Karthikeyan et al., 2020). The source of electromagnetic radiation can either be the Sun (passive remote sensing) or the satellite's light source (active remote sensing) (Sishodia et al., 2020). Remote sensing technology has a wide range of applications in different areas such as agriculture (crop growth, crop losses, irrigation), forestry and vegetation, coastal and ocean applications, hazard assessment, and natural source management (Fig. 8.7) (Karthikeyan et al., 2020; Li et al., 2010; NOAA, 2021).

8.2.2 Nonspectral sensors

8.2.2.1 Biosensors

A biosensor is composed of three parts, which are a bio-probe or bioreceptor (which is an element that reacts with the target analyte), a transducer that generates an electrical signal in response to that reaction and an electronic system which includes a signal amplifier, signal processor and display (please see Fig. 8.8 for the basic working principle of biosensors). They are employed for high-precision detection of pathogens and bacterial toxins, hygiene monitoring, pesticide and animal drug residue detection, phenolic compounds detection, and heavy metal detection (Fellows, 2009). In food processing, glucose, fructose, sucrose, lactose, lactic, malic, acetic, ascorbic, citric, and amino acids, ethanol, glycerol, and triglycerides are commonly monitored components with the use of biosensors. The key factors contributing to biosensor design are robustness, stability, and noninterference (Monosik et al., 2012). The most used bio-probes are nucleic acids, proteins and enzymes, antibodies, and phages (Nnachi et al., 2022; Petrenko & Vodyanoy, 2003). During the interaction between bio-probe and analyte, the process of signal generation (in the form of light, heat, pH, charge or mass change, plant or animal tissue, and microbial products) happens, which is known as *biorecognition* (Naresh & Lee, 2021). Electrochemical transducers, optical transducers, thermal transducers, and acoustic wave devices are the four basic types of transducers used in biosensors. The biosensor's specificity is determined by the proper selection and immobilization of the bio-probe element whereas its sensitivity is affected by the transducer (Li et al., 2010).

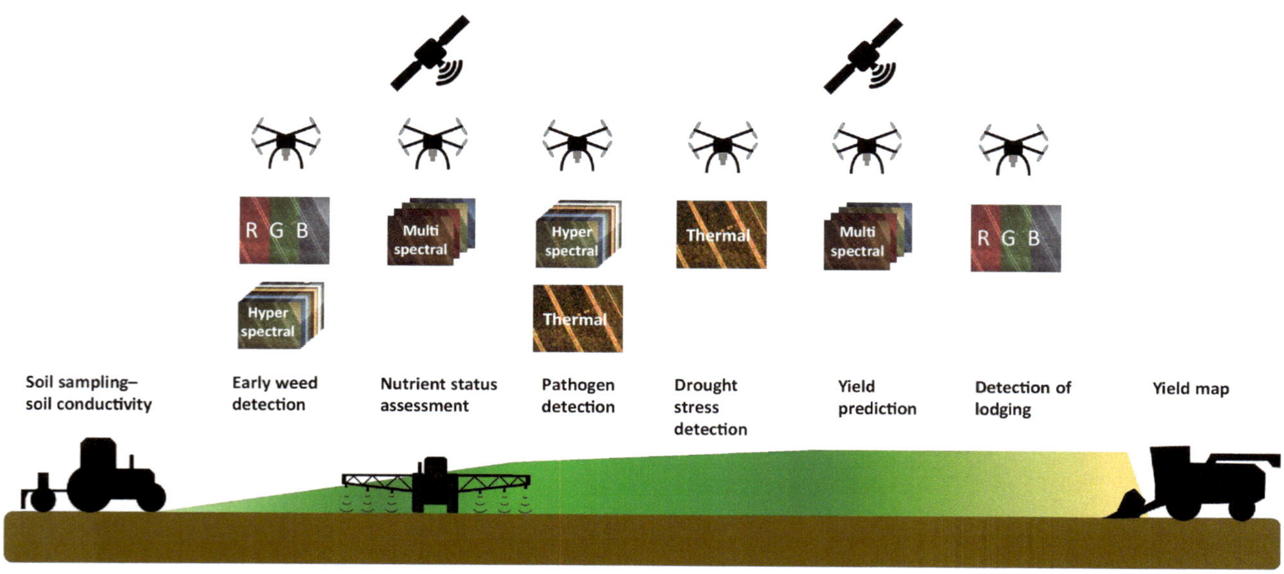

FIGURE 8.7 Schematic representation of remote sensors. *From Maes W.H., & Steppe, K. (2019). Perspectives for remote sensing with unmanned aerial vehicles in porecision agriculture.* Trends in Plant Science, 24*(2), 152–164. doi: 10.1016/j.tplants.2018.11.007.*

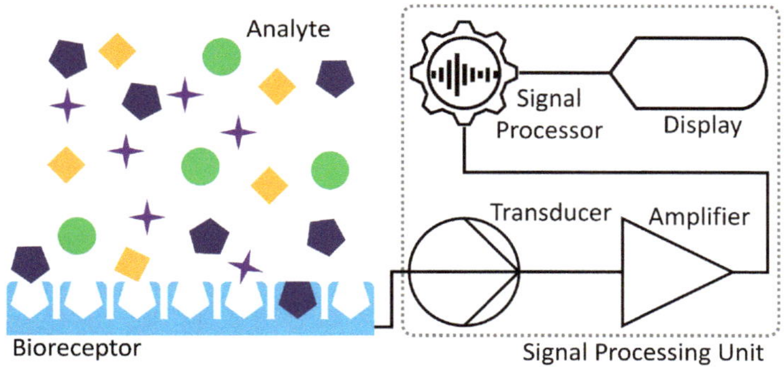

FIGURE 8.8 Schematic representation of the basic working principle of biosensors.

8.2.2.2 Electronic sensors (E-nose, E-tongue, and E-eye)

The electronic eyes (E-eye) are bionics-based detection instruments. They recognize and analzse visual information such as color and shape (YongXia et al., 2019). In other words, E-eye can simply be defined as the computer vision system built for specific purposes such as quality detection/measurement of food and crops. Besides that, E-nose and E-tongue are electronic devices equipped with gas and liquid sensors, respectively. All in all, these three sensors are built to mimic the olfactory, gustatory, and vision systems of the human being. They could provide fast and noninvasive detection and global information about the sample, instead of particular components (Xu et al., 2019).

8.2.2.3 Radio-frequency identification

Radio-frequency identification (RFID) is a partly remote technology. With the use of low-cost RFID tags and the incorporation of processors and sensors in them, it has been a key part of smart sensors (Rajaraman, 2017). RFID is a radio-frequency-based technology that is used for the autoidentification of objects. RFID reader (or interrogator) and RFID tag are the major parts of an RFID system. The working principle of the RFID system is based on the received radio-frequency by the interrogator with its antenna. The interrogator continuously sends radio waves at a specific frequency that activates the RFID chip in the tag. As the tag is activated, the ID code, which is a unique identifier of the data, is sent back to the interrogator (Cui et al., 2019). The RFID system is mainly operated at low-frequency (LF, 125 or 134 kHz), high-frequency (HF, 13.56 MHz), ultra-high-frequency (UHF, 860–960 MHz) bands, or microwave. The range of these RFID systems is up to 10 cm, 10–100 cm, and 10–15 m, respectively (Elyasizadeh, 2016; Dabbene et al., 2016). Three different kinds of RFID tags, namely passive, semipassive, and active tags, are available. Active tags have an internal power source for sending the response to the interrogator. Passive tags send the response relying on the radio waves received from the interrogator. The semipassive tags have an internal power supply, as well, but they behave like passive tags for responses (Rajaraman, 2017). Since the passive tags do not require power such as a battery, the sensing technique of RFID sensors becomes much more flexible and convenient, and its applications can be expanded into a broader field (Cui et al., 2019). RFID technology has a wide area of use including process control, warehouse and logistics, smart packaging, and cold-chain monitoring (Dabbene et al., 2016). However, the principal objective of RFID is not to perform measurements, but to wirelessly transport data gathered by sensors in a confined region and to make the sensor remote. A schematic representation of the use of RFID tags is given in Fig. 8.9.

8.2.2.4 Soft sensors

The name of *soft sensors (also called soft sensing technologies*, *software sensors*, *virtual sensors*, *intelligent sensors*, or *inferential estimators*) originates from the combination of software and sensor words (Luttmann et al., 2012). Soft sensing refers to techniques and algorithms that are used to estimate/predict particular properties of the process or processed material (Jiang et al., 2021). In other words, soft sensors are mathematical models (mechanistic, statistical, or empirical) used to predict the behavior of real systems (Cregan et al., 2017; Graziani et al., 2020). The generic steps of soft sensor development can be listed as data collection, variable selection, model development and validation, model evaluation and optimization, and implementation (Jiang et al., 2021; Zhu et al., 2020). Once a soft sensor is developed, it can be used for many different purposes such as process monitoring, process control, and optimization, obtaining prior knowledge for system design, etc. The most important advantages of soft sensors can be summarized as being very economical, being developable using many different methods, ease of adaptation to different processes and process conditions,

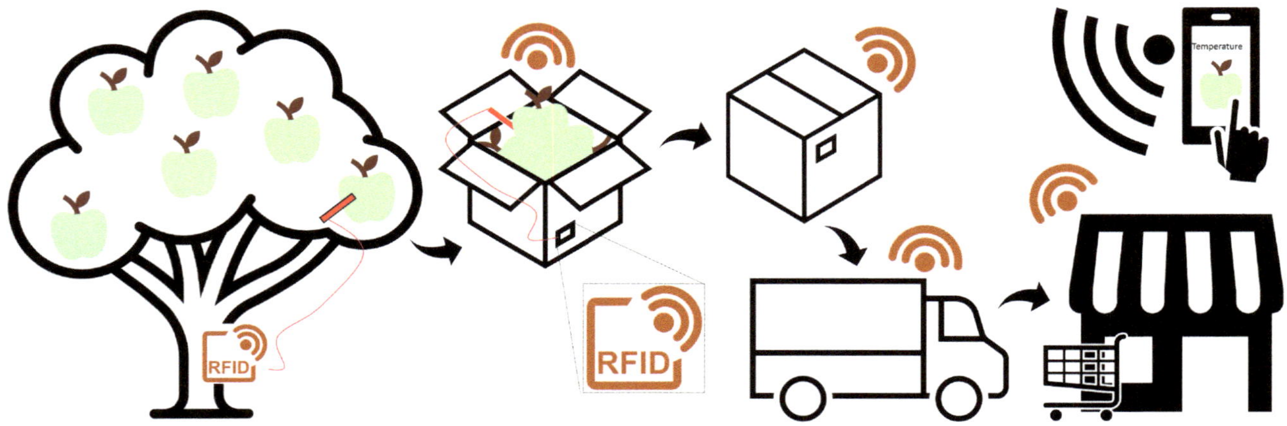

FIGURE 8.9 Schematic representation of the basic working principle of radio-frequency identification tags.

both in real-time and offline usage possibility, being usable in situations/positions where it is not possible with physical sensors, and the possibility to use an endless number of sensors to generate a huge amount of data in a very short time, which is not possible with physical sensors, as well.

8.3 Smart sensor applications in production

In the following sections, frequently used smart sensor applications are summarized covering the most recent studies in the available literature to draw a perspective on the potential use of these sensors through the agrofood chain. These studies are also listed in Table 8.1.

8.3.1 Cultivation period

One of the most used smart sensor technologies for the pre- and postharvesting phases of food items is remote sensors. Remote sensing is the use of sensors to detect, measure, and analyze physical features of an area or item from a distance. Remote sensing can be used in agriculture to acquire data about crops, soil, and other characteristics of the farming environment. This information may be utilized in precision agriculture for preseason planning, field preparation, maximizing agricultural yields, tracking crop health, and optimizing irrigation and fertilization procedures as the preharvest solution (Khanal et al., 2020; Maes & Steppe, 2019; Sishodia et al., 2020). In terms of postharvest use, remote sensors are helpful for crop residue assessment (Khanal et al., 2020). In agriculture, several different types of remote sensing technologies are commonly used including visible and infrared cameras, radar sensors, lidar sensors, and multispectral and hyperspectral sensors.

Visible and infrared cameras, as well as multispectral and hyperspectral cameras, are the most used sensors for remote sensing in agricultural production. They are mostly utilized to assess crop health, identify plant species and diseases, as well as to monitor soil moisture levels and detect problems such as pests. Besides that, multispectral and hyperspectral sensors, with the ability to scan a wide range of wavelengths, can identify specific materials and substances that might be more relevant for food safety, quality, and sustainability problems than the use of other sensors. For example, identification of hair fescue in wild blueberry fields with high accuracy (Bilodeau et al., 2023), determination of plant count and phenotyping variables by training a neural network model (Ariza-Sentís et al., 2023), automation of crop irrigation regime to minimize water usage (Tschand, 2023), detection of different phenological phase and growth stages of the vineyards (Comba et al., 2018), detection and counting palm trees (Putra et al., 2022), and development of a drone-based automated shepherd system (Yaxley et al., 2023) are examples of studies conducted recently using the images captured in the visible light spectrum. Detection of *Magnaporthe oryzae* infection to the rice crops (Tian et al., 2023) or classification of different rice varieties (Meng et al., 2022), detection of nitrogen level in maize crop (Gabriel et al., 2017), determination of leaf area index and chlorophyll content of mandarins as indicators of plants health and nutrition status (Ali & Imran, 2020), characterization of seasonal and diurnal dynamics of crop photosynthetic activity (Chang et al., 2020), plant disease identification using 3D information (Nagasubramanian et al., 2019) were performed using multispectral and hyperspectral data. Sometimes, it is also possible to combine the data obtained from different sources to increase the predictive ability of the developed model. For example, RGB, multispectral, and thermal sensors were used to estimate soybean yield with multimodal data fusion (Maimaitijiang et al., 2020).

TABLE 8.1 Literature overview of the studies about the use of smart sensors in the agriculture and food industry.

Application field	Sensor type	Aim	Source
Spectral sensors			
Agriculture (preharvest)	FT-NIR	On-site quality analysis of olive oil	Adame-Siles et al. (2018)
Agriculture (preharvest)	NIR	Identification of grain damage	Ba et al. (2023)
Food safety and quality	Portable NIR	Determination of chemical traits of meats	Cáceres-Nevado, Garrido-Varo, De Pedro-Sanz, and Pérez-Marín (2021)
Food safety and quality	FT-NIR and portable NIR	Frozen state determination of meat	Cáceres-Nevado, Garrido-Varo, De Pedro-Sanz, Tejerina-Barrado, et al. (2021)
Food safety and quality	FT-NIR	Frying oil quality evaluation	Calero et al. (2018)
Agriculture	NIR	Determination of water pollution for irrigation practices	Chen et al. (2020)
Food safety and quality	NIR and portable NIR	Composition evaluation of dairy milk	de la Roza-Delgado et al. (2017)
Agriculture	Vis/NIR	Soil property diagnosis	Delgadillo-Duran et al. (2022)
Food safety and quality	Portable NIR	Squash nitrate content and quality prediction	Entrenas et al. (2019)
Food safety and quality	FT-NIR and portable NIR	Spinach nitrate content and quality prediction	Entrenas et al. (2020)
Food safety and quality	NIR	Virgin olive oil quality assessment	Garrido-Varo et al. (2017)
Agriculture (postharvest)	Vis/NIR	Composition determination of citrus fruits	Kim et al. (2023)
Food safety and quality	NIR	Food powder classification	Ozturk et al. (2023)
Agriculture (postharvest)	Portable NIR	Quality determination of oranges and storage time assessment	Pérez-Marín, Calero, et al. (2019)
Agriculture (preharvest)	Portable NIR	Classification of spinach concerning quality and safety	Pérez-Marín, Torres, et al. (2019)
Food authenticity	Portable NIR	Identification of Iberian pig carcasses	Pérez-Marín, Fearn, et al. (2021)
Food authenticity and quality	Portable NIR	Authentication and quality assessment of bell pepper	Sánchez et al. (2019)
Food safety and quality	NIR	Determination of vinegar composition	Sánchez, Márquez, et al. (2020)
Agriculture (preharvest)	Portable NIR	Ripening stage evaluation of Lamuyo pepper	Sánchez, Pintado, et al. (2020)
Agriculture (preharvest)	Portable NIR	Quality analysis of citrus fruit	Torres et al. (2017)
Agriculture (preharvest)	Portable NIR	Chemical quality assessment of citrus species	Torres, Sánchez, de la Haba, et al. (2019)
Food safety and quality	FT-NIR and portable NIR	Summer squashes quality assessment	Torres, Sánchez, Entrenas, et al. (2019)
Food authenticity	Portable NIR	Fraud detection in sweet almond	Torres, Sánchez, Vega-Castellote, and Pérez-Marín (2021)
Food safety and quality	UV/Vis	Black tea recipe estimation	Turgut, Küçüköner, and Karacabey (2021)
Food safety and quality	NIR	Black tea quality estimation	Turgut et al. (2022)
Food safety and quality	NIR	Prediction of spinach nitrate content	Pérez-Marín, et al. (2021b)
Food safety and quality	Portable NIR	Fatty acid composition in-shell and shelled almonds	Pérez-Marín, et al. (2021a)

(Continued)

TABLE 8.1 (Continued)

Application field	Sensor type	Aim	Source
Food authenticity	FT-NIR and portable NIR	Detecting noncompliant batches of sweet almonds	Vega-Castellote, Sánchez, et al. (2021)
Agriculture (preharvest)	Portable NIR	Watermelon maturity assessment	Vega-Castellote et al. (2022)
Food quality and safety	Raman	Virus detection	Huang et al. (2021)
Authentication	Raman and FTIR	Authentication of turmeric powder with Sudan red and white turmeric	Chao et al. (2020)
Authentication	Raman	Adulteration in olive oils with other vegetable oils	Duraipandian et al. (2019)
Authentication	Raman	Food colorants	Gukowsky et al. (2018)
Food quality and safety	Surface-enhanced Raman	Determination of lead in food	Guo et al. (2022)
Authentication	Raman	Rapid detection of edible oil type and adulteration	Zhao et al. (2022)
Food quality and safety	Raman	Detection of foodborne pathogens	Huayhongthong et al. (2019)
Food quality and safety	Surface-enhanced Raman	Patulin detection in fruits	Wu et al. (2019)
Authentication	Raman	Parmigiano Reggiano cheese authenticity	Li Vigni et al. (2020)
Food quality and safety	Raman	Determination of titanium dioxide concentration in food	Lim et al. (2018)
Authentication	Raman	Detection of adulterated soft sugar with glucose in plastic packaging	Liu et al. (2021)
Authentication	FT-Raman	Authentication of the provenience of wine	Mandrile et al. (2016)
Authentication	Raman	Butter adulteration with margarine	Nedeljković et al. (2016)
Authentication	Raman	Detection of lard in butter	Taylan et al. (2020)
Food safety and quality	Laser speckle	Apple composition determination	Arefi et al. (2023)
Food quality	LIBS	Salt analysis in bakery products	Bilge et al. (2015)
Food quality	LIBS	Ash analysis in flour	Bilge, Sezer, Eseller, Berberoglu, Koksel, et al. (2016)
Authentication	LIBS	Determination of whey adulteration in milk powder	Bilge, Sezer, Eseller, Berberoglu, Topcu, et al. (2016)
Food quality	LIBS	Ca, K, and Mg detection in cassava flour samples	Costa et al. (2018)
Food quality	LIBS	Quantification of rubidium in beef	Dixit et al. (2017)
Food quality	LIBS	Determination of pesticides in spinach and rice contamination	Kim et al. (2012)
Authentication	LIBS	Rice geographic origin classification	Yang et al. (2018)
Authentication	LIBS	Fructose corn syrup and rape honey addition to the acacia honey	Peng et al. (2020)
Authentication	LIBS-Raman	Identification of fish species	Ren et al. (2023)
Authentication	LIBS	Identification of beef, chicken, and pork in fermented sausage and salami products	Sezer et al. (2022)
Authentication	LIBS	Coffee Arabica adulteration with wheat, corn, and chickpea	Sezer, Apaydin, et al. (2018)
Authentication	LIBS	Identification of bovine, caprine and ovine milk, and adulteration ratio in bovine milk	Sezer, Durna, et al. (2018)

(Continued)

TABLE 8.1 (Continued)

Application field	Sensor type	Aim	Source
Authentication	LIBS	Discrimination of olive oils based on the olive cultivar origin	Stefas et al. (2021b)
Authentication	LIBS	Determination of offal adulteration in beef meat	Velioglu et al. (2018)
Authentication	LIBS	Edible gelatine adulteration with industrial gelatine	Zhang et al. (2019)
Food quality	1H NMR	Identification of chemical aging markers in green coffee samples according to different packages	Borém et al. (2023)
Authentication	1H NMR	Authentication of saffron	Gunning et al. (2023)
Authentication	1H NMR	Cheese characterization and authentication	Haddad et al. (2022)
Food packaging	NMR	Noninvasive "through-package" assessment of the microstructural quality of a model food emulsion	Haiduc et al. (2007)
Authentication	1H NMR	Hen egg authentication by profiling of triacylglycerols	Hajjar et al. (2021)
Food quality	TD-NMR	Rapid monitoring of an edible oil's oxidation degree	Osheter et al. (2022)
Food packaging	TD-NMR	Through-package fat determination in commercial samples of mayonnaise and salad dressing	Pereira et al. (2015)
Food safety and quality	Hyperspectral imaging	Determination of coffee bean quality	Caporaso et al. (2018)
Food safety and quality	Hyperspectral imaging	Freeze damage evaluation in pork meat	Cheng et al. (2018)
Food safety and quality	Hyperspectral imaging	Identification of foreign material on chicken meat	Chung and Yoon (2021)
Food safety and quality	Hyperspectral imaging	Disease detection in sheep organs	Coombs et al. (2023)
Agriculture (preharvest)	Hyperspectral imaging	Potato disease diagnosis	Polder et al. (2019)
Agriculture (postharvest)	Hyperspectral imaging	Strawberry infection diagnosis	Siedliska et al. (2018)
Food safety and quality	Hyperspectral imaging	Fatty acid composition determination of almond	Pérez-Marín, et al. (2021)
Food safety and quality	Hyperspectral imaging	Classification of almonds according to bitterness	Torres-Rodríguez et al. (2022)
Food safety and quality	Hyperspectral imaging	Quality determination in chicken meat	Yang et al. (2018)
Agriculture (postharvest)	Hyperspectral imaging	Pear quality estimation	Yu et al. (2018)
Agriculture (preharvest)	Computer vision	Weed determination among food crops	Alrowais et al. (2022)
Food safety and quality	Computer vision	Date fruit sorting	Nasiri et al. (2019)
Agriculture (preharvest)	Computer vision	Disease diagnosis in sugar beet leaves	Ozguven and Adem (2019)
Agriculture (postharvest)	Computer vision	Papaya ripening identification	Santos Pereira et al. (2018)
Agriculture (preharvest)	Computer vision	Disease diagnosis in citrus fruits and leaves	Sharif et al. (2018)

(Continued)

TABLE 8.1 (Continued)

Application field	Sensor type	Aim	Source
Agriculture (preharvest)	Computer vision	Food crops disease diagnosis	Wongchai et al. (2022)
Aquaculture	Computer vision	Fish disease diagnosis	Li et al. (2022)
Agriculture (preharvest)	Computer vision	Food crop classification	Zheng et al. (2019)
Agriculture (preharvest)	Remote sensor	Spinach seed yield estimation	Ariza-Sentís et al. (2023)
Agriculture (preharvest)	Remote sensor	Identification of hair fescue on blueberry field	Bilodeau et al. (2023)
Agriculture	Remote sensor	Vineyard detection	Comba et al. (2018)
Agriculture (preharvest)	Remote sensor	Soybean yield estimation	Maimaitijiang et al. (2020)
Agriculture (preharvest)	Remote sensor	Rice variety classification	Meng et al. (2022)
Agriculture (preharvest)	Remote sensor	Variability management potato cultivation field	Mohammad EL-Basioni et al. (2022)
Agriculture	Remote sensor	Palm tree detection and counting	Putra et al. (2022)
Agriculture (preharvest)	Remote sensor	Rice infection identification	Tian et al. (2023)
Agriculture	Remote sensor	Water irrigation network development	Tschand (2023)
Agriculture	Remote sensor	Sheep shepherding	Yaxley et al. (2023)
Nonspectral sensors			
Food safety	Biosensor	Detection of *Escherichia coli* in dairy products	Arreguin-Campos et al. (2023)
Agriculture (preharvest)	Joint sensors	Wheat crop irrigation management	Elsherbiny et al. (2022)
Food packaging	Colorimetric sensor	Colorimetric sensor for storage condition tracking	Escobedo et al. (2023)
Food safety	Biosensor	Rapid detection of *Salmonella typhimurium* in food	Fulgione et al. (2018)
Food safety	Biosensor	Genetically modified organism detection	Huang et al. (2015)
Food safety and quality	Biosensor	Smartphone base *E. coli* detection	Zheng et al. (2019)
Agriculture (preharvest)	Joint sensors	Wheat crop storeroom management	Lydia et al. (2022)
Food packaging	Biosensor	Glucose biosensor as Time-Temperature Integrator	Mijanur Rahman et al. (2018)
Food safety and quality	Biosensor	Fish freshness and spoilage identification	Moradi et al. (2019)
Food safety	Biosensor	On-site monitoring of biogenic amines in foods	Qin et al. (2023)
Food safety	Biosensor	Detection of food allergens (peanut, hazelnut, almond, milk, wheat, and soybean) in biscuits, seafood substitutes, and probiotic foods	Sena-Torralba et al. (2023)
Food safety	Biosensor	Monitoring of *Salmonella enterica* in chicken broth	Soares et al. (2020)
Food safety	Biosensor	Hydrogen peroxide detection in dairy products	Vasconcelos et al. (2023)
Food packaging	Flexible sensor	Packaged lamb meat loss minimization	Xu et al. (2023)
Food safety	Biosensor	*Staphylococcus aureus* ST398 detection in pure culture	Zhou et al. (2023)
Food authenticity	E-nose, E-tongue, E-eye	Olive oil characterization and shelf life assessment	Buratti et al. (2018)
Food authenticity	E-nose, E-tongue	Discrimination of irradiated shrimps	Chung et al. (2019)

(*Continued*)

TABLE 8.1 (Continued)

Application field	Sensor type	Aim	Source
Food safety and quality	E-nose, E-tongue	Mushroom flavor and composition assessment	Donglu et al. (2017)
Food safety and quality	E-nose, E-tongue, E-eye	Fish freshness prediction during frozen storage	Li et al. (2023)
Food authenticity and quality	E-nose, E-tongue	Cultivation, odor, taste, and flavor compounds determination of tangerine peel	Li et al. (2022)
Food safety and quality	E-nose, E-tongue, E-eye	Tea quality assessment	Xu et al. (2019)
Food safety and quality	E-nose, E-tongue	Volatile compound analysis in rice bran	Li et al. (2023)
Food safety and quality	E-nose, E-tongue, E-eye	Chinese soybean paste flavor quality assessment	Yu et al. (2023)
Food safety and quality	E-nose, E-tongue	Microbial growth detection in fruit juice	Zhang et al. (2022)
Agriculture	Radio frequency	Soil moisture mapping	Balivada et al. (2022)
Food safety and quality	Radio frequency	Food temperature detection in a microwave oven	Banerjee and Srinivasan (2021)
Agriculture (preharvest)	Radio frequency	Tomato vegetation tracking	Cama-Pinto et al. (2020)
Food packaging	Radio frequency	Meat freshness estimation	Cao and Chung, (2019)
Food packaging	Radio frequency	Radiation exposure measurement	Heredia-Rivera et al. (2022)
Food packaging	Radio frequency	Cold-chain violation monitoring	Hussein et al. (2021)
Agriculture	Radio frequency	Optimization of energy efficiency in the agricultural field	Lu et al. (2021)
Agriculture (postharvest)	Radio frequency	Classification of avocado fruit according to the ripening	Occhiuzzi et al. (2022)
Food safety and quality	Radio frequency	Food spoilage detection and livestock farming management	Zhang et al. (2021)

These studies in which visible, hyperspectral, or multispectral cameras were used in agricultural production are not different from computer vision and hyperspectral imaging in the first principles, but the images to be processed are obtained from greater distances than in food processing lines.

Radar and lidar sensors measure the reflection or scattering of emitted microwave energy and laser pulses, respectively, of the Earth's surface. Radar sensors are more likely to be used to measure the height and structure of crops, as well as to detect water and soil moisture levels. On the other hand, lidar sensors are used to create high-resolution maps of the Earth's surface. In that way, the structure and height of crops are analyzed (Khanal et al., 2020). For example, erosion and deposition processes in a wasteland in the Province of Siena (Tuscany) (Neugirg et al., 2016), quality control and characterization of crops (sizes, orientation, and growth stages) (Lin & Habib, 2021), topographic variables from agricultural land in southern of Ontario to estimate crop yield (Eyre et al., 2021), absorption and distribution of light by avocado orchards (Westling et al., 2018) are some successful smart agriculture applications of lidar sensors. Moreover, the determination of drainage pipes for agricultural subsurface mapping (Allred et al., 2020), mapping the tile-drain network to improve the understanding of nutrient and water transport and management (Williamson et al., 2019), classification of different crops and agricultural surface types (Marais Sicre et al., 2020), and monitoring agricultural soil organic carbon for sustainable agricultural management (Nguyen et al., 2022) are among the recent studies that involved the use of radar sensors.

For remote sensing, the mentioned sensors and others can be placed on satellites, UAVs, aircrafts, tractors, or other farm machines depending on the application area. However, among these options, the most preferred sensor-carrying platform seems to be UAVs due to its advantages such as flexibility in providing on-demand information at a required spatial scale with the use of various types of sensors, getting continuous or frequent data, being low-cost, under cloud flight ability, options of automation, etc., which are limited in other platforms, especially for satellite images (Ehsani & Mari Maja, 2013; Huang et al., 2018; Primicerio et al., 2012). In some cases, manual handheld sensors are also used. For instance, NIR sensors are an extremely useful tool to determine the optimal harvest time linked to quality parameters and the possible industrial destination of fruits and vegetables, for systematic control of the stored product. The use of NIR sensors for pre- and postharvest applications increased in recent years since they have undergone a revolution, especially thanks to the development of instruments based on technologies such as linear variable filter, diode array, and micro-electro-mechanical systems (Pérez-Marín et al., 2021). Characterization of the parameters that determine the internal and external quality of nectarines during on-tree ripening (Pérez-Marín et al., 2009), determination of orange quality and optimum harvest time (Sánchez et al., 2013), on-site prediction of two of the most important quality parameters (dry matter and Brix) in intact bell peppers (Sánchez et al., 2019), quality prediction, and maturity classification in watermelons (Vega-Castellote et al., 2022) are among the studies showing the feasibility of the NIR spectroscopy technology to be used as a preharvest decision support tool for farmers. Besides that, the determination of quality attributes in intact Raf tomatoes (Torres et al., 2015), orange shelf life determination in cold storage (Pérez-Marín, Calero, et al., 2019), and fatty acid profile in intact in-shell and shelled almonds when they are received by the industry (Vega-Castellote, Pérez-Marín, et al., 2021a) are some examples of the postharvest use of NIR sensors.

RFID technologies are also frequently used in agricultural fields. These technologies can be categorized as partly remote since the data measured by the sensors are transmitted wirelessly over distances of meters. RFID technology is used in agriculture with similar purposes to remote sensors such as tracking and managing livestock (location tracking, health monitoring, and movement recording), and monitoring and optimization of crop production (monitoring soil moisture, temperature, and other environmental factors that affect crop growth). With this information, optimization of breeding and feeding regimes, prevention/reduction of disease transmission between animals, optimization/automation of irrigation and fertilization practices, as well as identification of the areas requiring additional attention could be possible (Balivada et al., 2022; Cama-Pinto et al., 2020; Hardie, 2020; Kulkarni et al., 2020; Lu et al., 2021; Mosadegh Sedghy, 2018; Xu et al., 2022). Overall, RFID technology can help farmers to improve the efficiency and productivity of their operations with minimal cost.

All in all, remote sensing can provide several benefits to the food and agriculture industry by monitoring and managing large areas of land efficiently, as well as detecting potential problems before they become too severe. In this way, both economic losses and food waste can be prevented with the early measures to be taken. Remote sensing can also be used to optimize irrigation and fertilization practices, helping to reduce water and chemical use and improve crop yields, which undoubtedly helps to reduce the stress on the environment and water resources and to achieve sustainable production with the economical use of natural resources and less use of pollutants. Overall, remote sensing is a strong tool for enhancing agricultural efficiency and sustainability. It has the potential to transform crop cultivation and management, thereby ensuring a reliable and sustainable food supply in the future.

8.3.2 Food production chain

Smart sensors are used in food production to capture data about various aspects of the process, such as temperature, humidity, pressure, pH levels, composition, shape, size and color, and sensory properties. Most of the time, the data is collected in real time, and may well be utilized to improve and regulate different areas of the manufacturing process, such as quality control, process optimization, safety and hygiene, and predictive maintenance. These measures assist manufacturers in improving efficiency, cost reduction, product quality and safety, and sustainability. Recent examples of the use of smart sensors in food packaging, food quality and safety, food authentication, and process control and optimization are reviewed later.

8.3.2.1 Food packaging

Smart sensors are extensively studied by embedding them into food packaging to increase traceability and detect and measure various physical, chemical, or biological parameters of the packaging material or the food product itself. Not all the sensors that were mentioned earlier are proper to be used for packaging applications because of limited portability. Therefore in the context of packaging applications, the majority of smart sensor examples focus on utilizing microprocessors, wireless communication, and biosensors to effectively collect and transmit data about the condition of both the food and the package (Fathi et al., 2020; Han et al., 2018; Mohammadian et al., 2020; Yousefi et al., 2019). In this way, not only modern sensor technologies but also traditional sensors (temperature, humidity, etc.) are the subject of smart applications thanks to the increasing use of IoT, wireless communication, and data fusion. The smart technologies used in food packaging can be categorized as time-temperature indicators (TTIs), gas indicators, microwave doneness indicators, RFID, and others (Han et al., 2018).

Smart sensors can be used in food packaging to monitor a variety of parameters that have impacts on the quality and safety of the food product. Microfluidics (Pou et al., 2022), fluorescent biosensors (Dong et al., 2017), electrochemical biosensors (Vargas et al., 2016), and gas biosensors (Shen et al., 2018) are frequently used to achieve this goal. Temperature sensors, for example, are used to monitor the temperature of a food product and ensure that it is stored and delivered at the proper temperature to avoid spoilage and foodborne health risks (Escobedo et al., 2021; Fathi et al., 2020; Kritchenkov et al., 2019). Especially, RFID tags are utilized to remotely transmit temperature data despite the present challenges, such as the high cost of these sensors' integrity due to the packaging material. TTI sensors measure physical changes in the food matrix as a function of temperature and time (Pereira Jr et al., 2015). They are, however, only suitable for cold storage of food products. Humidity sensors may be used to monitor the humidity of food and packaging, which is critical for the overall quality of many foods (Benabdellah et al., 2016; Rahman & Chowdhury, 2022). Moreover, to analyze the chemical and physical qualities of food and beverages, electronic sensors (E-nose) are also used to detect/analyze volatile organic compounds released by food to identify the individual chemicals contained in a sample, providing information about the food's quality, freshness, and spoilage (Wijaya et al., 2018). Oxygen sensors can be used to monitor the concentration of oxygen in food packaging (which is vital for limiting the growth of aerobic microorganisms) (Elisseeva et al., 2020; Kelly et al., 2020), and CO_2 sensors can also be used to monitor CO_2 concentrations, which is important to preserve the quality of carbonated beverages and other foodstuffs (Emanuel & Sandhu, 2019; Poghossian et al., 2019; Puligundla et al., 2012). Gas sensors are also particularly useful for determining the freshness of red meat and poultry (Arvanitoyannis & Stratakos, 2012). The pH sensors and color-changing indicators are also utilized extensively (Azman et al., 2022; Kuswandi et al., 2011; Mohammadian et al., 2020).

Spectroscopic methods are also utilized for nondestructive quality assessment of food packages. The analysis of packaged products is usually more challenging because of the barrier that the packaging can pose to light, considering also that the penetration capability of light is limited and that sometimes the packaging material can interfere with the spectral signal of the analyzed product. In addition, a key point in the application of spectral methods in this type of packaged product lies in the heterogeneity of packaging materials, since not all industries use the same, neither in terms of materials nor in terms of thickness, so it is an aspect that still requires a lot of research work to try to overcome these difficulties. Some research works carried out in recent years have focused on the analysis of vacuum-packed agrifood products in reasonably thin and totally or partially transparent plastics, mainly meat products. A fiber-optic contact probe coupled to a NIR spectrometer to categorize Iberian dry-cured *salchichon* and predict main quality characteristics (Ortiz et al., 2021), and a portable NIR sensor for the qualitative and quantitative characterization of vacuum-packed sliced of Iberico cured ham (Pérez-Marín & Garrido-Varo, 2022) were recently presented to the literature. Hyperspectral imaging and computer vision are also used to detect impurities or defects, such as pests, mold, and foreign and to automate processes, such as sorting and grading (Abraham, 2022), and to discriminate the type of plastic-based packages in recycling plants (Ulrici et al., 2013). This information can be utilized to not only improve the safety and quality of food products but also to reduce waste and increase efficiency. Spatially offset Raman spectroscopy (SORS) allows the signal from packaged foods to be recorded. For example, a fast independent component analysis was improved to separate the Raman signal of packaged food (FastICA) (Liu et al., 2022). Another study used the line-scan SORS technique to detect butter adulteration with margarine in packaged products. Partial Least Squares Regression (PLSR) scores for SORS data showed correlation coefficient values of 0.95 and 0.92, with prediction errors of 3.2% and 3.9% for two different covers (Lohumi et al., 2018).

Smart sensors may also be used to detect the movement and handling of food packaging, which can help with product monitoring and return in the case of a safety risk. GPS sensors, for example, help to record package movement history (Haroon et al., 2019; La Scalia et al., 2017), while accelerometers can determine how responsive the item is carried (Fernando et al., 2020; Ompusunggu et al., 2019; Paternoster et al., 2018). This information helps to ensure that the food is handled correctly and to identify potential physical damage during the distribution chain. Furthermore, smart sensors may be used to monitor the state of the packaging itself, such as detecting leaks of liquids and gases, fractures and tears that could threaten the food's safety and quality (Kuswandi et al., 2011; Yousefi et al., 2019).

In brief, the use of smart sensors in food packaging allows for real-time monitoring and control throughout the whole distribution chain. This helps producers to ensure the quality and safety of the food and improve the efficiency and effectiveness of the supply chain by taking early actions and minimizing losses.

8.3.2.2 Food quality and safety

Smart sensors are frequently used in food safety and quality in several ways including temperature, humidity, oxygen, and CO_2 monitoring, tracking the food supply chain, detection of contaminants such as pathogens, toxins, and pesticides, prediction of shelf life, and classification of defected or fraudulent products. Among these, processes such as temperature, oxygen, CO_2, and product tracking are more related to packaging materials, as discussed in the previous section. Regarding food quality and safety applications, more attention has been paid to at-line and offline applications related to the prediction of components and quality characteristics in foods. However, with the technological leap in the last decade, the use of these smart systems, especially spectral sensors rather than classical process control equipment, has become widespread for online and inline duties.

For example, LIBS, as a multielemental atomic emission spectroscopy technique, is one of the current spectral methods used for this purpose. Solid (dos Santos Augusto et al., 2017), liquid (Cama-Moncunill et al., 2020), and gas (Dong et al., 2012) samples can all be analyzed, but liquid and gas analyses may necessitate the use of special sampling techniques (Sezer et al., 2017). This adaptability is advantageous for various food matrix analyses. It has been previously used to analyze essential elements, for example, calcium ions concentration in liquid infant formula, and the results were confirmed using atomic absorption spectroscopy (Cama-Moncunill et al., 2020), and to analyze heavy metals (Cd concentrations), which is important due to toxicity and regulations, in rice roots (Wang et al., 2021). As the LIBS technique has the limitation of matrix dependence, a calibration graph is required for elemental quantification, and it is not appropriate for every sample matrix. This problem is overcome by the calibration-free LIBS (CF-LIBS) technique, which provides matrix-independent analysis. The technique calculates electron density and plasma temperature from the plasma plume produced by laser shutting, and the final concentration is calculated theoretically. CF-LIBS was used to identify iron, magnesium, sodium, potassium, calcium, aluminum, and zinc as essential nutrients and chromium, copper, manganese, nickel, and lead as toxic metals in black tea leaves (Rehan et al., 2022). The results obtained using CF-LIBS and inductively coupled plasma techniques showed excellent agreement, demonstrating that LIBS is an excellent tool for the spectro-chemical analysis of food products. Aside from elemental analysis, many foods can be analyzed using chemometric methods to correlate elemental compositions with quality parameters. *Salmonella enterica* serovar Typhimurium was successfully detected using LIBS based on Mg, Ca, Na, K, P, and C bands in milk, chicken broth, and brain heart infusion samples (Barnett et al., 2011). On the other hand, it has been reported that *E. coli* O157:H7 and *S. enterica* were identified using spectra collected directly from foods and food-contact surfaces (Multari et al., 2013).

Regarding Raman spectroscopy, its unique properties have made it a very appealing technique as a quality control detection tool. It can analyze quality control parameters and components in a variety of food matrices, including meat, seafood, dairy products, beverages, crops, fruits and vegetables, and oil products. The analysis of essential components such as lipids, proteins, carbohydrates, and water is especially important in many food products and can be performed using Raman spectroscopy. It is a useful tool for determining the distribution of secondary structures in proteins by analyzing the frequencies of amide bands among protein structures. Under the excitation of light irradiation, the chemical groups of amino acids and peptides produce specific Raman bands. Protein secondary structure (e.g., α-helix, β-sheet, β-turn, and random coil) can be investigated using Raman spectroscopy with amide I ($1600-1700\ cm^{-1}$), amide II ($1510-1560\ cm^{-1}$), and amide III ($1230-1350\ cm^{-1}$) bands (Herrero, 2008). Protein structures of various foods can thus be analyzed using the correlation of these bands and the quality parameters. For instance, a significant positive correlation between β-sheet structure and textural properties was discovered in heated meat batter samples (Shao et al., 2011). Furthermore, the group approved a positive correlation between the α-helical structure and total expressible fluid, while the β-sheet structure and total expressible fluid had a significant negative correlation. Raman spectroscopy can also reveal details about lipid structures such as unsaturation level and fatty acid composition (Baeten et al., 1996). Previously, the Raman method (785 nm excitation) was tested to quantify carotenoids, collagen, and fat in fish muscle (Marquardt & Wold, 2004). In another study, it was used to predict tenderness and slice shear force in aged pork samples (Santos et al., 2018).

NMR is another versatile analytical technique for the assessment of food quality and control. Although early NMR uses in foods were restricted to low-resolution NMR analyses of moisture, NMR began involving high-resolution investigations of liquid and solid state matrices for a variety of applications at the end of the 20th century including classification and authentication (Dais et al., 2015). NMR analysis has enormous potential for analyzing multicomponent systems, such as foods, and offers tremendous opportunities to advance the field of food science due to its nondestructive nature, high accuracy, and reproducibility, which are frequently achieved without using any separation or purification steps. Additionally, NMR spectroscopy is a quantitative approach where, depending on the food and the information sought, a variety of nuclei may be utilized for analysis. Under some circumstances, the area of the NMR signal is directly proportional to the number of nuclei that produce the signal (Bharti & Roy, 2012). NMR spectroscopy is one of the few techniques that can provide information about a molecule's regio/stereo chemistry, along with X-ray technology, which performs only for compounds that can be crystallized. Each atom can be studied selectively in NMR, and NMR experiments are frequently performed under physiological conditions such as temperature, pH, and salt concentration, which is especially important in biological studies. Despite its many advantages and numerous food applications in the literature, the NMR is underutilized because of its high cost, remarkably low sensitivity, and lack of NMR expertise.

NIR technology has enormous potential in terms of its application in ensuring the quality and safety of agrifood products. The massive, multiproduct, and multianalyte analysis capacity of NIR is a differential factor, which, together with information and communication technologies, opens an extremely interesting range of possibilities for the use of these sensors linked to the concept of Industry 4.0. Since 1990, numerous research works have been carried out regarding the use of NIR spectroscopy technology in ensuring the quality and safety of agrifood products throughout the entire agrifood chain, from the field to the final product. For example, in 2017, quality parameters in olive oil were determined using an at-line NIR spectrophotometer (Garrido-Varo et al., 2017). The results showed the feasibility of NIR technology as a powerful quality control method to monitor the production and storage of olive oil in the industry, providing results in a matter of seconds without the need for prior sample preparation. NIR analysis can also be used to assess food safety issues, such as the prediction of the nitrate content of some vegetables, in addition to other quality parameters. For example, the use of portable spectrometers was proposed to predict quality and safety parameters in intact spinach (Fig. 8.10A) (Pérez-Marín, Torres, et al., 2019; Torres et al., 2021). The safety and quality of summer squashes on the plant (Entrenas et al., 2019) and online determination of nitrate content in spinach plants with FT-NIR equipment (Fig. 8.10B) (Vega-Castellote, Pérez-Marín, et al., 2021b) are other recent examples from the literature.

Similar to NIR systems, hyperspectral imaging and computer vision are technologies used to analyze the chemical and physical aspects of food and drinks in smart food quality and safety. Because hyperspectral imaging involves the acquisition of a sample's entire electromagnetic spectrum, it can be used to identify specific compounds present in a sample, such as pigments, flavors, and contaminants, as well as their changes over time, which can provide information about the product's quality and freshness (Wieme et al., 2022). Computer vision only interprets visual information from the environment, thus, it may identify changes in color, texture, and shape, which can aid in the detection of potential safety concerns or spoilage, as well as quality and freshness. Therefore the utilization of hyperspectral imaging and computer vision is critical. Determination of coffee bean quality (Caporaso et al., 2018), identification of foreign material on chicken meat (Chung & Yoon, 2021), disease detection in sheep organs (Coombs et al., 2023), fatty acid composition determination of almond (Torres et al., 2021), date fruit sorting (Nasiri et al., 2019), and classification of almonds according to bitterness (Torres-Rodríguez et al., 2022) are examples of recent studies found in the literature about the use of hyperspectral imaging and computer vision in the context of food safety and quality.

The electronic sensors namely E-nose, E-tongue, and E-eye are used in smart food quality and safety to monitor the chemical and physical aspects of food and beverages. As previously described, an E-nose can detect volatile organic compounds generated by food, while an E-tongue can respond chemical composition of liquid samples. Along with E-eye, these sensors can be used separately or in tandem as a sensor system to check the product's quality and freshness

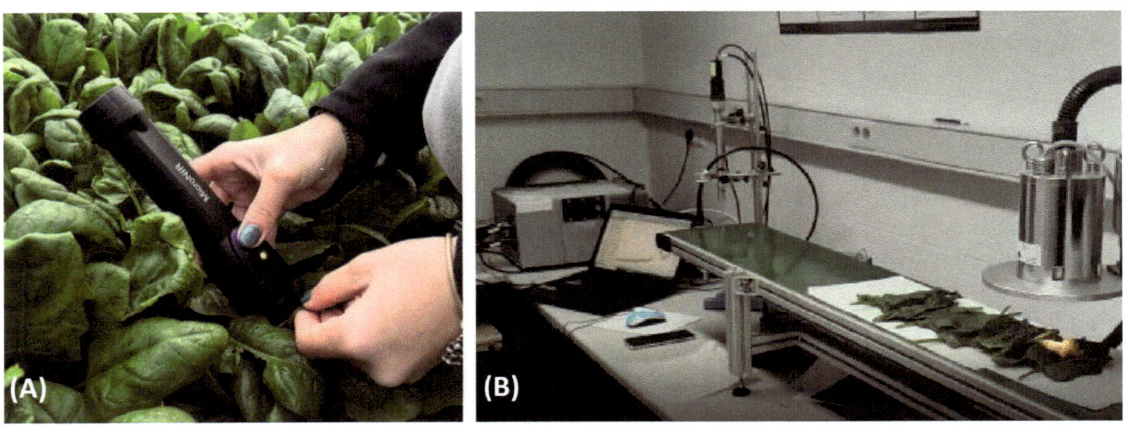

FIGURE 8.10 (A) On-site analysis of spinaches with a portable instrument (MicroNIR 1700) and (B) online analysis of spinach leaves over a conveyor belt using an FT-NIR instrument

of various food products such as olive oil (Buratti et al., 2018), shrimps (Chung et al., 2019), fish (Li et al., 2023), tangerine peel (Li et al., 2022), tea (Xu et al., 2019), rice bran (Li et al., 2023), soybean paste (Yu et al., 2023), and fruit juice (Zhang et al., 2022).

Instead of instrumental methods such as chromatographic, enzymatic, and spectral techniques, the food industry requires quick and affordable methods to assess the quality of food products. For that purpose, biosensors are too appealing alternatives in the food industry for rapid detection of pesticides (Zhao et al., 2018), pathogens (Habimana et al., 2018), toxins (Wang et al., 2019), and microorganisms (Xu et al., 2021). Recently, with the increased penetration of smartphones in every nook and corner of life worldwide, applications of smartphone combined analysis techniques have increased. A microfluidic chip, for example, based on fluorescence resonance energy transfer was developed for the simultaneous detection of multiplex foodborne bacteria ssDNA by smartphone. The biosensor detected *S. aureus*, *L. monocytogenes*, *E. coli*, and *S. enteritidis* ssDNA simultaneously within 5 minutes, with detection limits of 0.17, 0.18, 0.27, and 0.17 nM, respectively (Xing et al., 2023). The most important criteria in biosensors are low limits of detection and quantification and system reliability. Nanoparticles are frequently used materials to improve sensitivity, particularly in trace-amount material analyses. Aflatoxin-M1 in milk was successfully detected using quantum dot-gold hybrid nanoparticles. The graphene quantum dot composite with Au nanoparticles anchors the aptamer to the sensor surface and improves signal conductivity in this electrochemical microfluidic biosensor study (Ramalingam et al., 2023). The results showed that the designed biosensor could detect Aflatoxin-M1 in a range of 100 pM to 2 nM and with an detection limits of 0.3 nM. Online quantification of food components will become a future trend in the food industry.

8.3.2.3 Food authentication

Food authentication is validating the identity and origin of a food product to guarantee that it is what it claims to be and was manufactured and handled in accordance with specified quality and safety requirements. In this way, customers are safeguarded from fraudulent/mislabeled food items, and the food industry's reputation and integrity are preserved. With the adaptation of advanced computing technologies such as microprocessors, IoT, wireless communication, and blockchain, chemical and physical analyses are increasingly being replaced by smart sensors that produce rapid and robust results for commonly performed analyses. Traceability systems, in addition to the methods mentioned earlier, are used to verify the identity and origin of food products by tracing their movement along the supply chain (Abdo Hassoun, Alhaj Abdullah, et al., 2022).

Sensory analysis, chromatographic (gas and liquid chromatography), and various spectroscopic methods are routinely used to confirm the authenticity of foods (Danezis et al., 2016; Hassoun et al., 2020). For example, LIBS combined with multivariate data analysis techniques, is a powerful tool for detecting adulterations and approving the authenticity of various food matrixes. The melamine content of adulterated toddler milk powder was predicted using univariate and multivariate data analysis methods. It was calculated that LIBS spectra can be used to detect adulterated milk with a correlation coefficient of 0.99 (Moncayo et al., 2017). In another study, LIBS detected the addition of glucose syrup to honey capturing the differences in spectral band differences belonging to calcium, sodium, and potassium (Stefas et al., 2021a).

Raman spectroscopy has numerous advantages in food adulteration due to its rich spectral information. Meat and seafood, dairy products, alcoholic and nonalcoholic beverages, grains and products, oils and products, and honey are some application fields for Raman authenticity analyses. The Raman spectroscopy was used with machine learning to identify vegetable cold-pressed oils obtained from sesame, hemp, walnut, linseed, pumpkin, and sea buckthorn (Berghian-Grosan & Magdas, 2020). Another study used LIBS along with NIR and Raman spectroscopy to detect adulteration of animal feed with transformer and mineral oils. PLSR and Principal Component Regression (PCR) methods were used to process the data, but PLSR performed better than PCR (Graham et al., 2012). However, the low sensitivity of Raman and the requirement for high analyte concentrations limit the application fields, particularly for the analysis of trace chemical molecules. These issues were fortunately resolved by surface-enhanced Raman spectroscopy (SERS). Chemical molecules are absorbed through the surface of a rough nano-metal structure, resulting in an enhanced Raman signal (Fleischmann et al., 1974). For instance, a Fe_3O_4@Au core−shell SERS substrate was developed to identify acid orange II and brilliant blue in food samples (Xie et al., 2019).

Fraud in the agrifood chain takes advantage of the fact that product authentication inspections are carried out on a certain number of samples that are not representative of the complete batch to which they belong. NIR spectroscopy technology, being fast, noninvasive, and able to be applied to all products under inspection, arises as a very powerful tool in food fraud detection, although it needs its acceptance at the legislative level to be implemented as an official inspection method. The latest research works in this regard focus on the authentication of products belonging to a certain commercial category, or the inclusion of quantities of noncompliant products in the batches of the purchased product. In 2021 a portable NIR instrument was successfully used to evaluate its possible use for the authentication of the carcasses in the slaughterhouse reaching an overall percentage of correctly classified samples up to 92.7% (Dolores Pérez-Marín, Fearn, et al., 2021). Furthermore, the potential of NIR sensors was assessed to detect the adulteration of sweet almond batches with bitter almonds in different percentages and discrimination of sweet and adulterated almonds. The obtained results revealed the high potential of NIR-based technologies for authentication practices (Torres et al., 2021; Vega-Castellote, Pérez-Marín, et al., 2021a).

Data collected in the visible spectrum are not widely used for the detection of authenticity and adulteration because they only provide information on color, size, and shape but not on other chemical aspects. However, hyperspectral and multispectral imaging tools, which can provide information at different wavelengths, can be successfully combined with current image processing technologies due to their ability to carry information at the molecular composition. In addition, thanks to the spatial information obtained (unlike other nondestructive vibrational spectroscopy methods), hyperspectral imaging has become a very common tool in the agrifood sector due to its ability to provide both surface and subsurface information, rapid results, and flexible use (Huang et al., 2014). So it has been used for many applications, such as rice cultivar discrimination (Li et al., 2016), identification of coffee bean varieties (Zhang et al., 2018), authenticity evaluation of fresh and cooked mutton kebabs (Jiang et al., 2022), saffron authentication (Hashemi-Nasab & Parastar, 2022; Lu et al., 2020), quantitative detection of multiple adulterants in wheat flour (Lohumi et al., 2019) and in almond flour (Faqeerzada et al., 2020), and detection of meat and fish authenticity (Ghidini et al., 2019; Kamruzzaman, 2018).

The advantages of the presented techniques include accuracy, selectivity, and the capacity to identify the groups of chemicals found in the sample. On the other hand, they may necessitate a high cost of analysis and/or devices as well as long times for chemical or data analysis (Śliwińska et al., 2016). The progress made in electronic sensors (E-nose, E-tongue) and biosensors, decreased the time and cost required for the authenticity analysis (Bunney et al., 2017; Śliwińska et al., 2014). Biosensors are useful for determining the botanical or geographical origin of foods. It was previously demonstrated that a label-free DNA-based biosensor could detect wine authenticity. Because different *V. vinifera* varieties (leaf, must, and wine) can cause different characteristics in wine production, distinguishing these varieties is critical (Barrias et al., 2019). Moreover, biosensors were used in other applications, such as the detection of fraud/adulteration in meat species (Flauzino et al., 2022; Zhang et al., 2019), determination of protected designation of origin of foods (Apetrei & Ghasemi-Varnamkhasti, 2013), and detection of the adulteration in milk (Bougadi & Kalogianni, 2020) and coffee (Trantakis et al., 2012).

Another alternative for cost and time-saving food authenticity analysis is electronic sensors. The systems, called E-nose and E-tongue, can be described as equivalent to each other in relation to food authenticity. Both scan the fingerprint of odor and taste components in a complex food matrix. In the simplest definition, they generate an electrical signal by detecting distinguishable odor and taste compounds in the matrix (Wadood et al., 2020). With this method, many different foodstuffs can be analyzed in a very robust manner. Detection of adulteration in saffron samples using aroma fingerprint and artificial neural networks (Heidarbeigi et al., 2015), differentiation of coffee bean species with chemometric methods (Dong et al., 2017), adulteration of meat products by soy protein (Kalinichenko & Arseniyeva, 2020), evaluation of aroma rich honey products (Veloso et al., 2018), discrimination of cocoa liquors based on odor profile (Rottiers et al., 2019), and variety and geographical origin detection of apples (Wu et al., 2018) are among the up-to-date examples of the E-nose and E-tongue applications for food authenticity and adulteration.

RFID technology is one of the most practical applications when it comes to food traceability. This technology is utilized in smart food packaging to improve food quality and safety by enabling real-time monitoring and tracking of food items along the supply chain, as previously stated. RFID technology may also be used for food authentication, which is the process of tracing a food product's identity and provenance. The microchip with an antenna, which is programmed with specific information about a food product, such as its identity and origin, source, production date, and expiration date, is embedded in the food package. Then, it transmits information through a reader to a central database, where it can be accessed and analyzed by food producers, retailers, and regulatory agencies. Moreover, the digital record of the data can be stored in a decentralized network of computers (blockchain) at each step of the supply chain, creating a transparent and immutable record of the food's journey. In summary, RFID technology may be used to monitor foods from the farm or factory to the shop and even to the consumer's house (Deng & Feng, 2020; Abdo Hassoun, Alhaj Abdullah, et al., 2022). Extra virgin olive oil (Violino et al., 2020), tobacco (Yin et al., 2022), liquor (Sun et al., 2019), dairy products (Khanna et al., 2022), and halal products and meat (Cao et al., 2021; Shuib et al., 2021) are some of the food and agricultural products where RFID and blockchain technology have been successfully adapted.

8.3.2.4 Processes control and optimization

To ensure the desired product quality and food safety in food processes as well as agricultural production, process control and optimization are critical for achieving minimal energy use, and green and sustainable production targets within the context of combating global climate change and increasing population. To achieve this, it is important to collect and process real-time data and decide on the action to be taken to control and optimize processes. With emerging technologies during the transition to the Industry 4.0 era, it is more possible than ever before. The use of smart sensors alone or in conjunction with classical sensors has become widespread. In this respect, the fusion of the data acquired from different process equipment and different sensors is extensively employed in process optimization and control owing to the increasing use of IoT with the development of wireless communication technologies (Jagtap et al., 2021; Jin et al., 2020; Lv et al., 2017; Lydia et al., 2022; Tschand, 2023; Xu et al., 2023). Processes may be regulated in this manner with the use of intelligent decision-makers by interpreting the data in real time. For example, LIBS can be used for both online and inline control of the optimization of brining conditions in foods by mapping the salt diffusion in them (Dixit et al., 2018; Sezer et al., 2019). Raman spectroscopy can be used for inline monitoring of anethole and fenchone extraction from fennel seed (Sixt et al., 2018), and FT-Raman is used to monitor the distillation process of honey spirit online (Anjos et al., 2020).

Biosensors are also useful tools for reliable and fast analysis of foods and processes. Unlike conventional methods, it is possible to make evaluations at the molecular level with biosensors. For example, the volatile organic compounds in kombucha tea extract during fermentation were monitored using a nano colorimetric sensor (Kang et al., 2022). In another study, researchers used a screen-printed microchip to track L-glutamate levels during the microbial fermentation process (Yang et al., 2023). Another method for simultaneous prediction of food composition is hyperspectral imaging. Hyperspectral imaging and computer vision are among the most suitable approaches to future process control when compared with different methods. Images obtained in different spectral ranges can be processed in a very short time by advanced machine learning methods and used for purposes such as classification, defect detection, composition determination, color, texture, and shape analysis (Caporaso et al., 2018; Chung & Yoon, 2021; Huang et al., 2021; Mavani et al., 2022; Nasiri et al., 2019; Ndisya et al., 2021; Nyalala et al., 2019; Taheri-Garavand et al., 2019; Torres et al., 2021; Torres-Rodríguez et al., 2022).

The use of soft sensors in place of classical approaches has become more common for a variety of reasons, including the fact that the desired process parameters are physically impossible to obtain under certain conditions, cost, failure to obtain the results simultaneously, or the characteristic of interest can only be measured at the end of the process (Jiang et al., 2021). Despite the existence of data-driven and machine-learning-based soft sensors (Onwude et al., 2020; Peñaloza et al., 2021; Pinheiro, 2019), they may lack sufficient explanatory information about system dynamics compared to those developed using mechanistic modeling techniques and require a large amount of experimental data for accurate training. Therefore it is believed that soft sensors or digital twins developed using first principles are preferable by engineers for intelligent process control and optimization.

Soft sensor platforms built using mechanistic modeling techniques, which consider the conservation of heat, mass, and momentum, are indeed computerized clones of the physical system. Because of the complexity and high computational requirements, a targeted step is considered instead of modeling the entire process. Temperature, mass, flow rate, and pressure changes that occur in such a system can be acquired simultaneously with the process or before it using some recent examples of the soft sensors present in the literature (Defraeye et al., 2019; Kannapinn et al., 2022; Kielmann et al., 2022; Onwude et al., 2022; Shoji et al., 2022; Turgut et al., 2021; Turgut et al., 2022). In addition, by strengthening the mechanistic soft sensors with data-driven statistical and/or kinetic models, the results of the quality and yield of the product such as oxidation, degradation, color, volume, and texture change in foods can be estimated before and/or during the processing (Blikra et al., 2020; Rabeler et al., 2019; Touffet et al., 2021; Turgut et al., 2020). Then, data from the required number of soft probes on the digital twin can be fed back for online process control and management of system conditions (Verboven et al., 2020). The use of mechanistic models, and hence soft sensors, has focused more on the control of food processing processes. However, there is also a relatively small number of studies on agricultural production, leaving a lot of room for future research (Baker et al., 2018; Morelli et al., 2022; Schnepf & Leitner, 2009; Wang et al., 2021).

8.3.3 Joint technologies

A range of collaborative technologies may be utilized in smart food and agricultural production to increase efficiency and productivity while also improving product safety and quality. Joint technologies can be defined as the coordinated use of sensors with different technical characteristics or electronic components that enable sensors to be used more efficiently. For example, the IoT is one such technology, which refers to a linked network of physical devices, sensors, and other items embedded with computer and communication capabilities. IoT devices may be used in agriculture to monitor and regulate numerous elements of agricultural and livestock production, such as physical and chemical properties including irrigation, fertilizer, temperature, humidity, and disease (Alrowais et al., 2022; Balivada et al., 2022; Elsherbiny et al., 2022; Lydia et al., 2022; Mohammad EL-Basioni et al., 2022; Tschand, 2023; Wongchai et al., 2022; Xu et al., 2023). Other joint technologies that are commonly used in smart food and agricultural production include RFID tags, robotics, and UAVs, which can be used to monitor and control various aspects of the production process. More examples are given in previous sections and Table 8.1.

Not only combining electronics but also the combination of obtained data from various sources is also a joint technology. This is called data fusion modeling and is very popular in smart food and agricultural production since it helps to obtain more accurate estimates. Generally, a combination of data obtained from two or more different sensors (generally with different working principles) is done. Besides, a combination of smart sensor data with data from conventional methods such as chromatographic analysis or sensory evaluation is also very useful. For example, the use of E-nose and E-tongue in combination (Zakaria et al., 2011), joining spectral and textural results (Yang et al., 2018), combining visible near-infrared and short-wave infrared images (Chung & Yoon, 2021), a fusion of RGB images with motion influence map (Li et al., 2022), use of data taken by a UAV equipped with RGB, multispectral and thermal sensors by fusion (Maimaitijiang et al., 2020), data from multiple electronic sensors (Li et al., 2023), a fusion of data obtained from electronic sensors and texture analysis (Yu et al., 2023), and fusion of E-nose, E-tongue and E-eye combined with chemometrics (Xu et al., 2019) are examples of different fusion types utilized for different purposes through the agrofood chain. Overall, the use of joint technologies in smart food and agricultural production can assist to improve efficiency, reduce costs, and improve product quality and safety.

8.4 Conclusion

Smart sensors have the potential to transform the food and agricultural industries in the future by allowing for real-time monitoring and control of numerous elements of food production, processing, and distribution. Some examples of smart sensor applications in the food and agricultural industries include:

- Precision agriculture: To improve irrigation, fertilization, and pest control procedures, smart sensors may be utilized to monitor numerous environmental and plant-related data in fields, such as soil moisture, temperature, and nutrient levels.
- Livestock monitoring: Smart sensors may be used to monitor livestock health and well-being, such as recognizing symptoms of disease or stress, and improving feeding and breeding methods.
- Food safety, quality, and authenticity: Smart sensors can identify impurities and monitor temperature, composition, color, shape, and origin to monitor the quality and safety of food items during manufacturing, processing, and storage.
- Supply chain traceability: Smart sensors may be used to track the movement and handling of food items across the supply chain, from the farm to the fork, to increase efficiency and maintain food quality and safety.

Smart sensors are used at every step of the agrofood chain in many ways and for many different purposes. It is predicted that smart sensor applications will become more diverse and widespread in the future. With advances in Industry 4.0 enabling technologies and with electronic components getting cheaper and easier to access, smart sensors will likely become even more accurate and affordable. Moreover, with the development of wireless communication technologies, the diversity of data storage and processing possibilities, and the smaller flexibility of devices, as more data are collected and analyzed, it will become possible to develop more detailed, accurate, and generalizable models of agrofood chain assessment.

Overall, the use of smart sensors is expected to lead to analysis and control actions of almost every step in food and agriculture processes. The application of smart sensors in food and agricultural production has the potential to increase efficiency, minimize waste, and improve the quality and safety of food items, while also assisting in the optimization of the use of resources such as water and fertilizers to ensure green and sustainable food production.

References

Abraham, J. (2022). Future of food packaging: Intelligent packaging. In *Nanotechnology in intelligent food packaging* (pp. 383–417). <https://onlinelibrary.wiley.com/doi/abs/10.1002/9781119819011.ch16/>.

Adame-Siles, J. A., Sánchez-Müller, F., Maroto-Molina, F., Riccioli, C., Garrido-Varo, A., & Pérez-Marín, D. (2018). A fiber-optic probe and FT-NIR for onsite quality analysis of olive oils. *NIR News*, *29*(4), 4–8. Available from https://doi.org/10.1177/0960336018765584.

Akhtar, M. N., Shaikh, A. J., Khan, A., Awais, H., Bakar, E. A., & Othman, A. R. (2021). Smart sensing with edge computing in precision agriculture for soil assessment and heavy metal monitoring: A review. *Agriculture*, *11*(6), 475. Available from https://doi.org/10.3390/agriculture11060475.

Ali, A., & Imran, M. (2020). Evaluating the potential of red edge position (REP) of hyperspectral remote sensing data for real time estimation of LAI & chlorophyll content of kinnow mandarin (Citrus reticulata) fruit orchards. *Scientia Horticulturae*, *267*, 109326. Available from https://doi.org/10.1016/j.scienta.2020.109326.

Allred, B., Martinez, L., Fessehazion, M. K., Rouse, G., Williamson, T. N., Wishart, D., Koganti, T., Freeland, R., Eash, N., Batschelet, A., & Featheringill, R. (2020). Overall results and key findings on the use of UAV visible-color, multispectral, and thermal infrared imagery to map agricultural drainage pipes. *Agricultural Water Management*, *232*, 106036. Available from https://doi.org/10.1016/j.agwat.2020.106036.

Alrowais, F., Asiri, M. M., Alabdan, R., Marzouk, R., Hilal, A. M., alkhayyat, A., & Gupta, D. (2022). Hybrid leader based optimization with deep learning driven weed detection on internet of things enabled smart agriculture environment. *Computers and Electrical Engineering*, *104*, 108411. Available from https://doi.org/10.1016/j.compeleceng.2022.108411.

Anjos, O., Santos, R., Estevinho, L. M., & Caldeira, I. (2020). FT-RAMAN methodology for the monitoring of honeys' spirit distillation process. *Food Chemistry*, *305*, 125511.

Antonacci, A., Arduini, F., Moscone, D., Palleschi, G., & Scognamiglio, V. (2018). Nanostructured (Bio)sensors for smart agriculture. *TrAC Trends in Analytical Chemistry*, *98*, 95–103. Available from https://doi.org/10.1016/j.trac.2017.10.022.

Apetrei, C., & Ghasemi-Varnamkhasti, M. (2013). Biosensors in Food PDO Authentication. In M. de la Guardia, & A. Gonzálvez (Eds.), *Comprehensive Analytical Chemistry* (60, pp. 279–297). Elsevier. Available from https://doi.org/10.1016/B978-0-444-59562-1.00011-6.

Arefi, A., Sturm, B., Raut, S., von Gersdorff, G., & Hensel, O. (2023). NIR laser-based imaging techniques to monitor quality attributes of apple slices during the drying process: Laser-light backscattering & biospeckle imaging techniques. *Food Control*, *143*, 109289. Available from https://doi.org/10.1016/j.foodcont.2022.109289.

Ariza-Sentís, M., Valente, J., Kooistra, L., Kramer, H., & Mücher, S. (2023). Estimation of spinach (Spinacia oleracea) seed yield with 2D UAV data and deep learning. *Smart Agricultural Technology*, *3*, 100129. Available from https://doi.org/10.1016/j.atech.2022.100129.

Arreguin-Campos, R., Frigoli, M., Caldara, M., Crapnell, R. D., Ferrari, A. G.-M., Banks, C. E., Cleij, T. J., Diliën, H., Eersels, K., & van Grinsven, B. (2023). Functionalized screen-printed electrodes for the thermal detection of *Escherichia coli* in dairy products. *Food Chemistry*, *404*, 134653.

Arvanitoyannis, I. S., & Stratakos, A. Ch (2012). Application of modified atmosphere packaging and active/smart technologies to red meat and poultry: a review. *Food and Bioprocess Technology*, *5*(5), 1423–1446.

Azman, N. H., Khairul, W. M., & Sarbon, N. M. (2022). A comprehensive review on biocompatible film sensor containing natural extract: Active/intelligent food packaging. *Food Control*, *141*, 109189. Available from https://doi.org/10.1016/j.foodcont.2022.109189.

Ba, W., Jin, X., Lu, J., Rao, Y., Zhang, T., Zhang, X., Zhou, J., & Li, S. (2023). Research on predicting early *Fusarium* head blight with asymptomatic wheat grains by micro-near infrared spectrometer. *Spectrochimica Acta Part A: Molecular and Biomolecular Spectroscopy*, *287*, 122047. Available from https://doi.org/10.1016/j.saa.2022.122047.

Babich, N. (2020). Computer vision allows us to create products that can identify and process objects in images and videos in the same way that we do. Learn about the benefits of this technology at Adobe XD Ideas. Ideas What is Computer Vision & How Does it Work? An Introduction. <https://xd.adobe.com/ideas/principles/emerging-technology/what-is-computer-vision-how-does-it-work/>.

Baeten, V., Meurens, M., Morales, M. T., & Aparicio, R. (1996). Detection of virgin olive oil adulteration by Fourier transform Raman spectroscopy. *Journal of Agricultural and Food Chemistry*, *44*(8), 2225–2230.

Baker, R. E., Peña, J.-M., Jayamohan, J., & Jérusalem, A. (2018). Mechanistic models versus machine learning, a fight worth fighting for the biological community? *Biology Letters*, *14*, 20170660. Available from https://doi.org/10.1098/rsbl.2017.0660.

Balivada, S., Grant, G., Zhang, X., Ghosh, M., Guha, S., & Matamala, R. (2022). A wireless underground sensor network field pilot for agriculture and ecology: Soil moisture mapping using signal attenuation. *Sensors*, *22*(10), 3913. Available from https://doi.org/10.3390/s22103913.

Banerjee, A., & Srinivasan, K. (2021). RFTemp: Monitoring microwave oven leakage to estimate food temperature. *Proceedings of the ACM on Interactive, Mobile, Wearable and Ubiquitous Technologies*, *5*(4), 1–25. Available from https://doi.org/10.1145/3494967.

Barnett, C., Bell, C., Vig, K., Akpovo, A. C., Johnson, L., Pillai, S., & Singh, S. (2011). Development of a LIBS assay for the detection of *Salmonella enterica* serovar Typhimurium from food. *Analytical and Bioanalytical Chemistry*, *400*(10), 3323–3330.

Barreiro Elorza, P., Cristina Hernando, E. C., Arranz Saiz, F. J., Diezma Iglesias, B., Ruiz García, L., Villarroel, M., Robla Villalba, J. I., & García-Hierro, J. (2011). *Smart sensing applications in agriculture and food industry. Smart sensors and sensing technology*, 41.

Barrias, S., Fernandes, J. R., Eiras-Dias, J. E., Brazão, J., & Martins-Lopes, P. (2019). Label free DNA-based optical biosensor as a potential system for wine authenticity. *Food Chemistry*, *270*, 299–304.

Benabdellah, N., Bourhaleb, M., Nasri, M.', Benazzi, N., & Dahbi, S. (2016). Design of temperature and humidity sensors for an electronic nose used in rotten food. In *2016 International conference on electrical and information technologies (ICEIT)* (pp. 505–509). Available from https://doi.org/10.1109/EITech.2016.7519652.

Berghian-Grosan, C., & Magdas, D. A. (2020). Raman spectroscopy and machine-learning for edible oils evaluation. *Talanta*, *218*, 121176.

Bhargava, A., & Bansal, A. (2021). Fruits and vegetables quality evaluation using computer vision: A review. *Journal of King Saud University - Computer and Information Sciences*, *33*(3), 243–257. Available from https://doi.org/10.1016/j.jksuci.2018.06.002.

Bharti, S. K., & Roy, R. (2012). Quantitative 1H NMR spectroscopy. *TrAC Trends in Analytical Chemistry*, *35*, 5–26.

Bilge, G., Sezer, B., Eseller, K. E., Berberoglu, H., Topcu, A., & Boyaci, I. H. (2016). Determination of whey adulteration in milk powder by using laser induced breakdown spectroscopy. *Food Chemistry*, *212*, 183–188.

Bilge, G., Sezer, B., Eseller, K. E., Berberoglu, H., Koksel, H., & Boyaci, I. H. (2016). Ash analysis of flour sample by using laser-induced breakdown spectroscopy. *Spectrochimica Acta Part B: Atomic Spectroscopy*, *124*, 74–78.

Bilge, G., Boyacı, İ. H., Eseller, K. E., Tamer, U., & Çakır, S. (2015). Analysis of bakery products by laser-induced breakdown spectroscopy. *Food Chemistry*, *181*, 186–190.

Bilodeau, M. F., Esau, T. J., MacEachern, C. B., Farooque, A. A., White, S. N., & Zaman, Q. U. (2023). Identifying hair fescue in wild blueberry fields using drone images for precise application of granular herbicide. *Smart Agricultural Technology*, *3*, 100127. Available from https://doi.org/10.1016/j.atech.2022.100127.

Blikra, M. J., Hodnefjell, Å. V., Feyissa, A. H., & Skipnes, D. (2020). Dimensional change and cook loss during heating of fish: Problem formulation and semi-empirical modeling approach. *Journal of Food Engineering*, *281*, 110004. Available from https://doi.org/10.1016/j.jfoodeng.2020.110004.

Borém, F. M., de Abreu, G. F., Ferreira, A. G., da Silva Santos, M., da Conceição Alves, T., & de Carvalho Alves, A. P. (2023). 1H NMR spectroscopy applied to identify chemical aging markers in green coffee (*Coffea arabica* L.). *Food Chemistry*, *405*, 134667.

Bougadi, E. Th, & Kalogianni, D. P. (2020). Paper-based DNA biosensor for food authenticity testing. *Food Chemistry*, *322*, 126758. Available from https://doi.org/10.1016/j.foodchem.2020.126758.

Bouzembrak, Y., Klüche, M., Gavai, A., & Marvin, H. J. P. (2019). Internet of things in food safety: Literature review and a bibliometric analysis. *Trends in Food Science & Technology*, *94*, 54–64. Available from https://doi.org/10.1016/j.tifs.2019.11.002.

Bumbrah, G. S., & Sharma, R. M. (2016). Raman spectroscopy – Basic principle, instrumentation and selected applications for the characterization of drugs of abuse. *Egyptian Journal of Forensic Sciences*, *6*(3), 209–215. Available from https://doi.org/10.1016/j.ejfs.2015.06.001.

Bunney, J., Williamson, S., Atkin, D., Jeanneret, M., Cozzolino, D., & Chapman, J. (2017). The use of electrochemical biosensors in food analysis. *Current Research in Nutrition and Food Science Journal*, *5*(3), 183–195.

Buratti, S., Malegori, C., Benedetti, S., Oliveri, P., & Giovanelli, G. (2018). E-nose, e-tongue and e-eye for edible olive oil characterization and shelf life assessment: A powerful data fusion approach. *Talanta*, *182*, 131–141. Available from https://doi.org/10.1016/j.talanta.2018.01.096.

Bureau, S., Cozzolino, D., & Clark, C. J. (2019). Contributions of Fourier-transform mid infrared (FT-MIR) spectroscopy to the study of fruit and vegetables: A review. *Postharvest Biology and Technology*, *148*, 1–14. Available from https://doi.org/10.1016/j.postharvbio.2018.10.003.

Burgett, G. (2019). *OmniVision has created the world's smallest commercially-available image sensor*. <https://www.dpreview.com/news/4272574802/omnivision-has-created-the-world-s-smallest-commercially-available-image-sensor/>.

Burns, D.A., & Ciurczak,E.W. (2007). *Handbook of near-infrared analysis*. <https://www.taylorfrancis.com/books/edit/10.1201/9781420007374/handbook-near-infrared-analysis-donald-burns-emil-ciurczak/>.

Cáceres-Nevado, J. M., Garrido-Varo, A., De Pedro-Sanz, E., Tejerina-Barrado, D., & Pérez-Marín, D. C. (2021). Non-destructive near infrared spectroscopy for the labelling of frozen Iberian pork loins. *Meat Science*, *175*, 108440. Available from https://doi.org/10.1016/j.meatsci.2021.108440.

Cáceres-Nevado, J. M., Garrido-Varo, A., De Pedro-Sanz, E., & Pérez-Marín, D. C. (2021). NIR handheld miniature spectrometer to increase the efficiency of Iberian pig selection schemes based on chemical traits. *Spectrochimica Acta. Part A, Molecular and Biomolecular Spectroscopy*, *258*, 119865. Available from https://doi.org/10.1016/j.saa.2021.119865.

Calero, A. M., Muñoz, E., Pérez-Marin, D., Riccioli, C., Pérez, L., & Garrido-Varo, A. (2018). Evolution of frying oil quality using Fourier transform near-infrared (FT-NIR) spectroscopy. *Applied Spectroscopy*, *72*(7), 1001–1013. Available from https://doi.org/10.1177/0003702818764125.

Cama-Moncunill, X., Markiewicz-Keszycka, M., Cullen, P. J., Sullivan, C., & Casado-Gavalda, M. P. (2020). Direct analysis of calcium in liquid infant formula via laser-induced breakdown spectroscopy (LIBS). *Food Chemistry*, *309*, 125754.

Cama-Pinto, D., Damas, M., Holgado-Terriza, J. A., Arrabal-Campos, F. M., Gómez-Mula, F., Lao, J. A. M., & Cama-Pinto, A. (2020). Empirical model of radio wave propagation in the presence of vegetation inside greenhouses using regularized regressions. *Sensors (Switzerland)*, *20*(22), 6621. Available from https://doi.org/10.3390/s20226621.

Cao, S., Powell, W., Foth, M., Natanelov, V., Miller, T., & Dulleck, U. (2021). Strengthening consumer trust in beef supply chain traceability with a blockchain-based human-machine reconcile mechanism. *Computers and Electronics in Agriculture*, *180*, 105886. Available from https://doi.org/10.1016/j.compag.2020.105886.

Cao, X.-T., & Chung, W.-Y. (2019). An enhanced multiplication of RF energy harvesting efficiency using relay resonator for food monitoring. *Sensors (Switzerland)*, *19*(9). Available from https://doi.org/10.3390/s19091963.

Caporaso, N., Whitworth, M. B., Fowler, M. S., & Fisk, I. D. (2018). Hyperspectral imaging for non-destructive prediction of fermentation index, polyphenol content and antioxidant activity in single cocoa beans. *Food Chemistry*, *258*, 343–351. Available from https://doi.org/10.1016/j.foodchem.2018.03.039.

Chang, C. Y., Zhou, R., Kira, O., Marri, S., Skovira, J., Gu, L., & Sun, Y. (2020). An unmanned aerial system (UAS) for concurrent measurements of solar-induced chlorophyll fluorescence and hyperspectral reflectance toward improving crop monitoring. *Agricultural and Forest Meteorology*, *294*, 108145. Available from https://doi.org/10.1016/j.agrformet.2020.108145.

Chao, K., Dhakal, S., Schmidt, W. F., Qin, J., Kim, M., Peng, Y., & Huang, Q. (2020). Raman and IR spectroscopic modality for authentication of turmeric powder. *Food Chemistry*, *320*, 126567.

Chen, H., Chen, A., Xu, L., Xie, H., Qiao, H., Lin, Q., & Cai, K. (2020). A deep learning CNN architecture applied in smart near-infrared analysis of water pollution for agricultural irrigation resources. *Agricultural Water Management*, *240*, 106303. Available from https://doi.org/10.1016/j.agwat.2020.106303.

Chen, S., Brahma, S., Mackay, J., Cao, C., & Aliakbarian, B. (2020). The role of smart packaging system in food supply chain. *Journal of Food Science*, *85*(3), 517–525. Available from https://doi.org/10.1111/1750-3841.15046.

Cheng, W., Sun, D.-W., Pu, H., & Wei, Q. (2018). Heterospectral two-dimensional correlation analysis with near-infrared hyperspectral imaging for monitoring oxidative damage of pork myofibrils during frozen storage. *Food Chemistry*, *248*, 119–127. Available from https://doi.org/10.1016/j.foodchem.2017.12.050.

Chowdhury, S., Dey, P., Joel-Edgar, S., Bhattacharya, S., Rodriguez-Espindola, O., Abadie, A., & Truong, L. (2022). Unlocking the value of artificial intelligence in human resource management through AI capability framework. *Human Resource Management Review*, 100899. Available from https://doi.org/10.1016/j.hrmr.2022.100899.

Chung, N., Ameer, K., Jo, Y., & Kwon, J.-H. (2019). Comparison of electronic sensing techniques for screening dried shrimps irradiated using three types of approved radiation with standard analytical methods. *Food Chemistry*, *286*, 395–404. Available from https://doi.org/10.1016/j.foodchem.2019.02.038.

Chung, S., & Yoon, S.-C. (2021). Detection of foreign materials on broiler breast meat using a fusion of visible near-infrared and short-wave infrared hyperspectral imaging. *Applied Sciences*, *11*(24), 11987. Available from https://doi.org/10.3390/app112411987.

Comba, L., Biglia, A., Aimonino, D. R., & Gay, P. (2018). Unsupervised detection of vineyards by 3D point-cloud UAV photogrammetry for precision agriculture. *Computers and Electronics in Agriculture*, *155*, 84–95. Available from https://doi.org/10.1016/j.compag.2018.10.005.

Coombs, C. E. O., Allman, B. E., Morton, E. J., Gimeno, M., Horadagoda, N., Tarr, G., & González, L. A. (2023). A preliminary investigation into the automatic detection of diseased sheep organs using hyperspectral imaging sensors. *Smart Agricultural Technology*, *3*, 100122. Available from https://doi.org/10.1016/j.atech.2022.100122.

Correa-Hernando, E., Arranz, F. J., Diezma, B., Juliá, E., Robla, J. I., Ruiz-Garcia, L., García-Hierro, J., & Barreiro, P. (2011). Development of model based sensors for the supervision of a solar dryer. *Computers and Electronics in Agriculture*, *78*(2), 167–175. Available from https://doi.org/10.1016/j.compag.2011.07.004.

Costa, V. C., de Babos, D. V., de Aquino, F. W. B., Virgílio, A., Amorim, F. A. C., & Pereira-Filho, E. R. (2018). Direct determination of Ca, K and Mg in cassava flour samples by laser-induced breakdown spectroscopy (LIBS). *Food Analytical Methods*, *11*(7), 1886–1896.

Cracknell, A. P. (2018). The development of remote sensing in the last 40 years. *International Journal of Remote Sensing*, *39*(23), 8387–8427. Available from https://doi.org/10.1080/01431161.2018.1550919.

Cregan, V., Lee, W. T., & Clune, L. (2017). A soft sensor for the Bayer process. *Journal of Mathematics in Industry*, *7*(1), 7. Available from https://doi.org/10.1186/s13362-017-0037-9.

Cui, L., Zhang, Z., Gao, N., Meng, Z., & Li, Z. (2019). Radio frequency identification and sensing techniques and their applications—A review of the state-of-the-art. *Sensors*, *19*(18), 4012. Available from https://doi.org/10.3390/s19184012.

Dabbene F., Gay,P., & Tortia,C. (2016). Radio-frequency identification usage in food traceability. In M. Espiñeira, F. J. Santaclara (Eds.), *Advances in food traceability techniques and technologies* (pp. 67–89). <https://www.sciencedirect.com/science/article/pii/B9780081003107000053/>.

Dais, P., Misiak, M., & Hatzakis, E. (2015). Analysis of marine dietary supplements using NMR spectroscopy. *Analytical Methods*, *7*(12), 5226–5238.

Danezis, G. P., Tsagkaris, A. S., Camin, F., Brusic, V., & Georgiou, C. A. (2016). Food authentication: Techniques, trends and emerging approaches. *TrAC - Trends in Analytical Chemistry*, *85*, 123–132. Available from https://doi.org/10.1016/j.trac.2016.02.026, http://www.elsevier.com/locate/trac.

de la Roza-Delgado, B., Garrido-Varo, A., Soldado, A., González Arrojo, A., Cuevas Valdés, M., Maroto, F., & Pérez-Marín, D. (2017). Matching portable NIRS instruments for in situ monitoring indicators of milk composition. *Food Control*, *76*, 74–81. Available from https://doi.org/10.1016/j.foodcont.2017.01.004.

Defraeye, T., Tagliavini, G., Wu, W., Prawiranto, K., Schudel, S., Assefa Kerisima, M., Verboven, P., & Bühlmann, A. (2019). Digital twins probe into food cooling and biochemical quality changes for reducing losses in refrigerated supply chains. *Resources, Conservation and Recycling*, *149*, 778–794. Available from https://doi.org/10.1016/j.resconrec.2019.06.002.

Delgadillo-Duran, D. A., Vargas-García, C. A., Varón-Ramírez, V. M., Calderón, F., Montenegro, A. C., & Reyes-Herrera, P. H. (2022). Vis–NIR spectroscopy and machine learning methods to diagnose chemical properties in Colombian sugarcane soils. *Geoderma Regional*, *31*. Available from https://doi.org/10.1016/j.geodrs.2022.e00588.

Deng, M., & Feng, P. (2020). A food traceability system based on blockchain and radio frequency identification technologies. *Journal of Computer and Communications*, *8*(9), 17–27. Available from https://doi.org/10.4236/jcc.2020.89002.

Dixit, Y., Casado-Gavalda, M. P., Cama-Moncunill, R., Markiewicz-Keszycka, M., Cama-Moncunill, X., Cullen, P. J., & Sullivan, C. (2017). Quantification of rubidium as a trace element in beef using laser induced breakdown spectroscopy. *Meat Science*, *130*, 47–49.

Dixit, Y., Casado-Gavalda, M. P., Cama-Moncunill, R., Cama-Moncunill, X., Markiewicz-Keszycka, M., Jacoby, F., Cullen, P. J., & Sullivan, C. (2018). Introduction to laser induced breakdown spectroscopy imaging in food: Salt diffusion in meat. *Journal of Food Engineering*, *216*, 120–124.

Dong, J. X., Gao, Z. F., Zhang, Y., Li, B. L., Li, N. B., & Luo, H. Q. (2017). A selective and sensitive optical sensor for dissolved ammonia detection via agglomeration of fluorescent Ag nanoclusters and temperature gradient headspace single drop microextraction. *Biosensors and Bioelectronics*, *91*, 155–161.

Dong, M., Mao, X., Gonzalez, J. J., Lu, J., & Russo, R. E. (2012). Time-resolved LIBS of atomic and molecular carbon from coal in air, argon and helium. *Journal of Analytical Atomic Spectrometry*, *27*(12), 2066–2075.

Dong, W., Zhao, J., Hu, R., Dong, Y., & Tan, L. (2017). Differentiation of Chinese robusta coffees according to species, using a combined electronic nose and tongue, with the aid of chemometrics. *Food Chemistry*, *229*, 743–751. Available from https://doi.org/10.1016/j.foodchem.2017.02.149.

Donglu, F., Wenjian, Y., Kimatu, B. M., Liyan, Z., Xinxin, A., & Qiuhui, H. (2017). Comparison of flavour qualities of mushrooms (*Flammulina velutipes*) packed with different packaging materials. *Food Chemistry*, *232*, 1–9. Available from https://doi.org/10.1016/j.foodchem.2017.03.161.

dos Santos Augusto, A., Barsanelli, P. L., Pereira, F. M. V., & Pereira-Filho, E. R. (2017). Calibration strategies for the direct determination of Ca, K, and Mg in commercial samples of powdered milk and solid dietary supplements using laser-induced breakdown spectroscopy (LIBS). *Food Research International*, *94*, 72–78.

Duraipandian, S., Petersen, J. C., & Lassen, M. (2019). Authenticity and concentration analysis of extra virgin olive oil using spontaneous Raman spectroscopy and multivariate data analysis. *Applied Sciences*, *9*(12), 2433.

Ehsani, R., & Maja, J. M. (2013). The rise of small UAVs in precision agriculture. *Engineering and Technology for Sustainable World*, *20*(4), 18–19.

Elisseeva, S., Kelly, C., Cruz-Romero, M., Zhdanov, A. V., Kerry, J. P., & Papkovsky, D. B. (2020). The use of optical oxygen sensing and respirometry to quantify the effects of antimicrobials on common food spoilage bacteria and food samples. *Sensors and Actuators B: Chemical*, *322*, 128572. Available from https://doi.org/10.1016/j.snb.2020.128572.

El-Mesery, H. S., Mao, H., & Abomohra, A. El-Fatah (2019). Applications of non-destructive technologies for agricultural and food products quality inspection. *Sensors*, *19*(4), 846. Available from https://doi.org/10.3390/s19040846.

Elsherbiny, O., Zhou, L., He, Y., & Qiu, Z. (2022). A novel hybrid deep network for diagnosing water status in wheat crop using IoT-based multimodal data. *Computers and Electronics in Agriculture*, *203*, 107453. Available from https://doi.org/10.1016/j.compag.2022.107453.

Elyasizadeh, S. (2016). *Automated Monitoring System-Designing a Laboratory Equipment Tracking System*. Lahti University of Applied Sciences.

Emanuel, N., & Sandhu, H. K. (2019). Food packaging development: Recent perspective. *Journal of Thin Films Coating Science Technology and Application.*, *6*, 13–29.

Entrenas, J.-A., Pérez-Marín, D., Torres, I., Garrido-Varo, A., & Sánchez, M.-T. (2020). Simultaneous detection of quality and safety in spinach plants using a new generation of NIRS sensors. *Postharvest Biology and Technology*, *160*, 111026. Available from https://doi.org/10.1016/j.postharvbio.2019.111026.

Entrenas, J.-A., Pérez-Marín, D., Torres, I., Garrido-Varo, A., & Sánchez, M.-T. (2019). Safety and quality issues in summer squashes using handheld portable NIRS sensors for real-time decision making and for on-vine monitoring. *Journal of the Science of Food and Agriculture*, *99*(15), 6768–6777. Available from https://doi.org/10.1002/jsfa.9959.

Escobedo, P., Ramos-Lorente, C. E., Ejaz, A., Erenas, M. M., Martínez-Olmos, A., Carvajal, M. A., García-Núñez, C., de Orbe-Payá, I., Capitán-Vallvey, L. F., & Palma, A. J. (2023). QRsens: Dual-purpose quick response code with built-in colorimetric sensors. *Sensors and Actuators B: Chemical*, *376*, 133001. Available from https://doi.org/10.1016/j.snb.2022.133001.

Escobedo, P., Bhattacharjee, M., Nikbakhtnasrabadi, F., & Dahiya, R. (2021). Flexible strain and temperature sensing NFC tag for smart food packaging applications. *IEEE Sensors Journal*, *21*(23), 26406–26414. Available from https://doi.org/10.1109/JSEN.2021.3100876.

Eyre, R., Lindsay, J., Laamrani, A., & Berg, A. (2021). Within-field yield prediction in cereal crops using LiDAR-derived topographic attributes with geographically weighted regression models. *Remote Sensing*, *13*(20), 4152. Available from https://doi.org/10.3390/rs13204152.

FAO. (2009). How to feed the World 2050: Global agriculture towards 2050. Rome. <https://www.fao.org/fileadmin/templates/wsfs/docs/Issues_papers/HLEF2050_Global_Agriculture.pdf/>.

FAO. (2022). Thinking about the future of food safety – A foresight report. Rome. Unpublished content. Available from https://doi.org/10.4060/cb8667en, 978-92-5-135783-5.

Faqeerzada, M. A., Lohumi, S., Kim, G., Joshi, R., Lee, H., Kim, M. S., & Cho, B.-K. (2020). Hyperspectral shortwave infrared image analysis for detection of adulterants in almond powder with one-class classification method. *Sensors*, *20*(20), 5855. Available from https://doi.org/10.3390/s20205855.

Fathi, P., Karmakar, N. C., Bhattacharya, M., & Bhattacharya, S. (2020). Potential chipless RFID sensors for food packaging applications: A review. *IEEE Sensors Journal*, *20*(17), 9618–9636. Available from https://doi.org/10.1109/JSEN.2020.2991751.

Fellows, P.J. (2009). Properties of food processing. In *Food processing technology (third edition)* (pp. 11–95). <https://www.sciencedirect.com/science/article/pii/B9781845692162500025/>.

Fernando, I., Fei, J., Stanley, R., & Rouillard, V. (2020). Evaluating packaging performance for bananas under simulated vibration. *Food Packaging and Shelf Life*, *23*, 100428. Available from https://doi.org/10.1016/j.fpsl.2019.100428.

Flauzino, J. M. R., Alves, L. M., Rodovalho, V. R., Madurro, J. M., & Brito Madurro, A. G. (2022). Application of biosensors for detection of meat species: A short review. *Food Control*, *142*, 109214. Available from https://doi.org/10.1016/j.foodcont.2022.109214.

Fleischmann, M., Hendra, P. J., & McQuillan, A. J. (1974). Raman spectra of pyridine adsorbed at a silver electrode. *Chemical Physics Letters*, *26*(2), 163–166.

Fulgione, A., Cimafonte, M., Ventura, B. D., Iannaccone, M., Ambrosino, C., Capuano, F., Proroga, Y. T. R., Velotta, R., & Capparelli, R. (2018). QCM-based immunosensor for rapid detection of Salmonella Typhimurium in food. *Scientific Reports*, *8*(1), 1–8.

Gabriel, J. L., Zarco-Tejada, P. J., López-Herrera, P. J., Pérez-Martín, E., Alonso-Ayuso, M., & Quemada, M. (2017). Airborne and ground level sensors for monitoring nitrogen status in a maize crop. *Biosystems Engineering*, *160*, 124–133. Available from https://doi.org/10.1016/j.biosystemseng.2017.06.003.

Galvan, D., Aquino, A., Effting, L., Mantovani, A. C. G., Bona, E., & Conte-Junior, C. A. (2022). E-sensing and nanoscale-sensing devices associated with data processing algorithms applied to food quality control: a systematic review. *Critical Reviews in Food Science and Nutrition*, *62*(24), 6605–6645. Available from https://doi.org/10.1080/10408398.2021.1903384.

Gao, H., Yan, C., Wu, W., & Li, J. (2020). Application of microfluidic chip technology in food safety sensing. *Sensors*, *20*(6), 1792. Available from https://doi.org/10.3390/s20061792.

Garrido-Varo, A., Sánchez, M.-T., De la Haba, M.-J., Torres, I., & Pérez-Marín, D. (2017). Fast, low-cost and non-destructive physico-chemical analysis of virgin olive oils using near-infrared reflectance spectroscopy. *Sensors*, *17*(11), 2642. Available from https://doi.org/10.3390/s17112642.

Gaudiuso, R., Dell'Aglio, M., De Pascale, O., Senesi, G. S., & De Giacomo, A. (2010). Laser induced breakdown spectroscopy for elemental analysis in environmental, cultural heritage and space applications: A review of methods and results. *Sensors*, *10*(8), 7434–7468. Available from https://doi.org/10.3390/s100807434.

Ghidini, S., Varrà, M. O., & Zanardi, E. (2019). Approaching authenticity issues in fish and seafood products by qualitative spectroscopy and chemometrics. *Molecules*, *24*(9), 1812. Available from https://doi.org/10.3390/molecules24091812.

Graham, S. F., Haughey, S. A., Ervin, R. M., Cancouët, E., Bell, S., & Elliott, C. T. (2012). The application of near-infrared (NIR) and Raman spectroscopy to detect adulteration of oil used in animal feed production. *Food Chemistry*, *132*(3), 1614–1619.

Graziani, S., Xibilia,M.G., Pedrycz,W., & Chen,S.-M.. (2020). Deep learning for soft sensor design. In *Development and analysis of deep learning architectures* (pp. 31–59). https://doi.org/10.1007/978-3-030-31764-5_2.

Gukowsky, J. C., Xie, T., Gao, S., Qu, Y., & He, L. (2018). Rapid identification of artificial and natural food colorants with surface enhanced Raman spectroscopy. *Food Control*, *92*, 267–275.

Gunning, Y., Davies, K. S., & Kemsley, E. K. (2023). Authentication of saffron using 60 MHz 1H NMR spectroscopy. *Food Chemistry*, *404*, 134649.

Guo, Z., Chen, P., Yin, L., Zuo, M., Chen, Q., El-Seedi, H. R., & Zou, X. (2022). Determination of lead in food by surface-enhanced Raman spectroscopy with aptamer regulating gold nanoparticles reduction. *Food Control*, *132*, 108498.

Ha, N., Xu, K., Ren, G., Mitchell, A., & Ou, J. Z. (2020). Machine learning-enabled smart sensor systems. *Advanced Intelligent Systems*, *2*(9), 2000063. Available from https://doi.org/10.1002/aisy.202000063.

de Dieu Habimana, J., Ji, J., & Sun, X. (2018). Minireview: trends in optical-based biosensors for point-of-care bacterial pathogen detection for food safety and clinical diagnostics. *Analytical Letters*, *51*(18), 2933–2966.

Haddad, L., Francis, J., Rizk, T., Akoka, S., Remaud, G. S., & Bejjani, J. (2022). Cheese characterization and authentication through lipid biomarkers obtained by high-resolution 1H NMR profiling. *Food Chemistry*, *383*, 132434.

Haiduc, A. M., Trezza, E. E., van Dusschoten, D., Reszka, A. A., & van Duynhoven, J. P. M. (2007). Non-invasive 'through-package'assessment of the microstructural quality of a model food emulsion by the NMR MOUSE. *LWT-Food Science and Technology*, *40*(4), 737–743.

Hajjar, G., Haddad, L., Rizk, T., Akoka, S., & Bejjani, J. (2021). High-resolution 1H NMR profiling of triacylglycerols as a tool for authentication of food from animal origin: Application to hen egg matrix. *Food Chemistry*, *360*, 130056.

Halonen, N., Pálvölgyi, P. S., Bassani, A., Fiorentini, C., Nair, R., Spigno, G., & Kordas, K. (2020). Bio-based smart materials for food packaging and sensors – A review. *Frontiers in Materials*, *7*. Available from https://www.frontiersin.org/articles/10.3389/fmats.2020.00082.

Han, J.-W., Ruiz-Garcia, L., Qian, J.-P., & Yang, X.-T. (2018). Food packaging: A comprehensive review and future trends. *Comprehensive Reviews in Food Science and Food Safety*, *17*(4), 860–877. Available from https://doi.org/10.1111/1541-4337.12343.

Hardie, M. (2020). Review of novel and emerging proximal soil moisture sensors for use in agriculture. *Sensors*, *20*(23), 6934. Available from https://doi.org/10.3390/s20236934.

Haroon, A., Basharat, M., Khattak, A.M., & Ejaz, W. (2019). Internet of things platform for transparency and traceability of food supply chain. In *2019 IEEE 10th annual information technology, electronics and mobile communication conference (IEMCON)* (pp. 0013–0019). Available from https://doi.org/10.1109/IEMCON.2019.8936158.

Hashemi-Nasab, F. S., & Parastar, H. (2022). Vis-NIR hyperspectral imaging coupled with independent component analysis for saffron authentication. *Food Chemistry*, *393*, 133450. Available from https://doi.org/10.1016/j.foodchem.2022.133450.

Hassan, S. I., Alam, M. M., Illahi, U., Al Ghamdi, M. A., Almotiri, S. H., & Mohd Su'ud, M. (2021). A systematic review on monitoring and advanced control strategies in smart agriculture. *IEEE Access*, *9*, 32517–32548. Available from https://doi.org/10.1109/ACCESS.2021.3057865.

Hassoun, A., Måge, I., Schmidt, W. F., Temiz, H. T., Li, L., Kim, H. Y., Nilsen, H., Biancolillo, A., Aït-Kaddour, A., Sikorski, M., Sikorska, E., Grassi, S., & Cozzolino, D. (2020). Fraud in animal origin food products: Advances in emerging spectroscopic detection methods over the past five years. *Foods*, *9*(8). Available from https://doi.org/10.3390/foods9081069, https://www.mdpi.com/2304-8158/9/8/1069.

Hassoun, A., Aït-Kaddour,A., Abu-Mahfouz,A.M., Rathod,N.B., Bader,F., Barba,F.J., Biancolillo,A., Cropotova,J., Galanakis,C.M., Jambrak,A.R., Lorenzo,J.M., Måge,I., Ozogul,F., & Regenstein,J.. (2022). The fourth industrial revolution in the food industry—Part I: Industry 4.0 technologies. Available from https://doi.org/10.1080/10408398.2022.2034735. <https://www.tandfonline.com/doi/abs/10.1080/10408398.2022.2034735/>.

Hassoun, A., El Din Bekhit,A., Jambrak,A.R., Regenstein, J.M., Chemat,F., Morton,J.D., Gudjónsdóttir,M., Carpena,M., Prieto,M.A., Varela,P., Arshad,R.N., Aadil,R.M., Bhat,Z., & Ueland,Ø. (2022). The fourth industrial revolution in the food industry—Part II: Emerging food trends. Available from https://doi.org/10.1080/10408398.2022.2106472, <https://www.tandfonline.com/doi/abs/10.1080/10408398.2022.2106472/>.

Hassoun, A., Alhaj Abdullah,N., Aït-Kaddour,A., Ghellam,M., Beşir,A., Zannou,O., Önal,B., Aadil,R.M., Lorenzo,J.M., Mousavi Khaneghah,A., & Regenstein,J.M. (2022). Food traceability 4.0 as part of the fourth industrial revolution: key enabling technologies. Available from https://doi.org/10.1080/10408398.2022.2110033. <https://www.tandfonline.com/doi/abs/10.1080/10408398.2022.2110033/>.

Hassoun, A., Jagtap, S., Garcia-Garcia, G., Trollman, H., Pateiro, M., Lorenzo, J. M., Trif, M., Rusu, A. V., Aadil, R. M., Šimat, V., Cropotova, J., & Câmara, J. S. (2023). Food quality 4.0: From traditional approaches to digitalized automated analysis. *Journal of Food Engineering*, *337*, 111216. Available from https://doi.org/10.1016/j.jfoodeng.2022.111216.

He, S., Yuan, Y., Nag, A., Feng, S., Afsarimanesh, N., Han, T., Mukhopadhyay, S. C., & Organ, D. R. (2020). A review on the use of impedimetric sensors for the inspection of food quality. *International Journal of Environmental Research and Public Health*, *17*(14), 5220. Available from https://doi.org/10.3390/ijerph17145220.

Heidarbeigi, K., Mohtasebi, S. S., Foroughirad, A., Ghasemi-Varnamkhasti, M., Rafiee, S., & Rezaei, K. (2015). Detection of adulteration in saffron samples using electronic nose. *International Journal of Food Properties*, *18*(7), 1391–1401. Available from https://doi.org/10.1080/10942912.2014.915850.

Heredia-Rivera, U., Gopalakrishnan, S., Kadian, S., Nejati, S., Kasi, V., & Rahimi, R. (2022). A wireless chipless printed sensor tag for real-time radiation sterilization monitoring. *Journal of Materials Chemistry C*, *10*(26), 9813–9822. Available from https://doi.org/10.1039/D2TC00531J.

Herrero, A. M. (2008). Raman spectroscopy for monitoring protein structure in muscle food systems. *Critical Reviews in Food Science and Nutrition*, *48*(6), 512–523.

Huang, C.-C., Hsu, Z.-H., & Lai, Y.-S. (2021). Raman spectroscopy for virus detection and the implementation of unorthodox food safety. *Trends in Food Science & Technology*, *116*, 525–532.

Huang, H., Qureshi, J. U., Liu, S., Sun, Z., Zhang, C., & Wang, H. (2021). Hyperspectral imaging as a potential online detection method of microplastics. *Bulletin of Environmental Contamination and Toxicology*, *107*(4), 754–763. Available from https://doi.org/10.1007/s00128-020-02902-0.

Huang, H., Liu, L., & Ngadi, M. O. (2014). Recent developments in hyperspectral imaging for assessment of food quality and safety. *Sensors*, *14*(4), 7248–7276. Available from https://doi.org/10.3390/s140407248.

Huang, L., Zheng, L., Chen, Y., Xue, F., Cheng, L., Adeloju, S. B., & Chen, W. (2015). A novel GMO biosensor for rapid ultrasensitive and simultaneous detection of multiple DNA components in GMO products. *Biosensors and Bioelectronics*, *66*, 431–437.

Huang, W., Lu, J., Ye, H., Kong, W., Mortimer, A. H., & Shi, Y. (2018). Quantitative identification of crop disease and nitrogen-water stress in winter wheat using continuous wavelet analysis. *International Journal of Agricultural and Biological Engineering*, *11*(2), 145–152. Available from https://doi.org/10.25165/ijabe.v11i2.3467.

Huayhongthong, S., Khuntayaporn, P., Thirapanmethee, K., Wanapaisan, P., & Chomnawang, M. T. (2019). Raman spectroscopic analysis of food-borne microorganisms. *Lwt*, *114*, 108419.

Hussain Shah, S. K., Iqbal, J., Ahmad, P., Khandaker, M. U., Haq, S., & Naeem, M. (2020). Laser induced breakdown spectroscopy methods and applications: A comprehensive review. *Radiation Physics and Chemistry*, *170*, 108666. Available from https://doi.org/10.1016/j.radphyschem.2019.108666.

Hussein, H. M. E., Rinaldi, M., Onabajo, M., & Cassella, C. (2021). Capturing and recording cold chain temperature violations through parametric alarm-sensor tags. *Applied Physics Letters*, *119*(1). Available from https://doi.org/10.1063/5.0054022.

IBM. (2022). Computer vision is a field of artificial intelligence (AI) enabling computers to derive information from images, videos and other inputs. What is Computer Vision? | IBM <https://www.ibm.com/topics/computer-vision/>.

Issad, H. A., Aoudjit, R., & Rodrigues, J. J. P. C. (2019). A comprehensive review of data mining techniques in smart agriculture. *Engineering in Agriculture, Environment and Food*, *12*(4), 511–525. Available from https://doi.org/10.1016/j.eaef.2019.11.003.

Jagtap, S., Garcia-Garcia, G., & Rahimifard, S. (2021). Optimisation of the resource efficiency of food manufacturing via the internet of things. *Computers in Industry*, *127*, 103397. Available from https://doi.org/10.1016/j.compind.2021.103397.

Jiang, H., Yuan, W., Ru, Y., Chen, Q., Wang, J., & Zhou, H. (2022). Feasibility of identifying the authenticity of fresh and cooked mutton kebabs using visible and near-infrared hyperspectral imaging. *Spectrochimica Acta Part A: Molecular and Biomolecular Spectroscopy*, *282*, 121689. Available from https://doi.org/10.1016/j.saa.2022.121689.

Jiang, X., Sergienko, G., Schweer, B., Gierse, N., Hubeny, M., Kreter, A., Brezinsek, S., & Linsmeier, Ch (2017). Design and development of a LIBS system on linear plasma device PSI-2 for in situ real-time diagnostics of plasma-facing materials. *Nuclear Materials and Energy*, *12*, 1224–1230. Available from https://doi.org/10.1016/j.nme.2016.11.021.

Jiang, Y., Yin, S., Dong, J., & Kaynak, O. (2021). A review on soft sensors for monitoring, control, and optimization of industrial processes. *IEEE Sensors Journal*, *21*(11), 12868–12881. Available from https://doi.org/10.1109/JSEN.2020.3033153.

Jin, L., Li, T., Wu, B., Yang, T., Zou, D., Liang, X., Hu, L., Huang, G., & Zhang, J. (2020). Rapid detection of Salmonella in milk by nuclear magnetic resonance based on membrane filtration superparamagnetic nanobiosensor. *Food Control*, *110*, 107011. Available from https://doi.org/10.1016/j.foodcont.2019.107011.

Kakani, V., Nguyen, V. H., Kumar, B. P., Kim, H., & Pasupuleti, V. R. (2020). A critical review on computer vision and artificial intelligence in food industry. *Journal of Agriculture and Food Research*, *2*, 100033. Available from https://doi.org/10.1016/j.jafr.2020.100033.

Kalinichenko, A., & Arseniyeva, L. (2020). Electronic nose combined with chemometric approaches to assess authenticity and adulteration of sausages by soy protein. *Sensors and Actuators B: Chemical*, *303*, 127250. Available from https://doi.org/10.1016/j.snb.2019.127250.

Kamruzzaman, M. (2018). Applications of hyperspectral imaging for meat quality and authenticity. <https://www.taylorfrancis.com/chapters/edit/10.1201/9781315209203-13/applications-hyperspectral-imaging-meat-quality-authenticity-mohammed-kamruzzaman/>.

Kang, W., Lin, H., Yao-Say Solomon Adade, S., Wang, Z., Chen, Q., & Ouyang, Q. (2022). Advanced sensing of volatile organic compounds in the fermentation of tea extract enabled by nano-colorimetric sensor array based on density functional theory. *Food Chemistry*, 134193.

Kannapinn, M., Pham, M. K., & Schäfer, M. (2022). Physics-based digital twins for autonomous thermal food processing: Efficient, non-intrusive reduced-order modeling. *Innovative Food Science & Emerging Technologies*, *81*, 103143. Available from https://doi.org/10.1016/j.ifset.2022.103143.

Karthikeyan, L., Chawla, I., & Mishra, A. K. (2020). A review of remote sensing applications in agriculture for food security: Crop growth and yield, irrigation, and crop losses. *Journal of Hydrology*, *586*, 124905. Available from https://doi.org/10.1016/j.jhydrol.2020.124905.

Kelly, C. A., Santovito, E., Cruz-Romero, M., Kerry, J. P., & Papkovsky, D. P. (2020). Application of O2 sensor technology to monitor performance of industrial beef samples packaged on three different vacuum packaging machines. *Sensors and Actuators B: Chemical*, *304*, 127338. Available from https://doi.org/10.1016/j.snb.2019.127338.

Kempelis, A., Romanovs, A., & Patlins, A. (2022). Using computer vision and machine learning based methods for plant monitoring in agriculture: A systematic literature review. In *2022 63rd International scientific conference on information technology and management science of Riga Technical University (ITMS)* (pp. 1–6). Available from https://doi.org/10.1109/ITMS56974.2022.9937119.

Khanal, S., Kc, K., Fulton, J. P., Shearer, S., & Ozkan, E. (2020). Remote sensing in agriculture—Accomplishments, limitations, and opportunities. *Remote Sensing*, *12*(22), 3783. Available from https://doi.org/10.3390/rs12223783.

Khaneja, N. (Ed.). (2020). Nuclear Magnetic Resonance. IntechOpen. Avilable from https://doi.org/10.5772/intechopen.74899.

Khanna, A., Jain, S., Burgio, A., Bolshev, V., & Panchenko, V. (2022). Blockchain-enabled supply chain platform for Indian dairy industry: Safety and traceability. *Foods*, *11*(17), 2716. Available from https://doi.org/10.3390/foods11172716.

Kielmann, F., Knott, P., & Koch, W. (2022). Core temperature estimation of food items based on non-contact thermal and high frequency sensor data with an LSTM network. In *Sensor data gusion: Trends, solutions, applications (SDF)* (pp. 1–6). Available from https://doi.org/10.1109/SDF55338.2022.9931957.

Kim, G., Kwak, J., Choi, J., & Park, K. (2012). Detection of nutrient elements and contamination by pesticides in spinach and rice samples using laser-induced breakdown spectroscopy (LIBS). *Journal of Agricultural and Food Chemistry*, *60*(3), 718–724.

Kim, S.-Y., Hong, S.-J., Kim, E., Lee, C.-H., & Kim, G. (2023). Application of ensemble neural-network method to integrated sugar content prediction model for citrus fruit using Vis/NIR spectroscopy. *Journal of Food Engineering*, *338*, 111254. Available from https://doi.org/10.1016/j.jfoodeng.2022.111254.

Kondakci, T., & Zhou, W. (2017). Recent applications of advanced control techniques in food industry. *Food and Bioprocess Technology*, *10*(3), 522–542. Available from https://doi.org/10.1007/s11947-016-1831-x, http://www.springer.com/life + sci/food + science/journal/11947.

Kritchenkov, A. S., Egorov, A. R., Dubashynskaya, N. V., Volkova, O. V., Zabodalova, L. A., Suchkova, E. P., Kurliuk, A. V., Shakola, T. V., & Dysin, A. P. (2019). Natural polysaccharide-based smart (temperature sensing) and active (antibacterial, antioxidant and photoprotective) nanoparticles with potential application in biocompatible food coatings. *International Journal of Biological Macromolecules*, *134*, 480–486. Available from https://doi.org/10.1016/j.ijbiomac.2019.04.194.

Kulkarni, A. A., Dhanush, P., Chetan, B. S., Thamme Gowda, C. S., & Kumar Shrivastava, P. (2020). Applications of automation and robotics in agriculture industries; A review. *IOP Conference Series: Materials Science and Engineering*, *748*(1), 012002. Available from https://doi.org/10.1088/1757-899X/748/1/012002.

Kuswandi, B., Wicaksono, Y., Abdullah, A., Heng, L. Y., & Ahmad, M. (2011). Smart packaging: Sensors for monitoring of food quality and safety. *Sensing and Instrumentation for Food Quality and Safety*, *5*(3), 137–146.

La Scalia, G., Nasca, A., Corona, O., Settanni, L., & Micale, R. (2017). An innovative shelf life model based on smart logistic unit for an efficient management of the perishable food supply chain. *Journal of Food Process Engineering*, *40*(1), e12311. Available from https://doi.org/10.1111/jfpe.12311.

Lambert, J. B., Mazzola, E. P., & Ridge, C. D. (2019). Nuclear magnetic resonance spectroscopy: An introduction to principles, applications, and experimental methods (2nd ed, p. 480). UK: John Wiley & Sons, ISBN: 978-1-119-29523-5.

Lakhwani, K., Gianey, H., Agarwal, N., Gupta, S., Rathore, V.S., Worring, M., Mishra, D.K., Joshi, A., & Maheshwari, S. (2019). Singapore development of IoT for smart agriculture: A review. *Emerging trends in expert applications and security* (pp. 425–432). Available from https://doi.org/10.1007/978-981-13-2285-3_50, 9789811322853.

Li, B., Zhao, M., Zhou, Y., Hou, B., & Zhang, D. (2016). Detection of waxed rice using visible-near infrared hyperspectral imaging. *Journal of Food and Nutrition Research*, *4*(5), 267–275. Available from https://doi.org/10.12691/jfnr-4-5-1.

Li, H., Wang, Y., Zhang, J., Li, X., Wang, J., Yi, S., Zhu, W., Xu, Y., & Li, J. (2023). Prediction of the freshness of horse mackerel (*Trachurus japonicus*) using E-nose, E-tongue, and colorimeter based on biochemical indexes analyzed during frozen storage of whole fish. *Food Chemistry*, *402*, 134325. Available from https://doi.org/10.1016/j.foodchem.2022.134325.

Li, S., Simonian, A., & Chin, B. A. (2010). Sensors for agriculture and the food industry. *Electrochemical Society Interface*, *19*(4), 41–46. Available from https://doi.org/10.1149/2.F05104IF/XML.

Li Vigni, M., Durante, C., Michelini, S., Nocetti, M., & Cocchi, M. (2020). Preliminary assessment of parmigiano reggiano authenticity by handheld Raman spectroscopy. *Foods*, *9*(11), 1563.

Li, X., Hao, Y., zhang, P., Akhter, M., & Li, D. (2022). A novel automatic detection method for abnormal behavior of single fish using image fusion. *Computers and Electronics in Agriculture*, *203*, 107435. Available from https://doi.org/10.1016/j.compag.2022.107435.

Li, X., Yang, Y., Zhu, Y., Ben, A., & Qi, J. (2022). A novel strategy for discriminating different cultivation and screening odor and taste flavor compounds in Xinhui tangerine peel using E-nose, E-tongue, and chemometrics. *Food Chemistry*, *384*, 132519. Available from https://doi.org/10.1016/j.foodchem.2022.132519.

Li, Y., Gao, C., Wang, Y., Fan, M., Wang, L., & Qian, H. (2023). Analysis of the aroma volatile compounds in different stabilized rice bran during storage. *Food Chemistry*, *405*, 134753. Available from https://doi.org/10.1016/j.foodchem.2022.134753.

Lim, J.-H., Bae, D., & Fong, A. (2018). Titanium dioxide in food products: quantitative analysis using ICP-MS and Raman spectroscopy. *Journal of Agricultural and Food Chemistry*, *66*(51), 13533–13540.

Lin, Y.-C., & Habib, A. (2021). Quality control and crop characterization framework for multi-temporal UAV LiDAR data over mechanized agricultural fields. *Remote Sensing of Environment*, *256*, 112299. Available from https://doi.org/10.1016/j.rse.2021.112299.

Liu, Z., Huang, M., Zhu, Q., Qin, J., & Kim, M. S. (2022). A packaged food internal Raman signal separation method based on spatially offset Raman spectroscopy combined with FastICA. *Spectrochimica Acta Part A: Molecular and Biomolecular Spectroscopy*, *275*, 121154.

Liu, Z., Huang, M., Zhu, Q., Qin, J., & Kim, M. S. (2021). Detection of adulterated sugar with plastic packaging based on spatially offset Raman imaging. *Journal of the Science of Food and Agriculture*, *101*(15), 6281–6288.

Lohumi, S., Lee, H., Kim, M. S., Qin, J., & Cho, B.-K. (2018). Through-packaging analysis of butter adulteration using line-scan spatially offset Raman spectroscopy. *Analytical and Bioanalytical Chemistry*, *410*(22), 5663–5673.

Lohumi, S., Lee, H., Kim, M. S., Qin, J., & Cho, B.-K. (2019). Raman hyperspectral imaging and spectral similarity analysis for quantitative detection of multiple adulterants in wheat flour. *Biosystems Engineering*, *181*, 103–113. Available from https://doi.org/10.1016/j.biosystemseng.2019.03.006.

Lu, W., Xu, X., Huang, G., Li, B., Wu, Y., Zhao, N., & Yu, F. R. (2021). Energy efficiency optimization in SWIPT enabled WSNs for smart agriculture. *IEEE Transactions on Industrial Informatics*, *17*(6), 4335–4344. Available from https://doi.org/10.1109/TII.2020.2996672.

Lu, X., Xia, Z., Qu, F., Zhu, Z., & Li, S. (2020). Identification of authenticity, quality and origin of saffron using hyperspectral imaging and multivariate spectral analysis. *Spectroscopy Letters*, *53*(2), 76–85. Available from https://doi.org/10.1080/00387010.2019.1693403.

Lu, X. (2017). Sensing techniques for food safety and quality control. <https://pubs.rsc.org/en/content/ebook/978-1-78262-664-0/>.

Luttmann, R., Bracewell, D. G., Cornelissen, G., Gernaey, K. V., Glassey, J., Hass, V. C., Kaiser, C., Preusse, C., Striedner, G., & Mandenius, C.-F. (2012). Soft sensors in bioprocessing: A status report and recommendations. *Biotechnology Journal*, *7*(8), 1040–1048. Available from https://doi.org/10.1002/biot.201100506.

Lv, W., Zhang, M., Bhandari, B., Li, L., & Wang, Y. (2017). Smart NMR method of measurement of moisture content of vegetables during microwave vacuum drying. *Food and Bioprocess Technology*, *10*(12), 2251–2260. Available from https://doi.org/10.1007/s11947-017-1991-3.

Lydia, J., Leones Sherwin Vimalraj, S., Monisha, R., & Murugan, R. (2022). Automated food grain monitoring system for warehouse using IOT. *Measurement: Sensors*, *24*, 100472. Available from https://doi.org/10.1016/j.measen.2022.100472.

Macomber, R.S. (1998). *A complete introduction to modern NMR spectroscopy* (p. 400). Wiley. ISBN: 978-0-471-15736-6.

Maes, W. H., & Steppe, K. (2019). Perspectives for remote sensing with unmanned aerial vehicles in porecision agriculture. *Trends in Plant Science*, *24*(2), 152–164. Available from https://doi.org/10.1016/j.tplants.2018.11.007.

Maimaitijiang, M., Sagan, V., Sidike, P., Hartling, S., Esposito, F., & Fritschi, F. B. (2020). Soybean yield prediction from UAV using multimodal data fusion and deep learning. *Remote Sensing of Environment*, *237*, 111599. Available from https://doi.org/10.1016/j.rse.2019.111599.

Mandrile, L., Zeppa, G., Giovannozzi, A. M., & Rossi, A. M. (2016). Controlling protected designation of origin of wine by Raman spectroscopy. *Food chemistry*, *211*, 260–267.

Marais Sicre, C., Fieuzal, R., & Baup, F. (2020). Contribution of multispectral (optical and radar) satellite images to the classification of agricultural surfaces. *International Journal of Applied Earth Observation and Geoinformation*, *84*, 101972. Available from https://doi.org/10.1016/j.jag.2019.101972.

Marquardt, B. J., & Wold, J. P. (2004). Raman analysis of fish: a potential method for rapid quality screening. *LWT-Food Science and Technology*, *37*(1), 1–8.

Mavani, N. R., Ali, J. M., Othman, S., Hussain, M. A., Hashim, H., & Rahman, N. A. (2022). Application of artificial intelligence in food industry-a guideline. *Food Engineering Reviews*, *14*(1), 134–175. Available from https://doi.org/10.1007/s12393-021-09290-z.

Meng, Y., Ma, Z., Ji, Z., Gao, R., & Su, Z. (2022). Fine hyperspectral classification of rice varieties based on attention module 3D-2DCNN. *Computers and Electronics in Agriculture*, *203*, 107474. Available from https://doi.org/10.1016/j.compag.2022.107474.

Michael, A.P. (2021). Exactly how fast is technology growing? Technology grows by 2x every 18 months in processing capacity. Over 89% of Big Data has been generated in the past two years. According to the past 50 years technology is growing at a steady rate. Technology statistics: How fast is Tech advancing? <https://mediapeanut.com/how-fast-is-technology-growing-statistics-facts/>.

Mijanur Rahman, A. T. M., Kim, D. H., Jang, H. D., Yang, J. H., & Lee, S. J. (2018). Preliminary study on biosensor-type time-temperature integrator for intelligent food packaging. *Sensors*, *18*(6), 1949.

Mohammad EL-Basioni, B. M., Mohamed, E. S., Belal, A., Jalhoum, M. E. M., Abd EL-Kader, S. M., & Zahran, M. B. (2022). A case study of a real-time internet of things system for site-specific potato crop management in El-Salhia area-Egypt. *Scientific Reports*, *12*(1). Available from https://doi.org/10.1038/s41598-022-22690-3.

Mohammadian, E., Alizadeh-Sani, M., & Jafari, S. M. (2020). Smart monitoring of gas/temperature changes within food packaging based on natural colorants. *Comprehensive Reviews in Food Science and Food Safety*, *19*(6), 2885–2931. Available from https://doi.org/10.1111/1541-4337.12635.

Moncayo, S., Manzoor, S., Rosales, J. D., Anzano, J., & Caceres, J. O. (2017). Qualitative and quantitative analysis of milk for the detection of adulteration by laser induced breakdown spectroscopy (LIBS). *Food Chemistry*, *232*, 322–328.

Monosik, R., Stredansky, M., Tkac, J., & Sturdik, E. (2012). Application of enzyme biosensors in analysis of food and beverages. *Food Analytical Methods*, *5*(1), 40–53.

Moradi, M., Tajik, H., Almasi, H., Forough, M., & Ezati, P. (2019). A novel pH-sensing indicator based on bacterial cellulose nanofibers and black carrot anthocyanins for monitoring fish freshness. *Carbohydrate Polymers*, *222*, 115030. Available from https://doi.org/10.1016/j.carbpol.2019.115030.

Morelli, S., Cossio, F., Monarca, D., Marucci, A., Selli, S., Pierini, D., & Carlini, M. (2022). Parametric sweep simulation for greenhouse temperature field optimization: An Italian case study. *Energy Reports*, *8*, 881–895. Available from https://doi.org/10.1016/j.egyr.2022.07.105.

Mosadegh Sedghy, B. (2018). Evolution of radio frequency identification (RFID) in agricultural cold chain monitoring: A literature review. <https://mpra.ub.uni-muenchen.de/94448/>.

Mukherjee, A., Panja,A.K., Dey,N., Mukherjee,A., Panja,A.K., & Dey,N. (2020). Introduction to sensors and systems. In *A beginner's guide to data agglomeration and intelligent sensing* (pp. 1–27). <https://www.sciencedirect.com/science/article/pii/B9780128203415000011/>.

Multari, R. A., Cremers, D. A., Dupre, J. A. M., & Gustafson, J. E. (2013). Detection of biological contaminants on foods and food surfaces using laser-induced breakdown spectroscopy (LIBS). *Journal of Agricultural and Food Chemistry*, *61*(36), 8687–8694.

Mustafa, F., & Andreescu, S. (2018). Chemical and biological sensors for food-quality monitoring and smart packaging. *Foods*, *7*(10), 168. Available from https://doi.org/10.3390/foods7100168.

Nagasubramanian, K., Jones, S., Singh, A. K., Sarkar, S., Singh, A., & Ganapathysubramanian, B. (2019). Plant disease identification using explainable 3D deep learning on hyperspectral images. *Plant Methods*, *15*(1)98. Available from https://doi.org/10.1186/s13007-019-0479-8.

Naresh, V., & Lee, N. (2021). A review on biosensors and recent development of nanostructured materials-enabled biosensors. *Sensors (Basel, Switzerland)*, *21*(4), 1109. Available from https://doi.org/10.3390/s21041109.

Nasiri, A., Taheri-Garavand, A., & Zhang, Y.- D. (2019). Image-based deep learning automated sorting of date fruit. *Postharvest Biology and Technology*, *153*, 133–141. Available from https://doi.org/10.1016/j.postharvbio.2019.04.003.

Ndisya, J., Gitau, A., Mbuge, D., Arefi, A., Bădulescu, L., Pawelzik, E., Hensel, O., & Sturm, B. (2021). Vis-NIR hyperspectral imaging for online quality evaluation during food processing: A case study of hot air drying of purple-speckled cocoyam (*Colocasia esculenta* (L.) Schott). *Processes*, *9*(10), 1804. Available from https://doi.org/10.3390/pr9101804.

Nedeljković, A., Roesch, P., Popp, J., Miočinović, J., Radovanović, M., & Pudja, P. (2016). Raman spectroscopy as a rapid tool for quantitative analysis of butter adulterated with margarine. *Food Analytical Methods*, *9*(5), 1315–1320.

Neugirg, F., Stark, M., Kaiser, A., Vlacilova, M., Della Seta, M., Vergari, F., Schmidt, J., Becht, M., & Haas, F. (2016). Erosion processes in calanchi in the Upper Orcia Valley, Southern Tuscany, Italy based on multitemporal high-resolution terrestrial LiDAR and UAV surveys. *Geomorphology*, *269*, 8–22. Available from https://doi.org/10.1016/j.geomorph.2016.06.027.

Nguyen, T. T., Pham, T. D., Nguyen, C. T., Delfos, J., Archibald, R., Dang, K. B., Hoang, N. B., Guo, W., & Ngo, H. H. (2022). A novel intelligence approach based active and ensemble learning for agricultural soil organic carbon prediction using multispectral and SAR data fusion. *Science of The Total Environment*, *804*, 150187. Available from https://doi.org/10.1016/j.scitotenv.2021.150187.

Nnachi, R. C., Sui, N., Ke, B., Luo, Z., Bhalla, N., He, D., & Yang, Z. (2022). Biosensors for rapid detection of bacterial pathogens in water, food and environment. *Environment International*, *166*, 107357. Available from https://doi.org/10.1016/j.envint.2022.107357.

NOAA (2021). Remote sensing is the science of obtaining information about objects or areas from a distance, typically from aircraft or satellites What is remote sensing? <https://oceanservice.noaa.gov/facts/remotesensing.html/>.

Nyalala, I., Okinda, C., Nyalala, L., Makange, N., Chao, Q., Chao, L., Yousaf, K., & Chen, K. (2019). Tomato volume and mass estimation using computer vision and machine learning algorithms: Cherry tomato model. *Journal of Food Engineering*, *263*, 288–298. Available from https://doi.org/10.1016/j.jfoodeng.2019.07.012.

Occhiuzzi, C., Camera, F., D'Orazio, M., D'Uva, N., Amendola, S., Bianco, G. M., Miozzi, C., Garavaglia, L., Martinelli, E., & Marrocco, G. (2022). Automatic monitoring of fruit ripening rooms by UHF RFID sensor network and machine learning. *IEEE Journal of Radio Frequency Identification*, *6*, 649–659. Available from https://doi.org/10.1109/JRFID.2022.3174272.

Ompusunggu, A.P., Karel,J., & Vladimir,D. (2019). Long-term vibration monitoring of induction motors in the food industry with low-cost MEMS accelerometers. In *Surveillance, Vishno and AVE conferences*, Lyon, France. <https://hal.archives-ouvertes.fr/hal-02188706/>.

Onwude, D. I., Chen, G., Eke-emezie, N., Kabutey, A., Khaled, A. Y., & Sturm, B. (2020). Recent advances in reducing food losses in the supply chain of fresh agricultural produce. *Processes*, *8*(11), 1431. Available from https://doi.org/10.3390/pr8111431.

Onwude, D., Bahrami, F., Shrivastava, C., Berry, T., Cronje, P., North, J., Kirsten, N., Schudel, S., Crenna, E., Shoji, K., & Defraeye, T. (2022). Physics-driven digital twins to quantify the impact of pre- and postharvest variability on the end quality evolution of orange fruit. *Resources, Conservation and Recycling*, *186*, 106585. Available from https://doi.org/10.1016/j.resconrec.2022.106585.

Ortiz, A., León, L., Contador, R., & Tejerina, D. (2021). Near-infrared spectroscopy (NIRS) as a tool for classification of pre-sliced iberian salchichón, modified atmosphere packaged (MAP) according to the official commercial categories of raw meat. *Foods*, *10*(8), 1865. Available from https://doi.org/10.3390/foods10081865.

Osheter, T., Campisi-Pinto, S., Resende, M. T., Linder, C., & Wiesman, Z. (2022). 1H LF-NMR self-diffusion measurements for rapid monitoring of an edible oil's food quality with respect to its oxidation status. *Molecules*, *27*(18), 6064.

Otles, S., Ozyurt,V.H., Cheung,P.C. K., & Mehta,B.M. (2015). Instrumental food analysis. In *Handbook of food chemistry* (pp. 165–187). <https://doi.org/10.1007/978-3-642-36605-5_18/>.

Ozaki, Y., Christy,A.A., & McClure,W.F. (2006). Near-infrared spectroscopy in food science and technology. <https://www.wiley.com/en-us/Near + Infrared + Spectroscopy + in + Food + Science + and + Technology-p-9780471672012/>.

O'Sullivan, M.G. (2020). Salt, fat and sugar reduction: Sensory approaches for nutritional reformulation of foods and beverages. Woodhead Publishing. Avilable from https://doi.org/10.1016/C2019-0-00720-0.

Özdoğan, G., Lin, X., & Sun, D.-W. (2021). Rapid and noninvasive sensory analyses of food products by hyperspectral imaging: Recent application developments. *Trends in Food Science & Technology*, *111*, 151–165. Available from https://doi.org/10.1016/j.tifs.2021.02.044.

Ozguven, M. M., & Adem, K. (2019). Automatic detection and classification of leaf spot disease in sugar beet using deep learning algorithms. *Physica A: Statistical Mechanics and its Applications*, *535*, 122537. Available from https://doi.org/10.1016/j.physa.2019.122537.

Ozturk, S., Bowler, A., Rady, A., & Watson, N. J. (2023). Near-infrared spectroscopy and machine learning for classification of food powders during a continuous process. *Journal of Food Engineering*, *341*, 111339. Available from https://doi.org/10.1016/j.jfoodeng.2022.111339.

Paternoster, A., Vanlanduit, S., Springael, J., & Braet, J. (2018). Measurement and analysis of vibration and shock levels for truck transport in Belgium with respect to packaged beer during transit. *Food Packaging and Shelf Life*, *15*, 134–143. Available from https://doi.org/10.1016/j.fpsl.2017.12.007.

Peñaloza, E. A. G., Oliveira, V. A., & Cruvinel, P. E. (2021). Using soft sensors as a basis of an innovative architecture for operation planning and quality evaluation in agricultural sprayers. *Sensors*, *21*(4), 1269. Available from https://doi.org/10.3390/s21041269.

Peng, J., Xie, W., Jiang, J., Zhao, Z., Zhou, F., & Liu, F. (2020). Fast quantification of honey adulteration with laser-induced breakdown spectroscopy and chemometric methods. *Foods*, *9*(3), 341.

Pereira Jr, V. A., de Arruda, I. N. Q., & Stefani, R. (2015). Active chitosan/PVA films with anthocyanins from *Brassica oleraceae* (Red Cabbage) as time–temperature Indicators for application in intelligent food packaging. *Food Hydrocolloids*, *43*, 180–188.

Pereira, F. M. V., Rebellato, A. P., Lima Pallone, J. A., & Colnago, L. A. (2015). Through-package fat determination in commercial samples of mayonnaise and salad dressing using time-domain nuclear magnetic resonance spectroscopy and chemometrics. *Food Control*, *48*, 62–66.

Pérez-Marín, D., & Garrido-Varo,A. (2022). *Innovation proposal in the growing labelling of vacuum-packed sliced Iberian pork ham using NIRS portable sensors*. Sólo Cerdo Ibérico.

Pérez-Marín, D., Garrido-Varo,A., Cifuentes,A. (2021). *NIR sensors for the in-situ assessment of Iberian ham comprehensive foodomics* (pp. 340–345). <https://www.sciencedirect.com/science/article/pii/B9780081005965228606/>.

Pérez-Marín, D., Torres, I., Entrenas, J.-A., Vega, M., & Sánchez, M.-T. (2019). Pre-harvest screening on-vine of spinach quality and safety using NIRS technology. *Spectrochimica Acta Part A: Molecular and Biomolecular Spectroscopy*, *207*, 242–250. Available from https://doi.org/10.1016/j.saa.2018.09.035.

Pérez-Marín, D., Calero, L., Fearn, T., Torres, I., Garrido-Varo, A., & Sánchez, M.-T. (2019). A system using in situ NIRS sensors for the detection of product failing to meet quality standards and the prediction of optimal postharvest shelf-life in the case of oranges kept in cold storage. *Postharvest Biology and Technology*, *147*, 48–53. Available from https://doi.org/10.1016/j.postharvbio.2018.09.009.

Pérez-Marín, D., Sánchez, M.-T., Paz, P., Soriano, M.-A., Guerrero, J.-E., & Garrido-Varo, A. (2009). Non-destructive determination of quality parameters in nectarines during on-tree ripening and postharvest storage. *Postharvest Biology and Technology*, *52*(2), 180–188. Available from https://doi.org/10.1016/j.postharvbio.2008.10.005.

Pérez-Marín, D., Fearn, T., Riccioli, C., De Pedro, E., & Garrido, A. (2021). Probabilistic classification models for the in situ authentication of iberian pig carcasses using near infrared spectroscopy. *Talanta*, *222*, 121511. Available from https://doi.org/10.1016/j.talanta.2020.121511.

Petrenko, V. A., & Vodyanoy, V. J. (2003). Phage display for detection of biological threat agents. *Journal of Microbiological Methods*, *53*(2), 253–262. Available from https://doi.org/10.1016/S0167-7012(03)00029-0.

Pinheiro, F.C. (2019). In-line real-time water activity prediction based on soft sensors: A case study in a pet food industry. <https://repositorio.ufpe.br/handle/123456789/33997/>.

Poghossian, A., Geissler, H., & Schöning, M. J. (2019). Rapid methods and sensors for milk quality monitoring and spoilage detection. *Biosensors and Bioelectronics*, *140*, 111272. Available from https://doi.org/10.1016/j.bios.2019.04.040.

Polder, G., Blok, P. M., de Villiers, H. A. C., van der Wolf, J. M., & Kamp, J. (2019). Potato virus Y detection in seed potatoes using deep learning on hyperspectral images. *Frontiers in Plant Science*, *10*. Available from https://doi.org/10.3389/fpls.2019.00209.

Pou, K. R. J., Raghavan, V., & Packirisamy, M. (2022). Microfluidics in smart packaging of foods. *Food Research International*, 111873.

Primicerio, J., Di Gennaro, S. F., Fiorillo, E., Genesio, L., Lugato, E., Matese, A., & Vaccari, F. P. (2012). A flexible unmanned aerial vehicle for precision agriculture. *Precision Agriculture*, *13*(4), 517–523. Available from https://doi.org/10.1007/s11119-012-9257-6.

Puligundla, P., Jung, J., & Ko, S. (2012). Carbon dioxide sensors for intelligent food packaging applications. *Food Control*, *25*(1), 328–333. Available from https://doi.org/10.1016/j.foodcont.2011.10.043.

Putra, Y. C., Wijayanto, A. W., & Chulafak, G. A. (2022). Oil palm trees detection and counting on Microsoft Bing Maps very high resolution (VHR) satellite imagery and unmanned aerial vehicles (UAV) data using image processing thresholding approach. *Ecological Informatics*, *72*, 101878. Available from https://doi.org/10.1016/j.ecoinf.2022.101878.

Qin, Y., Ke, W., Faheem, A., Ye, Y., & Hu, Y. (2023). A rapid and naked-eye on-site monitoring of biogenic amines in foods spoilage. *Food Chemistry*, *404*, 134581.

Rabeler, F., Skytte, J. L., & Feyissa, A. H. (2019). Prediction of thermal induced color changes of chicken breast meat during convective roasting: A combined mechanistic and kinetic modelling approach. *Food Control*, *104*, 42–49. Available from https://doi.org/10.1016/j.foodcont.2019.04.018.

Rahman, S., & Chowdhury, D. (2022). Guar gum-sodium alginate nanocomposite film as a smart fluorescence-based humidity sensor: A smart packaging material. *International Journal of Biological Macromolecules*, *216*, 571–582. Available from https://doi.org/10.1016/j.ijbiomac.2022.07.008.

Raja P.M. V., & Barron,A.R. (2016). Chemical speciation. In *Physical methods in chemistry and nano science*. <https://chem.libretexts.org/Bookshelves/Analytical_Chemistry/Physical_Methods_in_Chemistry_and_Nano_Science_(Barron)/04%3A_Chemical_Speciation/4.02%3A_IR_Spectroscopy/>.

Rajaraman, V. (2017). Radio frequency identification. *Resonance*, *22*(6), 549–575. Available from https://doi.org/10.1007/s12045-017-0498-6.

Ramalingam, S., Elsayed, A., & Singh, A. (2023). An Aflatoxin-M1 biochip using graphene quantum dot-gold hybrid nanoparticles. *Food Chemistry*, *403*, 134302.

Rehan, I., Gondal, M. A., Aldakheel, R. K., Almessiere, M. A., Rehan, K., Khan, S., Sultana, S., & Khan, M. Z. (2022). Determination of nutritional and toxic metals in black tea leaves using calibration free LIBS and ICP: AES technique. *Arabian Journal for Science and Engineering*, *47*(6), 7531–7539.

Ren, L., Tian, Y., Yang, X., Wang, Q., Wang, L., Geng, X., Wang, K., Du, Z., Li, Y., & Lin, H. (2023). Rapid identification of fish species by laser-induced breakdown spectroscopy and Raman spectroscopy coupled with machine learning methods. *Food Chemistry*, *400*, 134043.

Režek Jambrak, A., Nutrizio, M., Djekić, I., Pleslić, S., & Chemat, F. (2021). Internet of nonthermal food processing technologies (IoNTP): Food Industry 4.0 and sustainability. *Applied Sciences*, *11*(2), 686. Available from https://doi.org/10.3390/app11020686.

Rottiers, H., Tzompa Sosa, D. A., Van de Vyver, L., Hinneh, M., Everaert, H., De Wever, J., Messens, K., & Dewettinck, K. (2019). Discrimination of cocoa liquors based on their odor fingerprint: A fast GC electronic nose suitability study. *Food Analytical Methods*, *12*(2), 475–488. Available from https://doi.org/10.1007/s12161-018-1379-7.

Saha, D., & Manickavasagan, A. (2021). Machine learning techniques for analysis of hyperspectral images to determine quality of food products: A review. *Current Research in Food Science*, *4*, 28–44. Available from https://doi.org/10.1016/j.crfs.2021.01.002.

Sánchez, M.-T., Pintado, C., de la Haba, M.-J., Torres, I., García, M., & Pérez-Marín, D. (2020). In situ ripening stages monitoring of Lamuyo pepper using a new-generation near-infrared spectroscopy sensor. *Journal of the Science of Food and Agriculture*, *100*(5), 1931–1939. Available from https://doi.org/10.1002/jsfa.10205.

Sánchez, M.-T., Torres, I., de la Haba, M.-J., Chamorro, A., Garrido-Varo, A., & Pérez-Marín, D. (2019). Rapid, simultaneous, and in situ authentication and quality assessment of intact bell peppers using near-infrared spectroscopy technology. *Journal of the Science of Food and Agriculture*, *99*(4), 1613–1622. Available from https://doi.org/10.1002/jsfa.9342.

Sánchez, M.-T., De la Haba, M.-J., Serrano, I., & Pérez-Marín, D. (2013). Application of NIRS for nondestructive measurement of quality parameters in intact oranges during on-tree ripening and at harvest. *Food Analytical Methods*, *6*(3), 826–837. Available from https://doi.org/10.1007/s12161-012-9490-7.

Sánchez, M.-T., Márquez, R., Torres, I., De la Haba, M.-J., Pérez-Marín, D., & López, M.-I. (2020). Chemical characterization of wine vinegars belonging to the vinagre de Montilla-Moriles protected designation of origin, using near-infrared spectroscopy. *Food Analytical Methods*, *13*(3), 802–810. Available from https://doi.org/10.1007/s12161-019-01697-z.

Santos, C. C., Zhao, J., Dong, X., Lonergan, S. M., Huff-Lonergan, E., Outhouse, A., Carlson, K. B., Prusa, K. J., Fedler, C. A., & Yu, C. (2018). Predicting aged pork quality using a portable Raman device. *Meat Science*, *145*, 79–85.

Santos Pereira, L. F., Barbon, S., Valous, N. A., & Barbin, D. F. (2018). Predicting the ripening of papaya fruit with digital imaging and random forests. *Computers and Electronics in Agriculture*, *145*, 76–82. Available from https://doi.org/10.1016/j.compag.2017.12.029.

Schaller, C., Farmer,S., Kennepohl,D., & Reusch,W., (2015). Infrared spectroscopy. In *Organic chemistry*. <https://chem.libretexts.org/Bookshelves/Organic_Chemistry/Organic_Chemistry_(LibreTexts)/12%3A_Structure_Determination_-_Mass_Spectrometry_and_Infrared_Spectroscopy/12.06%3A_Infrared_Spectroscopy/>.

Schnepf, A., & Leitner, D. (2009). FEM simulation of below ground processes on a 3-dimensional root system geometry using dismesh and multiphysis.

Sena-Torralba, A., Smits, N. G. E., Blázquez, D., Albero-Pérez, C., Pallás-Tamarit, Y., Maquieira, Á., & Morais, S. (2023). Simultaneous quantification of six major allergens in commercial foods for children using a multiplex array on a digital versatile disc. *Food Chemistry*, *404*, 134570.

Sezer, B., Bjelak, A., Murat Velioglu, H., & Hakkı Boyaci, I. (2022). Identification of meat species in processed meat products by using protein based laser induced breakdown spectroscopy assay. *Food Chemistry*, *372*, 131245.

Sezer, B., Bilge, G., & Boyaci, I. H. (2017). Capabilities and limitations of LIBS in food analysis. *TrAC Trends in Analytical Chemistry*, *97*, 345–353.

Sezer, B., Bilge, G., Eseller, K. E., Berberoglu, H., & Boyaci, I. H. (2019). Laser induced breakdown spectroscopy based diffusion modelling in cheese matrix. *Journal of Food Engineering*, *263*, 320–325.

Sezer, B., Apaydin, H., Bilge, G., & Boyaci, I. H. (2018). Coffee arabica adulteration: Detection of wheat, corn and chickpea. *Food Chemistry*, *264*, 142–148.

Sezer, B., Durna, S., Bilge, G., Berkkan, A., Yetisemiyen, A., & Boyaci, I. H. (2018). Identification of milk fraud using laser-induced breakdown spectroscopy (LIBS). *International Dairy Journal*, *81*, 1–7.

Shao, J.-H., Zou, Y.-F., Xu, X.-L., Wu, J.-Q., & Zhou, G.-H. (2011). Evaluation of structural changes in raw and heated meat batters prepared with different lipids using Raman spectroscopy. *Food Research International*, *44*(9), 2955–2961.

Sharif, M., Khan, M. A., Iqbal, Z., Azam, M. F., Lali, M. I. U., & Javed, M. Y. (2018). Detection and classification of citrus diseases in agriculture based on optimized weighted segmentation and feature selection. *Computers and Electronics in Agriculture*, *150*, 220–234. Available from https://doi.org/10.1016/j.compag.2018.04.023.

Sharma, P., & Sharma,A.. (2022). In *Proceedings of the international conference on innovative computing & communication (ICICC) 2022*. Available at SSRN. Available from https://doi.org/10.2139/ssrn.4024154.

Shen,, J.-Y., Wang,, M.-D., Wang,, Y.-F., Hu,, J.-Y., Zhu,, Y., Zhang,, Y. X., Li,, Z.-J., & Yao,, H.-C. (2018). Iron and carbon codoped WO3 with hierarchical walnut-like microstructure for highly sensitive and selective acetone sensor. *Sensors and Actuators B: Chemical*, *256*, 27–37.

Shoji, K., Schudel, S., Shrivastava, C., Onwude, D., & Defraeye, T. (2022). Optimizing the postharvest supply chain of imported fresh produce with physics-based digital twins. *Journal of Food Engineering*, *329*, 111077. Available from https://doi.org/10.1016/j.jfoodeng.2022.111077.

Shuib, A., Irwani,I., & Hazman,Y.F. (2021). Traceability system of halal chicken supply chain. In *2021 IEEE 12th control and system graduate research colloquium (ICSGRC)* (pp. 144–149). Available from https://doi.org/10.1109/ICSGRC53186.2021.9515246.

Siedliska, A., Baranowski, P., Zubik, M., Mazurek, W., & Sosnowska, B. (2018). Detection of fungal infections in strawberry fruit by VNIR/SWIR hyperspectral imaging. *Postharvest Biology and Technology*, *139*, 115–126. Available from https://doi.org/10.1016/j.postharvbio.2018.01.018.

Sishodia, R. P., Ray, R. L., & Singh, S. K. (2020). Applications of remote sensing in precision agriculture: A review. *Remote Sensing*, *12*(19), 3136. Available from https://doi.org/10.3390/rs12193136.

Sixt, M., Gudi, G., Schulz, H., & Strube, J. (2018). In-line Raman spectroscopy and advanced process control for the extraction of anethole and fenchone from fennel (*Foeniculum vulgare* L. MILL.). *Comptes Rendus Chimie*, *21*(2), 97–103.

Śliwińska, M., Wiśniewska,P., Dymerski,T., Wardencki,W., Namieśnik,J., & Downey,G. (2016). Advances in electronic noses and tongues for food authenticity testing. In *Advances in food authenticity testing* (pp. 201–225). <https://www.sciencedirect.com/science/article/pii/B9780081002209000084/>.

Śliwińska, M., Wiśniewska, P., Dymerski, T., Namieśnik, J., & Wardencki, W. (2014). Food analysis using artificial senses. *Journal of Agricultural and Food Chemistry*, *62*(7), 1423–1448. Available from https://doi.org/10.1021/jf403215y.

Soares, R. R. A., Hjort, R. G., Pola, C. C., Parate, K., Reis, E. L., Soares, N. F. F., McLamore, E. S., Claussen, J. C., & Gomes, C. L. (2020). Laser-induced graphene electrochemical immunosensors for rapid and label-free monitoring of *Salmonella enterica* in chicken broth. *ACS sensors*, *5*(7), 1900–1911.

Stefas, D., Gyftokostas, N., Kourelias, P., Nanou, E., Kokkinos, V., Bouras, C., & Couris, S. (2021a). Discrimination of olive oils based on the olive cultivar origin by machine learning employing the fusion of emission and absorption spectroscopic data. *Food Control*, *130*, 108318.

Stefas, D., Gyftokostas, N., Kourelias, P., Nanou, E., Kokkinos, V., Bouras, C., & Couris, S. (2021b). A laser-based method for the detection of honey adulteration. *Applied Sciences*, *11*(14), 6435.

Suma, V. (2021). Internet of things (IoT) based smart agriculture in India: An overview. *Journal of ISMAC*, *3*(1), 1–15. Available from https://doi.org/10.36548/jismac.2021.1.001.

Sun, W., Zhu, X., Zhou, T., Su, Y., & Mo, B. (2019). Application of blockchain and RFID in anti-counterfeiting traceability of liquor. In *2019 IEEE 5th international conference on computer and communications (ICCC)* (pp. 1248–1251). Available from https://doi.org/10.1109/ICCC47050.2019.9064152.

Sun, D.-W. (2009). Infrared spectroscopy for food quality analysis and control (p. 424) Burlington, MA: Academic Press, ISBN: 978-0-12-374136-3. Available from https://doi.org/10.1016/B978-0-12-374136-3.X0001-6.

Swain, A., Abdellatif, E., Mousa, A., & Pong, P. W. T. (2022). Sensor technologies for transmission and distribution systems: A review of the latest developments. *Energies*, *15*(19), 7339. Available from https://doi.org/10.3390/en15197339.

Taheri-Garavand, A., Fatahi, S., Omid, M., & Makino, Y. (2019). Meat quality evaluation based on computer vision technique: A review. *Meat Science*, *156*, 183–195. Available from https://doi.org/10.1016/j.meatsci.2019.06.002.

Tang, F., Vasas, M., Hatzakis, E., Spyros, A., & Webb, G. A. (2019). Chapter five - Magnetic resonance applications in food analysis. *Annual Reports on NMR Spectroscopy*, *98*. Available from https://www.sciencedirect.com/science/article/pii/S006641031930016X.

Taylan, O., Cebi, N., Tahsin Yilmaz, M., Sagdic, O., & Bakhsh, A. A. (2020). Detection of lard in butter using Raman spectroscopy combined with chemometrics. *Food Chemistry*, *332*, 127344.

Tian, L., Wang, Z., Xue, B., Li, D., Zheng, H., Yao, X., Zhu, Y., Cao, W., & Cheng, T. (2023). A disease-specific spectral index tracks Magnaporthe oryzae infection in paddy rice from ground to space. *Remote Sensing of Environment*, *285*, 113384. Available from https://doi.org/10.1016/j.rse.2022.113384.

Torres, I., Pérez-Marín, D., Haba, la M.-J. De, & Sánchez, M.-T. (2015). Fast and accurate quality assessment of Raf tomatoes using NIRS technology. *Postharvest Biology and Technology*, *107*, 9–15. Available from https://doi.org/10.1016/j.postharvbio.2015.04.004.

Torres, I., Pérez-Marín, D., De la Haba, M.-J., & Sánchez, M.-T. (2017). Developing universal models for the prediction of physical quality in citrus fruits analysed on-tree using portable NIRS sensors. *Biosystems Engineering*, *153*, 140–148. Available from https://doi.org/10.1016/j.biosystemseng.2016.11.007.

Torres, I., Sánchez, M.-T., Entrenas, J.-A., Garrido-Varo, A., & Pérez-Marín, D. (2019). Monitoring quality and safety assessment of summer squashes along the food supply chain using near infrared sensors. *Postharvest Biology and Technology*, *154*, 21–30. Available from https://doi.org/10.1016/j.postharvbio.2019.04.015.

Torres, I., Sánchez, M.-T., de la Haba, M.-J., & Pérez-Marín, D. (2019). LOCAL regression applied to a citrus multispecies library to assess chemical quality parameters using near infrared spectroscopy. *Spectrochimica Acta Part A: Molecular and Biomolecular Spectroscopy*, *217*, 206–214. Available from https://doi.org/10.1016/j.saa.2019.03.090.

Torres, I., Sánchez, M.-T., Vega-Castellote, M., & Pérez-Marín, D. (2021). Fraud detection in batches of sweet almonds by portable near-infrared spectral devices. *Foods*, *10*(6), 1221. Available from https://doi.org/10.3390/foods10061221.

Torres, I., Sánchez, M.-T., Vega-Castellote, M., Luqui-Muñoz, N., & Pérez-Marín, D. (2021). Routine NIRS analysis methodology to predict quality and safety indexes in spinach plants during their growing season in the field. *Spectrochimica Acta Part A: Molecular and Biomolecular Spectroscopy*, *246*, 118972. Available from https://doi.org/10.1016/j.saa.2020.118972.

Torres, I., Dolores, P.-M., Miguel, V.-C., Ana, G.-V., & María-Teresa, S. (2021). Hyperspectral imaging for the determination of the main unsaturated fatty acid distribution in shelled almonds analysed in bulk. In *Sensing for agriculture and food quality and safety XIII* (pp. 37–43). Available from https://doi.org/10.1117/12.2585699. <https://www.spiedigitallibrary.org/conference-proceedings-of-spie/11754/117540B/Hyperspectral-imaging-for-the-determination-of-the-main-unsaturated-fatty/10.1117/12.2585699.full 11754/>.

Torres-Rodríguez, I., Sánchez, M.-T., Entrenas, J.-A., Vega-Castellote, M., Garrido-Varo, A., & Pérez-Marín, D. (2022). Hyperspectral imaging for the detection of bitter almonds in sweet almond batches. *Applied Sciences*, *12*(10), 4842. Available from https://doi.org/10.3390/app12104842.

Touffet, M., Allouche, M. H., Ariane, M., & Vitrac, O. (2021). Coupling between oxidation kinetics and anisothermal oil flow during deep-fat frying. *Physics of Fluids*, *33*(8), 085105. Available from https://doi.org/10.1063/5.0055873.

Trantakis, I. A., Spaniolas, S., Kalaitzis, P., Ioannou, P. C., Tucker, G. A., & Christopoulos, T. K. (2012). Dipstick test for DNA-based food authentication. application to coffee authenticity assessment. *Journal of Agricultural and Food Chemistry*, *60*(3), 713–717. Available from https://doi.org/10.1021/jf203180h.

Tschand, A. (2023). Semi-supervised machine learning analysis of crop color for autonomous irrigation. *Smart Agricultural Technology*, *3*, 100116. Available from https://doi.org/10.1016/j.atech.2022.100116.

Turgut, S. S., Feyissa, A. H., Baltacıoğlu, C., Küçüköner, E., & Karacabey, E. (2020). Extraction simulation of porous media by CFD: Recovery of trans-resveratrol from grape cane by pressurised low polarity water system. *Chemical Engineering and Processing - Process Intensification*, *148*, 107779. Available from https://doi.org/10.1016/j.cep.2019.107779.

Turgut, S. S., Küçüköner, E., Feyissa, A. H., & Karacabey, E. (2021). A novel drying system – Simultaneous use of ohmic heating with convectional air drying: System design and detailed examination using CFD. *Innovative Food Science & Emerging Technologies*, *72*, 102727. Available from https://doi.org/10.1016/J.IFSET.2021.102727.

Turgut, S. S., Küçüköner, E., & Karacabey, E. (2021). TeaPot: A chemometric tool for tea blend recipe estimation. *Applied Food Research*, *1*(1), 100006. Available from https://doi.org/10.1016/j.afres.2021.100006.

Turgut, S. S., Entrenas, J. A., Taşkın, E., Garrido-Varo, A., & Pérez-Marín, D. (2022). Estimation of the sensory properties of black tea samples using non-destructive near-infrared spectroscopy sensors. *Food Control*, *142*, 109260. Available from https://doi.org/10.1016/j.foodcont.2022.109260.

Turgut, S. S., Siamos, E., & Feyissa, A. H. (2022). Simulation and parametric sensitivity evaluation for ohmic heating of chicken breast. *Proceedings of the 8th International Food Operations and Processing Simulation Workshop (FoodOPS 2022), 009*. Available from https://doi.org/10.46354/i3m.2022.foodops.009.

Uğuz, S., & Uysal, N. (2021). Classification of olive leaf diseases using deep convolutional neural networks. *Neural Computing and Applications, 33* (9), 4133–4149. Available from https://doi.org/10.1007/s00521-020-05235-5.

Ulrici, A., Serranti, S., Ferrari, C., Cesare, D., Foca, G., & Bonifazi, G. (2013). Efficient chemometric strategies for PET–PLA discrimination in recycling plants using hyperspectral imaging. *Chemometrics and Intelligent Laboratory Systems, 122*, 31–39. Available from https://doi.org/10.1016/j.chemolab.2013.01.001.

UN. (2015). *United Nations Sustainable Development Goals - Time for global action for people and planet*. United Nations Sustainable Development Goals. <https://www.un.org/sustainabledevelopment/>.

UNCTAD. (2017). *The role of science, technology and innovation in ensuring food security by 2030*. United Nations Commission on Science, & Technology for Development. Unpublished content. <https://unctad.org/system/files/official-document/dtlstict2017d5_en.pdf/ >

Vargas, E., Ruiz, M. A., Ferrero, F. J., Campuzano, S., Ruiz-Valdepeñas Montiel, V., Reviejo, A. J., & Pingarrón, J. M. (2016). Automatic bionalyzer using an integrated amperometric biosensor for the determination of L-malic acid in wines. *Talanta, 158*, 6–13.

Vasconcelos, H., Matias, A., Mendes, J., Araújo, J., Dias, B., Jorge, P. A. S., Saraiva, C., de Almeida, J. M. M. M., & Coelho, L. C. C. (2023). Compact biosensor system for the quantification of hydrogen peroxide in milk. *Talanta, 253*, 124062.

Vega-Castellote, M., Sánchez, M.-T., Torres, I., & Pérez-Marín, D. (2021). An innovative non-targeted control system based on NIR spectral information for detecting non-compliant batches of sweet almonds. *Spectrochimica Acta Part A: Molecular and Biomolecular Spectroscopy, 250*, 119407. Available from https://doi.org/10.1016/j.saa.2020.119407.

Vega-Castellote, M., Sánchez, M.-T., Torres, I., de la Haba, M.-J., & Pérez-Marín, D. (2022). Assessment of watermelon maturity using portable new generation NIR spectrophotometers. *Scientia Horticulturae, 304*, 111328. Available from https://doi.org/10.1016/j.scienta.2022.111328.

Vega-Castellote, M., Pérez-Marín, D., Torres, I., & Sánchez, M.-T. (2021a). Non-destructive determination of fatty acid composition of in-shell and shelled almonds using handheld NIRS sensors. *Postharvest Biology and Technology, 174*, 111459. Available from https://doi.org/10.1016/j.postharvbio.2020.111459.

Vega-Castellote, M., Pérez-Marín, D., Torres, I., & Sánchez, M.-T. (2021b). Online NIRS analysis for the routine assessment of the nitrate content in spinach plants in the processing industry using linear and non-linear methods. *LWT, 151*, 112192. Available from https://doi.org/10.1016/j.lwt.2021.112192.

Velioglu, H. M., Sezer, B., Bilge, G., Baytur, S. E., & Boyaci, I. H. (2018). Identification of offal adulteration in beef by laser induced breakdown spectroscopy (LIBS). *Meat Science, 138*, 28–33.

Veloso, A. C. A., Sousa, M. E. B. C., Estevinho, L., Dias, L. G., & Peres, A. M. (2018). Honey evaluation using electronic tongues: An overview. *Chemosensors, 6*(3), 28. Available from https://doi.org/10.3390/chemosensors6030028.

Verboven, P., Defraeye, T., Datta, A. K., & Nicolai, B. (2020). Digital twins of food process operations: the next step for food process models? *Current Opinion in Food Science, 35*, 79–87. Available from https://doi.org/10.1016/j.cofs.2020.03.002.

Violino, S., Pallottino, F., Sperandio, G., Figorilli, S., Ortenzi, L., Tocci, F., Vasta, S., Imperi, G., & Costa, C. (2020). A full technological traceability system for extra virgin olive oil. *Foods, 9*(5), 624.

Wadood, S. A., Boli, G., Xiaowen, Z., Hussain, I., & Yimin, W. (2020). Recent development in the application of analytical techniques for the traceability and authenticity of food of plant origin. *Microchemical Journal, 152*, 104295. Available from https://doi.org/10.1016/j.microc.2019.104295.

Wang, C., Xiao, R., Wang, S., Yang, X., Bai, Z., Li, X., Rong, Z., Shen, B., & Wang, S. (2019). Magnetic quantum dot based lateral flow assay biosensor for multiplex and sensitive detection of protein toxins in food samples. *Biosensors and Bioelectronics, 146*, 111754.

Wang, H., Zhang, M., Yang, Z., Wang, Z., Liu, X., Lu, Y., Ji, L., Wang, Z. L., & Cheng, J. (2021). Energy from greenhouse plastic films. *Nano Energy, 89*, 106328. Available from https://doi.org/10.1016/j.nanoen.2021.106328.

Wang, Wei, Kong, W., Shen, T., Man, Z., Zhu, W., He, Y., & Liu, F. (2021). Quantitative analysis of cadmium in rice roots based on LIBS and chemometrics methods. *Environmental Sciences Europe, 33*(1), 1–14.

Wang, W.-T., Zhang, H., Yuan, Y., Guo, Y., & He, S. (2018). Research progress of Raman spectroscopy in drug analysis. *AAPS PharmSciTech, 19* (7), 2921–2928. Available from https://doi.org/10.1208/s12249-018-1135-8.

Westling, F., Underwood, J., & Örn, S. (2018). Light interception modelling using unstructured LiDAR data in avocado orchards. *Computers and Electronics in Agriculture, 153*, 177–187. Available from https://doi.org/10.1016/j.compag.2018.08.020.

Wieme, J., Mollazade, K., Malounas, I., Zude-Sasse, M., Zhao, M., Gowen, A., Argyropoulos, D., Fountas, S., & Van Beek, J. (2022). Application of hyperspectral imaging systems and artificial intelligence for quality assessment of fruit, vegetables and mushrooms: A review. *Biosystems Engineering, 222*, 156–176. Available from https://doi.org/10.1016/j.biosystemseng.2022.07.013.

Wijaya, D. R., Sarno, R., & Zulaika, E. (2018). Electronic nose dataset for beef quality monitoring in uncontrolled ambient conditions. *Data in Brief, 21*, 2414–2420. Available from https://doi.org/10.1016/j.dib.2018.11.091.

Williams, P., & Norris, K. (Eds.), (1987). *Near-infrared technology in the agricultural and food industries* (330 p.). St. Paul: American Association of Cereal Chemists.

Williams, P., Antoniszyn, J., & Manley, M. (2019). Near infrared technology: Getting the best out of light. Stellenbosch, South Africa: African Sun Media, ISBN: 978-1-928480-30-3. Available from https://doi.org/10.18820/9781928480310.

Williamson, T. N., Dobrowolski, E. G., Meyer, S. M., Frey, J. W., & Allred, B. J. (2019). Delineation of tile-drain networks using thermal and multispectral imagery—Implications for water quantity and quality differences from paired edge-of-field sites. *Journal of Soil and Water Conservation, 74*(1), 1–11. Available from https://doi.org/10.2489/jswc.74.1.1.

Wongchai, A., Jenjeti, rao D., Priyadarsini, A. I., Deb, N., Bhardwaj, A., & Tomar, P. (2022). Farm monitoring and disease prediction by classification based on deep learning architectures in sustainable agriculture. *Ecological Modelling*, *474*, 110167. Available from https://doi.org/10.1016/j.ecolmodel.2022.110167.

Wu, H., Yue, T., & Yuan, Yahong (2018). Authenticity tracing of apples according to variety and geographical origin based on electronic nose and electronic tongue. *Food Analytical Methods*, *11*(2), 522–532. Available from https://doi.org/10.1007/s12161-017-1023-y.

Wu, L., Yan, H., Li, G., Xu, X., Zhu, L., Chen, X., & Wang, J. (2019). Surface-imprinted gold nanoparticle-based surface-enhanced Raman scattering for sensitive and specific detection of patulin in food samples. *Food Analytical Methods*, *12*(7), 1648–1657.

Xie, Y., Chen, T., Guo, Y., Cheng, Y., Qian, H., & Yao, W. (2019). Rapid SERS detection of acid orange II and brilliant blue in food by using Fe3O4@ Au core–shell substrate. *Food chemistry*, *270*, 173–180.

Xing, G., Li, N., Lin, H., Shang, Y., Pu, Q., & Lin, J.-M. (2023). Microfluidic biosensor for one-step detection of multiplex foodborne bacteria ssDNA simultaneously by smartphone. *Talanta*, *253*, 123980.

Xu, J., Yang, Z., Wang, Z., Li, J., & Zhang, X. (2023). Flexible sensing enabled packaging performance optimization system (FS-PPOS) for lamb loss reduction control in E-commerce supply chain. *Food Control*, *145*, 109394. Available from https://doi.org/10.1016/j.foodcont.2022.109394.

Xu, J., Gu, B., & Tian, G. (2022). Review of agricultural IoT technology. *Artificial Intelligence in Agriculture*, *6*, 10–22. Available from https://doi.org/10.1016/j.aiia.2022.01.001.

Xu, M., Wang, J., & Zhu, L. (2019). The qualitative and quantitative assessment of tea quality based on E-nose, E-tongue and E-eye combined with chemometrics. *Food Chemistry*, *289*, 482–489. Available from https://doi.org/10.1016/j.foodchem.2019.03.080.

Xu, X., Su, Y., Zhang, Y., Wang, X., Tian, H., Ma, X., Chu, H., & Xu, W. (2021). Novel rolling circle amplification biosensors for food-borne microorganism detection. *TrAC Trends in Analytical Chemistry*, *141*, 116293.

Yang, L., Bai, R., Xie, B., Zhuang, N., Lv, Z., Chen, M., Dong, W., Zhou, J., & Jiang, M. (2023). A biosensor based on oriented immobilization of an engineered l-glutamate oxidase on a screen-printed microchip for detection of l-glutamate in fermentation processes. *Food Chemistry*, *405*, 134792.

Yang, P., Zhu, Y., Yang, X., Li, J., Tang, S., Hao, Z., Guo, L., Li, X., Zeng, X., & Lu, Y. (2018). Evaluation of sample preparation methods for rice geographic origin classification using laser-induced breakdown spectroscopy. *Journal of Cereal Science*, *80*, 111–118.

Yang, Y., Wang, W., Zhuang, H., Yoon, S.-C., & Jiang, H. (2018). Fusion of Spectra and Texture Data of Hyperspectral Imaging for the prediction of the water-holding capacity of fresh chicken breast filets. *Applied Sciences*, *8*(4), 640. Available from https://doi.org/10.3390/app8040640.

Yaxley, K. J., Reid, A., Kenworthy, C., Hossny, M., Baxter, D. P., Allworth, M. B., McGrath, S. R., Joiner, K. F., & Abbass, H. (2023). Building a sky shepherd for the future of agriculture. *Smart Agricultural Technology*, *3*, 100137. Available from https://doi.org/10.1016/j.atech.2022.100137.

Yin, Z., Yang,L., Yongqiang,Y., & Cheng, C. (2022). Modeling tobacco traceability system based on blockchain and RFID technologies. In *2022 IEEE international conference on Smart Internet of Things (SmartIoT)*, Suzhou, China (pp. 241–245). Available from https://doi.org/10.1109/SmartIoT55134.2022.00045.

YongXia, C., RuiXin, L., ZhaoZhou, L., PengJu, C., LiLi, W., YanLi, W., & SuiQing, C. (2019). Quality evaluation based on color grading: quality discrimination of the Chinese medicine *Corni Fructus* by an E-eye. *Scientific Reports*, *9*(1), 17006. Available from https://doi.org/10.1038/s41598-019-53210-5.

Yousefi, H., Su, H.-M., Imani, S. M., Alkhaldi, K., Filipe, C. D. M., & Didar, T. F. (2019). Intelligent food packaging: A review of smart sensing technologies for monitoring food quality. *ACS Sensors*, *4*(4), 808–821. Available from https://doi.org/10.1021/acssensors.9b00440.

Yu, K., Ren, J., & Zhao, Y. (2020). Principles, developments and applications of laser-induced breakdown spectroscopy in agriculture: A review. *Artificial Intelligence in Agriculture*, *4*, 127–139. Available from https://doi.org/10.1016/j.aiia.2020.07.001.

Yu, S., Huang, X., Wang, L., Chang, X., Ren, Y., Zhang, X., & Wang, Y. (2023). Qualitative and quantitative assessment of flavor quality of Chinese soybean paste using multiple sensor technologies combined with chemometrics and a data fusion strategy. *Food Chemistry*, *405*, 134859. Available from https://doi.org/10.1016/j.foodchem.2022.134859.

Yu, X., Lu, H., & Wu, D. (2018). Development of deep learning method for predicting firmness and soluble solid content of postharvest Korla fragrant pear using Vis/NIR hyperspectral reflectance imaging. *Postharvest Biology and Technology*, *141*, 39–49. Available from https://doi.org/10.1016/j.postharvbio.2018.02.013.

Zakaria, A., Shakaff, A. Y. Md, Masnan, M. J., Ahmad, M. N., Adom, A. H., Jaafar, M. N., Ghani, S. A., Abdullah, A. H., Aziz, A. H. A., & Kamarudin, L. M. (2011). A biomimetic sensor for the classification of honeys of different floral origin and the detection of adulteration. *Sensors*, *11*(8), 7799–7822.

Zareef, M., Chen, Q., Hassan, Md. M., Arslan, M., Hashim, M. M., Ahmad, W., Kutsanedzie, F. Y. H., & Agyekum, A. A. (2020). An overview on the applications of typical non-linear algorithms coupled with NIR spectroscopy in food analysis. *Food Engineering Reviews*, *12*(2), 173–190. Available from https://doi.org/10.1007/s12393-020-09210-7.

Zhang, C., Zhang, X., Liao, G., Shang, Y., Ge, C., Chen, R., Wang, Y., & Xu, W. (2019). Species-specific TM-LAMP and trident-like lateral flow biosensor for on-site authenticity detection of horse and donkey meat. *Sensors and Actuators B: Chemical*, *301*, 127039. Available from https://doi.org/10.1016/j.snb.2019.127039.

Zhang, C., Liu, F., & He, Y. (2018). Identification of coffee bean varieties using hyperspectral imaging: influence of preprocessing methods and pixel-wise spectra analysis. *Scientific Reports*, *8*(1), 2166. Available from https://doi.org/10.1038/s41598-018-20270-y.

Zhang, H., Wang, S., Li, D., Zhang, Y., Hu, J., & Wang, L. (2019). Edible gelatin diagnosis using laser-induced breakdown spectroscopy and partial least square assisted support vector machine. *Sensors*, *19*(19), 4225.

Zhang, J.-W., Pan, L.-Q., & Tu, K. (2022). Growth prediction of the total bacterial count in freshly squeezed strawberry juice during cold storage using electronic nose and electronic tongue. *Sensors*, *22*(21). Available from https://doi.org/10.3390/s22218205.

Zhang, L., Yang, H., Tan, Q., Jing, L., Zhang, W., Xiong, J., & Chen, P.-Y. (2021). Wireless detection of biogenic amines using a split-ring resonator with silver nanoparticles-decorated molybdenum disulfide. *Sensors and Actuators B: Chemical*, *343*, 130155. Available from https://doi.org/10.1016/j.snb.2021.130155.

Zhao, F., Wu, J., Ying, Y., She, Y., Wang, J., & Ping, J. (2018). Carbon nanomaterial-enabled pesticide biosensors: design strategy, biosensing mechanism, and practical application. *TrAC Trends in Analytical Chemistry*, *106*, 62–83.

Zhao, H., Zhan, Y., Xu, Z., John Nduwamungu, J., Zhou, Y., Powers, R., & Xu, C. (2022). The application of machine-learning and Raman spectroscopy for the rapid detection of edible oils type and adulteration. *Food Chemistry*, *373*, 131471.

Zheng, L., Cai, G., Wang, S., Liao, M., Li, Y., & Lin, J. (2019). A microfluidic colorimetric biosensor for rapid detection of Escherichia coli O157:H7 using gold nanoparticle aggregation and smart phone imaging. *Biosensors and Bioelectronics*, *124–125*, 143–149. Available from https://doi.org/10.1016/j.bios.2018.10.006.

Zheng, Y.-Y., Kong, J.-L., Jin, X.-B., Wang, X.-Y., Su, T.-L., & Zuo, M. (2019). Cropdeep: The crop vision dataset for deep-learning-based classification and detection in precision agriculture. *Sensors (Switzerland)*, *19*(5). Available from https://doi.org/10.3390/s19051058.

Zhou, B., Ye, Q., Chen, M., Wang, C., Xiang, X., Li, Y., Zhang, J., Zhang, Y., Wang, J., & Wu, S. (2023). A label-free AuNP bioprobe-assisted CRISPR/Cas12a colorimetric platform for high-throughput detection of *Staphylococcus aureus* ST398. *Food Control*, *145*, 109451.

Zhu, X., Rehman, K. U., Wang, B., & Shahzad, M. (2020). Modern soft-sensing modeling methods for fermentation processes. *Sensors*, *20*(6), 1771. Available from https://doi.org/10.3390/s20061771.

Chapter 9

Artificial intelligence and Big Data revolution in the agrifood sector

Sigfredo Fuentes[1,2], Eden Tongson[1] and Claudia Gonzalez Viejo[1]
[1]*Digital Agriculture, Food and Wine Sciences Group, School of Agriculture, Food and Ecosystem Sciences, Faculty of Science, The University of Melbourne, Parkville, VIC, Australia,* [2]*Tecnologico de Monterrey, School of Engineering and Sciences, Monterrey, Nuevo Leon, Mexico*

9.1 Introduction

The potential implementation of digital technologies in the agrifood sector has been growing exponentially in the last decade due to the advances in computer science and electrical and electronic engineering. The latter is exemplified by the development and miniaturization of sensors and camera technologies utilized in advanced platforms such as unmanned aerial and terrestrial vehicles (UAV and UTV) that allow rapid and efficient remote sensing. That was considered expensive, bulky, and user-unfriendly. Computers also have increased in computer power to handle, analyze, and model Big Data that digital sensors have produced in the agrifood sector using the latest advances in artificial intelligence (AI).

Advanced computing power and sensor technologies to acquire, retrieve, and analyze digital data have been coined Agriculture and Food 4.0 or Agri-Food 4.0. Nowadays, Agri-Food 5.0 has been developed through the accumulation of data by different stakeholders within the Agri-Food sector worldwide, allowing the implementation of new and emerging technologies, such as within sensor technologies and communications, that is, 5G (Hilten & Wolfert, 2022), the Internet of Things (IoT) (Mahdad et al., 2022; Yadav, Choi, et al., 2022; Yadav, Luthra, et al., 2022), robotics (Hilten & Wolfert, 2022) coupled with AI in the form of advanced signal and Big Data analysis, computer vision, biometrics, and machine/deep learning modeling, among others (Gonzalez Viejo et al., 2019a).

It has been argued in recent years that one of the major utilities for AI and digital technologies (Renda, 2019) and Agri-Food 5.0 is to tackle the constraints of climate change in the Agri-Food sector (Ben Ayed et al., 2021; Tortorella et al., 2020) to secure sustainability. Specifically, some research has focused on implementing AI to increase efficiency in agrifood supply chains (Monteiro & Barata, 2021) and specifically after COVID-19 as a contingent effect of climate change that was not being foreseen or tested before (Di Vaio et al., 2020). The latter with current and future implications of novel pandemics affects countries differentially (i.e., Australia with the longest COVID-19 lockdowns and China with continuing lockdowns and new COVID-19 variants).

However, most of the developed and proposed Agri-Food 4.0 and 5.0 applications still start from products from the farmgate to the consumer level within the supply chain. Hence, it has been identified as a disconnection between the farm and the palate, including data, information, and models that can be applied as feedback for the agricultural product through modifications of management strategies (Fuentes et al., 2021).

The main general advantages of the implementation of AI in the agrifood sector are (1) gives transformational powers to the sector to become more predictive rather than as it has been historically reactive; (2) integration of the different sectors of the agrifood production chain from the farm to the palate including circular feedback of information and implementation of digital twins (DT) (Tebaldi et al., 2021) for efficient management purposes and to increase sustainability (Fuentes et al., 2021). The latter paper, published in Current Opinions in Food Science, also offers a solution to one of the most important questions for agrifood industry stakeholders: who owns the data and whether competitors can benefit from AI models DT developed for the industry. In short, the authors argue that data can be regarded as currency and access and handling of the data produced in the agrifood industries can be managed using blockchain technology to securely manage data throughout the food supply chain, securing the

Food Industry 4.0. DOI: https://doi.org/10.1016/B978-0-443-15516-1.00009-8

integrity and transparency between agrifood companies, government agencies, and technology providers. Furthermore, this technology and data handling can also be used to assess the provenance and traceability of products and to avoid counterfeiting.

Regarding the concerns of benefiting the competition through the application of Agri-Food 5.0 digital models based on AI and DT trained from data owned by the agrifood stakeholders data. The latter is based on the erroneous conception that AI models trained with data from farms can detect only the patterns and influences from the soil—plant—atmosphere continuum. In other words, there is the conception that AI models are deterministic and site-specific. However, machine and deep learning models also create patterns to analyze new data and continue learning, which reinforces the importance of monitoring the performance of AI models and not applying models blindly, especially those that can be bought. There is a need to validate AI models in deployed applications. Furthermore, the application of developed agrifood AI models will benefit the specific stakeholder applying them and, at the same time, the whole agrifood industry following the game theory (Gebser & Kaminski, 2022; Lu & Li, 2022), which is the basis of sustainability and response to climate change and food security issues.

The complexity of processes within the agrifood sector, from farm to the consumer, requires complex analysis beyond precision agriculture/food techniques and Agri-Food 4.0 tools. The implementation of AI, including digital technologies based on remote sensing, can identify emerging patterns due to unforeseen effects from climate change effects on crops and livestock to changes in consumer preferences.

This chapter is focused on the latest digital advances within agriculture, food, and beverages using combined digital sensors and AI, including computer vision, digital sensor technologies and signal analysis, machine/deep learning, and DT. Specific technologies and results will be covered in a table to increase the impact on readers from this chapter as a reference tool. However, more emphasis will be directed on the integrative approaches of the different AI agrifood digital technologies throughout the agrifood chains. Due to the explosion of research related to AI implemented in the agrifood sector in the last 10 years, it is only possible to include some of the body of research produced. Many terminologies used in this chapter related to AI and smart agriculture/food are currently in their fluid state by many researchers and the established media and social/professional media (Misra et al., 2022). Hence, within this chapter, the terminology will be explicitly described for its use in the context of the different digital technologies, applications, and deployments described.

9.2 Digital agriculture

9.2.1 State-of-the-art smart farming

There are many terminologies to describe the impact of digital technologies and AI on farming systems, from production, monitoring, and automation, ranging from precision farming, precision agriculture (PA), Agriculture 4.0, smart farming, and Agriculture 5.0 to name a few. PA has been around for more than 40 years, from the late 80s and early 90s of the 20th century, and it was focused on trying to make sense of the natural variability of farming systems, specifically related to soils with implications of farm management strategies, such as fertilization, water management, and yield variability. PA was possible by implementing Geographical Positioning Systems (GPS) from satellites to map these variabilities using different sensor technologies ranging from soil and plant analysis to remote sensing using passive sensors (measuring the reflectivity of radiations from soils and crops). Remote sensing platforms can vary from long-range (satellites), medium-range (airborne), low-range (UAV), and proximal (UTV, tractors, quad bikes), or just carrying sensors and cameras by users. The highest resolutions of data (pixel footprint) can be achieved based on the resolution of sensors/cameras within the medium to proximal ranges. Satellites still have the inconvenience of the timing of data recapture, which can range from 12 to 14 days for a new pass through the same site and lower resolutions of information with a pixel footprint of 0.5 m^2 for panchromatic information (i.e., Ikonos, Landsats), a couple of m^2 for multispectral information (i.e., WorldView 2 and 3) and up to 30 m^2 for Landsat as examples. Furthermore, passive remote sensors cannot obtain data on cloudy days (S. Fuentes & Gago, 2022). The main difference between PA and digital agriculture (DA) is that the former deals with methods, technologies, and sensors to acquire digital data and mapping differences in the field. The latter (DA) deals with implementing new and emerging technologies to analyze the data and generate more flexible and applicable models using AI tools.

State of the art for AI in DA, including livestock monitoring for welfare, traceability, and identification, has been recently published elsewhere with the latest papers on AI for digital livestock (Table 9.1) (Dac et al., 2022; Fuentes et al., 2022).

TABLE 9.1 Agriculture applications of the different technologies based on sensors, machine learning modeling, and Big Data.

Application	Product(s)	Assessment	Technologies	Accuracy	Deployment	References
Diseases	Citrus fruits	*Penicillium digitatum*	Whole-cell-based biosensor	NR	None	Chalupowicz et al. (2020)
	Potato	*Pectobacterium atrosepticum*	Electrochemical impedance immunosensor	NR	None	Hashemi Tameh et al. (2020)
	Potato	*Verticillium wilt*	NIR ANN—classification	94%	84%	Shin et al. (2023)
	Tomato	*Fusarium oxysporum*	E-nose ANN—classification	Plant: 94%–97%; soil: 95%–96%	None	Feng et al. (2022)
	Canola	Ascocospores and sclerotinia stem rot	Immuno-impedimetric biosensor	NR	None	Shoute et al. (2018)
	Tea plant	VOC; damaged plants by *Ectropis obliqua* and *Ectropis grisescens*	E-nose LPP and SVM—classification	100%	None	Sun, Wang, Sun, et al. (2019)
	Maize	Gray leaf spot disease	Multispectral sensors Random forest—classification	84%	None	Dhau et al. (2018)
	Carrot	*Candidatus Liberibacter solanacearum*	RobHortic—RBG, multispectral and hyperspectral cameras PLS-DA—classification	Laboratory: 66%; field: 60%	None	Cubero et al. (2020)
Pests	Wheat	Aphids	NIR E-nose ANN—classification	NIR: 97%–99%; E-nose: 94%–99%	None	Fuentes, Tongson, Unnithan, et al. (2021)
	Tomato	Whitefly	E-nose PCA and cluster analysis	Reported as accurate	None	Cui et al. (2021)
	Tomato	Aphids	E-nose PCA	Reported as accurate	None	Cui et al. (2019)
	Tea plants	*Ectropis obliqua* and *Ectropis grisescens*	E-nose SVM—classification	91%	None	Sun, Wang, Cheng, et al. (2019)

(*Continued*)

TABLE 9.1 (Continued)

Application	Product(s)	Assessment	Technologies	Accuracy	Deployment	References
Physiology	Wheat	Photosynthesis; stomatal conductance; transpiration	E-nose ANN—regression	$R = 0.86–0.94$	None	Fuentes et al. (2021)
	Potato	Photosynthesis; stomatal conductance; transpiration	NIR ANN—regression	$R = 0.89$	$R = 0.76$	Shin et al. (2023)
	Tomato	Photosynthesis; stomatal conductance; transpiration	E-nose ANN—regression	$R = 0.97–0.99$	None	Feng et al. (2022)
	Khasi mandarin orange plants	Water stress	E-nose SVM—bagging ensemble—classification	85%	None	Choudhury et al. (2019)
	Arugula	Crop water stress index	Infrared sensor RGB images + computer vision	86%–90%	None	Paulo et al. (2023)
	Olive trees	Water status	Thermal infrared sensors	$R^2 > 0.80$	None	Noguera et al. (2020)
	Grapevines	Drought stress	Hyperspectral images PLS-SVM—classification	>97%	None	Zovko et al. (2019)
	Grapevines	Berry cell death	NIR ANN—regression	$R = 0.94$	None	Fuentes, Tongson et al. (2020)
	Grapevines (Chardonnay and Shiraz)	Berry cell death	NIR Viticanopy—RBG images ANN—regression	NIR: $R = 0.87$; Viticanopy: $R = 0.98$	None	Fuentes et al. (2021)
Pesticides	Cherries	Diazinon	E-nose ANN—classification	100%	None	Nategh et al. (2021)
	Citrus fruits	Organophosphate, carbamate, and bipyridylium	Nanoplasmonic sensor array Linear discriminant analysis	$R^2 = 0.99$	None	Mirghafouri et al. (2020)
	Tea plants	Pesticide residues	E-nose Confocal Raman microspectroscopy ANN—regression	$R^2 = 0.94$	None	Sanaeifar et al. (2021)
Yield	Wheat	Yield (Mateo-Sanchis et al., 2019)	Multispectral sensor Weather data Random forest—regression	$R^2 = 0.74$	None	Ruan et al. (2022)
	Corn, soybean, wheat	Yield	Optical and microwave sensors Kernel ridge regression	$R^2 > 0.76$	None	Mateo-Sanchis et al. (2019)
	Soybean	Yield	Multispectral sensor RGB camera Random forest—classification	91%	None	Herrero-Huerta et al. (2020)
	Potato	Yield	Dual geometry sensor array Soil moisture sensor SVM—regression	RMSE = 4.62–6.60 t/ha	None	Abbas et al. (2020)

Quality of produces	Cacao trees	VAC of fermented cacao	Viticanopy—RGB images; ANN—regression	$R = 0.82$	None	Fuentes et al. (2019)
	Grapevines (Pinot noir)	Intensity of sensory descriptors in wine; physicochemical parameters in wine	Meteorological station ANN—regression	$R = 0.98–0.99$	None	Fuentes et al. (2020)
	Grapevines (Pinot noir)	Intensity of sensory descriptors in wine; physicochemical parameters in wine	Meteorological station ANN—regression	$R = 0.94–0.99$	None	Fuentes et al. (2019)
	Grapevines (Chardonnay and Shiraz)	Intensity of sensory descriptors in wine	Viticanopy—RGB images ANN—regression	$R = 0.99$	None	Fuentes et al. (2021)
	Grapevines (Pinot noir)	Intensity of sensory descriptors in wine	NIR ANN—regression	$R = 0.80$	None	Fuentes, Tongson et al. (2020)
	Dairy cows/milk	Milk quality and yield	Meteorological station data ANN—regression	$R = 0.87$	None	Fuentes, Gonzalez Viejo, et al. (2020)
	Dairy cows/milk	Milk quality and yield	RGB camera Computer vision ANN—regression	$R = 0.96$	$R = 0.93$	Fuentes et al. (2021)

NR, not reported; *NIR*, near-infrared spectroscopy; *E-nose*, electronic nose; *ANN*, artificial neural networks; *LPP*, locality preserving projections; *SVM*, support vector machine; *VOC*, volatile organic compounds; *VAC*, volatile aromatic compounds; *RGB*, reg, geen, blue; *PLS*, partial least squares; *DA*, discriminant analysis; *PCA*, principal components analysis.

9.2.2 Machine learning modeling, sensors, and Big Data

In general terms, there are supervised machine learning and unsupervised deep learning (ML and DL, respectively) algorithms for classification and regression AI modeling techniques that can be used for Agri-Food 5.0. These are subdisciplines of AI, which are combined with digital sensor technologies to acquire data and be used as inputs to the models. Specific targets can be modeled with existing data that, in the case of agriculture, can be Big Data accumulated through the years by automatic meteorological stations, soil moisture sensors, and plant-based sensors, which on many occasions have been used for more than 30 years. These types of data can give high temporal resolution (with data collected every 15 or 30 minutes through seasons) and high spatial resolution depending on the variability of the specific site where they are installed and the number of sensors distributed to represent this variability. Specific physiological processes, compounds of interest, and up-to-yield can be targeted using AI models. The basic premise that needs to be kept in mind for AI models is the parsimony principle, which in machine learning means that inputs need to be easier to obtain compared to targets, which is very relevant for automation when deploying models. This principle can also be applied in reverse to remote sensing from different platforms (satellite or UAV), which can be used as inputs to target ground truth data mentioned before, which allows increasing spatial resolution representation of a specific agricultural field.

A complete description of different ML/DL algorithms (over 20) can be found in the publication from Gonzalez Viejo et al. (2019a). However, many researchers settle on using a specific ML algorithm showing the accuracies and efficiencies without explaining why that algorithm was used or any comparison with other ML algorithms available.

9.2.3 Prediction stage using artificial intelligence

In the case of supervised ML models, the parameter engineering process is extremely important, which requires, in the case of agrifood application, knowledge of basic sciences, plant physiology, soil–plant–atmosphere interactions, chemistry, biochemistry, and mathematics, to name a few. Without this knowledge, models can arrive at highly accurate predictions by overfitting data or artifacts (Crawford, 2022; Rochewerger & Pang, 2021). There are many cases that can serve as examples of bias from the simplistic mistaken direct correlations between ice cream consumption on beaches for summer with the number of drowning cases, which presents a problem of causality. More complex bias from AI models in agrifood can be cited as the prediction of berry cell death (BCD) in grapevine berries using near-infrared spectroscopy (NIR)-based sensors in the range between 300 and 1200 nm, which includes color and sugar accumulation. BCD is a programmed process occurring around veraison to harvest, which coincides with sugar accumulation and color change (Fuentes, Tongson, et al., 2020). However, plant physiology and biochemistry knowledge would suggest tracking hydrogen peroxide (H_2O_2), which is in the range of 1600–1700 nm (Table 9.1). There are many other cases that can be cited when trying to develop AI models using complex biophysical organisms without the proper parameter engineering process.

A different scenario is presented with DL modeling in which it is up to the machine to uncover emerging patterns within the data provided. Hence, the shortcomings of these DL will be mainly related to the volume of data and variety of data used for the AI modeling. In the case of our grapevine berry example, even DL models will overestimate models by feeding inputs related to NIR between 300 and 1200 nm and targeting winemaking grapevine cultivar BCD. On the other hand, which is also part of parameter engineering and biology/plant physiology knowledge, by feeding DL models with NIR in the range of 300–2500 nm and targeting BCD for winemaking and table grape cultivars, the DL will make sense of the variability of the data. Table grapes achieve minimal BCD at harvest, and these cultivars have been selected for this purpose to end up to the consumer with crunchy, juicy, and sweet grapes. On the contrary, for winemaking cultivars, BCD levels are sought because it is directly related to the development of flavor and aroma profiles of final wines (Table 9.1).

9.2.4 Deployment stage

Due to the previous argument, the requirement for the deployment of AI models developed on different conditions/scenarios with new data (extra validation process) is extremely important to avoid problems of overfitting, bias, and artifacts to be included in the models. However, not much-published research offers data for deployment (Table 9.1). Furthermore, the adequate selection of inputs through the parsimony principle could secure the efficient deployment and automation of model deployments.

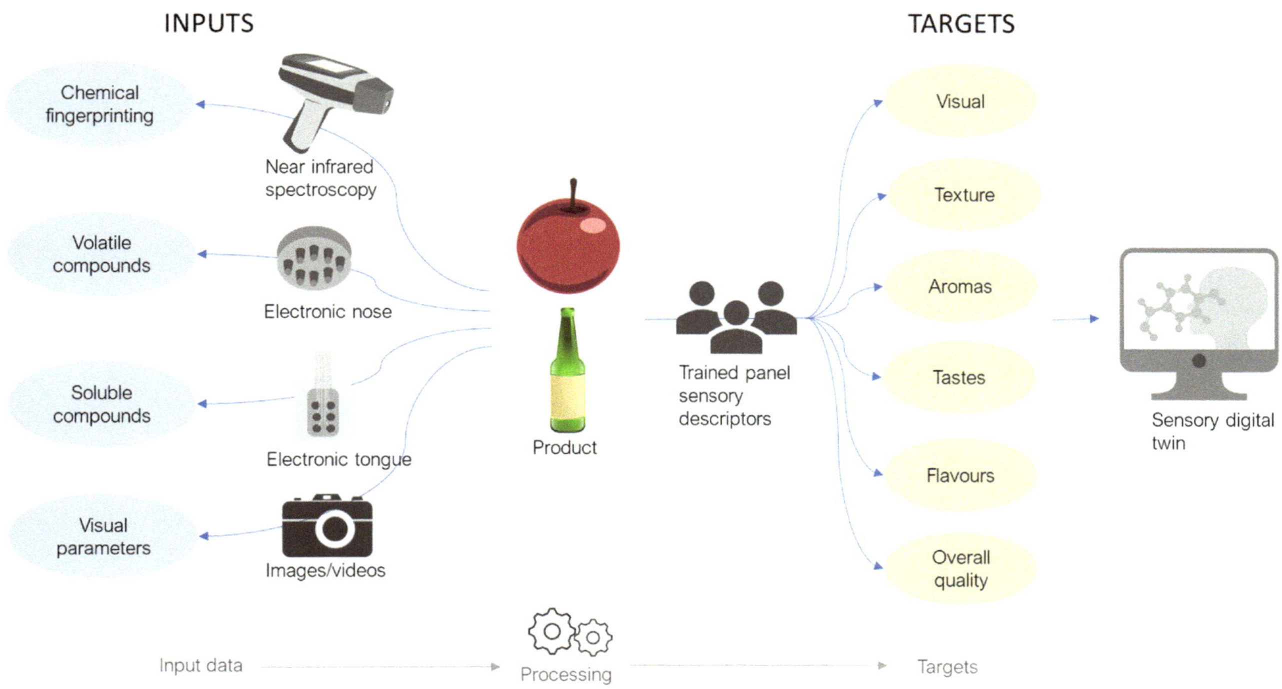

FIGURE 9.1 Example of digital twins applied to agrifood with inputs data, data processing by AI models and targets Targets can also be used to modify agronomical practices to tailor specific requirements from the agrifood industry or group of consumers.

The most efficient way for AI automation is the development of DT, which could be considered as a virtual replication of a process that automatically gathers data, analyzes it using AI models, and produces information for the automation of management strategies that, for agriculture, can be targeted to irrigation scheduling, fertilization, canopy management, and control of biotic and abiotic stresses, among others. The implementation of DT is considered the optimal way to deal with the complexities of the agrifood production and supply chains, securing the introductions of constraints presented by climate change and incorporating digital tools, such as blockchain for food security, traceability, and provenance, and to avoid counterfeiting. An example of a proposed digital twin for agricultural produce is depicted in Fig. 9.1. This DT example can be applied to any part of the Agri-Food processes, and it can even have feedback from every stage to increase efficiency and learn from emerging patterns that can be present or introduced due to climate change (Fig. 9.1).

9.3 Food and beverages

9.3.1 State-of-the-art smart food technology

The term "Smart Food" is unclear since it has been given different definitions within the industry. However, this chapter refers to Smart Foods as food produced using digital technologies that do not compromise food safety but enhance or improve food production, processing, distribution, and consumption to offer better products (Jofresa & Xirau, 2019). These technologies involve a broad range of tools that aid in efficiency and automation (Table 9.2) to secure high-quality and safe food and beverages for consumers. This is in response to the increasing demand for more premium, safe, and healthy products and also to aid in more cost-efficient and less time-consuming and labor-intensive methods in the food production chain from manufacturing to retailing (Gonzalez Viejo et al., 2019a; Wu et al., 2021).

9.3.2 Sensors, machine learning modeling, and Big Data

Sensors, machine learning modeling, and Big Data have become some of the most popular technologies in the food and beverage industry (Misra et al., 2022). There is a wide variety of sensors typically coupled with machine learning

TABLE 9.2 Digital technologies that comprise smart food and their main applications.

Technology	Applications	References
Sensors/biosensors	Food quality Food safety Packaging Authentication	Apetrei and Ghasemi-Varnamkhasti (2013), Fuentes et al. (2021), Gonzalez Viejo et al. (2020), Gonzalez Viejo et al. (2019a), and Thakur and Ragavan (2013)
Computer vision	Food quality Nutrition	Gonzalez Viejo et al. (2019a) and Sahoo et al. (2019)
Radio-frequency identification (RFID)	Packaging Traceability	Bhatia and Ahanger (2021), Fathi et al. (2020), Gonzalez Viejo, Gurr, et al. (2022), and Yiying et al. (2019)
Robotics	Processing automation Packaging	Gonzalez Viejo et al. (2019a), Iqbal et al. (2017), Khan et al. (2018), and Nayik et al. (2015)
Internet of Things (IoT)	Traceability	Bhatia and Ahanger (2021) and Krupitzer and Stein (2021)
Blockchain	Traceability Fraud Food safety	Abbas et al. (2020), Creydt and Fischer (2019), Galvez et al. (2018), Jofresa and Xirau (2019), Yiannas (2018), and Zhou et al. (2022)
3D printing	Product development Processing	Jofresa and Xirau (2019), Nachal et al. (2019), and Yang et al. (2017)
Machine learning	Processing Product development Packaging Food quality Food safety Traceability Authentication	Bhatia and Ahanger (2021), Gonzalez Viejo et al. (2019a), and Krupitzer and Stein (2021)
Artificial intelligence	Processing Packaging Food quality Food safety Traceability Authentication	Bhatia and Ahanger (2021), Gonzalez Viejo, Fuentes, et al. (2021), Gonzalez Viejo et al. (2019a), Kakani et al. (2020), and Zhou et al. (2022)
Big Data	Food safety Food security Food quality	Irani et al. (2018), Jeevanandam et al. (2022), Jin et al. (2020), and Zhou et al. (2022)

techniques, which are used in the food industry for different applications such as the detection of volatile compounds to assess aromas, detection of biomolecules like enzymes and allergens, food and beverage authentication, shelf life, sensory attributes and consumer acceptability, packaging integrity, and quality traits (Apetrei & Ghasemi-Varnamkhasti, 2013; Gonzalez Viejo, Gurr, et al., 2022; Gonzalez Viejo & Fuentes, 2020; Mustafa & Andreescu, 2018; Thakur & Ragavan, 2013). Big Data in the food industry refers to the use of large and complex data sets that cannot be handled using traditional methods to drive decision-making and improve operations (Sadiku et al., 2020). This involves collecting, analyzing, and interpreting data from different sources, such as sensors, production processes, market trends, and consumer needs and demands (Rejeb et al., 2022). This is to increase efficiency, reduce waste, ensure food safety, and improve consumers' overall experience. The main applications of Big Data in the food industry are for monitoring supply chain efficiency, designing and improving food safety and traceability systems, optimizing food packaging and distribution based on market trends and consumers demands, and personalizing food products based on consumers' needs and preferences (Vågsholm et al., 2020).

Within the last 5 years, most applications of sensors and machine learning technologies have been in the food and beverage quality assessment in a broad range of products (Table 9.3). Among the most popular sensors to use in food

TABLE 9.3 Food and beverage applications of the different technologies based on sensors and machine learning modeling.

Application	Product(s)	Assessment	Technologies	Accuracy	Deployment	References
Quality traits assessment	Rice	Moisture	NIR	$R^2 = 0.94$	None	Lin et al. (2019)
	Rice	VAC in raw and cooked rice	E-nose ANN—regression	$R = 0.95–0.96$	None	Aznan et al. (2022a)
	Sourdough bread	Type of wheat VAC	NIR E-nose ANN—classification ANN—regression	Type of wheat: 96%–99%; VAC: 0.97–0.99	None	Gonzalez Viejo, Harris, et al. (2022)
	Olive oil	Rancidity and type of olive oil VAC	NIR E-nose ANN—classification ANN—regression	Rancidity and type of olive oil: 87%–97%; VAC: 0.93–0.96	None	Gonzalez Viejo and Fuentes (2022b)
	Coffee	Roasting intensity VAC	E-nose ANN—classification ANN—regression	Intensity: 98%; VAC: $R = 0.99$	None	Gonzalez Viejo et al. (2021)
	Coffee	Chlorogenic acid	Electrochemical impedance spectroscopy	$R^2 = 0.99$	None	Salamanca-Neto et al. (2020)
	Green coffee	Sucrose	Multienzymatic biosensor	$R^2 = 0.97$	None	Stredansky et al. (2018)
	Carbonated water	Bubble size	RGB camera Computer vision	NR	None	Gonzalez Viejo et al. (2019)
	Beer	VAC	E-nose ANN—regression	$R = 0.93$	None	Gonzalez Viejo et al. (2020)
	Beer	Faults	NIR E-nose ANN—classification	$R = 96\%–99\%$	None	Gonzalez Viejo et al. (2021)
	Wine	Faults	NIR E-nose ANN—classification	90%–97%	None	Gonzalez Viejo and Fuentes (2022a)
	Wine	Smoke taint	E-nose ANN—classification	98%	90% Testing smoke taint amelioration success	Summerson et al. (2021)
	Wine	Smoke taint (glycoconjugates and volatile phenols) in berries (raw material) from wine (final product) measurements	NIR E-nose ANN—classification	NIR: 0.98; E-nose: $R = 0.98–0.99$	None	Fuentes, Summerson et al. (2020) and Summerson et al. (2020)
	Wine	Quercetin and carbendazim	Electrochemical sensors	NR	None	Raymundo-Pereira et al. (2020)

(Continued)

TABLE 9.3 (Continued)

Application	Product(s)	Assessment	Technologies	Accuracy	Deployment	References
Shelf life	Rice	Shelf life	E-nose k-NN—regression	$R^2 = 0.72$	None	Hanif et al. (2021)
	Rice	Shelf life	Colorimetric sensor LDA	98.75%	None	Lin et al. (2018)
	Milk	Hydrogen peroxide	Optical biosensor	NR	None	Vasconcelos et al. (2022)
	Tuna	Shelf life—cadaverine	Microcantilever biosensor	$R = 0.71-0.96$	None	Alexi et al. (2021)
	Pork meat	Shelf life—microbiology and sensory	E-nose	$R = 0.63-0.89$	None	Ramírez et al. (2018)
Authentication	Rice	Adulteration	E-nose NIR ANN—regression	E-nose ($R = 0.94-0.98$); NIR ($R = 0.95-0.98$)	None	Aznan et al. (2022b)
	Rice	Morpho-colorimetry; variety prediction	RGB images Computer vision ANN—classification	>91%	94%	Aznan et al. (2021)
	Coffee	Authentication	3D-printed electrochemical sensors	91%–95%	None	Mutz et al. (2023)
	Beer	Type of fermentation	RGB camera Computer vision Gas sensors ANN—classification	94%	Average: 95%	Gonzalez Viejo, Fuentes, Torrico, Lee et al. (2018) and Gonzalez Viejo et al. (2016)
	Wine	Vintage; sensory descriptors; VAC for authentication purposes	NIR through the bottle ANN—classification	Vintage: 97%; sensory descriptors: $R = 0.95$; VAC: $R = 0.88$	None	Harris et al. (2023)
	Tequila	Methanol	Fiber optic sensor	NR	None	Necochea-Chamorro et al. (2019)
	Fruit juices	Formaldehyde	Electrochemical biosensor	$R^2 = 0.98-0.99$	None	Kundu et al. (2019)
	Mozzarella and feta cheese	Adulteration with cow's milk	Silicon photonic immunosensor	NR	None	Angelopoulou et al. (2021)
	Halal beef	Adulteration with pork	E-nose SVM—classification	98%	None	Sarno et al. (2020)

Sensory evaluation	Sourdough bread	Aromas during sensory session	E-nose	NR	None	Farahmand et al. (2021)
	Spicy foods	Spicy compounds	E-tongue	72%–84%	None	Paup et al. (2019)
	Paneer cheese	Discrimination of different samples compared to consumers	E-tongue	93%	None	Schlossareck and Ross (2019)
	Beer	Intensity of sensory descriptors	E-nose ANN—regression	$R = 0.97$	None	Gonzalez Viejo et al. (2020)
	Beer	Consumers acceptability; intensity of sensory descriptors	NIR ANN—regression	$R = 0.93–0.99$	None	Gonzalez Viejo and Fuentes (2020)
	Beer	Consumers acceptability	RGB camera Computer vision Gas sensors ANN—regression	$R = 0.98$	None	Gonzalez Viejo et al. (2019b)
	Beer	Intensity of sensory descriptors	RGB camera E-nose Computer vision ANN—regression	$R = 0.91$	$R = 0.86$	Gonzalez Viejo et al. (2019b) and Gonzalez Viejo, Fuentes, Torrico, Howell et al. (2018)
	Wine	Consumers acceptability	E-nose ANN—regression	$R = 0.98$	None	Fuentes, Summerson et al. (2020)

NR, not reported; *NIR*, near-infrared spectroscopy; *E-nose*, electronic nose; *E-tongue*, electronic tongue; *ANN*, artificial neural networks; *k-NN*, k-nearest neighbors; *SVM*, support vector machine; *VAC*, volatile aromatic compounds; *RGB*, red, green, blue; *LDA*, linear discriminant analysis.

quality assessment are the electronic noses coupled with machine learning modeling to assess volatile aromatic compounds (VACs) as an alternative rapid and low-cost method to the costly and time-consuming gas chromatography–mass spectroscopy (GC–MS) equipment (Table 9.3) (Gonzalez Viejo et al., 2020). The assessment of VACs is important as an indicator of quality, which also leads to the successful prediction (accuracy $> 87\%$) of other quality traits, such as rancidity in olive oil (Gonzalez Viejo & Fuentes, 2022b), type of wheat in sourdough bread (Gonzalez Viejo, Harris, et al., 2022), roasting intensity in coffee (Gonzalez Viejo, Tongson, et al., 2021), and smoke taint in wines (Summerson et al., 2021), among others. Other sensors that have been used for the quality traits assessment in foods and beverages are NIR and electrochemical impedance spectroscopy sensors to detect specific compounds such as chloragenic acid in coffee (Salamanca-Neto et al., 2020). Shelf life is another important application that is directly related to food quality and safety but is assessed over time. Even though it is not the most explored application, there have been some recent developments using different types of sensors, such as optical biosensors to assess hydrogen peroxide in milk as an indicator of spoilage (Vasconcelos et al., 2022) or a microcantilever biosensor to evaluate cadaverine in tuna to determine shelf life (Alexi et al., 2021).

On the other hand, one of the most important applications of these digital technologies is for authentication; this is due to the high incidence of food fraud, as food and beverages account for 10% of the counterfeited products worldwide, including coffee, wine, milk, and meat within the most counterfeited products (Csapó et al., 2019). Therefore, researchers have used different types of sensors to identify adulterated foods (Table 9.3). Among some of these developments is the use of a low-cost and portable e-nose to assess adulterated rice; this method, coupled with artificial neural networks, showed to be highly accurate ($R > 0.94$) to assess the percentage of adulteration with other rice qualities or varieties (Aznan et al., 2022b). Similarly, an e-nose has been used along with support vector machine modeling to assess halal beef adulterated with pork with high accuracy (98%) (Sarno et al., 2020). A more recent development was the use of NIR through the wine bottle to assess vintage, sensory descriptors, and VAC of the wine as authentication indicators (Harris et al., 2023); this is important because wine is severely affected by counterfeiting with ~\$6B costs every year with 1/5 of wine sales being fake wines (Liska, 2022; Young, 2022). Other applications with sensors, such as electrochemical biosensors to assess formaldehyde in fruit juices (Kundu et al., 2019) and 3D-printed electrochemical sensors for coffee authentication (Mutz et al., 2023), have been successfully developed. Regarding the sensory descriptors and consumer acceptability of foods and beverages, e-noses and e-tongues have been very popular due to their ability to detect VACs and basic tastes, respectively. These have been applied to assess products such as bread, spicy foods, paneer cheese, beer, and wine (Table 9.3).

9.3.3 Prediction stage using artificial intelligence

The main AI tools used for the prediction stage in the food industry consist of integrating digital technologies such as robotics, sensors, computer vision, and machine learning models, among others, and more complex techniques such as DL and DT. As mentioned before, deep learning is a complex type of machine learning modeling that works using neural networks to process and analyze large amounts of data. It has given a wide range of applications within the food industry, such as food safety, food recognition and classification, quality control and assurance, and supply chain management (Zhou et al., 2019). On the other hand, DT in the food industry refers to a virtual simulation of the food production or assessment process developed using data from digital technologies such as sensors and ML models (Henrichs et al., 2022; Verboven et al., 2020). Some of the current applications of DT in the food industry are to simulate modifications for optimization within the processing line to assess the quality of the product before applying it in the real process or to personalize recommended formulas for consumers (Hinduja et al., 2020).

9.3.4 Deployment stage

The automated deployment stage using DT can be similar to the process described in Fig. 9.1 for agricultural applications when incorporating Agri-Food and the end consumer perspective. For increased efficiencies, continuous monitoring and AI modeling are proposed throughout the agrifood production and supply chain, including consumer acceptability and preference (Fig. 9.2).

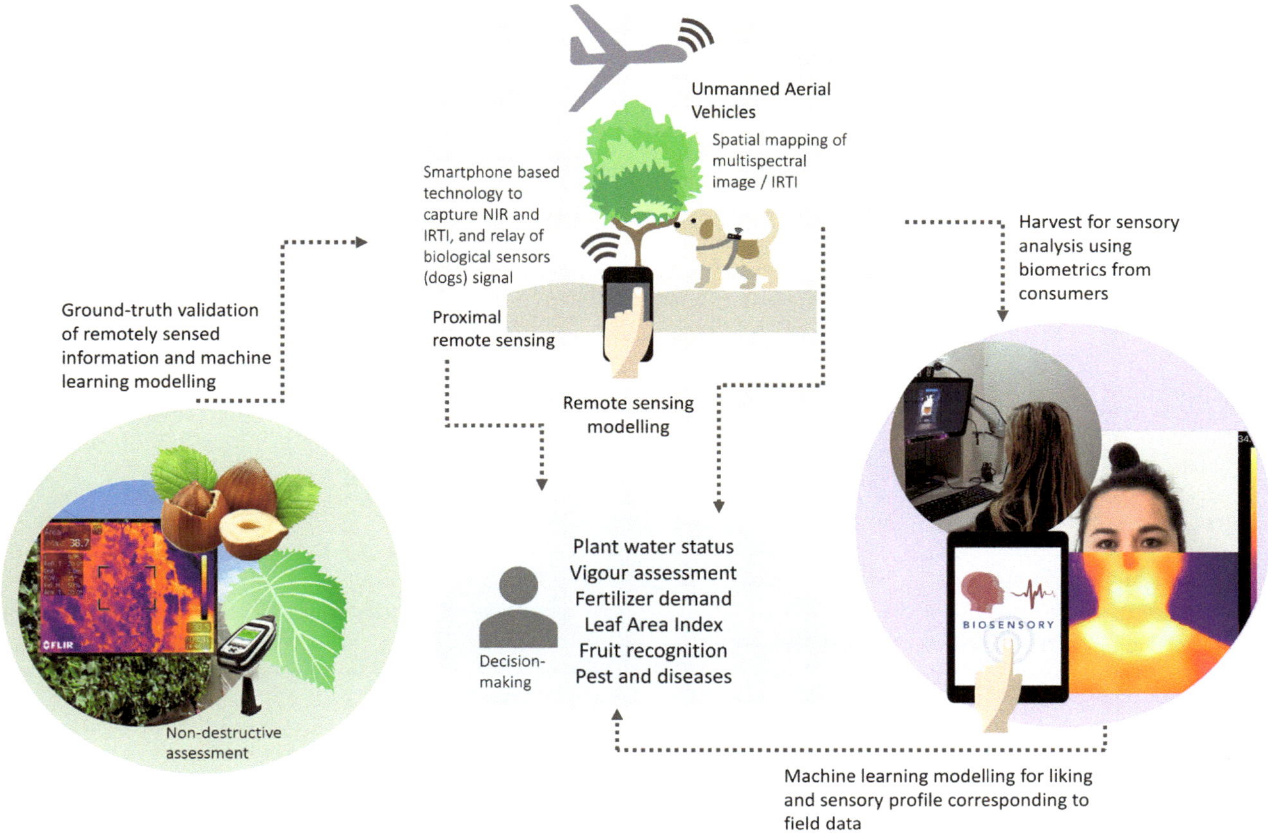

FIGURE 9.2 Implementation of digital technologies for the agrifood industry from the field to the consumer The information can be used back through the production and supply chain to tailor processes to achieve higher efficiency and reduce waste.

9.4 Conclusions

Artificial intelligence systems for Agri-Food 5.0 should consider implementation from field to fork and beyond. Current constraints of climate change and the reduction of arable soil for agricultural production around the world have pushed new and emerging technologies to maximize the efficiency of food production, such as vertical farming (VF). All aspects of production can be controlled in VF, from light, water and nutrient supply, seasonality, biotic and abiotic stresses, and yield. VF needs to implement tools from Agri-Food 5.0 described. Another area of implementation of Agri-Food 5.0 principles would be long-term space explorations soon with the NASA plans (Artemis program) for missions from the Earth to the Moon (2030) and from the Moon to Mars (2040). Agri-Food 5.0 offers the potential to enhance productivity, diversify space produce, boost nutrient availability, and produce materials and pharmaceuticals, while also fostering astronauts' well-being through plant-human interactions and monitored pick-and-eat activities using biometrics tailored for sensory applications.

References

Abbas, F., Afzaal, H., Farooque, A. A., & Tang, S. (2020). Crop yield prediction through proximal sensing and machine learning algorithms. *Agronomy*, *10*, 1046.

Alexi, N., Hvam, J., Lund, B. W., Nsubuga, L., de Oliveira Hansen, R. M., Thamsborg, K., Lofink, F., Byrne, D. V., & Leisner, J. J. (2021). Potential of novel cadaverine biosensor technology to predict shelf life of chilled yellowfin tuna (*Thunnus albacares*). *Food Control*, *119*. Available from https://doi.org/10.1016/j.foodcont.2020.107458, https://www.journals.elsevier.com/food-control.

Angelopoulou, M., Petrou, P. S., Raptis, I., Misiakos, K., Livaniou, E., Makarona, E., & Kakabakos, S. (2021). Rapid detection of mozzarella and feta cheese adulteration with cow milk through a silicon photonic immunosensor. *Analyst*, *146*(2), 529–537. Available from https://doi.org/10.1039/d0an01706j, http://pubs.rsc.org/en/journals/journal/an.

Apetrei, C., & Ghasemi-Varnamkhasti, M. (2013). Biosensors in food PDO authentication. *Comprehensive Analytical Chemistry*, *60*. Available from https://doi.org/10.1016/B978-0-444--59562-1.00011--6, http://www.elsevier.com/wps/find/bookseriesdescription.cws_home/BS_CAC/description#description.

Ben Ayed, R., Hanana, M., & Khan, R. (2021). Artificial intelligence to improve the food and agriculture sector. *Journal of Food Quality*, 1–7. Available from https://doi.org/10.1155/2021/5584754, ***2021***.

Aznan, A., Gonzalez Viejo, C., Pang, A., & Fuentes, S. (2022a). Rapid detection of fraudulent rice using low-cost digital densing devices and machine learning. *Sensors*, *22*(22), 8655.

Aznan, A., Gonzalez Viejo, C., Pang, A., & Fuentes, S. (2022b). Rapid assessment of rice quality traits using low-cost digital technologies. *Foods*, *11*. Available from https://doi.org/10.3390/foods11091181.

Aznan, A., Gonzalez Viejo, C., Pang, A., & Fuentes, S. (2021). Computer vision and machine learning analysis of commercial rice grains: A potential digital approach for consumer perception studies. *Sensors*, *21*(19), 6354. Available from https://doi.org/10.3390/s21196354.

Bhatia, M., & Ahanger, T. A. (2021). Intelligent decision-making in smart food industry: Quality perspective. *Pervasive and Mobile Computing*, *72*, 101304. Available from https://doi.org/10.1016/j.pmcj.2020.101304.

Chalupowicz, D., Veltman, B., Droby, S., & Eltzov, E. (2020). Evaluating the use of biosensors for monitoring of Penicillium digitatum infection in citrus fruit. *Sensors and Actuators B: Chemical*, *311*, 127896. Available from https://doi.org/10.1016/j.snb.2020.127896.

Choudhury*, R., Hazarika, S., & Sarma, U. (2019). Detection of water stress in Khasi Mandarin orange plants from volatile organic compound emission profile implementing electronic nose. *International Journal of Engineering and Advanced Technology*, *9*(1), 133–137. Available from https://doi.org/10.35940/ijeat.a1086.109119.

Crawford, K. (2022). Atlas of AI: Power, politics, and the planetary costs of artificial intelligence. *Perspectives on Science and Christian Faith*, *74*(1), 61–62. Available from https://doi.org/10.56315/pscf3--22crawford.

Creydt, M., & Fischer, M. (2019). Blockchain and more - Algorithm driven food traceability. *Food Control*, *105*, 45–51. Available from https://doi.org/10.1016/j.foodcont.2019.050.019.

Csapó, J., Némethy, S., & Albert, C. (2019). Food counterfeiting in general; counterfeiting of milk and dairy products. *Ecocycles*, *5*(1), 26–41. Available from https://doi.org/10.19040/ecocycles.v5i10.138.

Cubero, S., Marco-Noales, E., Aleixos, N., Barbé, S., & Blasco, J. (2020). RobHortic: A field robot to detect pests and diseases in horticultural crops by proximal sensing. *Agriculture*, *10*(7), 276. Available from https://doi.org/10.3390/agriculture10070276.

Cui, S., Cao, L., Acosta, N., Zhu, H., & Ling, P. P. (2021). Development of portable e-nose system for fast diagnosis of whitefly infestation in tomato plant in greenhouse. *Chemosensors*, *9*(11). Available from https://doi.org/10.3390/chemosensors9110297, https://www.mdpi.com/2227-9040/9/11/297/pdf.

Cui, S., Inocente, E. A. A., Acosta, N., Keener, H. M., Zhu, H., & Ling, P. P. (2019). Development of fast e-nose system for early-stage diagnosis of aphid-stressed tomato plants. *Sensors (Switzerland)*, *19*(16). Available from https://doi.org/10.3390/s19163480, https://www.mdpi.com/1424-8220/19/16/3480/pdf.

Dac, H. H., Gonzalez Viejo, C., Lipovetzky, N., Tongson, E., Dunshea, F. R., & Fuentes, S. (2022). Livestock identification using deep learning for traceability. *Sensors*, *22*(21). Available from https://doi.org/10.3390/s22218256, http://www.mdpi.com/journal/sensors.

Di Vaio, A., Boccia, F., Landriani, L., & Palladino, R. (2020). Artificial intelligence in the agri-food system: Rethinking sustainable business models in the COVID-19 scenario. *Sustainability*, *12*(12), 4851.

Dhau, I., Adam, E., Mutanga, O., Ayisi, K., Abdel-Rahman, E. M., Odindi, J., & Masocha, M. (2018). Testing the capability of spectral resolution of the new multispectral sensors on detecting the severity of grey leaf spot disease in maize crop. *Geocarto International*, *33*(11), 1223–1236. Available from https://doi.org/10.1080/10106049.2017.1343391, http://www.tandfonline.com/toc/tgei20/current.

Farahmand, E., Razavi, S. H., & Mohtasebi, S. S. (2021). Investigating effective variables to produce desirable aroma in sourdough using e-nose and sensory panel. *Journal of Food Processing and Preservation*, *45*(2). Available from https://doi.org/10.1111/jfpp.15157, http://onlinelibrary.wiley.com/journal/10.1111/(ISSN)1745-4549.

Fathi, P., Karmakar, N. C., Bhattacharya, M., & Bhattacharya, S. (2020). Potential chipless RFID sensors for food packaging applications: A review. *IEEE Sensors Journal*, *20*(17), 9618–9636. Available from https://doi.org/10.1109/JSEN.2020.2991751, http://ieeexplore.ieee.org/xpl/RecentIssue.jsp?punumber = 7361.

Feng, H., Gonzalez Viejo, C., Vaghefi, N., Taylor, P. W. J., Tongson, E., & Fuentes, S. (2022). Early detection of *Fusarium oxysporum* infection in processing tomatoes (*Solanum lycopersicum*) and pathogen–soil interactions using a low-cost portable electronic nose and machine learning modeling. *Sensors*, *22*(22), 8645. Available from https://doi.org/10.3390/s22228645.

Fuentes, S., Gonzalez Viejo, C., Tongson, E., Lipovetzky, N., & Dunshea, F. R. (2021). Biometric physiological responses from dairy cows measured by visible remote sensing are good predictors of milk productivity and quality through artificial intelligence. *Sensors*, *21*(20), 6844.

Fuentes, S., & Gago, J. (2022). *Modern approaches to precision and digital viticulture. Improving sustainable viticulture and winemaking practices* (pp. 125–145). Australia: Elsevier. Available from https://www.sciencedirect.com/book/9780323851503, 10.1016/B978-0-323-85150-3.00015-3.

Fuentes, S., Gonzalez Viejo, C., Cullen, B., Tongson, E., Chauhan, S. S., & Dunshea, F. R. (2020). Artificial intelligence applied to a robotic dairy farm to model milk productivity and quality based on cow data and daily environmental parameters. *Sensors*, *20*(10), 2975.

Fuentes, S., Gonzalez Viejo, C., Tongson, E., & Dunshea, F. R. (2022). The livestock farming digital transformation: Implementation of new and emerging technologies using artificial intelligence. *Animal Health Research Reviews*, *23*(1), 59–71. Available from https://doi.org/10.1017/S1466252321000177, http://journals.cambridge.org/action/displayJournal?jid = ahr.

Fuentes, S., Summerson, V., Gonzalez Viejo, C., Tongson, E., Lipovetzky, N., Wilkinson, K. L., Szeto, C., & Unnithan, R. R. (2020). Assessment of smoke contamination in grapevine berries and taint in wines due to bushfires using a low-cost E-nose and an artificial intelligence approach. *Sensors*, *20*(18), 5108.

Fuentes, S., Tongson, E., Chen, J., & Gonzalez Viejo, C. (2020). A digital approach to evaluate the effect of berry cell death on pinot noir wines' quality traits and sensory profiles using non-destructive near-infrared spectroscopy. *Beverages*, *6*(2), 39.

Fuentes, S., Tongson, E., & Gonzalez Viejo, C. (2021). Novel digital technologies implemented in sensory science and consumer perception. *Current Opinion in Food Science*, *41*, 99–106. Available from https://doi.org/10.1016/j.cofs.2021.030.014, http://www.journals.elsevier.com/current-opinion-in-food-science/.

Fuentes, S., Tongson, E., Torrico, D. D., & Gonzalez Viejo, C. (2019). Modeling pinot noir aroma profiles based on weather and water management information using machine learning algorithms: A vertical vintage analysis using artificial intelligence. *Foods*, *9*(1), 33.

Fuentes, S., Tongson, E., Unnithan, R. R., & Gonzalez Viejo, C. (2021). Early detection of aphid infestation and insect-plant interaction assessment in wheat using a low-cost electronic nose (E-Nose), near-infrared spectroscopy and machine learning modeling. *Sensors*, *21*(17), 5948.

Fuentes, S., Chacon, G., Torrico, D. D., Zarate, A., & Gonzalez Viejo, C. (2019). Spatial variability of aroma profiles of cocoa trees obtained through computer vision and machine learning modelling: A cover photography and high spatial remote sensing application. *Sensors*, *19*(14), 3054. Available from https://doi.org/10.3390/s19143054.

Fuentes, S., Gonzalez Viejo, C., Hall, C., Tang, Y., & Tongson, E. (2021). Berry cell vitality assessment and the effect on wine sensory traits based on chemical fingerprinting, canopy architecture and machine learning modelling. *Sensors*, *21*(21), 7312. Available from https://doi.org/10.3390/s21217312.

Fuentes, S., Torrico, D. D., Tongson, E., & Gonzalez Viejo, C. (2020). Machine learning modeling of wine sensory profiles and color of vertical vintages of pinot noir based on chemical fingerprinting, weather and management data. *Sensors*, *20*(13), 3618. Available from https://doi.org/10.3390/s20133618.

Galvez, J. F., Mejuto, J. C., & Simal-Gandara, J. (2018). Future challenges on the use of blockchain for food traceability analysis. *TrAC - Trends in Analytical Chemistry*, *107*, 222–232. Available from https://doi.org/10.1016/j.trac.2018.080.011, http://www.elsevier.com/locate/trac.

Gebser, K., & Kaminski, Y. (2022). *Essentials of game theory: A concise multidisciplinary introduction. Springer Nature*.

Gonzalez Viejo, C., & Fuentes, S. (2020). A digital approach to model quality and sensory traits of beers fermented under sonication based on chemical fingerprinting. *Fermentation*, *6*(3). Available from https://doi.org/10.3390/FERMENTATION6030073, https://www.mdpi.com/2311-5637/6/3/73.

Gonzalez Viejo, C., & Fuentes, S. (2022a). Digital assessment and classification of wine faults using a low-cost electronic nose, near-infrared spectroscopy and machine learning modelling. *Sensors*, *22*(6). Available from https://doi.org/10.3390/s22062303, https://www.mdpi.com/1424-8220/22/6/2303/pdf.

Gonzalez Viejo, C., & Fuentes, S. (2022b). Digital detection of olive oil rancidity levels and aroma profiles using near-infrared spectroscopy, a low-cost electronic nose and machine learning modelling. *Chemosensors*, *10*(5), 159.

Gonzalez Viejo, C., Fuentes, S., Godbole, A., Widdicombe, B., & Unnithan, R. R. (2020). Development of a low-cost e-nose to assess aroma profiles: An artificial intelligence application to assess beer quality. *Sensors and Actuators B: Chemical*, *308*, 127688.

Gonzalez Viejo, C., Fuentes, S., & Hernandez-Brenes, C. (2021). Smart detection of faults in beers using near-infrared spectroscopy, a low-cost electronic nose and artificial intelligence. *Fermentation*, *7*(3), 117.

Gonzalez Viejo, C., Fuentes, S., Li, G. J., Collmann, R., Condé, B., & Torrico, D. (2016). Development of a robotic pourer constructed with ubiquitous materials, open hardware and sensors to assess beer foam quality using computer vision and pattern recognition algorithms: RoboBEER. *Food Research International*, *89*, 504–513. Available from https://doi.org/10.1016/j.foodres.2016.080.045, http://www.elsevier.com/inca/publications/store/4/2/2/9/7/0.

Gonzalez Viejo, C., Fuentes, S., Torrico, D., Lee, M., Hu, Y., Chakraborty, S., & Dunshea, F. (2018). The effect of soundwaves on foamability properties and sensory of beers with a machine learning modeling approach. *Beverages*, *4*(3), 53.

Gonzalez Viejo, C., Fuentes, S., Torrico, D. D., Howell, K., & Dunshea, F. R. (2018). Assessment of beer quality based on a robotic pourer, computer vision, and machine learning algorithms using commercial beers. *Journal of Food Science*, *83*(5), 1381–1388. Available from https://doi.org/10.1111/1750-3841.14114, http://www3.interscience.wiley.com/journal/118509799/issueyear?year = 2008.

Gonzalez Viejo, C., Gurr, P. A., Dunshea, F. R., & Fuentes, S. (2022). *Food packaging: The smarter WayNovel packaging development, assessment and authentication using smart technologies. Non-invasive biometric sensory tools and artificial intelligence*. Springer.

Gonzalez Viejo, C., Harris, N. M., & Fuentes, S. (2022). Quality traits of sourdough bread obtained by novel digital technologies and machine learning modelling. *Fermentation. Fermentation*, *8*(10), 516.

Gonzalez Viejo, C., Tongson, E., & Fuentes, S. (2021). Integrating a low-cost electronic nose and machine learning modelling to assess coffee aroma profile and intensity. *Sensors*, *21*(6), 2016.

Gonzalez Viejo, C., Torrico, D., Dunshea, F., & Fuentes, S. (2019a). Development of artificial neural network models to assess beer acceptability based on sensory properties using a robotic pourer: A comparative model approach to achieve an artificial intelligence system. *Beverages*, *5*(2), 33.

Gonzalez Viejo, C., Torrico, D. D., Dunshea, F. R., & Fuentes, S. (2019b). Emerging technologies based on artificial intelligence to assess the quality and consumer preference of beverages. *Beverages*, *5*(4). Available from https://doi.org/10.3390/beverages5040062, https://www.mdpi.com/2306-5710/5/4/62/pdf.

Gonzalez Viejo, C., Torrico, D. D., Dunshea, F. R., & Fuentes, S. (2019). The effect of sonication on bubble size and sensory perception of carbonated water to improve quality and consumer acceptability. *Beverages*, *5*(3), 58. Available from https://doi.org/10.3390/beverages5030058.

Hanif, S., Wijaya, D.R., & Wikusna, W. (2021). Indonesia K-nearest neighbors algorithm for prediction shelf life of rice based on electronic nose dataset. In *Proceedings - 2021 IEEE Asia Pacific conference on wireless and mobile (APWiMob)* (pp. 116–120). Institute of Electrical and Electronics Engineers Inc. doi:10.1109/APWiMob51111.2021.9435229. http://ieeexplore.ieee.org/xpl/mostRecentIssue.jsp?punumber = 9435193.

Harris, N., Gonzalez Viejo, C., Barnes, C., & Fuentes, S. (2023). Non-invasive digital technologies to assess wine quality traits and provenance through the bottle. *Fermentation*, *9*(1), 10. Available from https://doi.org/10.3390/fermentation9010010.

Hashemi Tameh, M., Primiceri, E., Chiriacò, M. S., Poltronieri, P., Bahar, M., & Maruccio, G. (2020). Pectobacterium atrosepticum biosensor for monitoring blackleg and soft rot disease of potato. *Biosensors*, *10*(6), 64. Available from https://doi.org/10.3390/bios10060064.

Henrichs, E., Noack, T., Pinzon Piedrahita, A. M., Salem, M. A., Stolz, J., & Krupitzer, C. (2022). Can a byte improve our bite? An analysis of digital twins in the food industry. *Sensors*, *22*(1), 115. Available from https://doi.org/10.3390/s22010115.

Herrero-Huerta, M., Rodriguez-Gonzalvez, P., & Rainey, K. M. (2020). Yield prediction by machine learning from UAS-based mulit-sensor data fusion in soybean. *Plant Methods*, *16*(1). Available from https://doi.org/10.1186/s13007-020-00620-6, http://www.plantmethods.com/.

Hilten, S., & Wolfert. (2022). 5G in agri-food-a review on current status, opportunities and challenges. *Computers and Electronics in Agriculture*, *201*, 107291.

Hinduja, H., Kekkar, S., Chouraisa, S., & Chakrapani, H. B. (2020). Industry 4.0: Digital twin and its industrial applications. *International Journal of Science Engineering and Technology.*, *8*(4).

Iqbal, J., Khan, Z. H., & Khalid, A. (2017). Prospects of robotics in food industry. *Food Science and Technology*, *37*(2), 159–165. Available from https://doi.org/10.1590/1678-457x.14616.

Irani, Z., Sharif, A. M., Lee, H., Aktas, E., Topaloğlu, Z., van't Wout, T., & Huda, S. (2018). Managing food security through food waste and loss: Small data to big data. *Computers and Operations Research*, *98*, 367–383. Available from https://doi.org/10.1016/j.cor.2017.100.007, http://www.elsevier.com/inca/publications/store/3/0/0/.

Jeevanandam, J., Agyei, D., Danquah, M. K., & Udenigwe, C. (2022). *Food quality monitoring through bioinformatics and big data* (pp. 733–744). Elsevier BV. Available from 10.1016/b978-0-323-91001-9.00036-0.

Jin, C., Bouzembrak, Y., Zhou, J., Liang, Q., van den Bulk, L. M., Gavai, A., Liu, N., van den Heuvel, L. J., Hoenderdaal, W., & Marvin, H. J. P. (2020). Big Data in food safety – A review. *Current Opinion in Food Science*, *36*, 24–32. Available from https://doi.org/10.1016/j.cofs.2020.110.006, http://www.journals.elsevier.com/current-opinion-in-food-science/.

Jofresa, R.S., & Xirau, M.T. (2019). Bellaterra (Cerdanyola del Vallès): UAB Research Park Smart Food.

Kakani, V., Nguyen, V. H., Kumar, B. P., Kim, H., & Pasupuleti, V. R. (2020). A critical review on computer vision and artificial intelligence in food industry. *Journal of Agriculture and Food Research*, *2*. Available from https://doi.org/10.1016/j.jafr.2020.100033, http://www.journals.elsevier.com/journal-of-agriculture-and-food-research.

Khan, Z. H., Khalid, A., & Iqbal, J. (2018). Towards realizing robotic potential in future intelligent food manufacturing systems. *Innovative Food Science and Emerging Technologies*, *48*, 11–24. Available from https://doi.org/10.1016/j.ifset.2018.050.011, http://www.elsevier.com/locate/ifset.

Krupitzer, C., & Stein, A. (2021). Food informatics—Review of the current state-of-the-art, revised definition, and classification into the research landscape. *Foods*, *10*(11), 2889.

Kundu, M., Prasad, S., Krishnan, P., & Gajjala, S. (2019). A novel electrochemical biosensor based on hematite (α-Fe2O3) flowerlike nanostructures for sensitive determination of formaldehyde adulteration in fruit juices. *Food and Bioprocess Technology*, *12*(10), 1659–1671. Available from https://doi.org/10.1007/s11947-019-02318-7, http://www.springer.com/life + sci/food + science/journal/11947.

Lin, H., Man, Z. x, Kang, W. C., Guan, B. b, Chen, Q. s, & Xue, Z. l (2018). A novel colorimetric sensor array based on boron-dipyrromethene dyes for monitoring the storage time of rice. *Food Chemistry*, *268*, 300–306. Available from https://doi.org/10.1016/j.foodchem.2018.060.097, http://www.elsevier.com/locate/foodchem.

Lin, L., He, Y., Xiao, Z., Zhao, K., Dong, T., & Nie, P. (2019). Rapid-detection sensor for rice grain moisture based on NIR spectroscopy. *Applied Sciences*, *9*(8), 1654. Available from https://doi.org/10.3390/app9081654.

Liska, A. (2022). *Counterfeit wine, spirits, and cheese*. https://www.recordedfuture.com/counterfeit-wine-spirits-cheese.

Lu, Y., & Li, W. (2022). Techniques and paradigms in modern game AI systems. *Algorithms*, *15*(8), 282. Available from https://doi.org/10.3390/a15080282.

Mahdad, M., Hasanov, M., Isakhanyan, G., & Dolfsma, W. (2022). A smart web of firms, farms and internet of things (IOT): Enabling collaboration-based business models in the agri-food industry. *British Food Journal*, *124*(6), 1857–1874. Available from https://doi.org/10.1108/BFJ-07-2021-0756, https://www.emeraldinsight.com/loi/bfj.

Mateo-Sanchis, A., Piles, M., Muñoz-Marí, J., Adsuara, J. E., Pérez-Suay, A., & Camps-Valls, G. (2019). Synergistic integration of optical and microwave satellite data for crop yield estimation. *Remote Sensing of Environment*, *234*, 111460. Available from https://doi.org/10.1016/j.rse.2019.111460.

Mirghafouri, M. R., Abbasi-Moayed, S., Ghasemi, F., & Hormozi-Nezhad, M. R. (2020). Nanoplasmonic sensor array for the detection and discrimination of pesticide residues in citrus fruits. *Analytical Methods*, *12*(48), 5877–5884. Available from https://doi.org/10.1039/d0ay02039g, http://pubs.rsc.org/en/journals/journal/ay.

Misra, N. N., Dixit, Y., Al-Mallahi, A., Bhullar, M. S., Upadhyay, R., & Martynenko, A. (2022). IoT, Big Data, and artificial intelligence in agriculture and food industry. *IEEE Internet of Things Journal*, *9*(9), 6305–6324. Available from https://doi.org/10.1109/JIOT.2020.2998584, http://ieeexplore.ieee.org/servlet/opac?punumber = 6488907.

Monteiro, J., & Barata, J. (2021). Portugal artificial intelligence in extended agri-food supply chain: A short review based on bibliometric analysis. *Procedia Computer Science*, *192*, 3020–3029. Available from https://doi.org/10.1016/j.procs.2021.09.074.18770509, http://www.sciencedirect.com/science/journal/18770509192.

Mustafa, F., & Andreescu, S. (2018). Chemical and biological sensors for food-quality monitoring and smart packaging. *Foods*, *7*(10), 168. Available from https://doi.org/10.3390/foods7100168.

Mutz, Y. S., do Rosario, D., Silva, L. R. G., Galvan, D., Stefano, J. S., Janegitz, B. C., Weitz, D. A., Bernardes, P. C., & Conte-Junior, C. A. (2023). Lab-made 3D printed electrochemical sensors coupled with chemometrics for Brazilian coffee authentication. *Food Chemistry*, *403*. Available from https://doi.org/10.1016/j.foodchem.2022.134411, http://www.elsevier.com/locate/foodchem.

Nachal, N., Moses, J. A., Karthik, P., & Anandharamakrishnan, C. (2019). Applications of 3D Printing in food processing. *Food Engineering Reviews*, *11*(3), 123–141. Available from https://doi.org/10.1007/s12393-019-09199-8, http://www.springer.com/life + sci/food + science/journal/12393.

Nategh, N. A., Dalvand, M. J., & Anvar, A. (2021). Detection of toxic and non-toxic sweet cherries at different degrees of maturity using an electronic nose. *Journal of Food Measurement and Characterization*, *15*(2), 1213–1224. Available from https://doi.org/10.1007/s11694-020-00724-6, http://rd.springer.com/journal/11694.

Nayik, G. A., Muzaffar, K., & Gull, A. (2015). Robotics and food technology: A mini review. *Journal of Nutrition Food Sciences*, *5*, 1–11.

Necochea-Chamorro, J. I., Carrillo-Torres, R. C., Sánchez-Zeferino, R., & Álvarez-Ramos, M. E. (2019). Fiber optic sensor using ZnO for detection of adulterated tequila with methanol. *Optical Fiber Technology*, *52*. Available from https://doi.org/10.1016/j.yofte.2019.101982, http://www.journals.elsevier.com/optical-fiber-technology/.

Noguera, M., Millán, B., Pérez-Paredes, J. J., Ponce, J. M., Aquino, A., & Andújar, J. M. (2020). A new low-cost device based on thermal infrared sensors for olive tree canopy temperature measurement and water status monitoring. *Remote Sensing*, *12*(4). Available from https://doi.org/10.3390/rs12040723, https://res.mdpi.com/d_attachment/remotesensing/remotesensing-12-00723/article_deploy/remotesensing-12-00723-v2.pdf.

Paulo, R. L. D., Garcia, A. P., Umezu, C. K., Camargo, A. P. D., Soares, F. T., & Albeiro, D. (2023). Water stress index detection using a low-cost infrared sensor and excess green image processing. *Sensors*, *23*(3), 1318.

Paup, V. D., Barnett, S. M., Diako, C., & Ross, C. F. (2019). Detection of spicy compounds using the electronic tongue. *Journal of Food Science*, *84*(9), 2619–2627. Available from https://doi.org/10.1111/1750-3841.14709, http://onlinelibrary.wiley.com/journal/10.1111/(ISSN)1750-3841.

Ramírez, H. L., Soriano, A., Gómez, S., Iranzo, J. U., & Briones, A. I. (2018). Evaluation of the food sniffer electronic nose for assessing the shelf life of fresh pork meat compared to physicochemical measurements of meat quality. *European Food Research and Technology*, *244*(6), 1047–1055. Available from https://doi.org/10.1007/s00217-017-3021-0, http://link.springer.de/link/service/journals/00217/index.htm.

Raymundo-Pereira, P. A., Gomes, N. O., Carvalho, J. H. S., Machado, S. A. S., Oliveira, O. N., & Janegitz, B. C. (2020). Simultaneous detection of quercetin and carbendazim in wine samples using disposable electrochemical sensors. *ChemElectroChem*, *7*(14), 3074–3081. Available from https://doi.org/10.1002/celc.202000788, http://onlinelibrary.wiley.com/journal/10.1002/(ISSN)2196-0216.

Rejeb, A., Keogh, J. G., & Rejeb, K. (2022). Big data in the food supply chain: A literature review. *Journal of Data, Information and Management*, *4*(1), 33–47. Available from https://doi.org/10.1007/s42488-021-00064-0.

Renda, A. (2019). *The age of foodtech: Optimizing the agri-food chain with digital technologies. Achieving the sustainable development goals through sustainable food systems* (pp. 171–187). Belgium: Springer International Publishing. Available from http://www.springer.com/in/book/9783030239688, 10.1007/978-3-030-23969-5_10.

Rochewerger, A. S., & Pang, W. (2021). *Real world AI: A practical guide for responsible machine learning*. Lioncrest Publishing.

Ruan, G., Li, X., Yuan, F., Cammarano, D., Ata-UI-Karim, S. T., Liu, X., Tian, Y., Zhu, Y., Cao, W., & Cao, Q. (2022). Improving wheat yield prediction integrating proximal sensing and weather data with machine learning. *Computers and Electronics in Agriculture*, *195*, 106852. Available from https://doi.org/10.1016/j.compag.2022.106852.

Sadiku, M. N. O., Ashaolu, T. J., Ajayi-Majebi, A., & Musa, S. M. (2020). Big Data in food industry. *International Journal Of Scientific Advances*, *1*(3). Available from https://doi.org/10.51542/ijscia.v1i3.5.

Sahoo, D., Hao, W., Ke, S., Xiongwei, W., Le, H., Achananuparp, P., Lim, E.P., & Hoi, S.C. H. (2019). FoodAI: Food image recognition via deep learning for smart food logging. In *Proceedings of the ACM SIGKDD international conference on knowledge discovery and data mining* (pp. 2260–2268). Association for Computing Machinery. doi:10.1145/3292500.3330734.

Salamanca-Neto, C. A. R., Marcheafave, G. G., Scremin, J., Barbosa, E. C. M., Camargo, P. H. C., Dekker, R. F. H., Scarminio, I. S., Barbosa-Dekker, A. M., & Sartori, E. R. (2020). Chemometric-assisted construction of a biosensing device to measure chlorogenic acid content in brewed coffee beverages to discriminate quality. *Food Chemistry*, *315*. Available from https://doi.org/10.1016/j.foodchem.2020.126306, http://www.elsevier.com/locate/foodchem.

Sanaeifar, A., Li, X., He, Y., Huang, Z., & Zhan, Z. (2021). A data fusion approach on confocal Raman microspectroscopy and electronic nose for quantitative evaluation of pesticide residue in tea. *Biosystems Engineering*, *210*, 206–222. Available from https://doi.org/10.1016/j.biosystemseng.2021.080.016, http://www.elsevier.com/inca/publications/store/6/2/2/7/9/5/index.htt.

Sarno, R., Triyana, K., Sabilla, S. I., Wijaya, D. R., Sunaryono, D., & Fatichah, C. (2020). Detecting pork adulteration in beef for halal authentication using an optimized electronic nose system. *IEEE Access*, *8*, 221700–221711. Available from https://doi.org/10.1109/access.2020.3043394.

Schlossareck, C., & Ross, C. F. (2019). Electronic tongue and consumer sensory evaluation of spicy paneer Cheese. *Journal of Food Science*, *84*(6), 1563–1569. Available from https://doi.org/10.1111/1750-3841.14604, http://onlinelibrary.wiley.com/journal/10.1111/(ISSN)1750-3841.

Shin, M. Y., Gonzalez Viejo, C., Tongson, E., Wiechel, T., Taylor, P. W. J., & Fuentes, S. (2023). Early detection of verticillium wilt of potatoes using near-infrared spectroscopy and machine learning modeling. *Computers and Electronics in Agriculture*, *204*. Available from https://doi.org/10.1016/j.compag.2022.107567, http://www.elsevier.com/inca/publications/store/5/0/3/3/0/4.

Shoute, L. C. T., Anwar, A., MacKay, S., Abdelrasoul, G. N., Lin, D., Yan, Z., Nguyen, A. H., McDermott, M. T., Shah, M. A., Yang, J., Chen, J., & Li, X. S. (2018). Immuno-impedimetric biosensor for onsite monitoring of ascospores and forecasting of sclerotinia stem rot of canola. *Scientific Reports*, *8*(1). Available from https://doi.org/10.1038/s41598-018-30167-5, http://www.nature.com/srep/index.html.

Stredansky, M., Redivo, L., Magdolen, P., Stredansky, A., & Navarini, L. (2018). Rapid sucrose monitoring in green coffee samples using multienzymatic biosensor. *Food Chemistry*, *254*, 8–12. Available from https://doi.org/10.1016/j.foodchem.2018.010.171, http://www.elsevier.com/locate/foodchem.

Summerson, V., Gonzalez Viejo, C., Torrico, D. D., Pang, A., & Fuentes, S. (2021). Digital smoke taint detection in pinot grigio wines using an e-nose and machine learning algorithms following treatment with activated carbon and a cleaving enzyme. *Fermentation*, *7*(3). Available from https://doi.org/10.3390/fermentation7030119, https://www.mdpi.com/2311-5637/7/3/119/pdf.

Summerson, V., Viejo, C. G., Torrico, D. D., Pang, A., & Fuentes, S. (2020). Detection of smoke-derived compounds from bushfires in cabernet-sauvignon grapes, must, and wine using near-infrared spectroscopy and machine learning algorithms. *Oeno One*, *54*(4), 1105–1119. Available from https://doi.org/10.20870/OENO-ONE.2020.54.4.4501, https://oeno-one.eu/article/view/4501.

Sun, Y., Wang, J., Cheng, S., & Wang, Y. (2019). Detection of pest species with different ratios in tea plant based on electronic nose. *Annals of Applied Biology*, *174*(2), 209–218. Available from https://doi.org/10.1111/aab.12485, http://onlinelibrary.wiley.com/journal/10.1111/(ISSN)1744-7348.

Sun, Y., Wang, J., Sun, L., Cheng, S., & Xiao, Q. (2019). Evaluation of E-nose data analyses for discrimination of tea plants with different damage types. *Journal of Plant Diseases and Protection*, *126*(1), 29–38. Available from https://doi.org/10.1007/s41348-018-0193-1, http://link.springer.com/journal/41348.

Tebaldi, L., Vignali, G., & Bottani, E. (2021). Digital twin in the agri-food supply chain: A literature review. In *IFIP advances in information and communication technology* 1868422X (pp. 276–283). Springer Science and Business Media Deutschland GmbH, Italy. doi:10.1007/978-3-030-85910-7_29. http://www.springer.com/series/6102633.

Thakur, M. S., & Ragavan, K. V. (2013). Biosensors in food processing. *Journal of Food Science and Technology*, *50*(4), 625–641. Available from https://doi.org/10.1007/s13197-012-0783-z.

Tortorella, M. M., Di Leo, S., Cosmi, C., Fortes, P., Viccaro, M., Cozzi, M., Pietrapertosa, F., Salvia, M., & Romano, S. (2020). A methodological integrated approach to analyse climate change effects in agri-food sector: The TIMES water-energy-food module. *International Journal of Environmental Research and Public Health*, *17*(21), 1–21. Available from https://doi.org/10.3390/ijerph17217703, https://www.mdpi.com/1660-4601/17/21/7703/pdf.

Vasconcelos, H., Matias, A., Mendes, J., Araújo, J., Dias, B., Jorge, P.A. S., Saraiva, C., Coelho, L.C. C., & De Almeida, J.M. M.M. (2022). Optical biosensor for the detection of low concentrations of hydrogen peroxide in milk samples. In *Proceedings of SPIE – The International Society for Optical Engineering*, 1996756X SPIE 12139, Portugal. doi:10.1117/12.2621552. http://spie.org/x1848.xml.

Verboven, P., Defraeye, T., Datta, A. K., & Nicolai, B. (2020). Digital twins of food process operations: The next step for food process models? *Current Opinion in Food Science*, *35*, 79–87. Available from https://doi.org/10.1016/j.cofs.2020.030.002, http://www.journals.elsevier.com/current-opinion-in-food-science/.

Vågsholm, I., Arzoomand, N. S., & Boqvist, S. (2020). Food security, safety, and sustainability—getting the trade-offs right. *Frontiers in Sustainable Food Systems*, *4*. Available from https://doi.org/10.3389/fsufs.2020.00016, http://www.frontiersin.org/journals/sustainable-food-systems#.

Wu, W., Zhang, A., van Klinken, R. D., Schrobback, P., & Muller, J. M. (2021). Consumer trust in food and the food system: A critical review. *Foods*, *10*(10). Available from https://doi.org/10.3390/foods10102490, https://www.mdpi.com/2304-8158/10/10/2490/pdf.

Yadav, S., Choi, T. M., Luthra, S., Kumar, A., & Garg, D. (2022). Using internet of things (IoT) in agri-food supply chains: A research framework for social good with network clustering analysis. *IEEE Transactions on Engineering Management*, 1–10. Available from https://doi.org/10.1109/TEM.2022.3177188, https://ieeexplore.ieee.org/servlet/opac?punumber = 17.

Yadav, S., Luthra, S., & Garg, D. (2022). Internet of things (IoT) based coordination system in agri-food supply chain: Development of an efficient framework using DEMATEL-ISM. *Operations Management Research*, *15*(1–2), 1–27. Available from https://doi.org/10.1007/s12063-020-00164-x, http://www.springer.com/business/production/journal/12063.

Yang, F., Zhang, M., & Bhandari, B. (2017). Recent development in 3D food printing. *Critical Reviews in Food Science and Nutrition*, *57*(14), 3145–3153. Available from https://doi.org/10.1080/10408398.2015.1094732, http://www.tandf.co.uk/journals/titles/10408398.asp.

Yiannas, F. (2018). A new era of food transparency powered by blockchain. *Innovations: Technology, Governance, Globalization*, *12*(1–2), 46–56. Available from https://doi.org/10.1162/inov_a_00266.

Yiying, Z., Yuanlong, R., Fei, L., Jing, S., & Song, L. (2019). Research on meat food traceability system based on RFID technology. In *Proceedings of 2019 IEEE 3rd information technology, networking, electronic and automation control conference, ITNEC 2019* (pp. 2172–2175). Institute of Electrical and Electronics Engineers Inc., China. doi:10.1109/ITNEC.2019.8729408. http://ieeexplore.ieee.org/xpl/mostRecentIssue.jsp?punumber = 8717364.

Young, M. (2022). *Counterfeiting is a lingering stain on the wine industry*. https://wineindustryadvisor.com/2022/02/08/counterfeit-wine-investigations#:~:text = A%20European%20Union%20Intellectual%20Property,%242.61%20billion%20in%20tax%20revenues.

Zhou, L., Zhang, C., Liu, F., Qiu, Z., & He, Y. (2019). Application of deep learning in food: A review. *Comprehensive Reviews in Food Science and Food Safety*, *18*(6), 1793–1811. Available from https://doi.org/10.1111/1541-4337.12492, http://onlinelibrary.wiley.com/journal/10.1111/(ISSN)1541-4337.

Zhou, Q., Zhang, H., & Wang, S. (2022). Artificial intelligence, big data, and blockchain in food safety. *International Journal of Food Engineering*, *18*(1), 1–14. Available from https://doi.org/10.1515/ijfe-2021-0299, http://www.degruyter.com/view/j/ijfe.

Zovko, M., Žibrat, U., Knapič, M., Kovačić, M. B., & Romić, D. (2019). Hyperspectral remote sensing of grapevine drought stress. *Precision Agriculture*, *20*(2), 335–347. Available from https://doi.org/10.1007/s11119-019-09640-2, http://www.wkap.nl/journalhome.htm/1385-2256.

Chapter 10

Portability of miniaturized food analytical systems 4.0

Justyna Grabska, Krzysztof B. Beć and Christian W. Huck
Institute of Analytical Chemistry and Radiochemistry, University of Innsbruck, Innsbruck, Austria

10.1 Introduction

Near-infrared (NIR) spectroscopy (800–2500 nm; 12,500–4000 cm^{-1}) has been widely adopted in various fields due to its nondestructive analytical capabilities and its ability to provide simultaneous qualitative and quantitative characterization of a wide variety of samples. This tool has been particularly effective in analyzing whole foods and processed products and their components in industries such as the agrifood sector, pharmaceutical industry, and environmental monitoring (Beć et al., 2020, 2021a, 2022a, 2022b, 2023; Chapman et al., 2020, Cozzolino, 2016a, 2016b; Grabska et al., 2022a; Roberts et al., 2018). Industry 4.0 represents the latest stage of the Industrial Revolution, characterized by the wide-spreading autonomous "smart" devices and big data analytics. This new phase in industrial development brings opportunities for enhanced efficiency, cost savings, and the development of new products and services to the realm of instrumental analytical chemistry. With the miniaturization of NIR spectrometers, portable and handheld instruments offer several advantages for the food industry, including ease of use, affordability, and flexibility in a variety of analytical scenarios. These characteristics make miniaturized food analytical systems based on NIR spectroscopy a prime example of the integration of modern technologies of Industry 4.0, enabling the optimization of manufacturing processes, increased productivity, and more reliable product quality. However, the miniaturized NIR instruments face challenges when analyzing chemically complex and inhomogeneous samples, which are common in the food industry. As a result, there is ongoing research to evaluate the applicability and performance of these instruments to ensure their reliability in practical applications (Beć et al., 2022c; Grabska et al., 2022b; McVey, Elliott, et al., 2021; Müller-Maatsch et al., 2021).

For the past 40 years, NIR spectroscopy has become a particularly popular method for food analysis and quality control, as it provides key advantages in such applications (Cozzolino, 2022; Ozaki et al., 2018, 2021). In fact, the evolution of NIR spectroscopy has predominantly been driven by the needs of the agrifood sector, and the technique has undergone several advancements, which originated in this field of applications to later become adopted universally. Starting with the first analytical applications of NIR spectroscopy in the early 1950s, agrifood applications have remained the prime driver for the advancement of this technique, from the instrument design, through sampling strategies, to the data handling methods such as the development of the chemometric framework firstly by means of linear and multiple linear regression methods and then pioneering the standards still applicable nowadays, for example, partial least squares regression (PLSR).

In the regime of the instrumentation, the introduction of Fourier-transform (FT)-NIR spectrometers in the early 1990s should be seen as a major breakthrough (Ozaki, 2022). The development of portable and handheld instruments in the 2000s marked another key milestone, as it enabled analysis to be conducted on-site, bringing significant benefits to the agrifood sector (Beć et al., 2020, 2021b, 2018). In the past decade, portable NIR spectrometers have become essential tools in the food industry for monitoring raw and shelf product quality and ensuring food safety. With the advancement of miniaturization technology, these instruments can now be easily transported to production facilities and farms for on-site analysis. This not only saves time but also reduces transportation costs and eliminates the need for samples to be sent to laboratories for testing. This has been especially important in remote or resource-limited areas, where access to analytical laboratories may be limited.

Food Industry 4.0. DOI: https://doi.org/10.1016/B978-0-443-15516-1.00010-4

However, factors affecting the analytical performance of the portable systems in different scenarios need to be considered. Currently, there is a strong emphasis on evaluating the accuracy and reliability of various miniaturized spectrometers available in the market (Beć et al., 2020, 2021b; Crocombe, 2018). The significance of combining various tools into NIR analysis to improve its accuracy and reliability is emphasized. With the increasing volume of data generated by these instruments, new approaches for data analysis, implementing new developments in the field of machine learning and artificial intelligence, can be used to develop more robust predictive models and enhance the reliability of the analysis. Advanced calibration techniques, such as artificial neural networks (ANNs), enhance the performance of miniaturized instruments for analyzing complex samples, bringing their accuracy in line with benchtop spectrometers. The complexity of NIR spectra, particularly in the case of multiconstituent samples, is improved with the aid of theoretical simulations of NIR spectra, which provide a deeper understanding of the chemical information in the processed data-set and help interpret the differences between portable sensors based on distinct technologies. The effectiveness of sensor fusion approaches can also be noted, while calibration transfer allows sharing of trained calibration models between NIR systems.

The use of miniaturized NIR spectroscopy in food-related contexts addresses a range of issues where optimized analytical pathways are crucial for maximum results. This chapter examines the current state of the art, remaining challenges, and necessary research trends and evaluates the key factors that drive the progress of MicroNIR analysis for modern food analysis, quality control, and safety monitoring.

10.2 Modern NIR instrumentation—toward sensor ultra-miniaturization and integration

The distinction between stationary benchtops and handheld NIR spectrometers is crucial in understanding the specifics of the analytical frameworks involving miniaturized NIR sensors. Benchtop spectrometers, well-established instruments, typically use either a Michelson or a polarization interferometer, acquiring spectra based on the Fourier-transform principle. Conversely, handheld and miniaturized NIR instruments, employ various of optical and engineering principles, resulting in diversity in their operational characteristics, cost-effectiveness, and practicality (Beć et al., 2021b).

10.2.1 Design scheme of a miniaturized NIR spectrometer

10.2.1.1 Radiation source

Tungsten halogen incandescent lamps commonly serve as light sources in both benchtop and miniaturized NIR spectrometers. The emission profile of this source makes it well suited for the NIR region, providing a bright and stable light source. However, in miniaturized spectrometers, the energy efficiency and physical dimensions of the source must be optimized and its thermal stability is more challenging to be maintained, especially during field operation. In some designs, the source is supplied with power only during measurement to minimize heating, or a temperature correction function is implemented in the spectrometer software. LED sources offer low power consumption, compactness, and durability, but currently only emit within a narrow wavelength range that only partially covers the NIR region. Thus they are more suitable for use in visible/short-wavelength NIR (Vis/SW-NIR) spectrometers where compactness and cost-effectiveness are important (Beć et al., 2021b, 2022c).

10.2.1.2 Wavelength selector

The wavelength selection principle is a critical characteristic of a spectrometer and determines its design and operating parameters. Interferometer-based design is commonly used in benchtop spectrometers but is difficult to miniaturize for handheld devices. Therefore there is a diversity of designs present in the market and no uniform "best" concept has been established. Portable NIR spectrometers use various wavelength selectors, including Fabry–Pérot interferometer, Hadamard mask, or linear variable filter (LVF) integrated with an array detector. Microelectromechanical systems (MEMS) technology is used to miniaturize wavelength selectors, for example, digital micromirror device (DMD) or programmable Hadamard mask. The DMD is a type of wavelength selector using an array of mirrors made with MEMS technology that forms the wavelength-scanning element. It offers advantages in terms of mechanical robustness, size, and cost-effectiveness compared to traditional optical spectrometers. The principle of Hadamard transform has been implemented in multiple handheld NIR spectrometers and has practical advantages such as multiplex, frequency accuracy, and optical throughput advantages, with no reliance on moving parts. The Hadamard transform (HT) is performed using a programmable Hadamard mask implemented via MEMS technology. The Fabry–Pérot interferometer is another miniaturized wavelength selector that works based on two parallel mirrors separated by a constant or variable distance. It can be reconfigured for different spectral ranges and is possible to be made with MEMS technology as well (Beć et al., 2020, 2021b).

On the other hand, the FT-NIR principle has also been implemented in compact NIR instruments via Michelson interferometer and offers wide operational spectral regions and superior resolution. An alternative concept is an array detector with multiple independent photosensitive elements, which is expensive but efficient as a multichannel spectrometer. The multichannel sensor based on LVF is one of such solutions, featuring high optical performance and ultra-short "scanning" (i.e., integration interval) time in the range of microseconds. With no movement involved in the spectrometer, it is possible to acquire a spectrum consisting of thousands of averaged scans in a matter of seconds. The LVF element is very thin and allows the construction of short-path-length instruments with high optical performance (Beć et al., 2020, 2021b).

10.2.1.3 Detector

The miniaturized NIR spectrometers typically use indium gallium arsenide (InGaAs) or extended InGaAs detectors for adequate signal-to-noise ratio due to their high sensitivity, especially in the longer wavelengths of the NIR region. The typical wavelength range for optimal performance of InGaAs detectors is around 1000–1600 nm, but they can also perform well at adjacent wavelengths. They offer a fast response time, good quantum efficiency, and low dark current. The extended InGaAs detector is suitable for instruments operating at shorter wavelengths, but it may require cooling solutions and has lower sensitivity. On the other hand, photovoltaic Si detectors are suitable for cost-effective, compact spectrometers operating only in the Vis and SW-NIR region, but they require a wavelength-blocking filter to mask the detector from sunlight/stray light. These detectors are particularly suitable for spectrometers aimed at the consumer market due to their affordability (Beć et al., 2021b, 2022c).

10.2.1.4 Other elements

The main functional elements of a miniaturized spectrometer are supplemented by optical elements, power supply, control electronics, communication interface, and/or user interface. The optics include, for example, sample interface/detector window, and internal collimating lenses. Glass optics are compatible with NIR spectrometers and are cheap, and mechanically and chemically resistant, but high-quality fused silica may be required for the best performance in the long-wavelength part of the NIR region. Portable spectrometers may acquire power from an external source or be equipped with their own battery and have connectivity via USB or wireless (Wi-Fi, Bluetooth). The user control over the instrument can be achieved through a display screen and user-input device or externally through a PC or a smartphone application. Among consumer-oriented devices and analyzers intended for operation by nonexpert personnel, popular is becoming the scheme of operation in which the measured spectral data is uploaded to the cloud service and is subjected to automated analysis using precalibrated models. The end-user receives only the ultimate outcome of such analysis, reducing the complexity of NIR analysis to a "turn-key" scheme (Beć et al., 2022c).

10.2.2 Brief description of technical parameters of selected miniaturized NIR spectrometers

There are several companies that have developed handheld NIR instruments based on different principles and technologies. One of the first handheld NIR instruments was introduced by Polychromix, which is now the intellectual property of Thermo Fisher Scientific Inc. It employed a programmable microscale MEMS-based Hadamard mask, a low-power tungsten lamp source, and an InGaAs single-element detector, making it a robust, reliable, and reasonably compact instrument. Several successful products in the market are based on the NIRscan Digital Light Processor module from Texas Instruments. This solution is based on a DMD element manufactured with MEMS technology and is available in two variants: the High Performance and the Mobile Sensing Evaluation Modules. Spectral Engines implemented a Fabry–Pérot interferometer in a miniaturized NIR spectrometer NIRONE Sensor S is available in different variants with distinct operational wavelength ranges, signal-to-noise ratio (S/N), and resolution parameters while maintaining very compact dimensions. Hamamatsu has also recently unveiled a series of ultra-compact NIR sensors with narrow operational spectral regions, ranging from 1350 to 2150 nm. These advances in NIR spectrometer technology show promise for ultra-miniaturization in the future. Companies like Si-Ware Systems and Hefei SouthNest Technology offer compact FT-NIR spectrometers, such as the NeoSpectra device and the nanoFTIR NIR spectrometer, respectively (Beć et al., 2020, 2021b). These spectrometers are designed to be used in a variety of applications, ranging from online analysis to feed and forage assessment in the agrifood industry. In addition to the more traditional spectrometers, there is also a growing market for consumer-oriented NIR spectrometers, often referred to as "pocket food analyzers." These devices typically offer a cost-effective trade-off between performance and affordability, with the primary value being derived from the associated software and cloud services. One example of such a device is the Consumer Physics SCiO NIR microspectrometer, a compact unit that is intended for everyday consumer use in assessing food quality and nutritional value.

On the other hand, it is worth noting that many spectrometers are offered in a specialized variant as "turn-key analyzers." These devices are preconfigured for specific analyses and come equipped with a specialized software suite containing spectra-processing algorithms and precalibrated models for quantitative and qualitative analysis. These analyzers are preconfigured for a variety of applications, including animal feed analysis, plastics analysis in recycling, pharmaceutical analysis, and asbestos analysis. Finally, the idea of an NIR sensor fully integrated with a smartphone remains a concept that has been anticipated for over a decade (Michael, 2018; Reinig et al., 2018). While initial prototypes have shown promising results, no such solution has yet appeared on the market. However, recent advances in ultra-miniaturization, such as those demonstrated by Hamamatsu's MEMS-FPI spectrum sensors, suggest that this could change in the near future (McGonigle et al., 2018).

In Table 10.1, there are summarized characteristics parameters for the selected cases of benchtop and miniaturized spectrometers.

10.3 The versatile potential of miniaturized NIR spectrometers in the agrifood industry: an overview of applications

NIR spectroscopy has emerged as one of the most versatile analytical techniques with numerous applications in the food industry (Muller-Maatsch & van Ruth, 2021; Rodriguez-Saona et al., 2020). It is considered as a valuable tool for food quality control and analysis due to its ability to provide rapid and reliable results on many relevant quality parameters of foodstuff. The advent of miniaturized NIR spectrometers has particularly revolutionized the field of agriculture and food analysis by making it more portable and accessible (Beć et al., 2022c; Grabska et al., 2022b; McVey, Elliott, et al., 2021; Müller-Maatsch et al., 2021). The portability of these instruments enhances the conventional strengths of the NIR analytical framework and fits well with the challenges and specifics of food-related analysis. This is particularly important given the complex nature of the food production and delivery chain and the susceptibility of foods to quality loss, which requires a flexible analytical tool for monitoring and maintaining food quality (Beć et al., 2022c; Grabska et al., 2022b; McVey, Elliott, et al., 2021; Müller-Maatsch et al., 2021).

The use of portable NIR systems in the food industry displays high promise, as demonstrated by the wide range of applications for these instruments. The current landscape of miniaturized NIR spectroscopy in food analysis underscores the versatility and effectiveness of handheld instruments in solving a variety of problems (Abu-Khalaf & Hmidat, 2020; Amuah et al., 2019; Anyidoho et al., 2020; Aykas & Menevseoglu, 2021; Baca-Bocanegra et al., 2019; Barthet et al., 2020; Basri et al., 2017, 2018; Beć et al., 2021; Borba et al., 2021; Borghi et al., 2020; Cáceres-Nevado et al., 2021; Castrignanò et al., 2019; Coronel-Reyes et al., 2018; Correia et al., 2018; da Costa Fulgencio et al., 2023; Dashti et al., 2022; de la Roza-Delgado et al., 2017; de Lima et al., 2018; Dixit et al., 2020; dos Santos Pereira et al., 2021; Dumalisile et al., 2020; Edwards et al., 2020; Eskildsen et al., 2019; Giussani et al., 2021; Goi et al., 2022; Gonçalves et al., 2021; Grassi et al., 2018; Guelpa et al., 2017; Hao et al., 2019; Hasanzadeh et al., 2022; Henn et al., 2016, 2017; Horcada et al., 2020; Jiang et al., 2019; Karunathilaka et al., 2017, 2018; Kosmowski & Worku, 2018; Kucha & Ngadi, 2020; Lee et al., 2017; Liu, Aya Parra, et al., 2018; Liu, Parra, et al., 2018; Llano Suarez et al., 2018; Ma, Babu, et al., 2019; Ma, Zhang, et al., 2019; MacArthur et al., 2020; Malegori et al., 2017; Malvandi et al., 2022; Manuelian et al., 2022; Marinoni et al., 2017; Mayr, Beć, et al., 2021; Mayr, Schmelzer, et al., 2021; McGrath et al., 2020; McVey, Gordon, et al., 2021; McVey, McGrath, et al., 2021; Modroño et al., 2017; Muniz et al., 2020; Musa, 2022; Neves et al., 2019; Nieto-Ortega et al., 2022; Nolasco Perez et al., 2018a, 2018b; Oliveira et al., 2020; Parastar et al., 2020; Patel et al., 2021; Pennisi et al., 2021; Pérez-Marín et al., 2019, 2021; Pochanagone & Rittiron, 2019; Prieto et al., 2018; Reis et al., 2022; Risoluti et al., 2020; Riu et al., 2020; Rukundo et al., 2020, 2021; Sánchez et al., 2018; Schmutzler & Huck, 2016; Sciuto et al., 2021; Silva et al., 2020; Stocco et al., 2019; Tardaguila et al., 2017; Teye, Amuah, et al., 2019; Teye, Elliott, et al., 2019; Torres et al., 2017; van Ruth & Liu, 2019; Wang et al., 2019, 2020; Wei et al., 2016; Wiedemair & Huck, 2019; Wiedemair, Langore, Garsleitner, et al., 2019; Wiedemair et al., 2018; Wokadala et al., 2020; Yaman et al., 2022; Yan, van Stuijvenberg, et al., 2019; Yan, Xu, et al., 2019; You et al., 2017; Yu et al., 2020; Zareef et al., 2019). However, it should be noted that the potential applicability and performance of each spectrometer may differ, which requires systematic feasibility studies for each case. One of the challenges in food analysis is the wide chemical diversity of foodstuffs, which can be further complicated by complex matrices and high moisture content. Spectral signatures of food adulterants may also be difficult to access due to signal accessibility. Additionally, physical properties such as surface texture can interfere with the analysis, and nondestructive analysis of packaged products requires addressing the influence of the packaging material. These complexities require careful consideration when using miniaturized NIR spectrometers in food analysis. Despite these challenges, portable NIR spectroscopy has demonstrated remarkable success in the food industry and remains an active area of research.

TABLE 10.1 The operational characteristics of selected miniaturized NIR spectrometers available at the market in comparison with two exemplary benchtop FT-NIR devices.

Source	Spectrom. (vendor)		Key components		Operational wavelength region		Optical performance		Control/data transfer/power delivery	Weight (g)
	Wavelength selector	Detector	(nm)	(cm^{-1})	Resolution (at l) (nm)	S/N				
Benchtop	NIRFlex N-500 (Büchi)	TH (x2)	Polarization interferometer (FT)	InGaAs (TE-cooled)	800–2500	12,500–4000	avg. 1	10,000:1	PC/LAN/230 V	15,000
	Spectrum Two (PerkinElmer)	TH	Michels on interferometer	InGaAs (air-cooled)	680–4800	14,700–3800	0.8–6.4 (at 1000)	N/A	PC/LAN or USB/ 230 V	13,000
Miniaturized	microPHAZIR (Thermo Fisher Scientific)	TH	MOEMS Hadamard mask (HT)	InGaAs	1596–2396	6267–4173	11	N/A	Autonom./USB/ Li-ion cell	1250
	MicroNIR 1700 ES (VIAVI)	TH (x2)	LVF	InGaAs (array; 128 elements)	908–1676	11,013–5967	12.5 (at 1000) 25 (at 2000)	23,000:1	PC/USB/USB	58
	SCiO (Consumer Physics)	LED	Bandpass filter	Si photodiode (array, 12 elements)	740–1070	13,514–9346	N/A	N/A	Smartphone (Bluetooth)/ cloud/Li-ion cell	35
	NIRONE Sensor S (Spectral Engines)	TH (x2)	MOEMS Fabry–Pérot interferometer	S1.4: InGaAs S1.7–S2.5: "extended" InGaAs	1100–1350 (S1.4) 1350–1650 (S1.7) 1550–1950 (S2.0) 1750–2150 (S2.2) 2000–2450 (S2.5)	9090–7407 7407–6060 6451–5128 5714–4651 5000–4081	12–16 13–17 15–21 16–22 18–28	15,000:1–1500:1 (S1.4–S2.5)	Smartphone (Bluetooth)/ cloud/Li-ion cell	15
	NeoSpectra-Scanner (Si-Ware Systems)	TH	MEMS Michelson interferometer (FT)	InGaAs	1350–2500	7407–4000	16 (at 1550)	N/A	Smartphone (Bluetooth)/ cloud/Li-ion cell	1000
	nanoFTIR NIR (SouthNest Technology)	TH	Michelson interferometer (large mirror; FT)	InGaAs	800–2600	12,500–3846	2.5 (at 1000) 6 (at 1600) 13 (at 2400)	9000:1	PC/USB/USB	220
	Aurora NIR (Grain It)	Halogen	N/A	InGaAs	950–1650	10,526–6060	<10	11,000:1	USB/Li-ion cell	2000

(Continued)

TABLE 10.1 (Continued)

Source	Spectrom. (vendor)		Key components		Operational wavelength region		Optical performance		Control/data transfer/power delivery	Weight (g)
	Wavelength selector	**Detector**	**(nm)**	**(cm^{-1})**	**Resolution (at l) (nm)**	**S/N**				
	LabSpec 4 (ASD Inc.)	TH	Dispersive (reflective holographic diffraction grating)	3 detectors: Vis-NIR (350–1000 nm): Si (array, 512 elements) SW-NIR: InGaAs (1001–1800 nm) and (1801–2500 nm) TE-cooled	350–2500	28,571–4000	3 (at 700) 10 (at 1400/ 2100)	9000:1 (700 nm) 9000:1 (1400 nm) 4000:1 (2100 nm)	PC/USB/Acid-gel battery or 230V	5440
	LabSpec 2500 (ASD Inc.)	Halogen	Grating	Diode Array (Si, InGaAs)	350–1830	28,571–5464	3 (at 700) 10 (at 1400)	9000:1	PC	5600
	Link Square NIR (Stratio, Inc.)	LED	LVF	Si	700–1050	14,285–9523	3 (at 700) 4 (at 750) 5 (at 800) 7 (at 900) 10 (at 1000)	N/A	USB/Li-ion cell	57
	Link Square 1 (Stratio, Inc.)	LED, TH	LVF	Si	400–1000	25,000–10,000	1 (at 437) 6.5 (at 546) 12 (at 703) 44 (at 922)	N/A	USB/Li-ion cell	57

1. "at wavelength" parameter listed if available in the data-sheet provided by the vendor;
2. SCiO presents to the operator interpolated spectra with 1 nm data-spacing, but the real resolution is considerably lower;
3. depending on the sensor implementation/factory configuration.
TH, tungsten halogen; *TE*, thermo-electric; *x2*, duplicated element.
Unless stated otherwise, single-element detectors are used in the listed instruments. Multiple entries listed for some instruments correspond to different variants of the same model with specific factory settings
Source: Adapted (CC-BY 4.0. license) from Beć, K. B., Grabska, J. & Huck, C. W. (2022). Miniaturized NIR spectroscopy in food analysis and quality control: Promises, challenges, and perspectives. *Foods, 11*(10), 1465. https://doi.org/10.3390/foods11101465.

10.3.1 Milk

The utilization of miniaturized NIR spectroscopy for dairy product analysis has gained significant traction, with considerable research effort focusing specifically on developing methods for analyzing milk (de la Roza-Delgado et al., 2017; de Lima et al., 2018; dos Santos Pereira et al., 2021; Karunathilaka et al., 2018; Liu, Aya Parra, et al., 2018; Liu, Parra, et al., 2018; Llano Suarez et al., 2018; Muniz et al., 2020; Musa, 2022; Risoluti et al., 2020; Riu et al., 2020; van Ruth & Liu, 2019). Many studies have focused on qualitative assessments, such as differentiating organic and conventional milk (Liu, Parra, et al., 2018; van Ruth & Liu, 2019), checking authenticity and detecting adulteration (dos Santos Pereira et al., 2021; Musa, 2022) (Fig. 10.1), or identifying regular and lactose-free milk in the field (de Lima et al., 2018). Recent studies have shown that miniaturized NIR spectroscopy can provide rapid and quantitative predictions of key milk quality parameters, including fat and protein content and fatty acid composition (de la Roza-Delgado et al., 2017; Liu, Aya Parra, et al., 2018;

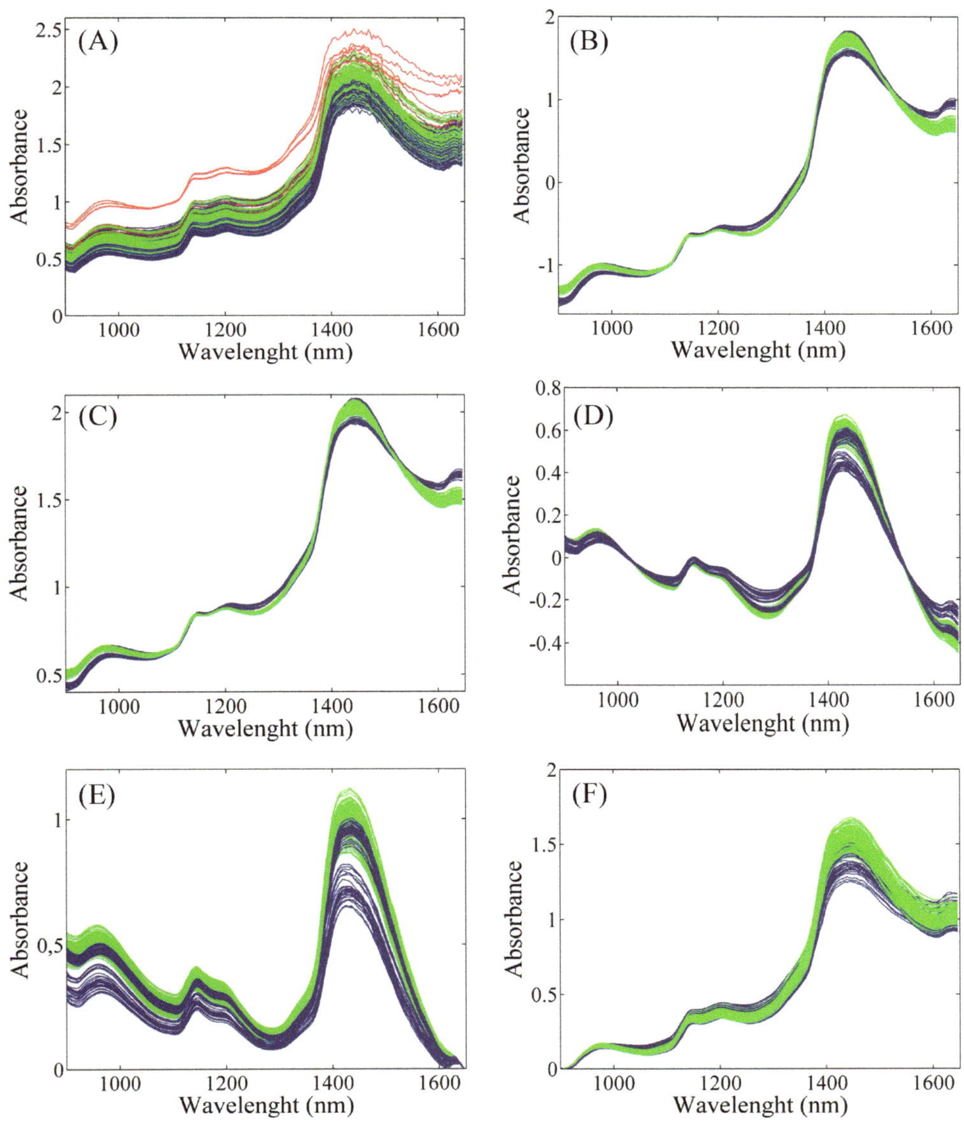

FIGURE 10.1 NIR spectra from goat (in blue) and cow (in red) milk and their mixtures (in green): raw (A) and preprocessed spectra with 7-point moving mean coupled with: standard normal variation (B), multiplicative scattering correction (C), linear baseline correction (D), linear baseline correction plus baseline offset (E), and baseline offset (F). As the pure cow milk samples were not included in the classification models, only their raw spectra are shown. *Reproduced with permission from dos Santos Pereira, E. V., de Sousa Fernandes, D. D., Ugulino de Araujo, M. C, Dias Diniz, H. G., & Sucupira Maciel, M. I. (2021). In-situ authentication of goat milk in terms of its adulteration with cow milk using a low-cost portable NIR spectrophotometer.* Microchemical Journal, Devoted to the Application of Microtechniques in all Branches of Science, 163, *105885. https://doi.org/10.1016/j.microc.2020.105885.*

Llano Suarez et al., 2018; Riu et al., 2020). These portable instruments have been found to offer similar prediction performance to benchtop NIR spectrometers, making real-time monitoring of cow milk quality control parameters feasible on the farm. Furthermore, ongoing efforts also target the analysis of goat milk, aiming to develop efficient calibration transfer methods for in situ analysis at the farm level using multiple portable instruments (dos Santos Pereira et al., 2021). This demonstrates the potential of miniaturized NIR spectroscopy as an effective tool for milk quality control and analysis. The key information from the reviewed studies is summarized and presented in a concise manner in Table 10.2.

10.3.2 Cheese

The use of miniaturized NIR spectrometers for the analysis of dairy products, specifically cheese, has gained widespread popularity as well (Eskildsen et al., 2019; Ma, Babu, et al., 2019; Manuelian et al., 2022; Marinoni et al., 2017; Reis et al., 2022; Stocco et al., 2019; Wiedemair, Langore, Garsleitner, et al., 2019; Yaman et al., 2022). The results of these studies have been summarized in Table 10.3 and show that these devices can deliver acceptable prediction

TABLE 10.2 Recent research activity oriented at miniaturized NIR spectroscopy for analysis of milk.

Ref.	Scope	Sample	Miniaturized NIR instr.	Data-analytical framework	General remarks (applicability/performance)
Liu, Parra, et al. (2018a)	Assessing the effectiveness of a portable NIR spectrometer for authentication of organic milk	87 samples (full-fat, pasteurized retail milk) including 7 organic retail milks and 50 nonorganic retail milks	MicroNIR 1700	PCA, PLS-DA	MicroNIR 1700 was able to accurately distinguish between organic and conventional milk, although it was not as successful in classifying pasture milk samples; however, in both instances, the performance of the MicroNIR was comparable to that of the benchtop NIR spectrometer
dos Santos Pereira et al. (2021)	Detecting adulteration of goat milk with cow milk for the purpose of authentication	200 samples (54 pure goat milk samples and 146 adulterated samples)	NIRscan nano	OC-PLS, PLS-DA, iSPA-PLS-DA	Miniaturized spectrometer effectively identified pure goat milk and detected one instance of adulteration with cow milk; all pure samples were accurately classified during test-set validation
Liu, Aya Parra, et al. (2018b)	Comparing miniaturized and benchtop NIRS with fatty acid profiling by gas chromatography for organic milk authentication and differentiation from other milk types	37 organic retail milk and 50 nonorganic retail milk (of which 36 conventional and 14 green "pasture" milk)	MicroNIR 1700	PCA, PLS-DA	MicroNIR 1700 accurately differentiated organic from conventional milk, but had difficulty with distinguishing pasture milk; fatty acid profiling by a benchtop NIR spectrometer was more effective in distinguishing organic from all nonorganic milk, including pasture milk

(Continued)

TABLE 10.2 (Continued)

Ref.	Scope	Sample	Miniaturized NIR instr.	Data-analytical framework	General remarks (applicability/performance)
de la Roza-Delgado et al. (2017)	Developing a method for handheld NIR spectrometer to collect raw milk spectra and analyze its protein, fat, and solids-nonfat (SNF) content, and transferring calibration models to another portable unit	542 fresh milk samples	microPHAZIR	MPLS	Successful calibration transfer was shown and sharing calibration models among multiple units was deemed essential for using portable instruments for in-situ analysis to give indicators of milk composition on the farm level
Muniz et al. (2020)	Improving monitoring in dairy facilities by classifying milk samples based on their quality	903 fresh cow milk samples	microPHAZIR	PLSR, ANN	microPHAZIR offers benefits at the milking stage by allowing real-time monitoring of quality control parameters for each individual cow milk sample
Riu et al. (2020)	Assessing the performance of two portable NIR instruments for rapid, low-cost, and simple quantitative determination of macronutrients in commercial milk	45 commercial milks	SCiO, NeoSpectra	PCA, PLSR	SCiO and NeoSpectra are capable of quickly and accurately analyzing fat in commercial milk, classifying it based on fat levels, and SCiO can also predict protein content and detect lactose presence/absence
de Lima et al. (2018)	Developing a method for in-field discrimination between regular and lactose-free UHT milks using benchtop FT-NIR and miniaturized NIR spectrometers	71 samples; 41 lactose-free UHT milk and 30 regular UHT milk	MicroNIR 1700	PLS-DA, GA-LDA, SPA-LDA	MicroNIR 1700 was found to be effective in differentiating between regular and lactose-free milk in field settings
Llano Suarez et al. (2018)	Developing a quick and easy-to-use on-site monitoring method for the fatty acid profile in raw milk at the farm level, using a miniaturized NIR spectrometer	108 raw milk samples	microPHAZIR	PLSR	microPHAZIR accurately classified milk by fatty acid composition at the farm level, successfully quantifying fatty acids and healthy indices in individual cow's milk, and predicting individual fatty acids, including saturated fatty acids
Risoluti et al. (2020)	Developing an on-site, contactless NIR analytical method for monitoring milk quality	17 milk specimens (commercially available in the Italian markets)	MicroNIR OnSite	PCA, PLSR	MicroNIR spectrometer accurately differentiated milk based on the distribution of fatty acids in a fast, nondestructive manner

(Continued)

TABLE 10.2 (Continued)

Ref.	Scope	Sample	Miniaturized NIR instr.	Data-analytical framework	General remarks (applicability/ performance)
van Ruth and Liu (2019)	Classification of organic milk	87 milk samples (37 organic retail milks and 50 conventional [nonorganic] retail milk)	SCiO	PLS-DA	SCiO showed separation between organic and conventional milk, but the difference was not as distinct as with the benchtop instrument
Karunathilaka et al. (2018)	Distinguishing between pure and adulterated milk powder	35 milk powder samples	microPHAZIR	PCA	microPHAZIR was successful in distinguishing pure milk powder from adulterated milk powder; the specificity of a nontargeted method is dependent on the type of adulterant, so further investigation using complementary techniques, such as Raman spectroscopy, may be necessary to fully cover all types of adulterants
Musa (2022)	Assessing the effectiveness of a portable NIR for detecting milk adulteration with water	114 samples (34 pure cow milk samples, 80 adulterated with water)	MicroNIR 1700	SIMCA, PLSR, PCA	MicroNIR 1700 was found suitable for the detection of milk adulteration with water

Abbreviations: ANN, artificial neural network; GA-LDA, genetic algorithm linear discriminant analysis; MPLS, modified partial least square; OC-PLS, one-class partial least squares; PCA, principal component analysis; PLSR, partial least squares regression; PLS-DA, partial least squares discriminant analysis; iSPA-PLS-DA, successive projections algorithm for interval selection in PLS-DA; SPA-LDA, successive projection algorithm linear discriminant analysis; SIMCA, soft independent modeling by class analogy.
Source: Adapted (CC-BY 4.0. license) from Beć, K. B., Grabska, J. & Huck, C. W. (2022). Miniaturized NIR spectroscopy in food analysis and quality control: Promises, challenges, and perspectives. *Foods, 11*(10), 1465. https://doi.org/10.3390/foods11101465.

performance even when using cost-effective sensors or cloud-based calibration methods. The Vis/SW-NIR region has been found to provide valuable information on the quality of cheese, making ultra-cost-effective sensors that operate in this range a good choice for these applications (Wiedemair & Langore, Garsleitner, et al., 2019; Yaman et al., 2022). For instance, Fig. 10.2 shows the PLS regressions for whole pieces of cheese delivered by "pocket" Vis/SW-NIR spectrometer SCiO compared with the result obtained with a benchtop instrument NIRFlex N-500, which operates in the wide conventional NIR region of 1000–2500 nm (10,000–4000 cm^{-1}) (Wiedemair, Langore, Garsleitner, et al., 2019). These compact sensors offer improved control in the cheese-making process by enabling early detection of deviations from target quality and monitoring of cheese aging during production (Manuelian et al., 2022; Yaman et al., 2022). The good performance of ultra-cost-effective sensors operating in the Vis/SW-NIR region, due to their construction with silicon photodetectors and LED sources, compared to conventional benchtop NIR spectrometers operating in the NIR spectral range, further supports the concluded application potential.

10.3.3 Meat

The literature showcases a substantial utilization of miniaturized NIR spectroscopy in the analysis of meat products (Table 10.4) (Cáceres-Nevado et al., 2021; Dashti et al., 2022; Dixit et al., 2020; Dumalisile et al., 2020; Edwards et al., 2020; Goi et al., 2022; Horcada et al., 2020; Kucha & Ngadi, 2020; Lee et al., 2017; Nolasco Perez et al., 2018a, 2018b;

TABLE 10.3 Recent research activity oriented at miniaturized NIR spectroscopy in the area of cheese applications.

Ref.	Scope	Sample	Miniaturized NIR instr.	Data-analytical framework	General remarks (applicability/performance)
Wiedemair, and Langore, Garsleitner, et al. (2019)	Comparing the performance of benchtop and miniaturized NIR spectrometers in determining the quality parameters of cheese	46 cheese samples (20 of hard cheese, 26 samples of semihard cheese, respectful to water content)	SciO	PLSR	Miniaturized, cost-effective NIR spectrometer accurately analyzes cheese quality parameters, even with consumer-aimed software
Eskildsen et al. (2019)	Assessing the capability of three different NIR instruments for online determination of fat and dry matter in cheese blocks	160 cheeses from 10 production batches	MicroNIR 1700	PLSR	Miniaturized NIR spectroscopy improves cheese production control by detecting deviations from target quality early in the process
Ma, Babu, et al. (2019)	Developing and validating a rapid quantification technique for intact casein and total protein content in cheddar cheese	49 white and yellow cheddar cheese samples	SciO	PLSR, iPLSR	SCiO effectively measured intact casein and total protein content in cheddar cheese, making it a cost-effective quality control tool for both cheddar and processed cheese production
Yaman et al. (2022)	Developing a rapid analytical method to determine key quality and ripening indicator compounds in cheese, and characterizing biochemical changes during ripening using vibrational spectroscopy	36 white cheese cubes were produced in 2 batches, making 72 cubes for analysis	NeoSpectra	PLSR	Handheld NIR spectrometer is suitable for fast, easy monitoring of cheese quality during aging and can detect deviations in the manufacturing process in real time
Manuelian et al. (2022)	Conducting a feasibility study on a low-cost NIR spectrometer to predict total nitrogen, soluble nitrogen, ripening index, major minerals, and fatty acids in cheese	104 ground cheese samples	SciO	MPLS	Miniaturized, cost-effective NIR spectrometer has comparable accuracy to benchtop devices in predicting targeted traits
Marinoni et al. (2017)	Predicting dry matter, fat, fat/dry matter, proteins, and proteins/dry matter in Grana Padano cheese, and conducting a feasibility study for screening production batches during fire-branding, warehousing, and packaging using cheese paste	195 samples of Grana Padano cheese	X-NIR (Dinamica Generale)	PLSR	Portable NIR spectrometer accurately predicted the chemical composition of Grana Padano cheese with performance similar to a benchtop FT-NIR instrument

(Continued)

TABLE 10.3 (Continued)

Ref.	Scope	Sample	Miniaturized NIR instr.	Data-analytical framework	General remarks (applicability/ performance)
Stocco et al. (2019)	Predicting chemical contents, pH, texture, and color of 37 cheese categories, and comparing the performance of three NIR instruments in reflectance and transmittance mode, using different wavelength intervals	1050 different cheeses from 104 cheese factories	LabSpec 2500	PLSR	Vis/NIR portable spectrometer was found to have better predictive performance than two laboratory NIR instruments in diffuse reflectance mode, regardless of using the full spectrum or selected intervals, and for both reflectance and transmittance modes. The portable instrument can analyze cheese's chemical composition in real time without sample preparation
Reis et al. (2022)	Comparing the performance of miniaturized NIR for estimation of fatty acids in cheese	68 cheese samples	QualitySpec Trek (ASD Inc.), NIRscan Nano	PLSR, SPLS	Comparable prediction performance for both instruments

Abbreviations: iPLSR, interval partial least squares regression; MPLS, modified partial least square; PCA, principal component analysis; PLSR, partial least squares regression; LS-SVM, least squares support-vector machines; RC, regression coefficients.

Parastar et al., 2020; Patel et al., 2021; Perez-Marín et al., 2021; Prieto et al., 2018; Silva et al., 2020; Wei et al., 2016; Wiedemair et al., 2018). The main focus there tends to be directed on rapid and nondestructive analysis of meat freshness and the detection of any mislabeling or adulteration, which are major concerns for meat consumption (Lee et al., 2017; Silva et al., 2020; Wei et al., 2016; Wiedemair et al., 2018). NIR spectral signature of meat shows reasonable levels of specificity associated with the cuts of meat that can be obtained from each muscle group (Fig. 10.3) (Dumalisile et al., 2020). It has also been shown that miniaturized NIR instruments can quickly and noninvasively analyze aging days, levels of chemical/microbial indicators such as thiobarbituric acid and volatile basic nitrogen, and bacteria levels (Kucha & Ngadi, 2020; Lee et al., 2017; Wei et al., 2016). Additionally, miniaturized NIR spectrometers have been used to classify chicken meat and distinguish between different parts, directly in the processing line (Edwards et al., 2020; Horcada et al., 2020; Nolasco Perez et al., 2018a, 2018b; Parastar et al., 2020; Perez-Marín et al., 2021).

Quantitative predictions in meat analysis often focus on fat content and microbiological status, as well as the detection of fat adulteration (Dumalisile et al., 2020). Miniaturized NIR spectroscopy has also been used to quantify specific types of meat in ground meat blends, such as the beef content in a chicken/beef, pork/beef, or chicken/beef/pork blend (Silva et al., 2020).

10.3.4 Fish

The escalation of food fraud in fish and seafood products has propelled the use of miniaturized NIR spectroscopy in seafood analysis (Table 10.5). These compact spectrometers have proven effective in identifying different fish species and predicting key chemical components in fish flesh, such as fat composition, protein, and lipids (Gonçalves et al., 2021). For instance, handheld NIR spectrometers have been utilized on the production line of cuttlefish and musky octopus, providing highly accurate results (as demonstrated in Fig. 10.4) (Pennisi et al., 2021). Miniaturized NIR instrumentation proves especially practical in challenging conditions commonly encountered in fish and seafood production sites, making it a valuable tool for maintaining quality standards and combatting food fraud (Gonçalves et al., 2021; Grassi et al., 2018; Nieto-Ortega et al., 2022; Pennisi et al., 2021; Pochanagone & Rittiron, 2019; Sciuto et al., 2021).

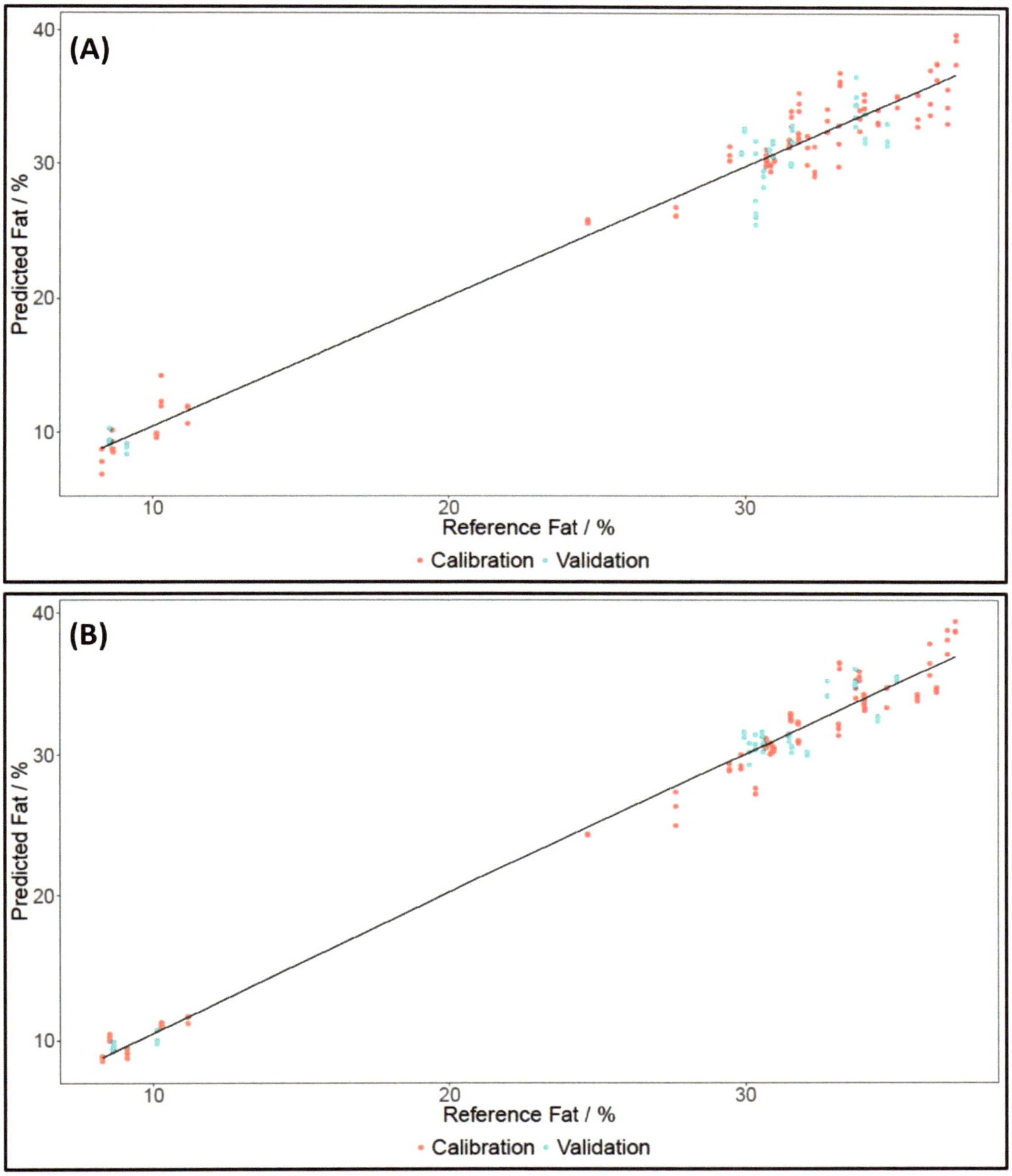

FIGURE 10.2 PLS regression of fat content of whole pieces of cheese, established using data of NIRFlex N-500 (A) and SCiO (B). *Reproduced (CC-BY 4.0 license) from Wiedemair, V., Langore, D., Garsleitner, R., Dillinger, K., & Huck, C. W. (2019). Investigations into the performance of a novel pocket-sized near-infrared spectrometer for cheese analysis.* Molecules (Basel, Switzerland), 24, *428. https://doi.org/10.3390/molecules24030428.*

10.3.5 Fruits and vegetables

Extensive research focuses on miniaturized NIR spectroscopy for the analysis of fruits and vegetables (Amuah et al., 2019; Baca-Bocanegra et al., 2019; Borba et al., 2021; Castrignanò et al., 2019; Hasanzadeh et al., 2022; Ma, Zhang, et al., 2019b; Malegori et al., 2017; Malvandi et al., 2022; Pérez-Marín et al., 2019; Sánchez et al., 2018; Schmutzler & Huck, 2016; Tardaguila et al., 2017; Torres et al., 2017; Wokadala et al., 2020; Yan, Xu, et al., 2019). The aim of these studies is to predict quality parameters, including moisture level, total sugar content (i.e., BRIX index), protein content, metrics of fruit maturity, soluble solids content (SSC), titratable acidity and ascorbic acid, and extractable polyphenols (Table 10.6).

TABLE 10.4 Recent research activity oriented at miniaturized NIR spectroscopy in the area of meat application.

Ref.	Scope	Sample	Miniaturized NIR instr.	Data-analytical framework	General remarks (applicability/ performance)
Wiedemair et al. (2018)	Comparing the performance of benchtop and miniaturized NIR spectrometers in detecting meat fraud	63 samples of different meat types (beef: 9, chicken: 10, mutton: 10, turkey: 10, pork: 10, horse meat: 14)	microPHAZIR	PCA, PLSR	Both benchtop and miniaturized spectroscopy can detect high-level meat adulterations ($>10\%$); however, the miniaturized instrument requires improvement with a larger sample set. Both types of instruments need improvement for detecting low-level meat adulterations ($<10\%$)
Dixit et al. (2020)	Evaluating the performance of a miniaturized NIR spectrometer (NIRscan) in predicting intramuscular fat, and comparing it with two portable and one Vis/ SW-NIR spectrometers	669 lamb meat samples: frozen (609 samples), fresh (60 samples)	LabSpec 5000 (ASD Inc.), LabSpec 4, NIRscan Nano	PLSR, VIP	Prediction performance of NIR spectroscopy on meat samples was not affected by sample temperature equilibration time. Frozen samples showed good performance with LabSpec 5000, LabSpec 4 instruments, but bias was observed for NIRscan Nano due to instrumental variations; for fresh meat, NIRscan Nano performed well and can be used as a good alternative for rapid and real-time classification of fresh lamb meat
Parastar et al. (2020)	Authenticating chicken meat using a miniaturized NIR spectrometer	153 fresh chicken fillet samples	MicroNIR 1700	PLS-DA, CP-ANN, SVM, RSDE	MicroNIR 1700 provides a cost-effective, quick (<20 s), and dependable method for checking the authenticity (and quality) of meat in the field
Cáceres-Nevado et al. (2021)	Evaluating a miniaturized NIR spectrometer as an on-site analyzer of meat quality traits in Iberian pig	Samples of Longissimus dorsi muscles were collected from 524 carcasses of Iberian pigs from "Sánchez Romero Carvajal Jabugo S.A."	MicroNIR 1700	MPLS	Method based on a handheld NIR device provides good accuracy in analyzing the chemical composition of intact pork loin directly at the industrial plant; this makes it a useful tool for monitoring meat authenticity and quality

(Continued)

TABLE 10.4 (Continued)

Ref.	Scope	Sample	Miniaturized NIR instr.	Data-analytical framework	General remarks (applicability/performance)
Wei et al. (2016)	Developing and evaluating a portable and low-cost Vis/NIR detection device for predicting TVB-N content and assessing pork meat freshness	58 pork samples with different freshness attributes	Self-developed LED-based portable Vis/NIR spectrometer (400–1100 nm)	MLR, PLSR	With the use of a miniaturized NIR spectrometer is possible to perform nondestructive detection of TVB-N (total volatile basic nitrogen) content in pork meat; further development of the instrument to simplify its structure and increase its cost-effectiveness is feasible
Lee et al. (2017)	Assessing the feasibility of using miniaturized NIR spectroscopy for on-site estimation of food freshness, including beef, pork, fish, and fruits such as tomato and watermelon	8 food items: meat (beef sirloin, beef eyeround, pork sirloin), fish (salmon, bass, corvina), vegetable (tomato), and fruit (watermelon)	SCiO	SVM	SCiO was successful in classifying food based on the aging day and chemical/microbial indicators such as thiobarbituric acid, volatile basic nitrogen, and bacteria levels; this method provides a high accuracy and is concluded to be fully satisfactory for point-of-need freshness assessment of meat, fish, vegetables, and fruits
Goi et al. (2022)	Assessing the feasibility of using miniaturized NIR spectroscopy to predict traits in Charolais cattle meat	40 rib cuts taken at the level of the 5th rib were collected from 40 Charolais beef cattle	SCiO	MPLS	SCiO can be used to predict certain quality traits of beef in real time during the production process; the use of this spectrometer eliminates the need for other expensive processes and materials, saving time and resources
Nolasco Perez et al. (2018a)	Using portable NIR to classify chicken parts based on their physical and chemical attributes, including pH and color features, and chemical composition such as protein, fat, moisture, and ash	137 chicken samples (52 breasts, 40 thighs, and 45 drumsticks) and 90 samples obtained by grinding the chicken parts (30 breasts, 30 thighs, and 30 drumsticks)	NIRscan Nano	LDA, RF, SVM	Portable NIR spectroscopy device was able to achieve good accuracy in classifying chicken meat, including different parts of the chicken during the processing line and also for authentication of shelf samples of processed products in the market

(Continued)

TABLE 10.4 (Continued)

Ref.	Scope	Sample	Miniaturized NIR instr.	Data-analytical framework	General remarks (applicability/performance)
Silva et al. (2020)	Investigating the potential of using miniaturized NIR spectroscopy to detect adulteration in ground meat	Cuts of cow, pig, and chicken breast (undisclosed number of samples)	MicroNIR 1700	PLSR, SVR	MicroNIR 1700 has satisfactory performance for the quantification of beef in ground meat blends, including blends with chicken and/or pork
Dumalisile et al. (2020)	Investigating the potential of a miniaturized NIR spectrometer to differentiate between different muscle types within each species of selected game animals and to classify species irrespective of the muscle type	42 animal samples (12 Impala) (*Aepyceros melampus*), 15 Eland (*Taurotragus oryx*), and 15 Ostrich (*Struthio camelus*)	MicroNIR OnSite	PCA, PLS-DA	The use of miniaturized NIR spectroscopy has been successful in authenticating different types of game meat, specifically impala, eland, and ostrich; was showed that the discrimination between species was less challenging than the identification of different muscles within each species
Patel et al. (2021)	Comparing three NIR instruments (a transportable Vis-NIR, a portable NIR, and a handheld NIR with varying sizes and characteristics) for their performance in predicting beef characteristics	178 beef samples (*Longissimus thoracis* muscle)	MicroNIR, Aurora NIR, LabSpec 2500	PLSR, LMS	Three different portable NIR spectrometers had similar accuracy in predicting 13 parameters of beef quality, with the most compact instrument (MicroNIR) tending to be the most precise; however, the wide spectrum and number of data points produced by NIR spectrometers can lead to data redundancy problems, which can affect their analytical performance
Kucha and Ngadi (2020)	Determining the freshness of pork by the use of a miniaturized NIR spectrometer	80 samples with four groups of 20 (storage in 4°C for 2, 4, and 8 days)	MicroNIR 2200	PLSR, MLR, SPA, RC	MicroNIR 2200 showed good performance in nondestructively monitoring the level of thiobarbituric acid reactive substances in minced pork, and it is considered suitable for this purpose

(*Continued*)

TABLE 10.4 (Continued)

Ref.	Scope	Sample	Miniaturized NIR instr.	Data-analytical framework	General remarks (applicability/performance)
Prieto et al. (2018)	Developing a method for handheld NIR instrument to predict fatty acids (FA) composition and iodine value (IV) of pig subcutaneous fat	357 pigs	LabSpec 4	PLSR	Portable NIR spectrometer can be used to predict pig fat quality; the miniaturized sensor technology was successfully implemented in a research abattoir, and spectra were collected directly on the carcass, allowing for carcass sorting based on fat composition or hardness for marketing purposes
Horcada et al. (2020)	Assessing the miniaturized NIR instrument for its feasibility in classifying individual Iberian pig carcasses into the four official quality categories	763 samples of Iberian pigs	microPHAZIR	PCA, PLS2-DA	microPHAZIR was successful in supporting the control of official quality category assignment in Iberian pig carcasses in commercial abattoirs by using subcutaneous fat samples
Edwards et al. (2020)	Differentiating South African game meat by miniaturized NIR spectrometer	307 samples from 52 animals (154 zebras, 133 springboks, 20 ostriches)	MicroNIR 1700	PCA, PLS-DA	NIR spectroscopy can be used as a rapid and nondestructive method for game meat authentication and can classify different types of meat using multilevel hierarchical modeling; the study shows that using NIR spectroscopy with different models and data pretreatment can lead to better classification results
Perez-Marín et al. (2021)	Assessing miniaturized NIR instrument for authentication of Iberian pig carcasses	568 samples (45 different Iberian pig producers were recorded in two consecutive years [66 samples in 2016 and 429 in 2017])	MicroNIR 1700	QDA, LDA	Portable NIR spectrometer enables the in situ inspection of individual Iberian pig carcasses at the speed of the slaughtering process; the best classification model, based on LDA, showed the device's potential for classifying the carcasses as "Premium" or "Non-Premium" based on the animal's feeding regime (acorn vs compound feeds)

(Continued)

TABLE 10.4 (Continued)

Ref.	Scope	Sample	Miniaturized NIR instr.	Data-analytical framework	General remarks (applicability/performance)
Nolasco Perez et al. (2018b)	Discriminating between chicken meat parts (breasts, thighs, and drumstick) with the use of miniaturized NIR spectrometer	137 samples (52 breasts, 40 thighs, and 45 drumsticks)	NIRscan Nano	LDA, SVM	Combination of portable NIR spectrophotometer and chemical analysis with machine learning algorithms (LDA, RF, and SVM) was successful in classifying different parts of chicken meat. The results were considered good, with a high accuracy rate
Dashti et al. (2022)	Combining two handheld spectrometers with chemometric methods for meat speciation, and its application for halal certification	48 beef muscle samples (fore and hind shank, chuck, brisket, and round), 41 lamb muscle samples (fore and hind shank), 40 chicken muscle samples (breast and drumstick), and 32 pork muscle samples (shoulder and leg)	Tellspec Enterprise, Link Square 1	PCA, SIMCA, KNN, PLS-DA, SVM	Handheld Vis-NIR and NIR sensors can simplify halal meat-species detection and screening. These sensors classify four meat species with promising classification for intact and ground meat. The Vis-NIR sensor is more accurate than NIR. They provide a fast-positive release and detect suspicious samples that can be confirmed with other methods

Abbreviations: CP-ANN, counter propagation artificial neural network; DS, direct standardization; MLR, multilinear regression; MPLS, modified partial least square; PCA, principal component analysis; PDS, piecewise direct standardization; PLSR, partial least squares regression; PLS-DA, partial least squares discriminant analysis; PLS2-DA, polynomial order of two partial least squares discriminant analysis; PLSR, partial least squares regression; RC, regression coefficients; RF, random forest; RSDE, random subspace discriminant ensemble; LDA, linear discriminant analysis; LSM, least squares means; QDA, quadratic discriminant analysis; SDW, spectral difference by wavelengths; SPA, successive projection algorithm; SVM, support vector machine; SVR, support vector machine regression; VIP, variable importance in projection.
Source: Adapted (CC-BY 4.0. license) from Beć, K. B., Grabska, J. & Huck, C. W. (2022). Miniaturized NIR spectroscopy in food analysis and quality control: Promises, challenges, and perspectives. *Foods, 11*(10), 1465. https://doi.org/10.3390/foods11101465.

Miniaturized NIR spectrometers have been shown to be feasible for performing these analyses directly under field conditions or during fruit ripening (Malegori et al., 2017). Portable spectrometers have expanded NIR spectroscopy's capabilities beyond quantitative analysis, enabling qualitative assessments in various scenarios. These instruments can identify food product variety and geographical origin, assess refrigerated storage duration, and detect mislabeling of conventionally produced fruits as organic (Amuah et al., 2019). The versatility of miniaturized NIR spectrometers makes them attractive analytical tools in the food industry, allowing for comprehensive quality assessment across various fruits and vegetables. For example, quality parameters of apple are also subject of the NIR study (Hasanzadeh et al., 2022; Schmutzler & Huck, 2016). Similar applications have been demonstrated for vegetables, such as examining spinach leaves directly on the plant (Pérez-Marín et al., 2019; Sánchez et al., 2018). However, mixed accuracy of analysis of the main quality parameters of spinach has been revealed in such case, with for example, a reliable prediction of SSC, but less adequate quality of the model fit in the case of nitrate or ascorbic acid contents (Fig. 10.5) (Pérez-Marín et al., 2019). The use of NIR spectroscopy technology has been found to be highly effective in optimizing cultural practices in agriculture, such as fertilization and irrigation. It can also be used to assess the quality of vegetables upon harvest, enabling farmers to make more informed decisions about the timing of their harvest and postharvest storage. These desirable features are fully in line with Agriculture 4.0 principles and cope well with the current era of precision agriculture and smart farming.

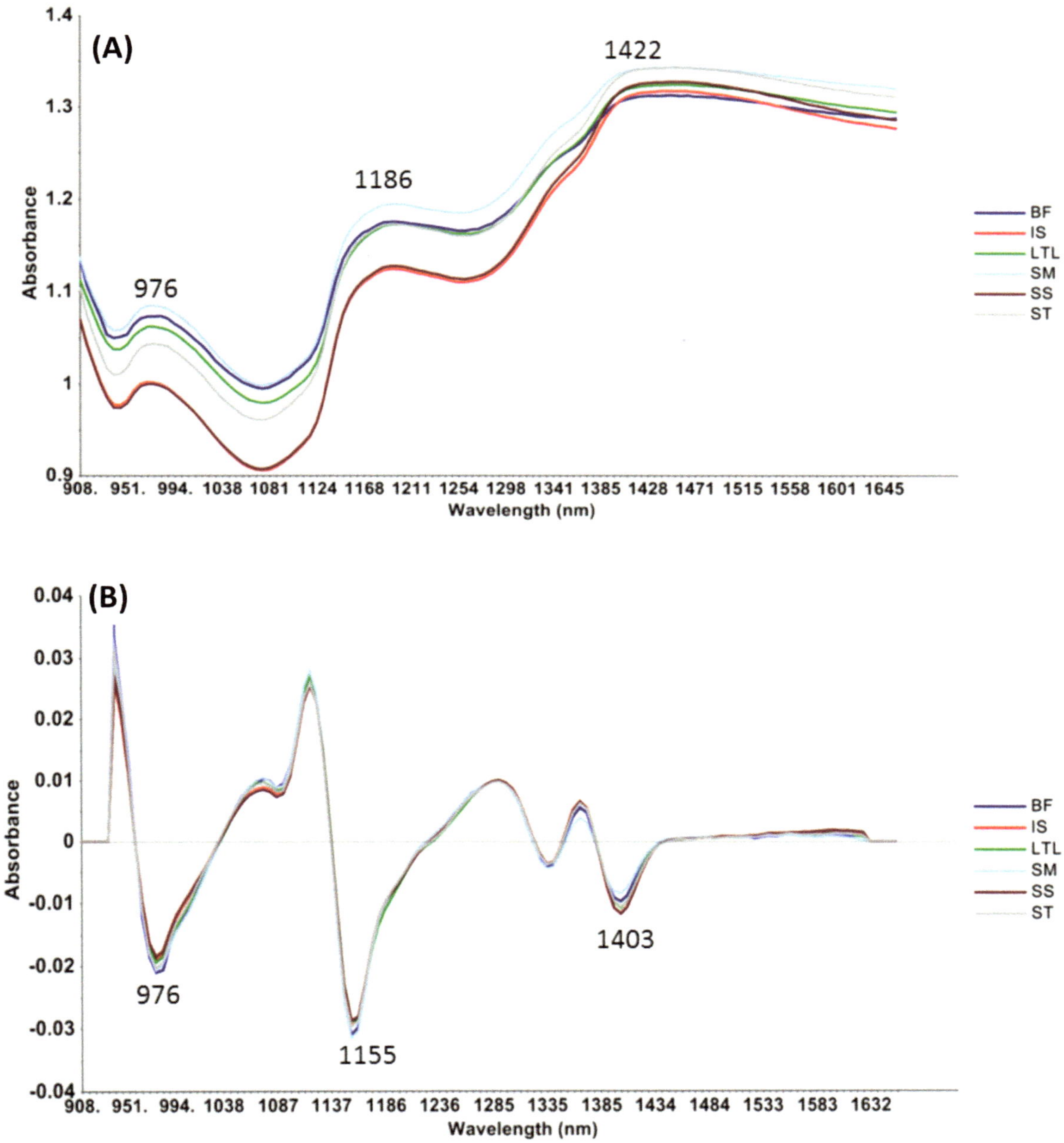

FIGURE 10.3 Mean spectra of impala selected muscles (BF, IS, LTL, SM, SS, and ST) showing the wavelength bands of (A) raw spectra, (B) SNV-Detrend and second derivative preprocessed spectra. Abbreviations: LTL, longissimus thoracis et lumborum; BF, biceps femoris; SM, semimembranosus; ST, semitendinosus; IS, infraspinatus; SS, supraspinatus. *Reproduced with permission from Dumalisile, P., Manley, M., Hoffman, L., & Williams, P. J. (2020). Discriminating muscle type of selected game species using near infrared (NIR) spectroscopy.* Food Control, 110, *106981. https://doi.org/10.1016/j.foodcont.2019.106981.*

10.3.6 Beverages and syrups

The literature indicates that miniaturized NIR spectroscopy is successfully employed in analyzing beverages and syrups (Correia et al., 2018; da Costa Fulgencio et al., 2023; Henn et al., 2016, 2017; Jiang et al., 2019; Mayr, Schmelzer, et al., 2021; Wang et al., 2019, 2020; Zareef et al., 2019). The technology can provide a rapid assessment of sugar content (as shown, e.g., in Fig. 10.6; Henn et al., 2016) and other chemical contents, such as polyphenols, amino acids, caffeine, and theanine, and can also detect the level of adulteration (Table 10.7). Miniaturized spectrometers have been utilized for both

TABLE 10.5 Recent research activity oriented at miniaturized NIR spectroscopy in the area of fish analysis.

Ref.	Scope	Sample	Miniaturized NIR instr.	Data-analytical framework	General remarks (applicability/performance)
Grassi et al. (2018)	Comparing the performance of a handheld NIR spectrometer and an FT-NIR benchtop spectrometer in distinguishing between fillets and patties of Atlantic cod and haddock	170 fresh fillets of *Gadus morhua* ($n = 80$) and *Melanogrammus aeglefinus* ($n = 90$)	MicroNIR OnSite	LDA	Handheld NIR device can be a simple, cost-effective, and reliable alternative to benchtop spectrometers for authenticating fish fillet and patty; the method was found to be suitable for use in commercial fraud detection and can be extended to authenticate fish species in processed products as well
Sciuto et al. (2021)	Assessing the potential for species identification in fish fillets using a pocket-sized NIR sensor	150 fish samples (9 fish species: *Merluccius merluccius*, *Pollachius virens*, *Epinephelus costae*, *Gadus morhua*, *Pleuronectes platessa*, *Sebastes norvegicus*, *Scomber scombrus*, *Chelidonichthys lucerna*, *Synaptura cadenati*)	SCiO	Pretreatment and analysis of spectra performed by the functions built-in in the proprietary cloud service	Handheld NIR device is a cost-effective and reliable alternative to benchtop spectrometers for authenticating fish fillet and patty; the accuracy of correctly identifying fish species was between 93.97%–96.58%, and the method was concluded to be a good screening approach to counter fish species fraud
Gonçalves et al. (2021)	Predicting nutritional values of fish, including protein, lipids, and moisture, and in distinguishing between different sources of fish (farmed vs. wild) and conditions (fresh or defrosted)	805 fish samples (133 Alaskan pollock) (*Gadus chalcogrammu*), 204 Atlantic cod (*G. morhua*), 22 European plaice (*Pleuronectes platessa*), 264 common sole (*Solea solea*) and 182 turbot (*Psetta maxima*)	Tellspec Enterprise	PCA, LDA, LSR, RF, LR, SVM, XGB	Tellspec sensor was found to have good to excellent performance in predicting the nutritional values of protein, lipids, and moisture, as well as in authenticating the source and condition of all the studied fish species
Nieto-Ortega et al. (2022)	Determining the fatty acid composition of industrial fish oils from fish by-products on-site	269 different mixtures of 8 fish oil samples	MicroNIR OnSite	PLSR	Miniaturized NIR spectrometer was successful in determining the fat composition of fish oil in a fast and nondestructive way, making it an attractive alternative to inefficient conventional methods of analysis

(*Continued*)

TABLE 10.5 (Continued)

Ref.	Scope	Sample	Miniaturized NIR instr.	Data-analytical framework	General remarks (applicability/ performance)
Pennisi et al. (2021)	Comparing the performance of three NIR instruments in identifying the storage conditions of fish products	50 fresh specimens of cuttlefish (*Sepia officinalis*) and musky octopus (*Eledone spp.*)	MicroNIR 1700, SCiO	PCA, PLS-DA	High accuracy of miniaturized sensors for classification purposes and the practical benefits of using the technology in a production line setting
Pochanagone and Rittiron (2019)	Determining the sodium chloride content in tuna fish variety skipjack	35 pieces of tuna fish	FQA-NIRGUN (FANTEC Research Institute)	MLR, PLSR	Successful analysis of sodium chloride in tuna fish using portable NIR instrument in detecting systematic changes of water content in samples

Abbreviations: MLR, multiple linear regression, PCA, principal component analysis; PLS-DA, partial least squares discriminant analysis; PLSR, partial least squares regression; RF, random forest; LDA, linear discriminant analysis; LR, logistic regression; SVM, support vector machine; XGB, xtreme gradient boosting.
Source: Adapted (CC-BY 4.0. license) from Beć, K. B., Grabska, J. & Huck, C. W. (2022). Miniaturized NIR spectroscopy in food analysis and quality control: Promises, challenges, and perspectives. *Foods, 11*(10), 1465. https://doi.org/10.3390/foods11101465.

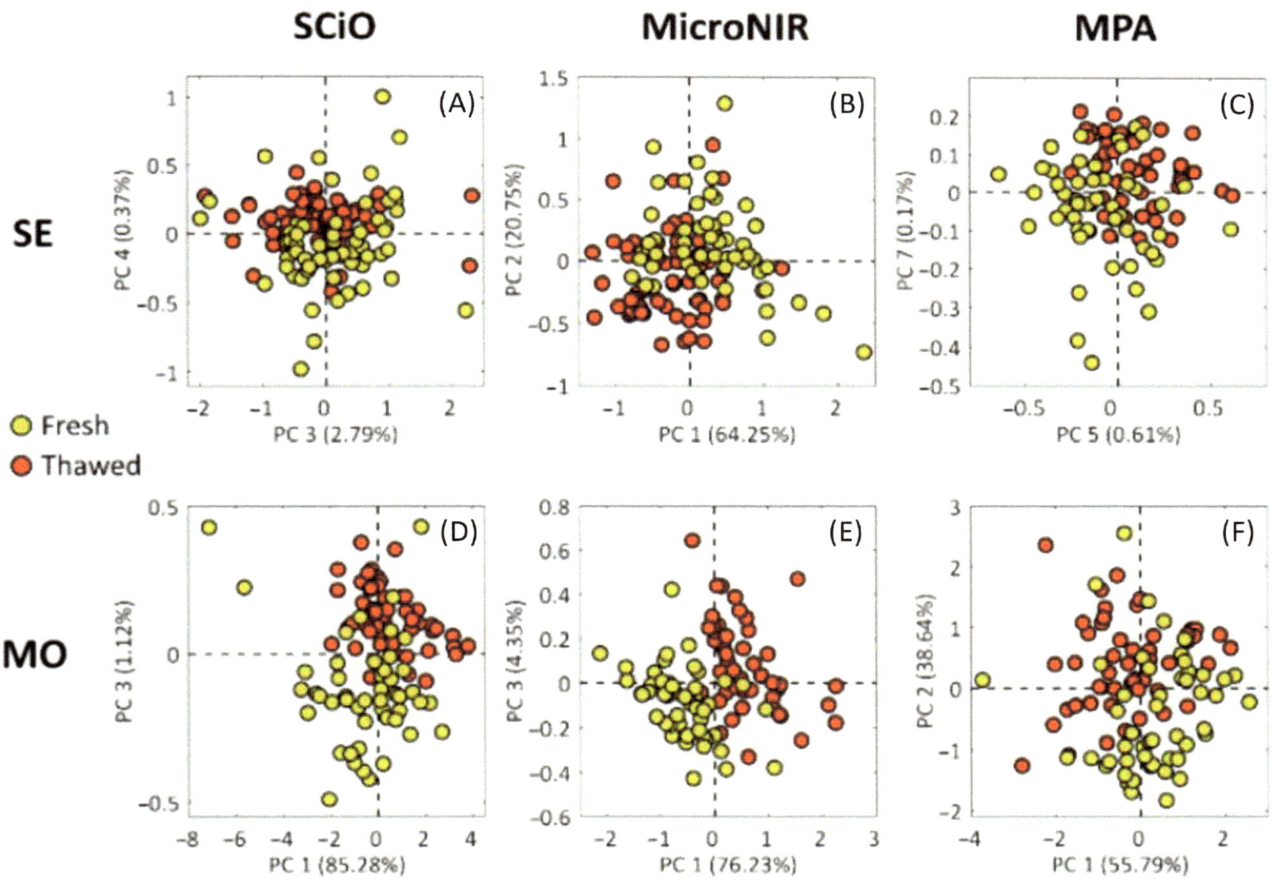

FIGURE 10.4 Most relevant PCA scores plots of cuttlefish' (SE; A–C) and musky octopus' (MO; D–F) samples. Fresh samples are depicted in yellow, while thawed samples are colored in orange. *Reproduced from (CC-BY 4.0 license) Pennisi, F., Giraudo, A., Cavallini, N., Esposito, G., Merlo, G., Geobaldo, F., Acutis, P. L., Pezzolato, M., Savorani, F., & Bozzetta, E. (2021). Differentiation between fresh and thawed cephalopods using NIR spectroscopy and multivariate data analysis.* Foods, 10, *528. https://doi.org/10.3390/foods10030528.*

TABLE 10.6 Recent research activity oriented at miniaturized NIR spectroscopy in the area of fruit and vegetable application.

Ref.	Scope	Sample	Miniaturized NIR instr.	Data-analytical framework	General remarks (applicability/performance)
Schmutzler and Huck (2016)	Using on-site analyses, a method is being developed for detecting the total antioxidant capacity and total soluble solids content of apples	92 apples of seven cultivars	microPHAZIR	PCA, PLSR	microPHAZIR was able to accurately predict the total sugar content of apples of different varieties, as well as the concentration of polyphenolic compounds in the peel of the fruits, through a nondestructive on-site analysis
Torres et al. (2017)	Evaluating the performance of different regression algorithms for predicting major physical-quality parameters in all citrus fruits using a handheld NIR spectrometer	611 samples belonging to the genus Citrus: 378 oranges (191 harvested in 2010 and 187 in 2011) and 233 mandarins	microPHAZIR	MPLS	Miniaturized NIR spectroscopy, combined with large databases and local regression algorithms, can effectively predict the quality of citrus fruits directly on the tree
Amuah et al. (2019)	Developing a method for the portable NIR spectrometer to simultaneously discriminate between organically produced pineapple fruits and conventionally produced ones (i.e., organic vs. inorganic) and predict total soluble solids	90 intact pineapple fruits	SCiO	KNN, PCA, LDA, PLSR, MSC-PCA-LDA	Portable NIR spectrometer combined with appropriate chemometric tools is suitable for rapidly performing nondestructive analysis of pineapple quality and is effective in detecting pineapple fraud, such as mislabeling of conventionally produced fruits as organic ones
Malegori et al. (2017)	Developing a method for in-field nondestructive analysis of titratable acidity and ascorbic acid content in acerola fruit during ripening	117 acerola fruit	MicroNIR 1700	PLSR, SVM	MicroNIR instrument demonstrated good ability in predicting chemical parameters of interest in acerola fruits during in-field monitoring
Baca-Bocanegra et al. (2019)	Developing in-field nondestructive analysis for titratable acidity and ascorbic acid content in acerola fruit during ripening	400 grape samples	MicroNIR 1700	PCA, LDA, MPLS, DPLS	MicroNIR instrument was successfully used for the in-vineyard screening of extractable polyphenols in red grape skins, however, challenges were identified in that environmental and physiological conditions interfere with sorting the berries according to their extractable polyphenol contents

(Continued)

TABLE 10.6 (Continued)

Ref.	Scope	Sample	Miniaturized NIR instr.	Data-analytical framework	General remarks (applicability/ performance)
Wokadala et al. (2020)	Evaluating the performance of a handheld Vis/NIR spectrometer in rapidly and nondestructively analyzing the moisture content of mangoes during solar drying	240 mango samples	F750 (Felix Instruments)	PLSR	Handheld Vis/NIR spectroscopy device has been proven to be a robust and effective method for quickly and nondestructively monitoring the moisture content of mangoes during solar drying; the device is able to provide accurate moisture information, making it a useful tool for the solar drying process
Malvandi et al. (2022)	Measuring and monitoring apple hardness nondestructively during ultrasonic contact drying through portable NIR spectroscopy in real time	148 samples from 8 fresh Gala apples	NIRscan Nano	PLSR, MLR, PLS-ANN	Portable NIR spectrometer was able for rapid and cost-effective, nondestructive measurement and real-time monitoring approach of apple hardness during ultrasonic contact drying
Hasanzadeh et al. (2022)	Measuring of quality parameters of apple fruit by using Vis/NIR spectroscopy	120 apple samples (60 Red Delicious and 60 Golden Delicious)	PS-100 (Apogee Instruments, Inc.)	PLSR	High ability to measure the pH, soluble solid contents, titratable acid and total phenol contenet parameters of red and yellow apple with the use of miniaturized device
Tardaguila et al. (2017)	Evaluating the feasibility of using portable NIR spectroscopy in-field for assessing the water status in diverse varieties of grapevine (grown under different environmental conditions) by miniaturized NIR	160 individual primary adult leaves (20 leaves per cultivar) of the mid-upper part of the shoot	microPHAZIR	PCA, MPLS	Nondestructive, on-site NIR spectroscopic analysis is a reliable method for assessing the grapevine water status under field conditions
Ma, Zhang, et al. (2019)	Quantifying water, protein, and soluble sugar in mulberry leaves using a miniaturized NIR spectrometer	83 mulberry leaves	MicroNIR 1700	PLSR	The use of handheld NIR spectrometers with wavelength optimization was found to be effective in predicting water content in fresh mulberry leaves and crude protein in dried mulberry leaves; however, the predictive performance was not ideal for the prediction of soluble sugar in mulberry leaves

(Continued)

TABLE 10.6 (Continued)

Ref.	Scope	Sample	Miniaturized NIR instr.	Data-analytical framework	General remarks (applicability/performance)
Yan, Xu, et al. (2019)	Developing a method for the authentication of fengdous and quantitative analysis of mulberry fruits using a miniaturized NIR spectrometer	434 mulberry fruits	MicroNIR 1700	GA, CARS, PLSR	Handheld NIR instrument has been successfully used for several plant analytical case studies and has demonstrated its capability to determine various nutritional parameters in a qualitative and quantitative manner
Pérez-Marín et al. (2019)	Developing a nondestructive and in situ quality evaluation of spinach plants using miniaturized NIR spectrometer and assessing spinach suitability for different uses once harvested	128 samples of spinach plants	microPHAZIR	MPLS, PLS-DA	microPHAZIR have the capability of monitoring important safety and quality parameters during the production of spinach
Sánchez et al. (2018)	Monitoring quality parameters in intact spinach using miniaturized NIR spectrometer in situ	149 spinach plants	microPHAZIR	MPLS	Miniaturized NIR spectroscopy is capable of analyzing the green color, texture, and dry matter of spinach leaves in situ on the plant; the results of these predictions can be used to optimize fertilization and irrigation strategies
Borba et al. (2021)	Developing a nondestructive method for assessing tomato quality attributes using a miniaturized NIR spectrometer	319 fresh market tomato samples	NeoSpectra	PLSR	Handheld NIR spectrometer can be used to quickly and effectively measure various quality characteristics of tomatoes, including multiple types of tomatoes
Castrignanò et al. (2019)	Determining quality attributes of tomato fruits using miniaturized NIR spectrometer	300 tomato fruits of San Marzano variety	MicroNIR 1700	PLSR	MicroNIR 1700 has been found to be very effective in real-time monitoring of the quality of San Marzano tomatoes. The device is considered to be cost-efficient, making it a practical solution for maintaining tomato quality

Abbreviations: CARS, competitive adaptive reweighted sampling; DPLS, discriminant partial least square analysis; GA, genetic algorithm; KNN, k-nearest neighbors; MPLS, modified partial least square; PCA, principal component analysis; PLS-ANN, partial least squares and artificial neural network; PLS-DA, partial least squares discriminant analysis; PLSR, partial least squares regression; LDA, linear discriminant analysis; SVM, support vector machine; MLR, multilinear regression; NIPALS, nonlinear iterative partial least squares.

Source: Adapted (CC-BY 4.0. license) from Beć, K. B., Grabska, J. & Huck, C. W. (2022). Miniaturized NIR spectroscopy in food analysis and quality control: Promises, challenges, and perspectives. *Foods, 11*(10), 1465. https://doi.org/10.3390/foods11101465.

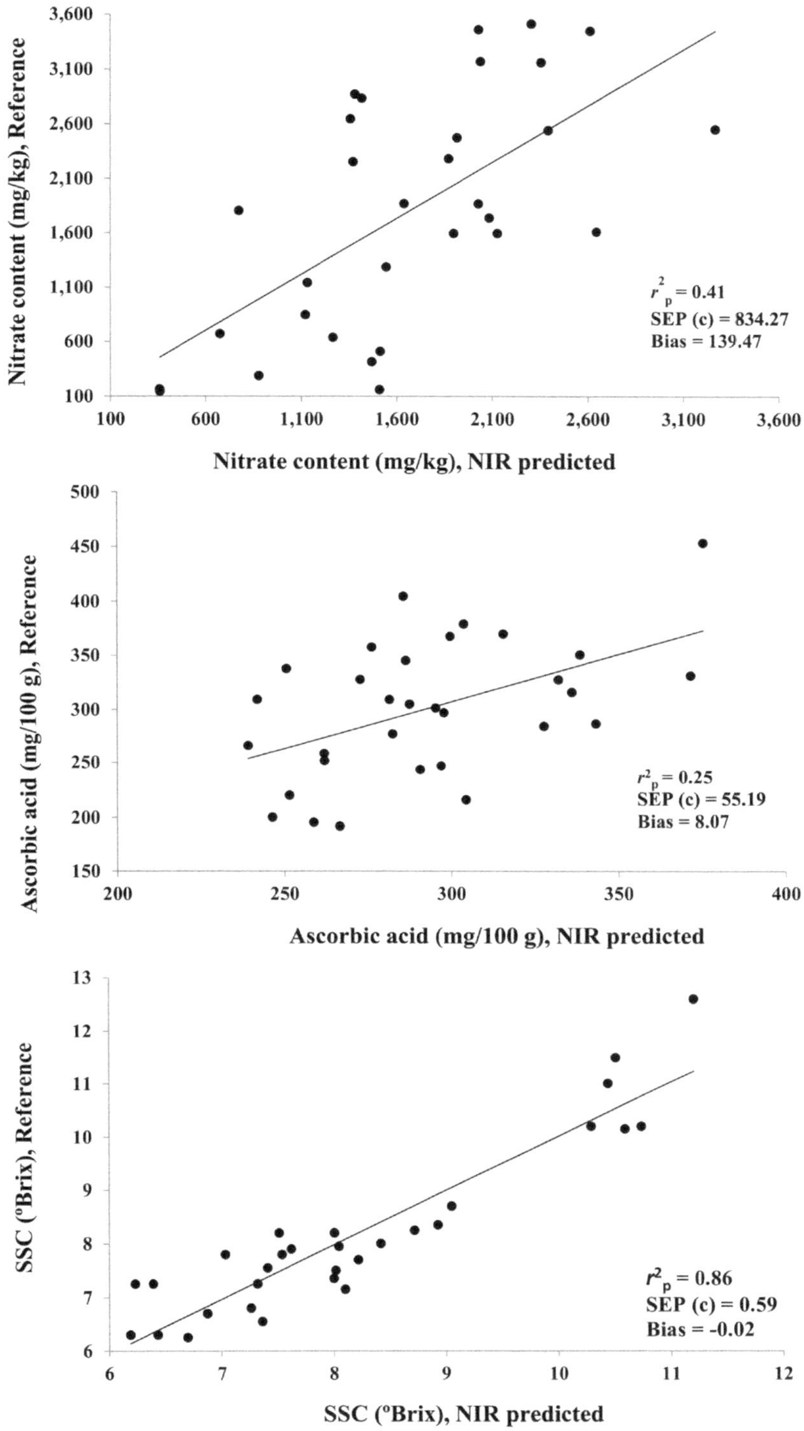

FIGURE 10.5 Reference and NIR predicted values for quality and safety parameters in intact spinach. *Reproduced with permission from Pérez-Marín, D., Torres, I., Entrenas, K. A., Vega, M., & Sánchez, M. T. (2019). Pre-harvest screening on-vine of spinach quality and safety using NIRS technology.* Spectrochimica Acta A, 207, *242–250. https://doi.org/10.1016/j.saa.2018.09.035.*

qualitative and quantitative analysis of various compounds in food formulations, including sugars in syrup (Henn et al., 2016, 2017). The investigations often focused on determining the content of key constituents in food items such as black tea (Mayr, Schmelzer, et al., 2021; Wang et al., 2020; Zareef et al., 2019). The study of quantification of alcohol content in four different fermented alcoholic beverages was performed as well (da Costa Fulgencio et al., 2023).

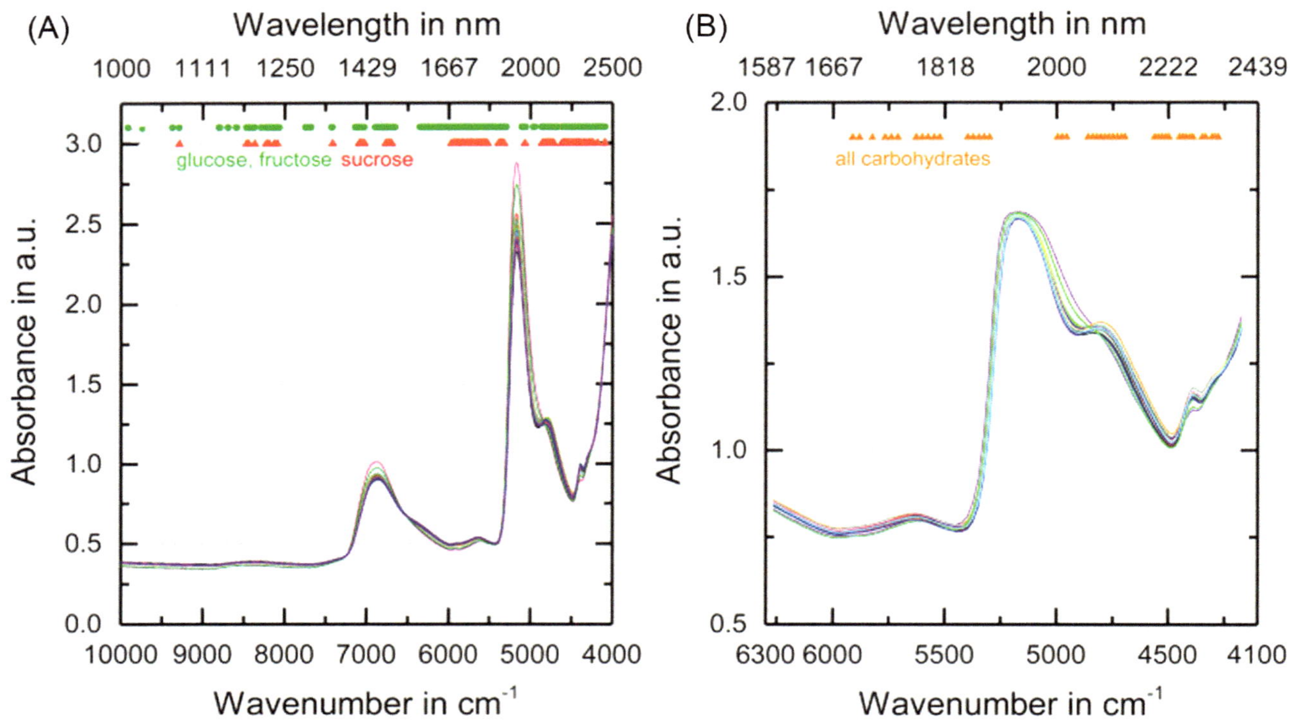

FIGURE 10.6 Raw spectra of standard syrups recorded with the benchtop (A) and portable (B) spectrometer with the used spectral ranges gained from significance test marked for the corresponding carbohydrates. *Reproduced with permission from Henn, R., Schwab, A., Huck, & C. W. (2016). Evaluation of benchtop versus portable near-infrared spectroscopic method combined with multivariate approaches for the fast and simultaneous quantitative analysis of main sugars in syrup formulations.* Food Control, 68, *97–104. https://doi.org/10.1016/j.foodcont.2016.03.037.*

TABLE 10.7 Recent research activity oriented at miniaturized NIR spectroscopy in the area of fruit and beverages and syrups application.

Ref.	Scope	Sample	Miniaturized NIR instr.	Data-analytical framework	General remarks (applicability/ performance)
Henn et al. (2016)	Comparing the performance of benchtop and portable NIR spectroscopy in qualitatively and quantitatively analyzing the main sugars in syrup formulations	116 samples (53 standard and 63 reformulated syrups)	microPHAZIR	PCA, PCR, SVMR, PLSR	microPHAZIR demonstrated good performance in the measurement of a wide range of sugar concentrations, making it suitable for use in quality control in industrial settings
Henn et al. (2017)	Predicting the main carbohydrate content in syrup and evaluating its potential for consumer use	116 syrups consisting of different flavor types	microPHAZIR	PLSR	Miniaturized NIR spectrometers by consumers is a promising technology, however, to make it more reliable and accessible, further development of robust spectral processing methods that require minimal or no supervision is needed

(Continued)

TABLE 10.7 (Continued)

Ref.	Scope	Sample	Miniaturized NIR instr.	Data-analytical framework	General remarks (applicability/performance)
Wang et al. (2019)	Evaluating the feasibility of using miniaturized NIR spectroscopy for analyzing the quality index of matcha tea	105 samples of matcha tea samples of different grades	NIRscan Nano	PLSR, Si-PLS, GA-PLS, CARS-PLS, RF-PLS	Model strategy based on portable NIR spectroscopy was developed and demonstrated potential for predicting and classifying the content of polyphenols and amino acids in matcha tea
Wang et al. (2020)	Assessing the feasibility of using miniaturized NIR spectroscopy for predicting the content of catechins and caffeine in green and black tea	270 tea samples (135 samples of black and 135 green tea)	NIR-S-R2; (InnoSpectra)	SVR	Miniaturized and cost-effective NIR spectrometer can be used to effectively analyze the quality of tea; results indicate that the NIR spectrometer provides a practical and accurate method for assessing tea quality
Mayr, Schmelzer, et al. (2021)	Evaluating the analytical performance of two miniaturized NIR spectrometers (in comparison with a benchtop one) for analyzing the caffeine and theanine content in black tea	65 samples (milled and ground) of black tea	microPHAZIR, MicroNIR 2200	PLSR	Accuracy of the predictions of caffeine and theanine content in tea differed for the two instruments; the differences in accuracy are said to be associated with the instruments' sensitivity to certain absorption bands that are unique to caffeine and theanine
Correia et al. (2018)	Using a miniaturized NIR spectrometer for quality control of Arabica coffee	125 blends of coffee	MicroNIR 1700	PCA, PLSR	MicroNIR spectrometer was successful in predicting adulterations in coffee samples and suggested its suitability for real-time quality control in the commercial coffee industry. The instrument was found to have the ability to detect adulterations at minimum quantification levels
Jiang et al. (2019)	Analyzing the sugar (sucrose) content in everyday drinks using miniaturized NIR spectroscopy	25 sucrose solutions	NIRScan Nano	OLS, SLR, MLR, SVM, RF, MPL	Miniaturized NIR spectrometer demonstrated successful analysis of sucrose content with reasonable performance
Zareef et al. (2019)	Developing a miniaturized NIR spectroscopic method for predicting the concentrations of cianidanol, ferulic acid, gallic acid, L-epicatechin, phloridzin, and rutin in congou black tea	140 sample of black tea from 7 batches	NIRQuest512 (Ocean Optics)	PLSR, CARS-PLS	Use of portable near-infrared spectroscopy in combination with multivariate chemometrics can be used as a nondestructive technique for rapid screening of phenolic compounds in congou black tea

(Continued)

TABLE 10.7 (Continued)

Ref.	Scope	Sample	Miniaturized NIR instr.	Data-analytical framework	General remarks (applicability/ performance)
da Costa Fulgencio et al. (2023)	Quantification of alcohol content in four different fermented alcoholic beverages by combining portable NIR spectroscopy and multivariate calibration	153 samples of alcoholic beverages (39 beer, 40 cider, 36 mead, 38 wine)	MicroNIR 1700	PLSR	New portable analytical platforms for quality control of beverages have incorporated a green analytical tool using NIR spectroscopy and a single multivariate calibration Multiproduct model, leading to better transparency in the food supply chain and improved quality inspection control of alcoholic beverages by regulatory agencies

Abbreviations: CARS-PLS, competitive adaptive reweighted sampling, partial least squares; GA-PLS, genetic algorithm partial least squares; MLR, multilinear regression; OLS, ordinary least square; PCA, principal component analysis; PCR, principal component regression; PLSR, partial least squares regression; RF, random forest; RF-PLS, random frog partial least square; SLR, single linear regression; Si-PLS, synergy interval partial least squares; SVM, support vector machine; SVMR, support vector machine regression; SVR, support vector machine regression; MPL, multilayer perceptron.
Source: Adapted (CC-BY 4.0. license) from Beć, K. B., Grabska, J. & Huck, C. W. (2022). Miniaturized NIR spectroscopy in food analysis and quality control: Promises, challenges, and perspectives. *Foods, 11*(10), 1465. https://doi.org/10.3390/foods11101465.

10.3.7 Oils

Application of miniaturized NIR spectroscopy has been extensively studied in the analysis of oils, particularly palm oil and olive oil (Abu-Khalaf & Hmidat, 2020; Basri et al., 2017; Basri et al., 2018; Borghi et al., 2020; Giussani et al., 2021; Karunathilaka et al., 2017; MacArthur et al., 2020; Teye, Elliott, et al., 2019; Yan, van Stuijvenberg, et al., 2019), for example, as demonstrated in Fig. 10.7. Table 10.8 presents a summary of key information from the reviewed studies. In general, these studies have found that the application of miniaturized NIR spectroscopy in the analysis of oils has proven to be a viable method for the detection of adulterants, as well as the quantification of adulteration levels (Basri et al., 2017, 2018; MacArthur et al., 2020; Teye, Elliott, et al., 2019). Interestingly, establishing an optimal sampling strategy has been identified as a meaningful factor in the analysis of edible oils, for example, as indicated by Giussani et al. (2021). Provided that this aspect is sufficiently addressed, the accuracy of the results obtained with miniaturized NIR spectroscopy has been concluded to be in line with, or even better than, those obtained with other conventional analytical methods. The findings of recent studies demonstrate that miniaturized NIR spectroscopy can offer a rapid and cost-effective alternative to conventional methods for analyzing oils, making it a promising option for the food industry. This technique can facilitate real-time quality control analysis and help to reduce production costs, allowing for greater efficiency in the industry.

10.3.8 Miscellaneous

Miniaturized NIR spectrometers showcase remarkable versatility in the field of food analysis. Several recent studies have proven the effectiveness of these instruments in analyzing a wide range of food samples, ranging from whole foods to highly processed products. The ability of miniaturized NIR spectrometers to rapidly analyze complex food samples in real time makes them a valuable tool for various industries, including agriculture and food processing (Anyidoho et al., 2020; Aykas & Menevseoglu, 2021; Barthet et al., 2020; Beć et al., 2021; Coronel-Reyes et al., 2018; Guelpa et al., 2017; Hao et al., 2019; Kosmowski & Worku, 2018; Mayr, Beć, et al., 2021; McGrath et al., 2020; McVey, Gordon, et al., 2021; McVey, McGrath, et al., 2021; Modroño et al., 2017; Neves et al., 2019; Oliveira et al., 2020; Rukundo et al., 2020, 2021; Teye, Amuah, et al., 2019; Wiedemair & Huck, 2019; You et al., 2017; Yu et al., 2020). The studies included in Table 10.9 offer a range of examples that illustrate the capabilities of miniaturized NIR

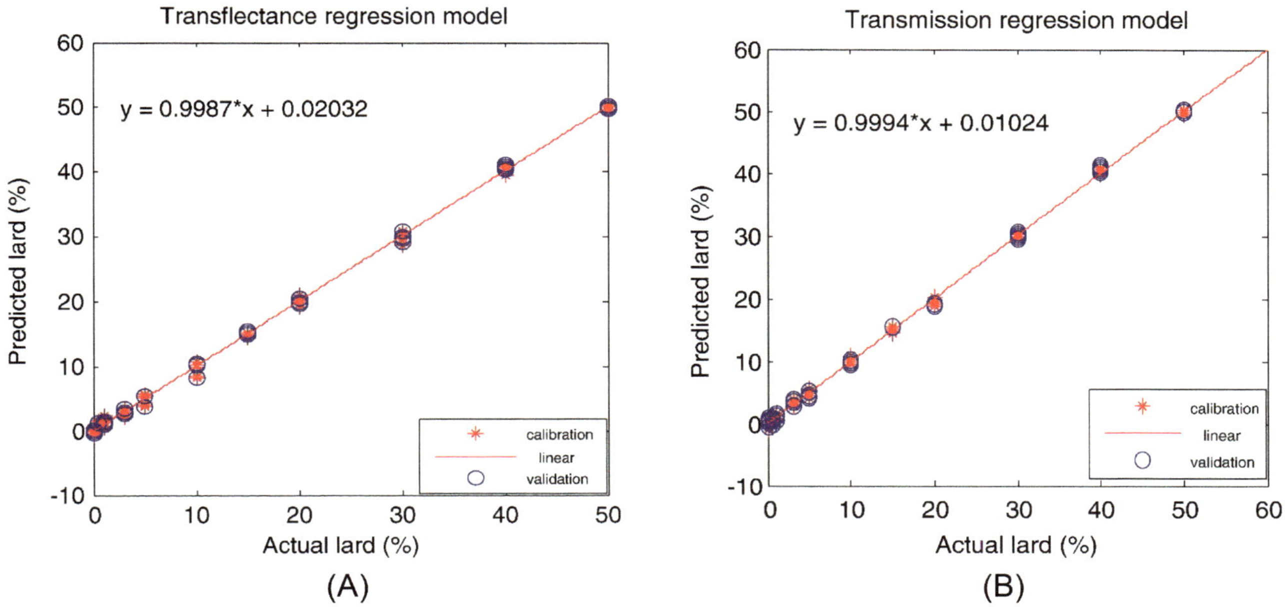

FIGURE 10.7 PLS regression model of palm oil using calibration and prediction data for (A) transflectance and (B) transmission mode. *Reproduced with permission from Basri, K. N., Hussain, M. N., Bakar, J., Sharif, Z., Khir, M. F. A., & Zoolfakar, A. S. (2017). Classification and quantification of palm oil adulteration via portable NIR spectroscopy.* Spectrochimica Acta A, 173, *335–342. https://doi.org/10.1016/j.saa.2016.09.028.*

TABLE 10.8 Recent research activity oriented at miniaturized NIR spectroscopy in the area of oil samples.

Ref.	Scope	Sample	Miniaturized NIR instr.	Data-analytical framework	General remarks (applicability/performance)
Basri et al. (2017)	Developing a MicroNIR-based analytical method to detect the presence of lard adulteration in palm oil, comparing the transmittance mode to the transflectance mode	Pure and adulterated palm oil samples (undisclosed sample count)	MicroNIR 2200	PCA, PLSR	MicroNIR was successfully used for the classification and quantification analysis of palm oil and was effective in discriminating between pure and adulterated samples. The transmittance mode was found to yield a better prediction model compared to the transflectance mode
Giussani et al. (2021)	Developing and optimizing various measurement strategies for two miniaturized NIR instruments to identify the optimal measurement conditions for rapid and cost-effective analysis of olive oils	66 samples of commercial oils	SCiO, NeoSpectra	LDA, PCA, PLS-DA	Analyzing olive oils using a portable NIR spectrometer can be challenging, especially with the NeoSpectra instrument; however, the study was able to successfully classify olive oils into different categories and distinguish between olive oil and sunflower oil without any sample pretreatment

(Continued)

TABLE 10.8 (Continued)

Ref.	Scope	Sample	Miniaturized NIR instr.	Data-analytical framework	General remarks (applicability/ performance)
Abu-Khalaf and Hmidat (2020)	Assessment of olive oil quality and differentiation between extra virgin, virgin, ordinary virgin, and lampante oils using miniaturized NIR spectrometer-based analysis	48 samples	Fieldspec spectrograph (Ocean Optics)	PCA	VIS/NIR spectroscopy in the 400–1100 nm range was effective in accurately predicting certain quality parameters such as acidity and peroxide in olive oil. The technology was able to differentiate between the four groups of olive oil (extra virgin, virgin, ordinary virgin, and lampante)
Teye, Elliott, et al. (2019)	Detection of Sudan dyes as indicators of palm oil adulteration using miniaturized NIR device	520 samples including spiked	IRIS Visum Palm NIR	LDA, SVM, PCA	Portable Visum palm NIR spectroscopy when coupled with the appropriate chemometrics can be used for quick and nondestructive analysis of fraudulent adulteration of palm oil with Sudan dyes (I, II, III, IV) for the purpose of authentication and identification
MacArthur et al. (2020)	Detection of Sudan IV dyes as indicators of palm oil adulteration using miniaturized NIR spectrometer	150 samples including spiked samples	SCiO	PCA, PLSR, SVM	Portable NIR spectroscopy can be effectively utilized to certify the absence of Sudan IV dye in palm oil, implying that it can be used as a tool for quality control in the palm oil industry
Yan, van Stuijvenberg, et al. (2019)	Distinguishing other olive oil grades from extra virgin olive oil (EVOO) with miniaturized NIR spectrometer	130 commercial olive oil samples (80 EVOO, 40 ROO, 10 POO)	MicroNIR 1700	KNN, SVM	MicroNIR 1700 shows promise in being able to distinguish between different grades of pure olive oil
Karunathilaka et al. (2017)	Developing method to determining of authenticity of extra virgin olive oil (EVOO)	24 reference oils, 35 reference set oils and 100 test-set oils	microPHAZIR	PCA	microPHAZIR was able to distinguish between authentic EVOO reference samples and commercial EVOO products, and that the results were better when a multivariate analysis (SIMCA) was used, as opposed to a univariate analysis (CI)

(Continued)

TABLE 10.8 (Continued)

Ref.	Scope	Sample	Miniaturized NIR instr.	Data-analytical framework	General remarks (applicability/ performance)
Borghi et al. (2020)	Classifying vegetable oils (soybean, corn, sunflower, and canola oil) as extra virgin olive oil	228 authentic and artificially spiked samples	MicroNIR 1700	PLS-DA	Portable NIR spectrometer in combination with multivariate data processing can be a reliable tool for identifying, classifying, and quantifying the content of different types of vegetable oils (soybean, corn, sunflower, and canola oil) in extra virgin olive oil
Basri et al. (2018)	Detecting the adulteration of palm oil with lard using NIR spectroscopy	165 samples	MicroNIR 1700	PLS-DA, LDA	LED-MicroNIR instrument was found to have the best performance compared to two other NIR spectrometers (FT-NIR and MicroNIR) in terms of classifying different types of vegetable oils; introduction of LED into the NIR spectrometer system reduced the number of variables, which improved the accuracy of the classification model

Abbreviations: PCA, principal component analysis; PLSR, partial least squares regression; PLS-DA, partial least squares discriminant analysis; LDA, linear discriminant analysis; KNN, k-nearest neighbors; SVM, support vector machine.

TABLE 10.9 Recent research activity oriented at miniaturized NIR spectroscopy in the area of variety of application.

Ref.	Scope	Sample	Miniaturized NIR instr.	Data-analytical framework	General remarks (applicability/ performance)
Oliveira et al. (2020)	Using a miniaturized NIR spectrometer for rapidly authenticating adulterated paprika powder	3 types of paprika (sweet, smoked, and spicy): 315 samples; spiked with potato starch and acacia gum (0%–36% w/w), and annatto (0%–18% w/ w)	NIRscan Nano	PLS-DA, PLSR	NIRscan Nano showed good accuracy in detecting adulterated samples and differentiating the types of adulterations, with lower accuracy only for annatto but still adequate for screening purposes

(Continued)

TABLE 10.9 (Continued)

Ref.	Scope	Sample	Miniaturized NIR instr.	Data-analytical framework	General remarks (applicability/ performance)
Beć et al. (2021)	Developing on-site quantitative analysis of protein content in handcrafted insect bars using miniaturized NIR spectrometers; employing low-level data fusion for simultaneous utilization of cost-effective Vis/NIR and NIR sensors	Insect-protein enhanced fitness 40 bars, 8 of each flavor (peanut-cranberry, hazelnut-cocoa, macadamia-salted caramel and cashew blueberry, and "Omas Apfelstrudel")	MicroNIR 1700, Tellspec Enterprise, SCiO	PCA, PLSR, GPR	GPR method used for calibration can accurately measure protein content with handheld instruments, with MicroNIR performing similarly to benchtop instruments; data fusion improved prediction performance for consumer-graded pocket food analyzers
Coronel-Reyes et al. (2018)	Predicting egg storage time at room temperature by ultra-cost-efficient miniaturized NIR spectrometry	30 shell-intact brown poultry eggs	SCiO	PLSR, ANN	Smatphone-connected, low-cost NIR spectrometer can be used effectively for assessing egg storage time, and its reliability can be improved when combined with traditional destructive techniques
Yu et al. (2020)	Evaluating the performance of miniaturized NIR spectrometer (compared to benchtop) in classifying high oleic acid peanuts (HOPs) and quantifying major fatty acids	150 different peanut varieties and strains from 10 main planting provinces	MicroNIR 1700	PCA, QDA, LDA, PLS-DA	Miniaturized NIR sensors successfully distinguished HOP from other oils and predicted the content of its main fatty acids with accuracy comparable to benchtop instruments
Mayr, Beć, et al. (2021)	Evaluating the performance of three miniaturized NIR instruments in quantifying piperine in black pepper	66 samples; whole and milled seeds of black pepper	MicroNIR 2200, microPHAZIR	PLSR	MicroNIR 2200 was able to provide a reliable prediction of whole seeds, but miniaturized spectrometers with a narrow spectral range had limited performance in quantifying piperine in black pepper; MicroNIR had better performance, as it acquires more meaningful absorption bands of piperine and offers prediction results comparable to benchtop instruments. The microPHAZIR had reduced applicability, as it only acquires C-H stretching bands of piperine

(Continued)

TABLE 10.9 (Continued)

Ref.	Scope	Sample	Miniaturized NIR instr.	Data-analytical framework	General remarks (applicability/ performance)
You et al. (2017)	Evaluating the performance of a miniaturized NIR spectrometer in classifying food powders	8 visually indistinguishable food powder: sugar, salt, cream, flour, corn, rice, bean and potato powder	Link Square 1	KNN, RF, SVM	Miniaturized NIR spectroscopy has been successful in classifying food powders
Neves et al. (2019)	Using a miniaturized NIR spectrometer in determining nutritional parameters of pasta/ sauce blends	commercial products: 5 pasta products, 5 sauce products; for each, five different pasta/sauce-type blend combinations (0%–100% (w/w) sauce addition)	MicroNIR 1700	PLSR	MicroNIR 1700 has demonstrated satisfactory accuracy in quantifying the content of energy, carbohydrate, fat, fiber, protein, and sugar in pasta/sauce meals in a realistic analytical scenario
Modroño et al. (2017)	Evaluating two handheld NIR spectrometers for on-site and real-time analysis of nutritive parameters in raw compound feed	100 samples of intact compound feeds (going from feed for dairy cows, piglets, laying hens, chicken, sheep, rabbits, horses and lambs using different presentation forms [crumbs, pellets, and meals])	microPHAZIR, MicroNIR 1700	PLSR	Handheld NIR instruments are effective in determining the composition changes of individual compound feeds on farms instantly, eliminating the need for transporting samples to the lab. Two popular miniaturized NIR instruments show similar performance
Rukundo et al. (2021)	Evaluating the feasibility and comparing the performance of two different miniaturized/ handheld NIR spectrometers for quantitatively analyzing crude protein (CP) content in mixed forage and feedstuff composed of Sweet Bran, distiller's grains, corn silage, and corn stalk	147 total—Sweet Bran, corn silage, corn stalks, and three kinds of corn distillers grains: wet distillers grain with solubles, modified distillers grain with solubles, and dry distillers grain with solubles	Tellspec Enterprise, QualitySpec Trek (ASD Inc.)	PCA, PLSR	Both evaluated handheld NIR instruments accurately measured the crude protein (CP) content in forage and feed, making them suitable for use in screening, quality control, and process control scenarios
Kosmowski and Worku (2018)	Using ultra-cost-efficient miniaturized NIR spectrometry for identifying cultivars of barley, chickpea, and sorghum in the context of Ethiopia	2650 grains of barley, chickpea, and sorghum cultivars	SCiO	SVM, PLS-DA	Barley, chickpea, and sorghum cultivars were accurately identified using miniaturized NIR spectrometers in a low-cost and rapid analysis

(Continued)

TABLE 10.9 (Continued)

Ref.	Scope	Sample	Miniaturized NIR instr.	Data-analytical framework	General remarks (applicability/performance)
Teye, Amuah, et al. (2019)	Estimating rice authenticity and quality in real time using ultra-cost-efficient miniaturized NIR spectrometer	520 rice samples from different quality grades	SCiO	PCA, KNN, SVM	Miniaturized NIR spectrometer enables rapid and nondestructive classification of rice samples based on quality grades, geographical origin, and production location with ultra-cost efficiency
McGrath et al. (2020)	Authenticating rice varieties, qualities, and grades by miniaturized NIR spectrometer	1399 samples	SCiO	OPLS-DA, SVM, KNN	Successful differentiation of various rice types, including high-value Basmati rice, geographically indicated rice from China, different qualities of rice in Ghana and Vietnam, and locally produced and imported rice in Ghana
Hao et al. (2019)	Identifying rice varieties using a miniaturized spectrometer	720 rice samples	MicroNIR 1700	PLS-DA, PCA, SVM	Portable NIR spectroscopy can provide a fast and accurate method for identifying rice types and transgenic characteristics, such as Bt63
McVey, Gordon, et al. (2021)	Investigating coriander seed authenticity by two miniaturized NIR spectrometers	290 coriander seed samples	Flame-NIR, SCiO	PLS-DA, OPLS-DA, RF	The accuracy of miniaturized NIR spectrometers (Flame-NIR and SCiO) is inferior compared to benchtop (iS50) NIR spectrometer in quantitative analysis, but they are still considered viable for screening purposes
Barthet et al. (2020)	Determining several quality parameters of canola seed by a miniaturized NIR spectrometer	181 intact whole canola seeds	MicroNIR OnSite-W	PLSR	MicroNIR OnSite-W has successfully predicted several quality parameters of canola seed such as oil, protein, oleic acid, and iodine. However, it was not accurate in predicting the chlorophyll content using the handheld instrument

(Continued)

TABLE 10.9 (Continued)

Ref.	Scope	Sample	Miniaturized NIR instr.	Data-analytical framework	General remarks (applicability/ performance)
Wiedemair and Huck (2019)	Analyzing total antioxidant capacity by Folin–Ciocalteu and NIR spectroscopy, with evaluation and comparison of the performance of three miniaturized NIR instruments and a benchtop FT-NIR spectrometer	77 samples comprising buckwheat, millet and oat	microPHAZIR, MicroNIR 2200, SCiO	PLSR, OSC	All the examined instruments were able to predict the total antioxidant capacity, but with varying accuracy
Guelpa et al. (2017)	Verifying authenticity and detecting fraud in South African honey with the use of miniaturized NIR spectrometers	84 samples (44 authentic South African honey, 40 nonauthentic and adulterated honey)	MicroNIR 1700, NIRQuest model 512–2.5 (Ocean Insight)	PLS-DA	NIR spectroscopy delivers high accuracy in classifying different types of honey food and its quality-related parameters; the accuracy of prediction varied between different NIR instruments, but all examined instruments shown the ability to provide useful information for authenticity verification
Anyidoho et al. (2020)	Determining the geographical origin of cocoa beans	330 samples	TellSpec Enterprise	LDA, SMV	Handheld NIR spectroscopy found to be effective for rapid and nondestructive classification of cocoa beans
Aykas and Menevseoglu (2021)	Detecting green pea and peanut adulteration in pistachio by NIR spectroscopy	60 samples	NeoSpectra	PCA, PLSR	NeoSpectra successfully determined the green pea and peanut adulteration in pistachios and showed better performance compared to FT-MIR benchtop spectrometer
Rukundo et al. (2020)	Detecting adulteration of turmeric powder with metanil yellow	120 samples including spiked samples	QualitySpec Trek (ASD Inc.)	PLSR	Miniaturized NIR device was found to be a cost-effective tool for assessing the presence of metanil yellow adulteration in turmeric powder, with acceptable and reliable results within a range of 0%–30% (w/w)

(Continued)

TABLE 10.9 (Continued)

Ref.	Scope	Sample	Miniaturized NIR instr.	Data-analytical framework	General remarks (applicability/ performance)
McVey, McGrath, et al. (2021)	Identifying coriander seed authenticity by miniaturized NIR spectrometer	290 samples including spiked	SCiO	PLS-DA, OPLS-DA	Calibration model based on data from a benchtop NIR spectrometer was successfully transferred to a handheld NIR spectrometer; SCiO was able to correctly predict adulterated samples 100% of the time and authentic samples 98.5% of the time with only minimal loss in prediction performance

Source: Adapted (CC-BY 4.0. license) from Beć, K. B., Grabska, J. & Huck, C. W. (2022). Miniaturized NIR spectroscopy in food analysis and quality control: Promises, challenges, and perspectives. *Foods, 11*(10), 1465. https://doi.org/10.3390/foods11101465.

spectrometers for qualitative and quantitative analysis. The technology has been used to predict critical qualitative and quantitative properties of food, such as the content of carbohydrates, fats, fiber, protein, and sugar, as well as specific chemical constituents like piperine in black pepper (Mayr, Beć, et al., 2021) or insect proteins in energy snacks (fitness bars) (Beć et al., 2021) (Fig. 10.8). Miniaturized NIR spectrometers can also be used for authentication or classification of different food products, like the quality grades and geographical origin of rice (Hao et al., 2019; McGrath et al., 2020; Teye, Amuah, et al., 2019). It is also possible to predict other quality parameters of a more general nature, such as the assessment of egg storage time (Coronel-Reyes et al., 2018). However, it's essential to note that miniaturized NIR instruments have limitations in their applicability, as seen in the inability to accurately predict chlorophyll content in canola seeds (Barthet et al., 2020).

10.4 Exploring the advancements and future potential of miniaturized NIR spectroscopy in the food industry

With its ability to rapidly and nondestructively analyze samples, NIR spectroscopy has proven to be a valuable tool in various industries. The recent development of portable spectrometers has further expanded the utility of this technique, particularly in the agrifood sector, where the analysis of diverse and complex samples is essential. The miniaturization of NIR spectroscopy allows for real-time monitoring of food quality throughout the entire production and supply chain, "from farm to fork." The prospects of further progress in this field are bright, with the advances brought by the progressing Industrial Revolution 4.0.

The food industry is shifting toward a more holistic approach to sustainability that emphasizes social, economic, and environmental aspects. As a nondestructive form of analysis that is highly efficient, cost-effective, and environmentally friendly, NIR spectroscopy meets the key demands of the modern world. Miniaturized NIR spectroscopy provides a versatile and accessible tool that combines unparalleled flexibility and analytical throughput. Effectively, portable NIR systems can be applied at any stage of the food production and supply chain, directly on-site. By enabling real-time monitoring throughout the production and supply chain, miniaturized NIR spectroscopy supports the goal of sustainable food production and consumption.

Some of the key areas for further refinement of the methods include addressing the sensor deterioration and the transferability of models to new generations of instruments in a highly practical manner. To address these challenges, the progress of the tools offered by Industry 4.0 seems critical, resulting in increased reliability and user-friendliness of

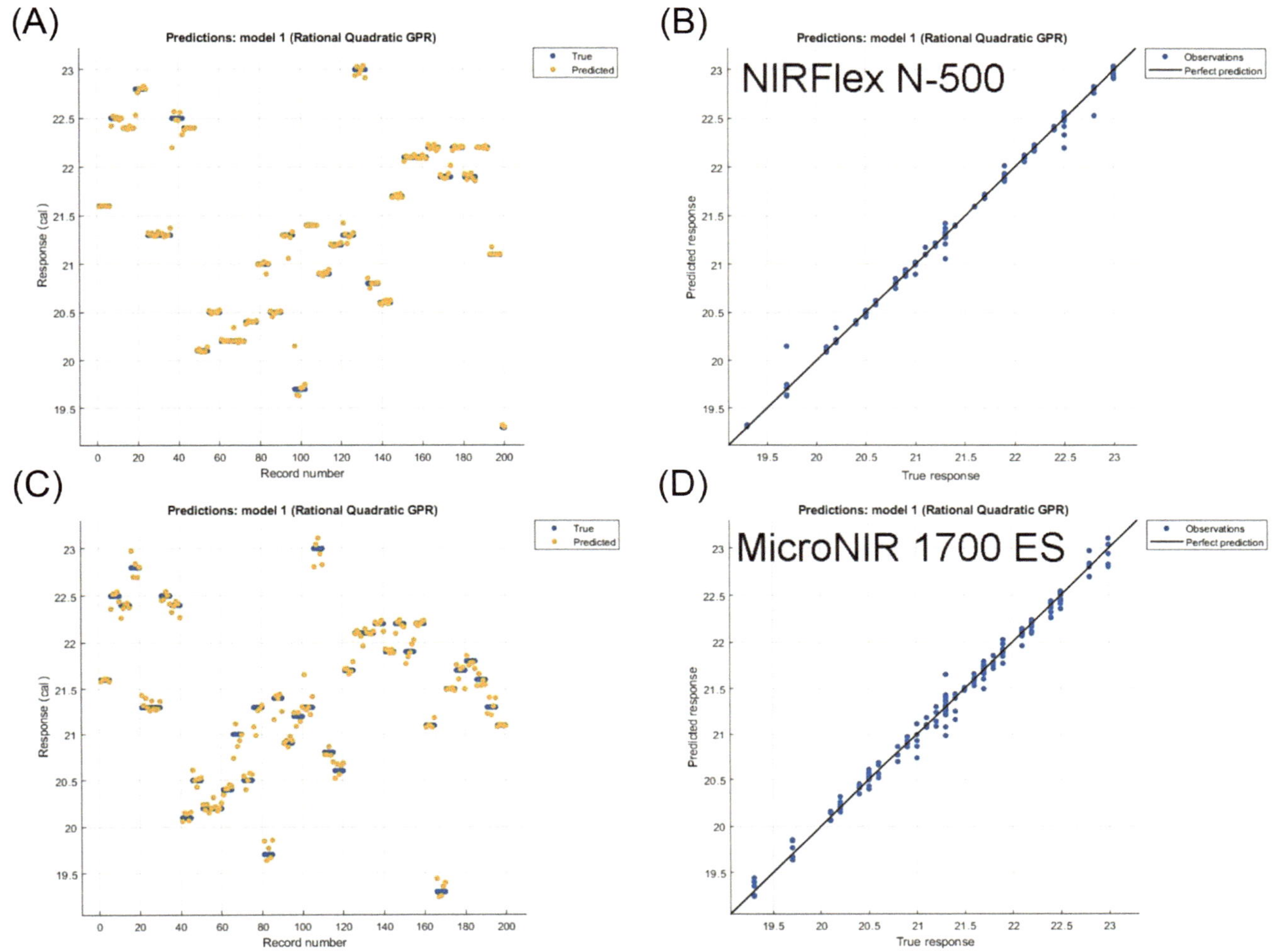

FIGURE 10.8 The fit of the GPR model to the calibration set of milled samples for (A, B) benchtop and (C, D) miniaturized MicroNIR 1700 ES spectrometer. (A, C) Response plot. (B, D) Predicted versus true response plot. *Reproduced from (CC-BY 4.0 license) Beć, K. B., Grabska, J., & Huck, C. W. (2021a). X-Ray, LIBS, NMR, and MS applications in food, feed, and agriculture. In R.A. Crocombre, P.E. Leary, B.W. Kammrath (Eds.),* Portable spectroscopy and spectrometry 2: Applications *(pp. 299–324). John Wiley & Sons, Ltd. https://doi.org/10.1002/9781119636489.ch36; Beć, K. B., Grabska, J., & Huck, C. W. (2021b). Principles and applications of miniaturized near-infrared (NIR) spectrometers.* Chemistry-A European Journal, *27, 1514–1532. https://doi.org/10.1002/chem.202002838.*

cloud-based services for processing and analyzing spectra, as well as the development of ultra-miniaturized spectrometers that are even more accessible to consumers. Despite the remaining challenges, the potential of miniaturized NIR spectroscopy to support sustainability in modern food monitoring systems is becoming increasingly apparent.

References

Abu-Khalaf, N., & Hmidat, M. (2020). Visible/near infrared (VIS/NIR) spectroscopy as an optical sensor for evaluating olive oil quality. *Computers and Electronics in Agriculture.*, *173*, 105445. Available from https://doi.org/10.1016/j.compag.2020.105445.

Amuah, C. L. Y., Teye, E., Lamptey, F. P., Nyandey, K., Opoku-Ansah, J., & Osei-Wusu Adueming, P. (2019). Feasibility study of the use of handheld NIR spectrometer for simultaneous authentication and quantification of quality parameters in intact pineapple fruits. *Journal of Spectroscopy*, *597*, 5461. Available from https://doi.org/10.1155/2019/5975461, 2019.

Anyidoho, E. K., Teye, E., & Agbemafle, R. (2020). Nondestructive authentication of the regional and geographical origin of cocoa beans by using a handheld NIR spectrometer and multivariate algorithm. *Analytical Methods*, *12*, 4150–4158. Available from https://doi.org/10.1039/d0ay00901f.

Aykas, D. P., & Menevseoglu, A. (2021). A rapid method to detect green pea and peanut adulteration in pistachio by using portable FT-MIR and FT-NIR spectroscopy combined with chemometrics. *Food Control*, *121*, 107670. Available from https://doi.org/10.1016/j.foodcont.2020.107670.

Baca-Bocanegra, B., Hernandez-Hierro, J. M., Nogales-Bueno, J., & Heredia, F. J. (2019). Feasibility study on the use of a portable micro near infrared spectroscopy device for the "in vineyard" screening of extractable polyphenols in red grape skins. *Talanta*, *192*, 353–359. Available from https://doi.org/10.1016/j.talanta.2018.09.057.

Barthet, V. J., Petryk, M. W. P., & Siemens, B. (2020). Rapid nondestructive analysis of intact canola seeds using a handheld near-infrared spectrometer. *Journal of the American Oil Chemists' Society*, *97*, 577–589. Available from https://doi.org/10.1002/aocs.12335.

Basri, K. N., Hussain, M. N., Bakar, J., Sharif, Z., Khir, M. F. A., & Zoolfakar, A. S. (2017). Classification and quantification of palm oil adulteration via portable NIR spectroscopy. *Spectrochimica Acta A*, *173*, 335–342. Available from https://doi.org/10.1016/j.saa.2016.09.028.

Basri, K. N., Laili, A. R., Tuhaime, N. A., Hussain, M. N., Bakar, J., Sharif, Z., Khir, M. F. A., & Zpplfakar, A. S. (2018). FT-NIR, MicroNIR and LED-MicroNIR for detection of adulteration in palm oil via PLS and LDA. *Analytical Methods*, *10*, 4143–4151. Available from https://doi.org/10.1039/c8ay01239c.

Beć, K. B., Grabska, J., & Huck, C. W. (2020). NIR spectral analysis of natural medicines supported by novel instrumentation, methods of data analysis and interpretation. *Journal of Pharmaceutical and Biomedical Analysis*, *193*, 113686. Available from https://doi.org/10.1016/j.jpba.2020.113686.

Beć, K. B., Grabska, J., & Huck, C. W. (2021). X-Ray, LIBS, NMR, and MS applications in food, feed, and agriculture. In R. A. ; Crocombre, P. E. ; Leary, & B. W. Kammrath (Eds.), *Portable spectroscopy and spectrometry 2: Applications* (pp. 299–324). John Wiley & Sons, Ltd. Available from http://doi.org/10.1002/9781119636489.ch36.

Beć, K. B., Grabska, J., & Huck, C. W. (2021). Principles and applications of miniaturized near-infrared (NIR) spectrometers. *Chemistry-A European Journal*, *27*, 1514–1532. Available from https://doi.org/10.1002/chem.202002838.

Beć, K. B., Grabska, J., & Huck, C. W. (2022). Infrared and near-infrared spectroscopic techniques for the quality control of herbal medicines. In P. K. Mukherjee (Ed.), *Evidence-based validation of herbal medicine* (2nd ed., pp. 603–627). Elsevier. Available from http://doi.org/10.1016/B978-0-323-85542-6.00018-4.

Beć, K. B., Grabska, J., & Huck, C. W. (2022). Miniaturized near-infrared spectroscopy in current analytical chemistry: from natural products to forensics. In V. P. Gupta (Ed.), *Molecular and laser spectroscopy - Advances and applications* (Vol. 3, pp. 141–188). Elsevier. Available from http://doi.org/10.1016/B978-0-323-91249-5.00009-0.

Beć, K. B., Grabska, J., & Huck, C. W. (2022). Miniaturized NIR spectroscopy in food analysis and quality control. Promises, challenges and perspectives. *Foods*, *11*, 1465. Available from https://doi.org/10.3390/foods11101465.

Beć, K. B., Grabska, J., & Huck, C. W. (2023). Near-infrared (NIR) sensors in environmental analysis. In R. Narayan (Ed.), *Encyclopedia of Sensors and Biosensors* (Vol. 4, pp. 484–503). Elsevier. Available from http://doi.org/10.1016/B978-0-12-822548-6.00093-5.

Beć, K. B., Grabska, J., Plewka, N., & Huck, C. W. (2021). Insect protein content analysis in handcrafted fitness bars by NIR spectroscopy. Gaussian process regression and data fusion for performance enhancement of miniaturized cost-effective consumer-grade sensors. *Molecules (Basel, Switzerland)*, *26*, 6390. Available from https://doi.org/10.3390/molecules26216390.

Beć, K. B., Grabska, J., Siesler, H. W., & Huck, C. W. (2020). Handheld near-infrared spectrometers: Where are we heading? *NIR News*, *31*, 28–35. Available from https://doi.org/10.1016/j.talanta.2020.121165.

Borba, K. R., Aykas, D. P., Milani, M. I., Colnago, L. A., Ferreira, M. D., & Rodriguez-Saona, L. E. (2021). Portable near infrared spectroscopy as a tool for fresh tomato quality control analysis in the field. *Applied Sciences*, *11*, 3209. Available from https://doi.org/10.3390/app11073209.

Borghi, F. T., Santos, P. C., Santos, F. D., Nascimento, M. H. C., Correa, T., Cesconetto, M., Pires, A. A., Ribeiro, A. V. F. N., Lacerda, V., Jr., Romao, W., & Filgueiras, P. R. (2020). Quantification and classification of vegetable oils in extra virgin olive oil samples using a portable near-infrared spectrometer associated with chemometrics. *Microchemical Journal, Devoted to the Application of Microtechniques in all Branches of Science*, *159*, 105544. Available from https://doi.org/10.1016/j.microc.2020.105544.

Cáceres-Nevado, J. M., Garrido-Varo, A., De Pedro-Sanz, E., & Pérez-Marín, D. C. (2021). NIR handheld miniature spectrometer to increase the efficiency of Iberian pig selection schemes based on chemical traits. *Spectrochimica. Acta A*, *258*, 119865. Available from https://doi.org/10.1016/j.saa.2021.119865.

Castrignanò, A., Buttafuoco, G., Malegori, C., Genorini, E., Iorio, R., Stipic, M., Girone, G., & Venezia, A. (2019). Assessing the feasibility of a miniaturized near-infrared spectrometer in determining quality attributes of San Marzano tomato. *Food Analytical Methods*, *12*, 1497–1510. Available from https://doi.org/10.1007/s12161-019-01475-x.

Chapman, J., Truong, V. K., Elbourne, A., Gangadoo, S., Cheeseman, S., Rajapaksha, P., Latham, K., Crawford, R. J., & Cozzolino, D. (2020). Combining chemometrics and sensors: toward new applications in monitoring and environmental analysis. *Chemical Reviews*, *120*, 6048–6069. Available from https://doi.org/10.1021/acs.chemrev.9b00616.

Coronel-Reyes, J., Ramirez-Morales, I., Fernandez-Blanco, E., Rivero, D., & Pazos, A. (2018). Determination of egg storage time at room temperature using a low-cost NIR spectrometer and machine learning techniques. *Computers and Electronics in Agriculture*, *145*, 1–10. Available from https://doi.org/10.1016/j.compag.2017.12.030.

Correia, R. M., Tosato, F., Domingos, E., Rodrigues, R. R. T., Aquino, L. F. M., Filgueiras, P. R., Lacerd, V., Jr, & Romão, W. (2018). Portable near infrared spectroscopy applied to quality control of Brazilian coffee. *Talanta*, *176*, 59–68. Available from https://doi.org/10.1016/j.talanta.2017.08.009.

Cozzolino, D. (2016). Near infrared spectroscopy and food authenticity. In M. Espiñeira, & F. J. Santaclara (Eds.), *Advances in food traceability techniques and technologies: Improving quality throughout the food chain* (pp. 119–136). Cambridge, MA, United States: Elsevier. Available from http://doi.org/10.1016/B978-0-08-100310-7.00007-7.

Cozzolino, D. (2016). Near infrared spectroscopy as a tool to monitor contaminants in soil, sediments and water - state of the art, advantages and pitfalls. *Trends in Environmental Analytical Chemistry*, *9*, 1–7. Available from https://doi.org/10.1016/j.teac.2015.10.001.

Cozzolino, D. (2022). Advantages, opportunities, and challenges of vibrational spectroscopy as tool to monitor sustainable food systems. *Food Analytical Methods*. Available from https://doi.org/10.1007/s12161-021-02207-w.

Crocombe, R. A. (2018). Portable spectroscopy. *Application Spectroscopy*, *72*, 1701–1751. Available from https://doi.org/10.1177/0003702818809719.

da Costa Fulgencio, A. C., Passos Resende, G. A., Fontoura Teixeira, M. C., Goncalves Botelho, B., & Matins Sena, M. (2023). Combining portable NIR spectroscopy and multivariate calibration for the determination of ethanol in fermented alcoholic beverages by a multi-product model. *Talanta Open*, *7*, 100180. Available from https://doi.org/10.1016/j.talo.2023.100180.

Dashti, A., Müller-Maatsch, J., Weesepoel, Y., Parastar, H., Kobarfard, F., Daraei, B., Shojaee AliAbadi, M. H., & Yazdanpanah, H. (2022). The feasibility of two handheld spectrometers for meat speciation combined with chemometric methods and its application for halal certification. *Foods*, *11*, 71. Available from https://doi.org/10.3390/foods11010071.

de la Roza-Delgado, B., Garrido-Varo, A., Soldado, A., Gonzalez Arrojo, A., Cuevas Valdes, M., Maroto, F., & Perez-Marin, D. (2017). Matching portable NIRS instruments for in situ monitoring indicators of milk composition. *Food Control*, *76*, 74–81. Available from https://doi.org/10.1016/j.foodcont.2017.01.004.

de Lima, G. F., Cardoso Andrade, S. A., da Silva, V. H., & Araujo, H. (2018). Multivariate classification of UHT milk as to the presence of lactose using benchtop and portable NIR spectrometers. *Food Analytical Methods*, *11*, 2699–2706. Available from https://doi.org/10.1007/s12161-018-1253-7.

dos Santos Pereira, E. V., de Sousa Fernandes, D. D., Ugulino de Araujo, M. C., Dias Diniz, H. G., & Sucupira Maciel, M. I. (2021). In-situ authentication of goat milk in terms of its adulteration with cow milk using a low-cost portable NIR spectrophotometer. *Microchemical Journal, Devoted to the Application of Microtechniques in all Branches of Science*, *163*, 105885. Available from https://doi.org/10.1016/j.microc.2020.105885.

Dixit, Y., Pham, H. Q., Realini, C. E., Agnew, M. P., Craigie, C. R., & Reis, M. M. (2020). Evaluating the performance of a miniaturized NIR spectrophotometer for predicting intramuscular fat in lamb: A comparison with benchtop and hand-held Vis-NIR spectrophotometers. *Meat Science*, *162*, 108026. Available from https://doi.org/10.1016/j.meatsci.2019.108026.

Dumalisile, P., Manley, M., Hoffman, L., & Williams, P. J. (2020). Discriminating muscle type of selected game species using near infrared (NIR) spectroscopy. *Food Control*, *110*, 106981. Available from https://doi.org/10.1016/j.foodcont.2019.106981.

Edwards, K., Manley, M., Hoffman, L. C., Beganovic, A., Kirchler, C. G., Huck, C. W., & Williams, P. J. (2020). Differentiation of South African game meat using near-infrared (NIR) spectroscopy and hierarchical modelling. *Molecules (Basel, Switzerland)*, *25*, 1845. Available from https://doi.org/10.3390/molecules25081845.

Eskildsen, C. E., Sanden, K. W., Wubshet, S. G., Andersenm, P. V., Oyaas, J., & Wold, J. P. (2019). Estimating dry matter and fat content in blocks of Swiss cheese during production using on-line near infrared spectroscopy. *Journal of Near Infrared Spectroscopy*, *27*, 293–301. Available from https://doi.org/10.1177/0967033519855436.

Giussani, B., Escalante-Quiceno, A. T., Boque, R., & Riu, J. (2021). Measurement strategies for the classification of edible oils using low-cost miniaturised portable NIR instruments. *Foods*, *10*, 2856. Available from https://doi.org/10.3390/foods10112856.

Goi, A., Hocquette, J.-F., Pellattiero, E., & De Marchi, M. (2022). Handheld near-infrared spectrometer allows on-line prediction of beef quality traits. *Meat Science*, *184*, 108694. Available from https://doi.org/10.1016/j.meatsci.2021.108694.

Gonçalves, D. B., Santos, C. S. P., Pinho, T., Queirós, R., Vaz, P. D., Bloore, M., Satta, P., Kovács, Z., Casal, S., & Hoffmann, I. (2021). Near infrared reflectance spectroscopy coupled to chemometrics as a cost-effective, rapid, and non-destructive tool for fish fraud control: monitoring source, condition, and nutritional value of five common whitefish species. *Journal of AOAC International*, *104*, 53–60. Available from https://doi.org/10.1093/jaoacint/qsaa114.

Grabska, J., Beć, K.B., & Huck, C.W. (2022a). Current and future applications of IR and NIR spectroscopy in ecology, environmental studies, wildlife and plant investigations. In D. Cozzolino (Ed.), *Comprehensive analytical chemistry* (Vol. 98, pp. 45–76). Elsevier. https://doi.org/10.1016/bs.coac.2020.08.002.

Grabska, J., Beć, K. B., & Huck, C. W. (2022). Miniaturized near-infrared Spectroscopy – The ultimate analytical tool in food and agriculture. In R. A. Meyers (Ed.), *Encyclopedia of analytical chemistry: Applications, theory and instrumentation*. John Wiley & Sons. Available from http://doi.org/10.1002/9780470027318.a9790.

Grassi, S., Casiraghi, E., & Alamprese, C. (2018). Handheld NIR device: A non-targeted approach to assess authenticity of fish fillets and patties. *Food Chemistry*, *243*, 382–388. Available from https://doi.org/10.1016/j.foodchem.2017.09.145.

Guelpa, A., Marini, F., du Plessis, A., Slabbert, R., & Manley, M. (2017). Verification of authenticity and fraud detection in South African honey using NIR spectroscopy. *Food Control*, *73*, 1388–1396. Available from https://doi.org/10.1016/j.foodcont.2016.11.002.

Hao, Y., Geng, P., Wu, W., Wen, Q., & Rao, M. (2019). Identification of rice varieties and transgenic characteristics based on near-infrared diffuse reflectance spectroscopy and chemometrics. *Molecules (Basel, Switzerland)*, *24*, 4568. Available from https://doi.org/10.3390/molecules24244568.

Hasanzadeh, B., Abbaspour-Gilandeh, Y., Soltani-Nazarloo, A., De La Cruz-Gámez, E., Hernández-Hernández, J. L., & Martínez-Arroyo, M. (2022). Non-destructive measurement of quality parameters of apple fruit by using visible/near-infrared spectroscopy and multivariate regression analysis. *Sustainability*, *14*, 14918. Available from https://doi.org/10.3390/su142214918.

Henn, R., Kirchler, C. G., & Huck, C. W. (2017). Miniaturized NIR spectroscopy for the determination of main carbohydrates in syrup. *NIR News*, *28*, 3–6. Available from https://doi.org/10.1177/0960336017695732.

Henn, R., Schwab, A., & Huck, C. W. (2016). Evaluation of benchtop versus portable near-infrared spectroscopic method combined with multivariate approaches for the fast and simultaneous quantitative analysis of main sugars in syrup formulations. *Food Control*, *68*, 97–104. Available from https://doi.org/10.1016/j.foodcont.2016.03.037.

Horcada, A., Valera, M., Juarez, M., & Fernandez-Cabanas, V. M. (2020). Authentication of Iberian pork official quality categories using a portable near infrared spectroscopy (NIRS) instrument. *Food Chemistry*, *318*, 126471. Available from https://doi.org/10.1016/j.foodchem.2020.126471.

Jiang, W., Marini, G., van Berkel, N., Sarsenbayeva, Z., Tan, Z., Luo, C., He, X., Dingler, T., Goncalves, J., Kawahara, Y., & Kostakos, V. (2019). Probing sucrose contents in everyday drinks using miniaturized near-infrared spectroscopy scanners. *Proceedings of the ACM on Interactive, Mobile, Wearable and Ubiquitous Technologies*, *3*, 1–25. Available from http://doi.org/10.1145/3369834.

Karunathilaka, S. R., Fardin-Kia, A. R., Srigley, C., Chung, J. K., & Mossoba, M. M. (2017). Rapid screening of commercial extra virgin olive oil products for authenticity: Performance of a handheld NIR device. *NIR News*, *28*, 9–14. Available from https://doi.org/10.1177/0960336016687521.

Karunathilaka, S. R., Yakes, B. J., He, K., Chung, J. K., & Mossoba, M. M. (2018). Non-targeted NIR spectroscopy and SIMCA classification for commercial milk powder authentication: a study using eleven potential adulterants. *Heliyon*, *4*, e00806. Available from https://doi.org/10.1016/j.heliyon.2018.e00806.

Kosmowski, F., & Worku, T. (2018). Evaluation of a miniaturized NIR spectrometer for cultivar identification: The case of barley, chickpea and sorghum in Ethiopia. *PLOS ONE*, *13*, e0193620. Available from https://doi.org/10.1371/journal.pone.0193620.

Kucha, C. T., & Ngadi, M. O. (2020). Rapid assessment of pork freshness using miniaturized NIR spectroscopy. *Journal of Food Measurement Characterization*, *14*, 1105–1115. Available from https://doi.org/10.1007/s11694-019-00360-9.

Lee, S., Noh, T.G., Choi, J.H., Han, J., Ha, J.Y., Lee, J.Y., & Park, Y. (2017). NIR spectroscopic sensing for point-of-need freshness assessment of meat, fish, vegetables and fruits. In *Sensing for agriculture and food quality and safety IX*, 1021708. International Society for Optics and Photonics. https://doi.org/10.1117/12.2261803.

Liu, N., Aya Parra, H., Pustjens, A., Hettinga, K., Mongondry, P., & van Ruth, S. M. (2018). Evaluation of portable near-infrared spectroscopy for organic milk authentication. *Talanta*, *184*, 128–135. Available from https://doi.org/10.1016/j.talanta.2018.02.097.

Liu, N., Parra, H. A., Pustjens, A., Hettinga, K., Mongondry, P., & van Ruth, S. M. (2018). Evaluation of portable near-infrared spectroscopy for organic milk authentication. *Talanta*, *184*, 128–135. Available from https://doi.org/10.1016/j.talanta.2018.02.097.

Llano Suarez, P., Soldado, A., Gonzalez-Arrojo, A., Vincente, F., & de la Roza-Delgado, B. (2018). Rapid on-site monitoring of fatty acid profile in raw milk using a handheld near infrared sensor. *Journal of Food Composition Analysis.*, *70*, 1–8. Available from https://doi.org/10.1016/j.jfca.2018.03.003.

Ma, Y., Zhang, G.-Z., & Rita-Cindy, S. A.-A. (2019). Quantification of water, protein and soluble sugar in mulberry leaves using a handheld near-infrared spectrometer and multivariate analysis. *Molecules (Basel, Switzerland)*, *24*, 4439. Available from https://doi.org/10.3390/molecules24244439.

Ma, Y. B., Babu, K. S., & Amamcharla, J. K. (2019). Prediction of total protein and intact casein in cheddar cheese using a low-cost handheld shortwave near-infrared spectrometer. *LWT-Food Science Technology*, *109*, 319–326. Available from https://doi.org/10.1016/j.lwt.2019.04.039.

MacArthur, R. L., Teye, E., & Darkwa, S. (2020). Predicting adulteration of Palm oil with Sudan IV dye using shortwave handheld spectroscopy and comparative analysis of models. *Vibrational Spectroscopy*, *110*, 103129. Available from https://doi.org/10.1016/j.vibspec.2020.103129.

Malegori, C., Nascimento Marques, E. J., de Freitas, S. T., Pimentel, M. F., Pasquini, C., & Casiraghi, E. (2017). Comparing the analytical performances of Micro-NIR and FT-NIR spectrometers in the evaluation of acerola fruit quality, using PLS and SVM regression algorithms. *Talanta*, *165*, 112–116. Available from https://doi.org/10.1016/j.talanta.2016.12.035.

Malvandi, A., Kapooor, R., Feng, H., & Kamruzzaman, M. (2022). Non-destructive measurement and real-time monitoring of apple hardness during ultrasonic contact drying via portable NIR spectroscopy and machine learning. *Infrared Physics & Technology*, *122*, 104077. Available from https://doi.org/10.1016/j.infrared.2022.104077.

Manuelian, C. L., Ghetti, M., De Lorenzi, C., Pozza, M., Franzoi, M., & De Marchi, M. (2022). Feasibility of pocket-sized near-infrared spectrometer for the prediction of cheese quality traits. *Journal of Food Composition Analysis*, *105*, 104245. Available from https://doi.org/10.1016/j.jfca.2021.104245.

Marinoni, L., Stroppa, A., Barzaghi, S., Cremonesi, K., Pricca, N., Meucci, A., Pedrolini, G.M., Gallia, A., & Cabassia, G. (2017, June 11–15). On site monitoring of Grana Padano cheese production using portable spectrometers. In *Proceedings of the 18th international conference on near infrared spectroscopy*, Copenhagen, Denmark (pp. 85–90). https://doi.org/10.1255/nir2017.085.

Mayr, S., Beć, K. B., Grabska, J., Schneckenreiter, E., & Huck, C. W. (2021). Near-infrared spectroscopy in quality control of *Piper nigrum*: A Comparison of performance of benchtop and handheld spectrometers. *Talanta*, *223*, 121809. Available from https://doi.org/10.1016/j.talanta.2020.121809.

Mayr, S., Schmelzer, J., Kirchler, C. G., Pezzei, C. K., Beć, K. B., Grabska, J., & Huck, C. W. (2021). Theae nigrae folium: Comparing the analytical performance of benchtop and handheld near-infrared spectrometers. *Talanta*, *221*, 121165. Available from https://doi.org/10.1016/j.talanta.2020.121165.

McGonigle, A. J. S., Wilkes, T. C., Pering, T. D., Willmott, J. R., Cook, J. M., Mims, F. M., III, & Parisi, A. V. (2018). Smartphone spectrometers. *Sensors*, *18*, 223–238. Available from https://doi.org/10.3390/s18010223.

McGrath, T. F., Shannon, M., Chevallier, O. P., Ch, R., Xu, F., Kong, F., Peng, H., Teye, E., Akaba, S., Wu, D., Wu, L., Cai, Q., Le Nguyen, D. D., Man Le, V. V., Pandor, S., Kapil, A. P., Zhang, G., McBride, M., & Elliott, C. T. (2020). Food fingerprinting: Using a two-tiered approach to monitor and mitigate food fraud in rice. *Journal of AOAC International*, *104*, 16–28. Available from https://doi.org/10.1093/jaoacint/qsaa109.

McVey, C., Elliott, C. T., Cannavan, A., Kelly, S. D., Petchkongkaew, A., & Haughey, S. A. (2021). Portable spectroscopy for high throughput food authenticity screening: Advancements in technology and integration into digital traceability systems. *Trends Food Science Technology*, *118*, 777–790. Available from https://doi.org/10.1016/j.tifs.2021.11.003.

McVey, C., Gordon, U., Haughey, S. A., & Elliott, C. T. (2021). Assessment of the analytical performance of three near-infrared spectroscopy instruments (benchtop, handheld and portable) through the investigation of coriander seed authenticity. *Foods*, *10*, 956. Available from https://doi.org/10.3390/foods10050956.

McVey, C., McGrath, T. F., Haughey, S. A., & Elliott, C. T. (2021). A rapid food chain approach for authenticity screening: The development, validation and transferability of a chemometric model using two handheld near infrared spectroscopy (NIRS) devices. *Talanta*, *222*, 121533. Available from https://doi.org/10.1016/j.talanta.2020.121533.

Michael, I. (2018). Smartphone NIR. *Spectroscopy Europe*. Available from https://www.spectroscopyeurope.com/news/smartphone-nir.

Modroño, S., Soldado, A., Martínez-Fernández, A., & De la Roza-Delgado, B. (2017). Handheld NIRS sensors for routine compound feed quality control: real time analysis and field monitoring. *Talanta, 162*, 597−603. Available from https://doi.org/10.1016/j.talanta.2016.10.075.

Muller-Maatsch., & van Ruth, S. M. (2021). Handheld devices for food authentication and their applications: a review. *Foods, 10*, 2901. Available from https://doi.org/10.3390/foods10122901.

Muniz, R., Cuevas-Valdes, M., & de la Roza-Delgado, B. (2020). Milk quality control requirement evaluation using a handheld near infrared reflectance spectrophotometer and a bespoke mobile application. *Journal of Food Composition Analysis, 86*, 103388. Available from https://doi.org/10.1016/j.jfca.2019.103388.

Musa, M. A. (2022). Evaluation of portable nir assay for detection of milk adulteration with water. *Journal of Advanced Diary, 10*, 1000599. Available from https://doi.org/10.35248/2329-888X.22.10.599.

Müller-Maatsch, J., Bertani, F. R., Mencattini, A., Gerardino, A., Martinelli, E., Weesepoel, Y., & van Ruth, S. (2021). The spectral treasure house of miniaturized instruments for food safety, quality and authenticity applications: A perspective. *Trends in Food Science & Technology, 110*, 841−848. Available from https://doi.org/10.1016/j.tifs.2021.01.091.

Neves, M. D. G., Poppi, R. J., & Siesler, H. W. (2019). Rapid determination of nutritional parameters of pasta/sauce blends by handheld near-infrared spectroscopy. *Molecules (Basel, Switzerland), 24*, 2029. Available from https://doi.org/10.3390/molecules24112029.

Nieto-Ortega, S., Olabarrieta, I., Saitua, E., Arana, G., Foti, G., & Melado-Herreros, A. (2022). Improvement of oil valorization extracted from fish by-products using a handheld near infrared spectrometer coupled with chemometrics. *Foods, 11*, 1092. Available from https://doi.org/10.3390/foods11081092.

Nolasco Perez, I. M., Badaro, A. T., Barbon, S., Jr, Barbon, A. P. A., Pollonio, M. A. R., & Barbin, D. F. (2018a). Classification of chicken parts using a portable near-infrared (NIR) spectrophotometer and machine learning. *Applications Spectroscopy, 72*, 1774−1780. Available from https://doi.org/10.1177/0003702818788878.

Nolasco Perez, I. M., Badaro, A. T., Barbon, S., Barbon, A. P. A., Pollonio, M. A. R., & Barbin, D. F. (2018b). Classification of chicken parts using a portable near-infrared (NIR) spectrophotometer and machine learning. *Applied Spectroscopy, 72*, 1774−1780. Available from https://doi.org/10.1177/0003702818788878.

Oliveira, M. M., Cruz-Tirado, J. P., Roque, J. V., Teófilo, R. F., & Barbin, D. F. (2020). Portable near-infrared spectroscopy for rapid authentication of adulterated paprika powder. *Journal of Food Composition Analysis, 87*, 103403. Available from https://doi.org/10.1016/j.jfca.2019.103403.

Ozaki, Y. (2022). NIR spectroscopy − What a wonderful world!. *NIR News, 33*, 10−17. Available from https://doi.org/10.1177/09603360211067093.

Ozaki, Y., Huck, C. W., & Beć, K. B. (2018). Near-IR spectroscopy and its applications. In V. P. Gupta (Ed.), *Molecular and laser spectroscopy. Advances and applications* (pp. 11−38). San Diego, Calif.: Elsevier. Available from http://doi.org/10.1016/B978-0-12-849883-5.00002-4.

Ozaki, Y., Huck, C. W., Tsuchikawa, S., & Engelsen, S. B. (Eds.), (2021). *Near-infrared spectroscopy*. Singapore: Springer.

Parastar, H., van Kollenburg, G., Weesepoel, Y., van den Doel, A., Buydens, L., & Jansen, J. (2020). Integration of handheld NIR and machine learning to "Measure & Monitor" chicken meat authenticity. *Food Control, 112*, 107149. Available from https://doi.org/10.1016/j.foodcont.2020.107149.

Patel, N., Toledo-Alvarado, H., & Bittante, G. (2021). Performance of different portable and hand-held near-infrared spectrometers for predicting beef composition and quality characteristics in the abattoir without meat sampling. *Meat Science, 178*, 106518. Available from https://doi.org/10.1016/j.meatsci.2021.108518.

Pennisi, F., Giraudo, A., Cavallini, N., Esposito, G., Merlo, G., Geobaldo, F., Acutis, P. L., Pezzolato, M., Savorani, F., & Bozzetta, E. (2021). Differentiation between fresh and thawed cephalopods using NIR spectroscopy and multivariate data analysis. *Foods, 10*, 528. Available from https://doi.org/10.3390/foods10030528.

Perez-Marín, D., Fearn, T., Riccioli, C., De Pedro, E., & Garrido, A. (2021). Probabilistic classification models for the in situ authentication of iberian pig carcasses using near infrared spectroscopy. *Talanta, 222*, 121511. Available from https://doi.org/10.1016/j.talanta.2020.121511.

Pérez-Marín, D., Torres, I., Entrenas, K. A., Vega, M., & Sánchez, M. T. (2019). Pre-harvest screening on-vine of spinach quality and safety using NIRS technology. *Spectrochimica Acta A, 207*, 242−250. Available from https://doi.org/10.1016/j.saa.2018.09.035.

Pochanagone, S., & Rittiron, R. (2019). Rapid detection of infrared inactive sodium chloride content in frozen tuna fish for determining commercial value using short wavelengths. *Journal of Near Infrared Spectroscopy, 27*, 424−431. Available from https://doi.org/10.1177/0967033519872541.

Prieto, N., Dugan, M. E. R., Juárez, M., López-Campos, Ó., Zijlstra, R. T., Aalhus, J. L., & Plaizier, J. (2018). Using portable near-infrared spectroscopy to predict pig subcutaneous fat composition and iodine value. *Candian Journal of Animal Science, 98*, 221−229. Available from https://doi.org/10.1139/cjas-2017-0033.

Reinig, R., Grüger, H., Knobbe, J., Pügner, T., & Meyer, S. (2018). Bringing NIR spectrometers into mobile phones. In *Proceedings of SPIE 10545, MOEMS and miniaturized systems XVII*, 105450F. https://doi.org/10.1117/12.2289931.

Reis, M. G., Agnew, M., Jacob, N., & Reis, M. M. (2022). Comparative evaluation of miniaturized and conventional NIR spectrophotometer for estimation of fatty acids in cheeses. *Spectrochimica Acta A, 279*, 121433. Available from https://doi.org/10.1016/j.saa.2022.121433.

Risoluti, R., Gullifa, G., & Materazi, S. (2020). Assessing the quality of milk using a multicomponent analytical platform MicroNIR/chemometric. *Frontiers in Chemistry., 8*, 614718. Available from https://doi.org/10.3389/fchem.2020.614718.

Riu, J., Gorla, G., Chakif, D., Boque, R., & Giussani, B. (2020). Rapid analysis of milk using low-cost pocket-size NIR spectrometers and multivariate analysis. *Foods, 9*, 1090. Available from https://doi.org/10.3390/foods9081090.

Roberts, J., Power, A., Chapman, J., Chandra, S., & Cozzolino, D. (2018). Vibrational spectroscopy methods for agro-food product analysis. *Comprehensive Analytical Chemistry., 80*, 51−68. Available from https://doi.org/10.1016/bs.coac.2018.03.002.

Rodriguez-Saona, L., Aykas, D. P., Rodrigues Borba, K., & Urtubia, A. (2020). Miniaturization of optical sensors and their potential for high-throughput screening of foods. *Current Opinion Food Science*, *31*, 136–150. Available from https://doi.org/10.1016/j.cofs.2020.04.008.

Rukundo, I. R., Danao, M. G. C., MacDonald, J. C., Wehling, R. L., & Weller, C. L. (2021). Performance of two handheld NIR spectrometers to quantify crude protein of composite animal forage and feedstuff. *AIMS Agriculture Food*, *6*, 462–477. Available from https://doi.org/10.3934/agrfood.2021027.

Rukundo, I. R., Danao, M. G. C., Weller, C. L., Wehling, R. L., & Eskridge, K. M. (2020). Use of a handheld near infrared spectrometer and partial least squares regression to quantify metanil yellow adulteration in turmeric powder. *Journal of Near Infrared Spectroscopy*, *28*, 81–92. Available from https://doi.org/10.1177/0967033519898889.

Sánchez, M. T., Entrenas, J. A., Torres, I., Vega, M., & Pérez-Marín, D. (2018). Monitoring texture and other quality parameters in spinach plants using NIR spectroscopy. *Computers and Electronics in Agriculture*, *155*, 446–452. Available from https://doi.org/10.1016/j.compag.2018.11.004.

Schmutzler, M., & Huck, C. W. (2016). Simultaneous detection of total antioxidant capacity and total soluble solids content by Fourier transform near-infrared (FT-NIR) spectroscopy: a quick and sensitive method for on-site analyses of apples. *Food Contaniers*, *66*, 27–37. Available from https://doi.org/10.1016/j.foodcont.2016.01.026.

Sciuto, S., Esposito, G., Dell'Atti, L., Rossi, F., Riina, M. V., Merlo, G., Magnani, L., Benso, A., Bozzetta, E. M., & Acutis, P. L. (2021). A New approach against food frauds: The portable near-infrared device for fish fillets identification. *Scholarly Journal of Food and Nutrition*, *4*, 442–447. Available from https://doi.org/10.32474/SJFN.2021.04.000177.

Silva, L. C. R., Folli, G. S., Santos, L. P., Barros, I. H. A. S., Oliveira, B. G., Borghi, F. T., Santos, F. D. d, Filgueiras, P. R., & Romão, W. (2020). Quantification of beef, pork, and chicken in ground meat using a portable NIR spectrometer. *Vibrational Spectroscopy*, *111*, 103158. Available from https://doi.org/10.1016/j.vibspec.2020.103158.

Stocco, G., Cipolat-Gotet, C., Ferragina, A., Berzaghi, P., & Bittante, G. (2019). Accuracy and biases in predicting the chemical and physical traits of many types of cheeses using different visible and near-infrared spectroscopic techniques and spectrum intervals. *Journal of Dairy Science*, *102*, 9622–9638. Available from https://doi.org/10.3168/jds.2019-16770.

Tardaguila, J., Fernández-Novales, J., Gutiérrez, S., & Paz Diago, M. (2017). Non-destructive assessment of grapevine water status in the field using a portable NIR spectrophotometer. *Journal of the Science of Food and Agriculture*, *97*, 3772–3780. Available from https://doi.org/10.1002/jsfa.8241.

Teye, E., Amuah, C. L. Y., McGrath, T., & Elliot, C. (2019). Innovative and rapid analysis for rice authenticity using hand-held NIR spectrometry and chemometrics. *Spectrochimica Acta A*, *217*, 147–154. Available from https://doi.org/10.1016/j.saa.2019.03.085.

Teye, E., Elliott, C., Sam-Amoah, L. K., & Mingle, C. (2019). Rapid and nondestructive fraud detection of palm oil adulteration with Sudan dyes using portable NIR spectroscopic techniques. *Food Additives & Contaminants. A*, *36*, 1589–1596. Available from https://doi.org/10.1080/19440049.2019.1658905.

Torres, I., Pérez-Marín, D., De la Haba, M. J., & Sánchez, M. T. (2017). Developing universal models for the prediction of physical quality in citrus fruits analysed on-tree using portable NIRS sensors. *Biosystem Engineering*, *153*, 140–148. Available from https://doi.org/10.1016/j.jfoodeng.2011.12.003.

van Ruth, S., & Liu, N. (2019). How organic is organic milk? Can we have a quick check? *NIR News*, *30*, 18–21. Available from https://doi.org/10.1177/0960336018823490.

Wang, J., Zareef, M., He, P., Sun, H., Chen, Q., Li, H., Ouyang, Q., Guo, Z., Zhang, Z., & Xu, D. (2019). Evaluation of matcha tea quality index using portable NIR spectroscopy coupled with chemometric algorithms. *Journal of the Science of Food and Agriculture*, *99*, 5019–5027. Available from https://doi.org/10.1002/jsfa.9743.

Wang, Y.-J., Li, T.-H., Li, L.-Q., Ning, J.-M., & Zhang, Z.-Z. (2020). Micro-NIR spectrometer for quality assessment of tea: comparison of local and global models. *Spectrochimica Acta A*, *237*, 118403. Available from https://doi.org/10.1016/j.saa.2020.118403.

Wei, W., Peng, Y., & Qiao, L. (2016). Development of hand-held nondestructive detection device for assessing meat freshness. In *SPIE commercial + scientific sensing and imaging*. International Society for Optics and Photonics (pp. 98640W–98640W). https://doi.org/10.1117/12.2223300.

Wiedemair, V., De Biasio, M., Leitner, R., Balthasar, D., & Huck, C. W. (2018). Application of design of experiment for detection of meat fraud with a portable near-infrared spectrometer. *Current Analytical Chemistry*, *14*, 58–67. Available from https://doi.org/10.2174/1573411013666170207121113.

Wiedemair, V., & Huck, C. W. (2019). Investigating the total antioxidant capacity of gluten-free grains with miniaturized near-infrared spectrometer. *NIR News*, *30*, 35–38. Available from https://doi.org/10.1177/0960336019861509.

Wiedemair, V., Langore, D., Garsleitner, R., Dillinger, K., & Huck, C. W. (2019). Investigations into the performance of a novel pocket-sized near-infrared spectrometer for cheese analysis. *Molecules (Basel, Switzerland)*, *24*, 428. Available from https://doi.org/10.3390/molecules24030428.

Wokadala, O. C., Human, C., Willemse, S., & Emmambux, N. M. (2020). Rapid non-destructive moisture content monitoring using a handheld portable Vis–NIR spectrophotometer during solar drying of mangoes (Mangifera indica L.). *Journal of Food Measutrement Characterization.*, *14*, 790–798. Available from https://doi.org/10.1007/s11694-019-00327-w.

Yaman, H., Aykas, D., Jiménez-Flores, R., & Rodriguez-Saona, L. E. (2022). Monitoring the ripening attributes of Turkish white cheese using miniaturized vibrational spectrometers. *Journal of Dairy Science*, *105*, 40–55. Available from https://doi.org/10.3168/jds.2021-20313.

Yan, H., Xu, Y.-C., Siesler, H. W., Han, B.-X., & Zhang, G.-Z. (2019). Hand-held near-infrared spectroscopy for authentication of fengdous and quantitative analysis of mulberry fruits. *Frontiers in Plant Science*, *10*, 1548. Available from https://doi.org/10.3389/fpls.2019.01548.

Yan, J., van Stuijvenberg, L., & van Ruth, S. M. (2019). Handheld near-infrared spectroscopy for distinction of extra virgin olive oil from other olive oil grades substantiated by compositional data. *European Journal of. Lipid Science and Technology*, *121*, 1–11. Available from https://doi.org/10.1002/ejlt.201900031.

You, H., Kim, Y., Lee, J.H., & Choi, S. (2017). Classification of food powders using handheld NIR spectrometer. In *9th international conference on ubiquitous and future networks (ICUFN)*, Milan (pp. 732–734). https://doi.org/10.1109/ICUFN.2017.7993887.

Yu, H., Liu, H., Wang, Q., & van Ruth, S. (2020). Evaluation of portable and benchtop NIR for classification of high oleic acid peanuts and fatty acid quantitation. *LWT*, *128*, 109398. Available from https://doi.org/10.1016/j.lwt.2020.109398.

Zareef, M., Chen, Q. S., Ouyang, Q., Arslan, M., Hassan, M. M., Ahmad, W., Viswadevarayalu, A., Wang, P., & Ancheng, W. (2019). Rapid screening of phenolic compounds in congou black tea (Camellia sinensis) during in vitro fermentation process using portable spectral analytical system coupled chemometrics. *Journal of Food Processing and Preservation*, *43*, e13996. Available from https://doi.org/10.1111/jfpp.13996.

Chapter 11

Emerging food trends: Cellular Agriculture—novel food production technology

Sissel Beate Rønning[1], Mona Elisabeth Pedersen[1] and Erik Bjørnerud[2]
[1]Nofima AS, Raw Materials and Process Optimization, Ås, Norway, [2]Høgskolen i Østfold, Faculty of Teacher Education and Languages, Department of Natural Sciences, Practical-Aesthetic, Social and Religious Studies, Halden, Norway

11.1 Introduction to Cellular Agriculture—two different technologies for novel food production

The world's livestock sector is developing at an extraordinary rate to meet the growing demand for high-value animal protein. Current estimates predict that global food production must increase by more than 60% by 2050. As a result, there is increasing pressure on the livestock sector to meet the growing demand for high-value animal protein. Livestock products are excellent protein sources with high nutritional value and valuable sources of essential micronutrients. There is also an increased recognition of protein's role in a healthy diet, especially in the growing elderly population. However, rising animal production to meet this demand is not sustainable. In fact, food production is responsible for about 25% of the annual greenhouse gas emissions, and meat/dairy products, especially from cows, are the biggest contributors (Poore & Nemecek, 2018). Also, poor animal welfare conditions are often a source of foodborne illnesses such as *Escherichia coli* and *Salmonella* adding to the nonbeneficial impact of traditional meat production. This calls for alternative, new and more environmentally friendly ways to produce animal proteins.

Cellular Agriculture (CellAg) is a new revolutionary way in how we produce food that can significantly change agricultural output and ownership, land use, policy-making, food habits, and not least ethical questions (Burton, 2019; White & Hall, 2017). This chapter focuses on two kinds of CellAg food production systems, focusing on technological aspects and reflections on ethical considerations (see Fig. 11.1 for an overview of the two technologies). The first is precision fermentation (PF), which in addition to traditional fermentation (e.g., beer, bread) includes the cultivation of a microbial species for either whole-cell biomass (biomass fermentation) or for the production of functional food ingredients by recombinant protein technology. The second type of CellAg is cultivated meat, or tissue-engineered cellular agriculture, which involves growing live cells from an animal and then directing these cells into a desired product (e.g., sausage, mince patty, meat). Although currently not cost-effective at scale, these are potentially disruptive technologies predicted to radically change the future landscape of agricultural food production. According to Wikipedia (n.d.), CellAg refers to the "production of agricultural products from cell cultures using a combination of biotechnology, tissue engineering, molecular biology, and synthetic biology." CellAg allows us to consume the same food products that we are accustomed to, like a dairy ice cream or a burger, but created without raising and slaughtering farmed livestock. If successful, this technology is predicted to lead to less land, water, and feedstock use, while the potential energy requirements are still debated (Lynch & Pierrehumbert, 2019) (Fig. 11.1).

11.2 Precision fermentation—the new golden standard for food production?

PF is well established in the biomedical sector and for enzyme production, and this technology has been around since the bacteria *E. coli* was modified to produce insulin in 1982 to help manage diabetes. The technology was developed further to include drugs for cancers and vaccines for hepatitis B. In the 1990s, the technology was further developed in

Food Industry 4.0. DOI: https://doi.org/10.1016/B978-0-443-15516-1.00011-6

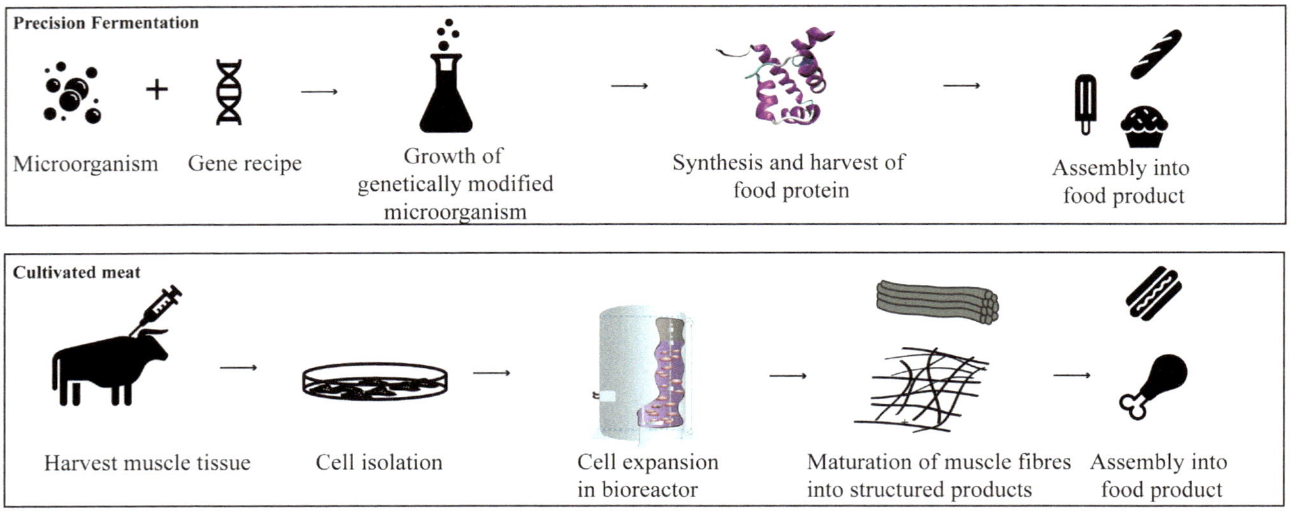

FIGURE 11.1 CellAg production systems Overview of the two kinds of CellAg food production systems described in this chapter. *Image by Sissel Beate Rønning/Nofima. Reproduced with permission from Nofima AS.*

the food segment, with the enzyme chymosin for rennet, an enzyme essential for cheese production. Today, more than 80% of the rennet used for cheese production comes from microorganisms rather than from the stomach of ruminant mammals (GFI, 2020). PF is based on genetically engineered microbial strains producing the proteins of interest during fermentation, and many species may be used as host cells. Foods on the market employing this technology include Impossible Foods' burgers, which contain leghemoglobin (from soybean) heterologous produced in the yeast *Pichia pastoris* (Impossible foods, 2023), and Perfect Day's ice cream, which contains bovine casein and whey proteins produced in the fungus *Trichoderma reseei* (Perfect Day, 2023).

However, further technological developments, including innovations that reduce production costs, are still needed for the technology to become feasible on a wider scale. The most significant advantage of CellAg proteins is the possibility to retain both the functional properties of the original animal-based food and the high nutritional value of these proteins. The resulting products thus resemble their animal-based counterparts with sensory and nutritional properties yet avoid the environmental and ethical issues connected to animal production. However, this is not straightforward. Ingredients such as egg, milk, and meat, used in many foods, are complex and consist of considerably more than the individual proteins produced by CellAg. Bovine milk is a nutritionally rich, chemically complex biofluid consisting of hundreds of different components. Whey proteins (including β-lactoglobulin and α-lactalbumin) constitute 20% of the milk proteins and have a well-defined tertiary structure that demonstrates high resistance to proteolytic enzymes. β-Lactoglobulin is the most abundant whey protein in ruminant milk, and this protein has been expressed in various host organisms in soluble and insoluble forms for decades. Casein is perhaps the most important protein in milk and the most difficult to make. It comprises up to 80% of cow's milk, contains the nine essential amino acids that humans need from the diet, and binds calcium and phosphorus, which are crucial for skeletal growth. Casein is a complex structure composed of four different caseins, which must be produced separately. In mammalians, these four caseins form, together with calcium phosphate clusters, a circular structure called a micelle, which occurs in the mammalian glands. It has proven difficult to replicate the formation of micelles using PF. The production of milk casein using PF is similarly costly due to the need for dedicated growth media, large-scale fermenters, and purification of the expressed protein. On top of that, milk is a complex product in which different proteins, minerals, sugars, and fats interreact: a single protein produced by PF will not act the same on its own compared to the "original," which is a complicated matrix.

Plant-based alternatives to milk cannot replicate the functionality of whey or casein proteins, for example, in the production of yogurt or cheese. Ovalbumin is the most abundant protein in egg whites; it comprises over 50% of egg white proteins and has been expressed in *E. coli* and *P. pastoris* mainly for research purposes. PF can also be used for the production of vitamins, minerals, and relevant food components. Vitamin B12 is one example, an essential vitamin involved in the metabolism of every cell in the human body; and only obtained by consuming animal-sourced foods, such as milk, meat, fish, and eggs. Vegans are advised to consume dietary supplements to avoid serious health consequences from lacking vitamins like B12. Industrial production of vitamin B12 can be achieved through PF; *Pseudomonas denitrificans*, *Sinorhizobium meliloti*, and *Propionibacterium shermanii* are common production systems.

However, these systems have long fermentation cycles, require expensive growth media, lack suitable tools for genetic engineering, and often have low B12 yields (Fang et al., 2017). Replacing these platforms with *E. coli* could solve some of the challenges, as *E. coli* is used extensively in the industrial production of similar products (e.g., chemicals, terpenoids) and is well established in metabolic engineering and synthetic biology strategies. Vitamin B12 biosynthesis is a complex metabolic process, and genetic modification is a promising strategy for improving biosynthesis and production efficiency. As another example, Vitamin K is a fat-soluble vitamin present in a variety of foods, including vegetables, fish, meat, cheese, and eggs. Vitamin K does not denote a single molecule but rather a group of closely related derivates with a 2-methyl-1,4-naphthoquinone structure as its common framework. Vitamin K2 is made of menaquinones (MKs), in which MK-4 has the shortest isoprenic side chain. Menaquinones are present in various amounts in animal products (such as meat and cheese) and plant-based fermented food (e.g., fermented cabbage and natto). Industrially, Vitamin K2 can be produced through organic synthesis or fermentation (Kang et al., 2022).

The list of relevant extracellular components produced by fermentation could continue with glutamate, which is produced via some *Micrococcus* species, amino acids (e.g., lysine, threonine, tryptophan), *Corynebacterium*, and, as technology develops, other sources as well (Kumar et al., 2014). However, many challenges must still be overcome to enable commercial low-cost production, including the lack of efficient fermentation systems (in terms of microorganisms, input, and equipment) and expensive growth media. It is important to identify key proteins/ingredients and highly specialized macromolecules that can be produced with suitable expression systems and optimize the fermentation process using food industry by-products to achieve economically viable production. Recombinant collagen production via plants or microorganisms like yeast and *E. coli* are promising approaches to producing animal-free collagen, but they still suffer from low yield, high cost, and challenges with posttranslational modification that limit the biological functionality of the recombinant collagen. However, new metabolic engineering processes are required to produce edible proteins. In modern industrial fermentation, a host microorganism's genes are often edited to produce desired macromolecules. However, the expression of fully functional novel collagen molecules is difficult due to the complex nature of their biosynthesis.

11.2.1 Gene editing is necessary

Production of CellAg food products using microbes, such as meat-like and dairy-like products, involves gene editing and reverse engineering and has the potential to tailor-make the texture, taste, and nutrients of food. Main technical barriers must be overcome, including identifying suitable expression systems for each target protein, metabolic engineering, process optimization to achieve scalable production, and incorporation and inclusion in food products to ensure high functional and nutritional quality. Recombinant proteins for these applications are mainly produced using prokaryotic and eukaryotic expression host systems such as mammalian cells, bacteria, yeast, insect cells, and transgenic plants at laboratory scale as well as in large-scale settings (Tripathi & Shrivastava, 2019). *Escherichia coli* offers high product yield and fast growth rate and has a well-known biochemistry and genetics. Disadvantages of *E. coli* include lack of posttranslational modifications, misfolding, inclusion body formation, and issues with endotoxin production (Georgiou & Valax, 1996; Tripathi & Shrivastava, 2019). Recently, the gene-editing tool CRISP/Cas9 has been used to successfully integrate functional genes in *E. coli* strain (Huang et al., 2019). The yeast systems *Saccharomyces cerevisiae* and *P. pastoris* (a.k.a. *Komagataella phaffi* or *K. pastori*) are good choices for expression systems as they provide posttranslational modifications, they have rapid growth, can easily be genetically manipulated, and the complete genome sequence is available. *P. pastoris* has qualified presumption of safety (QPS) status for food ingredient production in the EU regulatory system, and it can produce proteins with glycosylation and disulfide binding. As another benefit, *P. pastoris* often secretes the macromolecules it produces, as opposed to the proteins produced in *E. coli*, which can be trapped in inclusion bodies or the periplasm (the space just inside a cell's outer membrane or layer). Although mammalian cell lines can express large and complex recombinant proteins, they are expensive and mostly used for the production of recombinant therapeutics. Accurate replication of not only the amino acid sequence but also posttranslational modifications such as glycosylation, acetylation, or phosphorylation might be required to achieve identical protein secondary and tertiary structures and, thus, functional properties like those of animal-based protein(s). Also, the process's scalability and identifying edible and sustainable components of the growth medium are major technical problems still to be solved.

11.3 Cultivated meat from skeletal muscle stem (satellite) cells

The methodology for cultivated meat (CM) is inspired by techniques used for medical purposes like reconstructing the damaged muscle tissue of patients using their own cells or the large-scale production of biopharmaceuticals using mammalian cells. Efforts to produce food in vitro began with the aim to produce muscle-like protein for astronauts and space station residents (Benjaminson et al., 2002). CM from muscle satellite cells or fat cells is considered a new type of meat with great prospects due to the tremendous potential of these cells to multiply. The basic technology for cultivating meat is shown in the following steps (1) harvesting and isolating muscle or adipose stem cells, (2) increasing the number of cells in a bioreactor in a growth medium, (3) maturing the stem cells into muscle or fat cells and fibers using edible scaffolds, and (4) processing and assembling into a final meat product.

The cultivation of muscle or fat cells outside the living animal in a bioreactor promises a revolutionary new alternative to the traditional way of producing animal protein that bypasses animal production. Optimistic estimates suggest that, 10,000 kg of cultivated meat can be generated from as little as 1 g of beef muscle using this modern technology. According to calculations described by Melzener et al. (2021) only around 50 cattle would be needed to meet the beef demand in Maastricht for up to 2 years, and a herd size of 500 animals could supply cultivated meat to all of Europe. The first laboratory-grown burger was available for tasting at Prof. Mark Post's lab at Maastricht University in 2013, where it had been cultivated. This first proof-of-concept burger had a production cost of $300,000 and took more than 2 years to produce. Through constant process optimization, the price dropped but is still considered too expensive for a profitable business. The estimated production cost per kilogram of animal protein from a bioreactor is tentative as no current industrial process exists. In a technical assessment cost analysis performed by Delft in 2021, the calculations demonstrate the current cultivated meat production costs are 100–10,000 × higher than conventional meat products, and the cell culture medium is the most expensive. Nevertheless, as the industry scales and the cultivated meat production becomes more efficient beyond 2030, further cost is expected to decrease (Vergeer et al., 2021). The amount of input protein needed to produce cultivated meat is very low compared to traditional livestock production, which is why cultivated meat represents a new and sustainable alternative. For beef, 1.33 kg of protein (i.e., 7 kg of grain containing 19% protein) must be fed to produce 200 g of protein in meat. In contrast, the scaled-up clean meat procedure requires an estimated 225 g of nutrients (amino acids, glucose, etc.) to produce the same 200 g protein. Thus, cultivated meat production is predicted to be about six times more efficient than conventional beef production (Post, 2014).

Having already established that it is possible to create meat grown in a bioreactor, the current challenge is to succeed with a large-scale production that is economically feasible and environmentally friendly. It is challenging to grow high concentrations of viable muscle or adipose cells in a bioreactor because they are adherent, meaning they need to grow on a solid surface. To solve this, microcarriers are often used for their high surface-to-volume ratio. Furthermore, efficient cell growth depends on optimal growth conditions, including sufficient nutrients and an appropriate microenvironment, which is difficult to maintain at high cell concentrations. Contractile proteins, such as actin and myosin, are essential to achieve the functional properties of cultivated muscle as a food ingredient. For the cells to start producing these proteins, it is important to facilitate tension and anchorage during the differentiation phase so that they can self-organize into muscle fibers. This is achieved by letting the cells grow in a functional gel—for example, collagen. After achieving mature muscle fibers, muscle proteins with good functional properties (protein content, water binding ability, gel formation) can be harvested and assembled into food products. The potential of using these animal proteins as ingredients in food products is unlimited, not only as pure meat products, but also as protein-enriched nutraceuticals or vegan products needed optimization regarding water binding and gelling properties.

11.3.1 The cell source matters

The most promising cell source for cultivating meat is skeletal muscle satellite stem cells (MuSCs) (see Fig. 11.2 for the structure and different cell types present in a skeletal muscle).

In 1961, Mauro first suggested that these cells were involved in skeletal muscle regeneration after injury (Mauro, 1961). Under normal circumstances in the adult muscle, these cells are in a state of inactivity (quiescent), but the cells are activated upon exercise, injury, or disease. Following activation, these cells exhibit a remarkable ability to self-renew, expand, grow, or undergo differentiation to fuse and restore damaged muscle. These cells can be isolated in the lab and grown ex vivo. Comparative studies have shown that the MuSCs cultivated in the lab display morphological, metabolic, and biochemical similarities to the adult muscle (Aas et al., 2013). Such cultivated cells have been used for more than 30 years in medical research purposes, and these unique features allow us to take the step further in cultivating muscle cells to produce meat.

These cells were also the type of cells that Prof. Mark Post used when he and his colleagues produced the famous burger in 2013. MuSCs are an attractive cell source for the production of cultivated meat, but their utility is limited by the fact that they have low proliferative capacity; that is, these cells can only produce a limited amount of cell-grown meat, with 50–70 doublings in accordance with the Hayflick limit (Hayflick & Moorhead, 1961). Various approaches

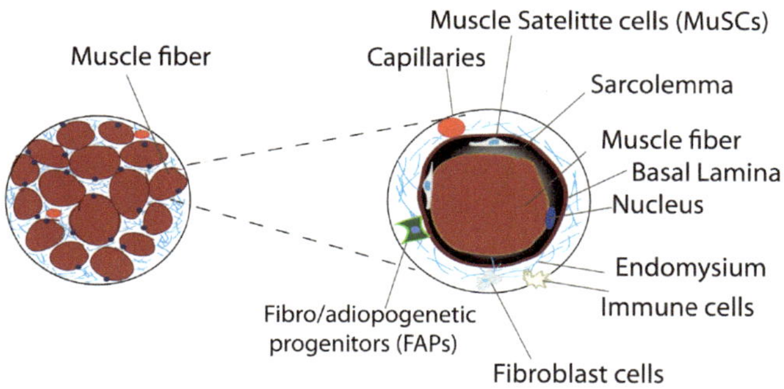

FIGURE 11.2 The structure of and the different cell types present in a mammalian skeletal muscle The mammalian skeletal muscle cells contain different cell types, including satellite cells, immune cells, fibro/adiopogenetic progenitors, and fibroblast cells. *Image by Sissel Beate Rønning/Nofima. Reproduced with permission from Nofima AS.*

have been explored to expand the proliferation capacity of the cells to improve manufacturing, including optimizing culture media and culture conditions (Ding et al., 2018). However, genetic modification is perhaps the most powerful tool for generating cell lines for cell-cultivated meat production.

One robust strategy for developing stable cell lines capable of omitting senescence, old age, and eventual cell death entails targeted approaches at the molecular level of the expression levels of telomerase, which affects the aging of the cells and key regulators driving the cell cycle. Establishing a muscle satellite cell line capable of indefinite population doubling is an essential first step in developing a commercial process for cell-cultivated meat production. Such a cell line can differentiate into 3D muscle tissue, thus mimicking the texture of conventional meat and eventually realizing the large-scale production of cultivated meat. Cells other than MuSCs can also be used as starting points to produce cultivated meat; for instance, embryonic stem cells and induced pluripotent stem cells represent an exciting opportunity (Post et al., 2020). These stem cells are isolated from the early stage of an embryo. They can either be used as a cell line or are cellularly reprogrammed to form myoblasts, embryonic precursors of muscle cells. The advantage of using stem cells is that they are potentially immortal and can be developed into all cell types. However, making a stable embryonic stem cell line is extremely difficult, as these cells are usually complicated to work with. Cellular reprogramming to obtain induced pluripotent stem cells can overcome this challenge. It can be done either by various advanced molecular techniques, such as viral-mediated overexpression of certain transcription factors or delivery of mRNA (Rackham et al., 2016). These types of cells have been extensively studied; however, they often require gene editing to proliferate in culture, they might suffer from low yields, and most importantly, they may not mimic the precise behavior of matured muscle cells.

Another approach is the use of other starting cells (e.g., fibroblasts) to be transdifferentiated into muscle or fat. In a recent paper, scientists induced spontaneous immortalization of fibroblasts derived from chicken embryos, reaching high densities of fat cells (Pasitka et al., 2023). As mentioned, most of the previous work has been done on mammalian or avian skeletal muscle cells, but fish, various seafood, and insect cells have also been reported (Rubio et al., 2019). Some start-up companies within cultivated seafood, like the San Francisco company Wildtype managed to produce enough lab-grown Coho salmon to serve it at a restaurant for invited guests. Finless Food has produced cells from bluefin tuna, although in low-capacity culture flasks with adherent growing cells, which at present limit upscaled production. Also, the production costs for lab-grown seafood are substantial and presently too high for a profitable business. Fish cultivation is performed at a different temperature scale than mammalian cells, and developers of efficient bioprocesses and bioreactor designs must take this into consideration. However, as a sustainable, potentially CO_2-neutral, and nutrient-rich protein source, lab-grown seafood promises numerous opportunities. Another interesting source of cultivated meat and fat is using insect cells. The advantages of this approach include low requirements of input resources, robust cells, and relatively simple projections for large-scale production (Letcher et al., 2022; Rubio et al., 2019).

11.3.2 Tissue engineering—use of scaffolds to produce a piece of cultivated meat

To create 3D structures, the cells must be surrounded by a scaffold (Seah et al., 2022). Muscle development is greatly influenced by its microenvironment; in nature, cells are surrounded by the extracellular matrix (ECM) protein scaffold that supports cell anchorage and tissue assembly. The ECM is composed of fibrous proteins such as collagens, glycoproteins, and proteoglycans. The ECM also contains carbohydrate epitopes, such as glycosaminoglycans (GAG), attached to the protein core of the proteoglycans. The GAGs and the fibrous proteins communicate with the cell receptors and regulate growth factor activities of importance for the fate of the skeletal muscle cells. Culturing bovine skeletal muscle cells on simple coating of collagen versus a mixed composition of glycoproteins and GAGs improves primary bovine skeletal muscle proliferation and fusion into myofibers (Rønning et al., 2013). Collagen is the most abundant ECM protein in skeletal muscle and accounts for roughly 1%−10% of the dry muscle matter. Scaffold design is critical to cultivated meat production, and an optimal scaffold must provide a suitable 3D environment on which the cells can attach, proliferate, and differentiate while maintaining optimal porosity to enable nutrient and oxygen supply.

Most importantly, the scaffold must be derived from inexpensive sources. There are a large number of possible options: as, alginate, chitosan, and collagen, all of which are natural ingredients found in algae, shellfish, plants, and animals. Alginate is a low-cost, nonmammalian carbohydrate polymer extracted from algae and used in tissue engineering, while chitosan, also of a carbohydrate nature, can be produced from mushrooms. However, alginate and chitosan do not possess the cell recognition sites necessary for muscle tissue cells to attach and migrate and must be modified with proteins or peptides to promote cell adhesion. Collagen and collagen-derived gelatins are attractive ingredients for edible scaffold production and are often used in both food and pharmaceutical industries. This material contains cell recognition sequences, and is biocompatible and biodegradable, and gelatin facilitates efficient cell attachment for most cell types. Collagen isolated from turkey tendon restrains fibrillar properties after extraction (Grønlien et al., 2019), but needs crosslinking-stabilization for mechanical properties needed as hydrogels or scaffolds. For optimal function, the scaffold should have optimal mechanical properties mimicking ECM in vivo, and with certain elastic properties (Kohane & Langer, 2008). This collagen modification should be edible, and UV-crosslinking by riboflavin or lumichrome is promising in that perspective (Grønlien et al., 2022). Eggshell membrane, consisting of fibrillar collagens together with a highly organized structured network of glycoproteins and GAGs, and with good mechanical properties, is an interesting collagen alternative, of edible nature and potential for use as a scaffold in meat production (Andreassen et al., 2022; Rønning et al., 2020). Ideally, the source of the scaffold should be animal-free. However, collagen and gelatin are derived from animal sources, which is a drawback, and other vegan sources as biomaterial is at present explored as scaffold material.

Recently, Ben-Arye and colleagues produced a meat-like product with high nutritional value and texture using textured soybean as a scaffold (Ben-Arye et al., 2020). This affordable, widely available, and sustainable plant-based scaffold has the potential to bring cultivated meat one step closer to realization. 3D-bioprinting, using edible bioink with skeletal muscle cells, is an alternative for culturing skeletal muscle in or on scaffolds, where the scaffold (a bioink) is printed together with skeletal muscle cells in advanced 3D-morphological structures or cell sheets. This contrasts to premade porous scaffolds produced by technologies such as electrospinning, freeze-dried, or extruder technology (Levi et al., 2022; Rahmati et al., 2021). Developing edible printable bioinks is in progress, but generating good viscosity properties with proper rheological properties is a challenge for natural biopolymers as collagen, and the use of collagen-alginate together with synthetic nonedible polymers as polyethylene glycol is often more successful. The viscosity of the bioink and gelling time is relevant for the resolution of the printed structure and for preserving its shape postprinting. However, bioinks with high polymer concentrations are more easily printable but usually not suitable for cells since they limit cell spreading, migration, proliferation, and matrix remodeling (Ostrovidov et al., 2019). The delivery of nutrients to cells is a key problem in fabricating thick tissues. In addition, maintaining cell viability and long-term functionality postprinting until remodeling and regeneration is complete represent key challenges with this technology (Ostrovidov et al., 2019). Cocultures of skeletal muscle cells with endothelial cells for vascularization of muscle, fibrin coating for stimulating angiogenesis, and perforation with nozzles to mimic channel structures are some examples of overcoming nutrition delivery problems. Bioprinting offers some advantages of inducing myotube fusion compared to common porous scaffolds following cell seeding, as cell density and cell−cell contact are increased (Fig. 11.3).

11.4 Scalability—a mind-blowing task?

To scale up cell culture to a sufficient level to feed the world is a mind-blowing task that requires substantial innovation and transdisciplinary collaboration. In a report by Humbird, several challenges were identified, including the need for large-scale bioreactors (Humbird, 2021). According to calculations from Post and Van Der Weele (2020), to feed the world population in 2050 with cultivated meat, one would need at least 600 million bioreactors, each of 25,000 L. Estimations by McKinsey (https://www.mckinsey.com/industries/agriculture/our-insights/cultivated-meat-out-of-the-lab-into-the-frying-pan, n.d.) suggest between 2.2×10^8 and 4.4×10^8 L bioreactors just to reach 0.5% of global meat production. To meet this demand, thousands of new large-scale bioreactors must be manufactured. Unfortunately, the worldwide fermentation capacity was reduced in 2021, not increased. Most of the production facilities are not developed for food but rather for pharmaceutical and bioethanol production. Finally, there is little public or private capital for companies to develop scale-up facilities. For cultivated meat, tissue culturing is mostly done on a small scale and/or

FIGURE 11.3 Skeletal muscle created using a 3D bioprinter Small pieces of cultivated meat are produced in the laboratory, using a 3D bioprinter, an edible bioink, and bovine skeletal muscle cells. *Image by Kjell Merok/Nofima. Reproduced with permission from Nofima AS.*

by rather costly means. Large-scale bioproduction of animal proteins for the purpose of cultivated meat production requires the suspension of cell cultures in bioreactors. Muscle/fat cells grow in suspensions and adhere to microcarrier beads that provide support. Nutrients and gases inside the bioreactor are mixed by an impeller, gas flow, or bioreactor rotation. Once the muscle cells have reached high densities, new beads are added to the reactor; this allows the cells to transfer and colonize new beads.

Cell proliferation (growth) and differentiation (maturation) can occur on the microcarriers, which helps avoid any additional handling of the cell culture. Stirred bioreactors have a significant advantage over static culture vessels as they provide a more homogenous environment for the growing cells. Still,the need for adequate oxygen supply must be balanced against the detrimental effects of hydrodynamic shear stress induced by the impeller. On-line monitoring of nutrition utilization ensures efficient production. The build-up of metabolic by-products and changes in pH may affect each cell type and phenotype stage (proliferation, maturation) differently. Also, adequate handling and removal of CO_2 are critical, as its accumulation will inhibit cell growth and protein production. The selection of appropriate microcarriers is fundamental because their structure, coating, and particle size influence cell attachment, growth, and detachment (Bodiou et al., 2020). Microcarriers are often made of polystyrene so they float easily in the medium; Cytodex1 is a typical microcarrier used in stirred tank bioreactors that works well with bovine muscle cells. However, they are not food grade, and the cells must thus be harvested from the carriers at the end of the production, which can be a challenging process. Therefore, it would be highly beneficial to develop edible microcarriers that can be included in the final food product. Collagen-based biodegradable beads and beads coated with extracellular matrix materials have shown potential usefulness in other tissue-engineering fields (e.g., for cartilage repair). As a by-product of animal production, they are biological in origin and exhibit properties that may not exist in synthetic materials—edibility and bioactivity, for example. Other cheap, abundant, edible, and degradable alternative materials include eggshell membranes, alginates, pectins, cellulose, and textured soy protein (Seah et al., 2022; Andreassen et al., 2022; Ben-Arye et al., 2020).

An optimal cell growth medium for CellAg must be affordable, contain only food-grade components, be available in massive quantities, and effectively promote cell growth and differentiation. Growth media generally contain glucose, amino acids, minerals, vitamins, and buffers. All these ingredients that can be generated by bacterial fermentation and obtained from plant sources. In addition, traditional cell culture media for cultivated meat application contain blood serum. This blood serum is harvested from bovine fetuses at the slaughterhouse and contains growth factors, enzymes, lipids, hormones, carbohydrates, and proteins crucial for growth and cell differentiation. However, serum supplementation is a limiting factor for sustainable large-scale production of cultivated meat for human consumption. Serum is extremely expensive; the cost can be up to 96% of the total cost of cell growth media, supply is often lower than demand and its composite, and undefined nature can induce variability in cell growth (GFI, 2019). A considerable amount of research aims to develop serum-free media (SFM) using recombinant proteins and has so far resulted in several different SFM formulations allowing high cell growth and differentiation capacity (Butler, 2013; Kolkmann et al., 2022; Post et al., 2020; Stout et al., 2022; van der Valk et al., 2010; Yamanaka et al., 2023). While only low concentrations of growth factors are needed in the serum, they are rather costly, and as such, reducing their cost is critical. Key drivers for high costs are fibroblast growth factor 2 and transforming growth factor-β. One approach to avoiding the high costs of serum is to adapt cells to require fewer growth factors or to produce cheap synthetic analogs. These proteins could also be produced through fermentation by genetically engineered microorganisms such as *E. coli* (Venkatesan et al., 2022).

Other promising constituents that can be included in a tailor-made growth medium include bioactive molecules derived from food industry by-products like carbohydrates, glycosaminoglycans, eggshell membranes, yeast extracts, and protein hydrolysates (Andreassen et al., 2020). In a recent techno-economic analysis for the production of cultivated meat, produced by CE Delft, one of the most promising measures for price reduction was to use soy hydrolysates as medium ingredients (Vergeer et al., 2021). There are several reports that show beneficial effects on cell culture performance when supplementing the media with hydrolysates (Aldén et al., 2007; Andreassen et al., 2020). Protein hydrolysates are produced from purified protein sources by heating with acid or, preferably, the addition of proteolytic enzymes for cleavage specificity, followed by downstream purification procedures. Each protein hydrolysate is a complex mixture of peptides of different chain lengths together with free amino acids. The hydrolysates have growth-promoting effects, and some also stimulate insulin-associated signaling pathways in mammalian cell culture (Lima et al., 2015). The hydrolysates contain analogs for hormones and active domains of specific proteins that stimulate cell signaling, such as growth factors (GFs). In addition, hydrolysates are low-cost, easy-to-obtain, food-safe ingredients containing a wide range of low molecular weight nutrients found in traditional basic cell culture mediums. They have potential GF-mimicking properties that can regulate cell growth or differentiation and antimicrobial and antioxidant properties that can abolish the need for antibiotics and protect cells from oxidative stress while increasing the stability of the cell media (Jakubczyk et al., 2020; Lima et al., 2015).

Many biomass sources can theoretically also serve as input for PF (Singh et al., 2017). By-products from food production, such as poultry carcasses, intestines, blood, and skin, alongside dairy by-products, are available in massive quantities, cannot be used as food in their current form, and contain proteins and other essential nutrients, making them promising carbon and nitrogen sources in CellAg production (Aspevik et al., 2017; Lindberg et al., 2017). A study performed at Nofima demonstrated that growth media based on food-grade by-products pretreated with enzymatic hydrolysis supported growth to a similar extent as commercial nonfood control media (Tzimorotas et al., 2018) for the production of fatty acids by fermentation. Also, plant-based waste streams have a high potential to be recycled as components of feedstocks in various types of fermentations (e.g., starch, an essential feedstock in the fermentation industry) (Xu et al., 2016).

11.5 Functional attributes, sensory and nutritional value of CellAg food

The biggest challenge may be to mimic the vast complexity of food products and unlock the mysteries of why we crave them. The biological constituents, protein properties structural organization, or postmortem processes are important for achieving this. This has been a topic for almost 50 years in traditional meat science. Conventional meat products are highly nutritious and contain all the essential amino acids, vitamins, and minerals necessary for humans. In addition to being complete protein sources, the structural muscle proteins myosin and actin hold various functional properties, such as forming gels, binding water and proteins, and forming emulsions. A typical cut of meat is made up primarily of skeletal muscle cells and connective tissue, while fat, arteries, and veins are also important constituents of a muscle. The lipids in the fat contribute to the flavor and juiciness of the cooked meat, and the marbling of a steak makes the meat juicier and more flavorful. All these properties may be taken into consideration when CellAg products are incorporated into new food products.

While tasting the first cultivated meat burger in 2013, the burger was deemed salty and dry. As such, improving the meat quality characteristics of CellAg meat products was and is necessary for successful production. Mitigation strategies include cultivating skeletal muscle cells and fat cells separately, and then combining them in the final product, as was demonstrated recently by Kang et al. (2022), who assembled differentiated muscle, adipose tissue, and blood capillary cell fibers into a steak-like meat using 3D bioprinting. Isolating primary cells for culturing bovine myoblasts and adipocytes separately has been investigated and proven successful, but attempts at coculturing these cells have proven difficult (Post et al., 2020). This is because each cell type grows and specializes in specific media. Although adipocytes can produce various saturated and polyunsaturated fatty acids, essential fatty acids, such as linoleic and α-linoleic acids, may still be missing in cultivated meat. Much of the flavor of the meat is released during heating, where an enormous number of volatiles are induced, mainly through Maillard reactions and lipid degradation reactions. Maillard reactions in conventional meat involve a reaction between free amino acids and reducing sugars, precursor products typically produced during postmortem processes. Whether these precursor molecules will be present in CellAg meat is unknown, as the postmortem metabolism needs to be better studied. Adding flavor compounds produced by PF, similar to what is generally done in plant-based meat analogs, could be an option to obtain good flavor.

Another vital quality attribute is color—the red color of conventional meat results from myoglobin, which is present in skeletal tissue. During muscle tissue formation in vitro, skeletal muscle cells produce limited amounts of myoglobin.

Strategies to increase myoglobin protein expression involve growing cells in hypoxic conditions or adding myoglobin protein to the growth media. Conventional meat is also an excellent source of minerals such as iron, zinc, and selenium. Iron is usually present in conventional meat as part of the heme group in myoglobin or stored in complexes with ferritin in a nonheme form. Iron is most bioavailable in the heme form, and increasing myoglobin content in cultivated meat is, therefore, not only crucial for color but also for nutrition. Zinc and selenium are typically absent in cell culture media and must be supplemented. Assessments of the functional, nutritional, and sensory values of CellAg food could benefit from the methods normally used to assess meat quality, including proteomics and monitoring degradation profiles and comparing them with the profiles of protein degradation and metabolic changes within conventional meat. Functional properties can be monitored using swelling, water-holding capacities, and Warner–Bratzler texture analysis. At the same time, nutritional values can be analyzed using both amino acid composition analysis and in vitro digestion models to monitor digestibility (Rieder et al., 2021).

11.6 CellAg: ethical perspectives

The eventual successful realization of CellAg, alongside the technical challenges discussed earlier, relates to how CellAg impacts ethical issues. An ethical analysis of CellAg should start with the values explicitly and implicitly endorsed by the industry and science of cultivated meat. As discussed at the outset of this chapter, these values are food security and food safety, animal well-being, environmental improvement, and stabilization or improvement of the global climate system.

11.6.1 Animals, existence, and well-being

There are several ethical issues of how humans treat their food animals. All measures should be taken to address the cruel treatment of animals. However, avoiding suffering through not bringing the animals into existence seems to be an ethical challenge. In the context of humans, David Benatar argues that as human life is full of suffering and the absence of suffering and pleasure connected with nonexistence are neutral, it is better not to bring more humans into the world (Benatar, 2007). From other perspectives, human and nonhuman animals striving to fulfill their life potentials are seen as good in themselves (Ebels-Duggan & Korsgaard, 2019; Nussbaum, 2023). From this perspective, to exist is good in and by itself. However, both Ebels-Duggan and Nussbaum are highly critical of current factory farming (Ebels-Duggan & Korsgaard, 2019; Nussbaum, 2023). The question regarding whether animals should cease to exist remains open. There are arguments on both sides and a prevailing sense that using animals for human nourishment is not inherently negative but rather hinges on how they are treated. However, this perspective holds even when reducing the population of nonhuman farm animals to minimize meat consumption and substituting animal proteins with plant-based ones (though notable ethical distinctions arise, which will be explored further).

Closely related to the mistreatment of animals are the aims of mitigating climate change, reducing land and energy use, and reducing pollution of the environment through CellAg. If we abandon the current ways of producing meat through animals, these laudable and positive aims will likely be realized. However, some studies indicate that cultivated meat might consume more energy—and also be resource intensive and have a higher environmental impact—than its nonmeat alternatives (Smetana et al., 2015). In a time when energy is increasingly becoming a scarce and expensive resource, a significant ethical and societal challenge for cultivated meat is reducing its energy use, which is currently estimated to be higher than for beef production (Rubio et al., 2020).

A challenging question is whether farmers should be considered as a party that should be taken into consideration by a possible transfer away from animals toward CellAg, that is, if they are a moral object in this respect. From an ethical point of view, the disappearance of jobs is significant to the extent that the jobs that disappear affect well-being, freedom, and/or central societal values. The farmer's freedom is affected by such an introduction, but an important consideration then becomes the extent to which this freedom is based on animals' unfreedom. It seems plausible that we cannot respect claims to freedom based on the subjugation of the liberty of others. The same argument is valid for well-being. A central image in most countries with a rural and farmer-based past is the image of animals at small-scale farms with ample opportunity to move, feed, play, socialize, etc. Claiming that such a farm life constitutes a value in these societies is uncontroversial. Nussbaum presents ideas that there is a threshold for when animals—also farmed animals—lead full lives and that the failure to provide for lives above this threshold violates these animals' rights (Nussbaum, 2023). Accordingly, it could be a point of departure for discussing whether farmers are moral objects depending on whether they treat their animals above a threshold or not. Most instances of factory farming would disqualify farmers as moral objects. This leads us to the question of burden of proof: Should we assume all farmers are morally acceptable unless evidence shows they mistreat animals? Alternatively, should no farmers be seen as morally acceptable until they demonstrate providing for animals beyond a certain threshold? This discussion over the burden of proof should primarily protect basic freedoms or basic well-being. Whereas the animals have no choice but to remain in the custody of their farm, the farmer has the capability of choosing other ways of acting than being a farmer. Consequently, it seems reasonable to place the animals first and the farmer second. For the discussion of the burden of proof, this prioritization means that farmers need to prove that they provide beyond a specific threshold to be considered as a moral object with regard to the introduction of cultivated meat.

11.6.2 Truthfulness, honesty, and openness are key ethical concerns that loom over the CellAg industry

A common argument is that CellAg is necessary since people are unlikely to embrace an all-vegetarian/vegan diet. However, consumer skepticism toward CellAg is high and there is currently little demand for this type of food. Accordingly, policy programs and educational efforts are pointed toward supporting the introduction of CellAg (Bryant et al., 2019; Post et al., 2020). From a utilitarian point of view, such public efforts to persuade citizens to change their preferences should be primarily targeted toward the alternatives to meat that are least harmful and/or most beneficial (with respect to animals, climate, humans, and the environment) (Alvaro, 2022), while from ethics that emphasizes freedom, such efforts should target the alternatives that humans are most likely to accept, prefer, or value. Currently, CellAg does not meet these criteria. Studies further suggest that knowledge of novel foods increases acceptance. Consequently, it could be argued that education should address the most beneficial and least harmful proteins. However, an interesting avenue to pursue in further research would be to investigate whether CellAg might be the compromise that realizes a combination of these values.

CellAg further aims to provide food security. While there are uncertainties regarding how much more food and proteins the world population needs by 2050 (van Dijk et al., 2021), the distribution of the need for food security is more certain. Population growth will occur mainly in Africa, to some extent in India and Pakistan—and to a lesser extent in other countries like Indonesia and the United States (United Nations Department of Economic and Social Affairs PD. World Population Prospects, 2022). Looking at the map of the CellAg company landscape, there is a discrepancy between the United States, European Union, and Southeast Asia's dominance in the industry with the future demand. The issues of technology transfer and access to novel medications might be relevant comparisons in this respect (WHO 2011). One initiative often discussed in this respect is to build knowledge capacity and technological infrastructure in these countries and regions. To achieve the goal of food security, such capacity building is vital to secure the potential benefits of CellAg as a climate mitigation strategy. Furthermore, the conditions for licensing of patented solutions need to be a central theme in the political support for CellAg to combat food inequity (Mahoney, 2022).

Moreover, the claim or the controversy over that CellAg is meat should also be closely examined from a nutritional point of view. It is a highly chemical discussion, but what matters is that the things we humans put into our mouths become nutritious and healthy through the human metabolism. If CellAg should be less beneficial in this respect than its nonlab counterparts, then the other qualities of CellAg need to outweigh this lack of benefits to the individual. As discussed by the EAT-Lancet report, meat consumption, and red meat in particular, should be reduced for health reasons and almost totally removed for climate reasons (Willett et al., 2019).

Finally, some ethical issues loom over the cultivated meat industry: truthfulness, honesty, and openness. Regarding truthfulness: every year, proponents of cultivated meat say that they will be ready for the market in 1 or 2 years—and nothing emerges (at least not in any meaningful quantity) (Abrell, 2023). The CellAg industry shares this promissory nature with other new and emerging technologies. The moral judgment over such a promise will depend on the perceived moral desirability of CellAg (Lucivero et al., 2011). Of course, these are not qualities of the CellAg in itself but qualities in the CellAg milieu.

Regarding honesty: The values discussed earlier, reduction of greenhouse gases, ending of animal suffering, and lesser use of energy, land, and water, are presented as a complete package that will be realized. This rhetorical strategy of realizing several goals through one solution places great responsibility on the industry. Accordingly, meat products that do not meet all criteria might raise doubts about the CellAg industry's honesty. In terms of openness, the CellAg industry appears structured so that the success of companies depends on them finding better, cheaper, and more efficient processes than their competitors. Thus, secrecy has become the preferred way (Guthman & Biltekoff, 2021). However, this lack of transparency might be detrimental to public trust. Opaqueness also prevents serious independent studies of the impacts and value potentials of cultivated meat. In all fairness, it should be added that the meat industry has promoted meat through sponsored research in a way that leaves much to be desired (Nestle, 2018).

11.7 Status in the world by 2023

In the commercial landscape, according to Good Food Institute State of the Industry reports from 2021 (GFI, 2019; GFI, 2021a; GFI, 2021b), the CellAg sector expanded both for PF and for cultivated meat. The growth in fermentation was seen with start-ups in the fermentation of alternative proteins, new suppliers of ingredients, biomass fermentation, and the development of business lines. The EVERY company launched in 2021 is the first animal-free egg protein, ClearEgg, while Perfect Days whey proteins became commercially available in a range of brands and products. The Israeli alternative dairy pioneer Remilk is building the world's largest PF facility in the eco-industrial park in Kalundborg, Denmark. Change Foods is also building a factory in the United Arab Emirates and claims to produce milk equivalent to 10,000 cows. They aim to produce cheese and will sell their products to the Asia-Pacific region. Also, the two companies, Cultivated Biosciences and Mycorena, announced that they are producing functional fat ingredients that can be used in new food products. Interestingly, big food brands are gaining interest; Nestle, for example, is now evaluating new, innovative technologies for CellAg ingredients. General Mills has established its own PF start-up, Renegade Creamery, to produce cheese. Also, Unilever considered PF a "significant trend," and estimated products in 1−2 years. The regulatory framework has advanced, and a number of companies have received a letter from the FDA claiming the products are generally recognized as safe (GRAS). Investment in fermentation was high in 2021, raising more than $1.69 billion in 2021.

According to the Good Food Institute's annual reports (GFI, 2021a; GFI, 2021b), more than 21 new companies dedicated to producing cultivated meat and seafood were announced in 2021, and there are now more than 107 companies in total. Of these, several were focused on cultivated meat (e.g., beef, chicken, pork, duck, horse, kangaroo, foie gras, bison, and mouse), some were focused on seafood (e.g., tuna, salmon, crustaceans, fish maw, and sturgeon). In contrast, others focused on serving the cultivated meat and seafood industry, offering, for example, cell culture media, cell lines, scaffolding, or bioreactors. Although most of these start-ups are localized in the United States, they also spread over more than 19 countries across five continents. In Europe, 13 cultivated protein start-ups have founded the "Cellular Agriculture Europe" foundation, to ensure transparency in the industry and create unity. There were a number of facility launches, including Future Meat in Israel, being able to produce 500 kg a day. WildType has a maximum capacity of producing approximately 91,000 kg of seafood per year, while Upside Foods opened a facility that can produce up to 182,000 kg of chicken per year. The cultivated meat industry is still in its early days regarding investments, and the most active investors are still US-based. However, Israel is accounting for a significant part of the investments. Total investment in the cultivated meat sector was $1.93 billion in 2021. Interestingly, JBS, the largest protein company in the world, announced a $100 million investment. Regarding the regulation, progress is slowly moving; with Singapore approving to Eat Just in 2020, China named cultivated proteins as part of its 5 years plan, and Upside Foods was the first company in the United States to receive a letter from the FDA in 2022. In June 2023, both Eat Just's Good Meat and Upside Foods completed the premarket regulatory review process from the US Department of Agriculture (USDA), allowing both companies to proceed with commercial production and sales of cultivated chicken. In the European Union, no one has submitted an application to the European Food Safety Authority.

11.8 Concluding remarks

CellAg has the potential to revolutionize how we produce food. It can significantly change agricultural output and ownership, land use, policy-making, and food habits and address ethical challenges while raising new ones. The numerous potential advantages of CellAg over traditional agriculture, including a smaller carbon footprint and more consistent products, have led to extensive exploration of its utility. Still, CellAg is in its infancy, and issues such as global scalability and distribution remain open. Much of the innovation and developments are still being performed "behind closed doors" in private companies. Still, governmental funding and increasing efforts in public research will help increase transparency, trust, and openness, in addition to identifying challenges and exploring mitigation strategies.

11.9 Funding

Funding from the Norwegian Fund for Research Fees for Agricultural Products ("Precision," no. 314111) and funding from the Norwegian Research Council ("ARRIVAL," no. 336136) is acknowledged. The funding sources were not involved in planning, conducting, or writing this chapter.

References

Aas, V., Bakke, S. S., Feng, Y. Z., Kase, E. T., Jensen, J., Bajpeyi, S., Thoresen, G. H., & Rustan, A. C. (2013). Are cultured human myotubes far from home? *Cell and Tissue Research*, *354*(3), 671–682. Available from https://doi.org/10.1007/s00441-013-1655-1.

Abrell, E. (2023). *The emty promises of cultured meat. The good it promises, the harm it does.* Oxford University Press.

Aldén, A., Gonzalez, L., Persson, A., Christensson, K., Holmqvist, O., & Ohlson, S. (2007). Porcine platelet lysate as a supplement for animal cell culture. *Cytotechnology*, *55*(1), 3–8. Available from https://doi.org/10.1007/s10616-007-9097-9.

Alvaro, C. (2022). A virtue-ethical approach to cultured meat. *Nature Food*, *3*(10), 788–790. Available from https://doi.org/10.1038/s43016-022-00601-z, https://www.nature.com/natfood/.

Andreassen, R. C., Pedersen, M. E., Kristoffersen, K. A., & Beate Rønning, S. (2020). Screening of by-products from the food industry as growth promoting agents in serum-free media for skeletal muscle cell culture. *Food and Function*, *11*(3), 2477–2488. Available from https://doi.org/10.1039/c9fo02690h, http://pubs.rsc.org/en/journals/journal/fo.

Andreassen, R. C., Rønning, S. B., Solberg, N. T., Grønlien, K. G., Kristoffersen, K. A., Høst, V., Kolset, S. O., & Pedersen, M. E. (2022). Production of food-grade microcarriers based on by-products from the food industry to facilitate the expansion of bovine skeletal muscle satellite cells for cultured meat production. *Biomaterials*, *286*. Available from https://doi.org/10.1016/j.biomaterials.2022.121602, http://www.journals.elsevier.com/biomaterials/.

Aspevik, T., Oterhals, Å., Rønning, S. B., Altintzoglou, T., Wubshet, S. G., Gildberg, A., Afseth, N. K., Whitaker, R. D., & Lindberg, D. (2017). Valorization of proteins from Co- and by-products from the fish and meat industry. *Topics in Current Chemistry*, *375*(3). Available from https://doi.org/10.1007/s41061-017-0143-6, http://www.springer.com/series/128.

Ben-Arye, T., Shandalov, Y., Ben-Shaul, S., Landau, S., Zagury, Y., Ianovici, I., Lavon, N., & Levenberg, S. (2020). Textured soy protein scaffolds enable the generation of three-dimensional bovine skeletal muscle tissue for cell-based meat. *Nature Food*, *1*(4), 210–220. Available from https://doi.org/10.1038/s43016-020-0046-5, https://www.nature.com/natfood/.

Benatar, D. (2007). *Better never to have been: The harm of coming into existence* (pp. 1–256). South Africa: Oxford University Press. Available from http://www.oxfordscholarship.com/view/10.1093/acprof:oso/9780199296422.001.0001/acprof-9780199296422, 10.1093/acprof:oso/9780199296422.001.0001.

Benjaminson, M. A., Gilchriest, J. A., & Lorenz, M. (2002). In vitro edible muscle protein production system (mpps): Stage 1, fish. *Acta Astronautica*, *51*(12), 879–889. Available from https://doi.org/10.1016/s0094-5765(02)00033-4.

Bodiou, V., Moutsatsou, P., & Post, M. J. (2020). Microcarriers for upscaling cultured meat production. *Frontiers in Nutrition*, *7*. Available from https://doi.org/10.3389/fnut.2020.00010, http://journal.frontiersin.org/journal/nutrition.

Bryant, C., Szejda, K., Parekh, N., Deshpande, V., & Tse, B. (2019). A survey of consumer perceptions of plant-based and clean meat in the USA, India, and China. *Frontiers in Sustainable Food Systems*, *3*. Available from https://doi.org/10.3389/fsufs.2019.00011.

Burton, R. J. F. (2019). The potential impact of synthetic animal protein on livestock production: The new "war against agriculture"? *Journal of Rural Studies*, *68*, 33–45. Available from https://doi.org/10.1016/j.jrurstud.2019.030.002, http://www.elsevier.com/inca/publications/store/3/4/8/.

Butler, M. (2013). Serum-free media: Standardizing cell culture system. *Pharmaceutical Bioprocessing*, *1*(4), 315–318. Available from https://doi.org/10.4155/pbp.13.45.

van Dijk, M., Morley, T., Rau, M. L., & Saghai, Y. (2021). A meta-analysis of projected global food demand and population at risk of hunger for the period 2010–2050. *Nature Food*, *2*(7), 494–501. Available from https://doi.org/10.1038/s43016-021-00322-9, https://www.nature.com/natfood/.

Ding, S., Swennen, G. N. M., Messmer, T., Gagliardi, M., Molin, D. G. M., Li, C., Zhou, G., & Post, M. J. (2018). Maintaining bovine satellite cells stemness through p38 pathway. *Scientific Reports*, *8*(1). Available from https://doi.org/10.1038/s41598-018-28746-7, http://www.nature.com/srep/index.html.

Wikipedia. (n.d.). Direct quote from Wikipedia.

Ebels-Duggan, K., & Korsgaard, C. M. (2019). Fellow creatures: Our obligations to the other animals Oxford: Oxford University Press, 2018 Pp. 272 ISBN 9780198753858 (hbk.) $24.95. *Kantian Review*, *24*(4), 653–659. Available from https://doi.org/10.1017/s1369415419000359.

Fang, H., Kang, J., & Zhang, D. (2017). Microbial production of vitamin B12: A review and future perspectives. *Microbial Cell Factories*, *16*(1). Available from https://doi.org/10.1186/s12934-017-0631-y, http://www.microbialcellfactories.com/home/.

Georgiou, G., & Valax, P. (1996). Expression of correctly folded proteins in *Escherichia coli*. *Current Opinion in Biotechnology*, *7*(2), 190–197. Available from https://doi.org/10.1016/S0958-1669(96)80012-7, http://www.elsevier.com/locate/copbio.

GFI. (2019). *Cultivated meat: 2019 State of the industry report*. Good Food Institute.

GFI. (2020). *Fermentation—an introduction to a pillar of the alternative protein industry*. Good Food Institute.

GFI. (2021a). Fermentation: 2021 State of the industry report.

GFI. (2021b). Cultivated meat: 2021 State of the industry report. Good Food Institute.

Grønlien, K. G., Pedersen, M. E., Rønning, S. B., Solberg, N. T., & Tønnesen, H. H. (2022). Tuning of 2D cultured human fibroblast behavior using lumichrome photocrosslinked collagen hydrogels. *Materials Today Communications*, *31*. Available from https://doi.org/10.1016/j.mtcomm.2022.103635, http://www.journals.elsevier.com/materials-today-communications/.

Grønlien, K. G., Pedersen, M. E., Sanden, K. W., Høst, V., Karlsen, J., & Tønnesen, H. H. (2019). Collagen from Turkey (Meleagris gallopavo) tendon: A promising sustainable biomaterial for pharmaceutical use. *Sustainable Chemistry and Pharmacy*, *13*. Available from https://doi.org/10.1016/j.scp.2019.100166, http://www.journals.elsevier.com/sustainable-chemistry-and-pharmacy.

Guthman, J., & Biltekoff, C. (2021). Magical disruption? Alternative protein and the promise of de-materialization. *Environment and Planning E: Nature and Space*, *4*(4), 1583–1600. Available from https://doi.org/10.1177/2514848620963125, http://journals.sagepub.com/home/ene.

Hayflick, L., & Moorhead, P. S. (1961). The serial cultivation of human diploid cell strains. *Experimental Cell Research*, *25*(3), 585–621. Available from https://doi.org/10.1016/0014-4827(61)90192-6.

https://www.mckinsey.com/industries/agriculture/our-insights/cultivated-meat-out-of-the-lab-into-the-frying-pan (n.d.).

Huang, C., Ding, T., Wang, J., Wang, X., Guo, L., Wang, J., Zhu, L., Bi, C., Zhang, X., Ma, X., & Huo, Y. X. (2019). CRISPR-Cas9-assisted native end-joining editing offers a simple strategy for efficient genetic engineering in *Escherichia coli*. *Applied Microbiology and Biotechnology*, *103* (20), 8497–8509. Available from https://doi.org/10.1007/s00253-019-10104-w, http://link.springer.de/link/service/journals/00253/index.htm.

Humbird, D. (2021). Scale-up economics for cultured meat. *Biotechnology and Bioengineering*, *118*(8), 3239–3250. Available from https://doi.org/10.1002/bit.27848, http://onlinelibrary.wiley.com/journal/10.1002/(ISSN)1097-0290.

Impossible foods. (2023). http://Impossiblefoods.com.

Jakubczyk, A., Karaś, M., Rybczyńska-Tkaczyk, K., Zielińska, E., & Zieliński, D. (2020). Current trends of bioactive peptides—New sources and therapeutic effect. *Foods*, *9*(7), 846. Available from https://doi.org/10.3390/foods9070846.

Kang, M. J., Baek, K. R., Lee, Y. R., Kim, G. H., & Seo, S. O. (2022). Production of vitamin K by wild-type and engineered microorganisms. *Microorganisms*, *10*(3). Available from https://doi.org/10.3390/microorganisms10030554, https://www.mdpi.com/2076-2607/10/3/554/pdf.

Kohane, D. S., & Langer, R. (2008). Polymeric biomaterials in tissue engineering. *Pediatric Research*, *63*(5), 487–491. Available from https://doi.org/10.1203/01.pdr.0000305937.26105.e7.

Kolkmann, A. M., Van Essen, A., Post, M. J., & Moutsatsou, P. (2022). Development of a chemically defined medium for in vitro expansion of primary bovine satellite cells. *Frontiers in Bioengineering and Biotechnology*, *10*. Available from https://doi.org/10.3389/fbioe.2022.895289, http://journal.frontiersin.org/journal/bioengineering-and-biotechnology#archive.

Kumar, R., Vikramachakravarthi, D., & Pal, P. (2014). Production and purification of glutamic acid: A critical review towards process intensification. *Chemical Engineering and Processing: Process Intensification*, *81*, 59–71. Available from https://doi.org/10.1016/j.cep.2014.040.012, http://www.elsevier.com/inca/publications/store/5/0/4/0/8/1/index.htt.

Letcher, S. M., Rubio, N. R., Ashizawa, R. N., Saad, M. K., Rittenberg, M. L., Mccreary, A., Ali, A., Calkins, O. P., Trimmer, B. A., & Kaplan, D. L. (2022). In vitro insect fat cultivation for cellular agriculture applications. *ACS Biomaterials Science and Engineering*, *8*(9), 3785–3796. Available from https://doi.org/10.1021/acsbiomaterials.2c00093, http://pubs.acs.org/journal/abseba.

Levi, S., Yen, F. C., Baruch, L., & Machluf, M. (2022). Scaffolding technologies for the engineering of cultured meat: Towards a safe, sustainable, and scalable production. *Trends in Food Science and Technology*, *126*, 13–25. Available from https://doi.org/10.1016/j.tifs.2022.050.011, http://www.elsevier.com/wps/find/journaldescription.cws_home/601278/description#description.

Lima, C. A., Campos, J. F., Filho, J. L. L., Converti, A., da Cunha, M. G. C., & Porto, A. L. F. (2015). Antimicrobial and radical scavenging properties of bovine collagen hydrolysates produced by Penicillium aurantiogriseum URM 4622 collagenase. *Journal of Food Science and Technology*, *52*(7), 4459–4466. Available from https://doi.org/10.1007/s13197-014-1463-y, http://www.springerlink.com/content/121580/.

Lindberg,D., Aaby,K., Borge,G.I. A., Haugen,J.-E., Nilsson,A., & Rødbotten,R., et al. (2017). Kartlegging av restråstoff fra jordbruket.

Lucivero, F., Swiestra, T., & Boenink, M. (2011). Assessing expectations: Towards a toolbox for an ethics of emerging technologies. *NanoEthics*, *5*(2), 129–141. Available from https://doi.org/10.1007/s11569-011-0119-x.

Lynch, J., & Pierrehumbert, R. (2019). Climate impacts of cultured meat and beef cattle. *Frontiers in Sustainable Food Systems*, *3*. Available from https://doi.org/10.3389/fsufs.2019.00005, http://www.frontiersin.org/journals/sustainable-food-systems#.

Mahoney, B. (2022). Let them eat cultured meat: Diagnosing the potential for meat alternatives to increase inequity. *Food Ethics*, *7*(2). Available from https://doi.org/10.1007/s41055-022-00108-0, https://www.springer.com/journal/41055.

Mauro, A. (1961). Satellite cell of skeletal muscle fibers. *The Journal of Biophysical and Biochemical Cytology*, *9*(2), 493–495. Available from https://doi.org/10.1083/jcb.9.20.493.

Melzener, L., Verzijden, K. E., Buijs, A. J., Post, M. J., & Flack, J. E. (2021). Cultured beef: From small biopsy to substantial quantity. *Journal of the Science of Food and Agriculture*, *101*(1), 7–14. Available from https://doi.org/10.1002/jsfa.10663, http://onlinelibrary.wiley.com/journal/10.1002/(ISSN)1097-0010.

Nestle, M. (2018). Unsavory truth: How food companies skew the science of what we eat (p. 310) New York: Basic Books.

Nussbaum, M. C. (2023). In S. Schuster (Ed.), *Justice for animals: Our collective responsibility* (p. 400). Simon and Schuster.

Ostrovidov, S., Salehi, S., Costantini, M., Suthiwanich, K., Ebrahimi, M., Sadeghian, R. B., et al. (2019). Bioprinting in skeletal muscle tissue engineering. *Small (Weinheim an der Bergstrasse, Germany)*, *15*(24).

Pasitka, L., Cohen, M., Ehrlich, A., Gildor, B., Reuveni, E., Ayyash, M., Wissotsky, G., Herscovici, A., Kaminker, R., Niv, A., Bitcover, R., Dadia, O., Rudik, A., Voloschin, A., Shimoni, M., Cinnamon, Y., & Nahmias, Y. (2023). Spontaneous immortalization of chicken fibroblasts generates stable, high-yield cell lines for serum-free production of cultured meat. *Nature Food*, *4*(1), 35–50. Available from https://doi.org/10.1038/s43016-022-00658-w, https://www.nature.com/natfood/.

Perfect Day. (2023). https://perfectdayfoods.com.

Poore, J., & Nemecek, T. (2018). Reducing food's environmental impacts through producers and consumers. *Science (New York, N.Y.)*, *360*(6392), 987–992. Available from https://doi.org/10.1126/science.aaq0216, http://science.sciencemag.org/content/360/6392/987/tab-pdf.

Post, M., & Van Der Weele, C. (2020). *Principles of tissue engineering for food principles of tissue engineering* (pp. 1355–1368). Netherlands: Elsevier. Available from https://www.elsevier.com/books/principles-of-tissue-engineering/lanza/978-0-12-398358-9, 10.1016/B978-0-12-818422-6.00074-5.

Post, M. J. (2014). An alternative animal protein source: Cultured beef. *Annals of the New York Academy of Sciences*, *1328*(1), 29–33. Available from https://doi.org/10.1111/nyas.12569, http://www.blackwellpublishing.com/0077-8923.

Post, M. J., Levenberg, S., Kaplan, D. L., Genovese, N., Fu, J., Bryant, C. J., Negowetti, N., Verzijden, K., & Moutsatsou, P. (2020). Scientific, sustainability and regulatory challenges of cultured meat. *Nature Food*, *1*(7), 403–415. Available from https://doi.org/10.1038/s43016-020-0112-z, https://www.nature.com/natfood/.

Rackham, O. J. L., Firas, J., Fang, H., Oates, M. E., Holmes, M. L., Knaupp, A. S., Suzuki, H., Nefzger, C. M., Daub, C. O., Shin, J. W., Petretto, E., Forrest, A. R. R., Hayashizaki, Y., Polo, J. M., & Gough, J. (2016). A predictive computational framework for direct reprogramming between human cell types. *Nature Genetics*, *48*(3), 331–335. Available from https://doi.org/10.1038/ng.3487, http://www.nature.com/ng/index.html.

Rahmati, M., Mills, D. K., Urbanska, A. M., Saeb, M. R., Venugopal, J. R., Ramakrishna, S., & Mozafari, M. (2021). Electrospinning for tissue engineering applications. *Progress in Materials Science*, *117*. Available from https://doi.org/10.1016/j.pmatsci.2020.100721, https://www.journals.elsevier.com/progress-in-materials-science.

Rieder, A., Afseth, N. K., Böcker, U., Knutsen, S. H., Kirkhus, B., Mæhre, H. K., Ballance, S., & Wubshet, S. G. (2021). Improved estimation of in vitro protein digestibility of different foods using size exclusion chromatography. *Food Chemistry*, *358*, 129830. Available from https://doi.org/10.1016/j.foodchem.2021.129830.

Rubio, N., Datar, I., Stachura, D., Kaplan, D., & Krueger, K. (2019). Cell-based fish: A novel approach to seafood production and an opportunity for cellular agriculture. *Frontiers in Sustainable Food Systems*, *3*. Available from https://doi.org/10.3389/fsufs.2019.00043, http://www.frontiersin.org/journals/sustainable-food-systems#.

Rubio, N. R., Fish, K. D., Trimmer, B. A., & Kaplan, D. L. (2019). In vitro insect muscle for tissue engineering applications. *ACS Biomaterials Science and Engineering*, *5*(2), 1071–1082. Available from https://doi.org/10.1021/acsbiomaterials.8b01261, http://pubs.acs.org/journal/abseba.

Rubio, N. R., Xiang, N., & Kaplan, D. L. (2020). Plant-based and cell-based approaches to meat production. *Nature Communications*, *11*(1). Available from https://doi.org/10.1038/s41467-020-20061-y, http://www.nature.com/ncomms/index.html.

Rønning, S. B., Berg, R. S., Høst, V., Veiseth-Kent, E., Wilhelmsen, C. R., Haugen, E., Suso, H. P., Barham, P., Schmidt, R., & Pedersen, M. E. (2020). Processed eggshell membrane powder is a promising biomaterial for use in tissue engineering. *International Journal of Molecular Sciences*, *21*(21), 1–14. Available from https://doi.org/10.3390/ijms21218130, https://www.mdpi.com/1422-0067/21/21/8130/pdf.

Rønning, S. B., Pedersen, M. E., Andersen, P. V., & Hollung, K. (2013). The combination of glycosaminoglycans and fibrous proteins improves cell proliferation and early differentiation of bovine primary skeletal muscle cells. *Differentiation; Research in Biological Diversity*, *86*(1-2), 13–22. Available from https://doi.org/10.1016/j.diff.2013.060.006.

Seah, J. S. H., Singh, S., Tan, L. P., & Choudhury, D. (2022). Scaffolds for the manufacture of cultured meat. *Critical Reviews in Biotechnology*, *42*(2), 311–323. Available from https://doi.org/10.1080/07388551.2021.1931803, http://www.tandfonline.com/loi/ibty20.

Singh, V., Haque, S., Niwas, R., Srivastava, A., Pasupuleti, M., & Tripathi, C. K. M. (2017). Strategies for fermentation medium optimization: An in-depth review. *Frontiers in Microbiology*, *7*. Available from https://doi.org/10.3389/fmicb.2016.02087, http://journal.frontiersin.org/article/10.3389/fmicb.2016.02087/full.

Smetana, S., Mathys, A., Knoch, A., & Heinz, V. (2015). Meat alternatives: Life cycle assessment of most known meat substitutes. *The International Journal of Life Cycle Assessment*, *20*(9), 1254–1267. Available from https://doi.org/10.1007/s11367-015-0931-6.

Stout, A. J., Mirliani, A. B., Rittenberg, M. L., Shub, M., White, E. C., Yuen, J. S. K., & Kaplan, D. L. (2022). Simple and effective serum-free medium for sustained expansion of bovine satellite cells for cell cultured meat. *Communications Biology*, *5*(1). Available from https://doi.org/10.1038/s42003-022-03423-8, https://www.springer.com/journal/11852.

Tripathi, N. K., & Shrivastava, A. (2019). Recent developments in bioprocessing of recombinant proteins: Expression hosts and process development. *Frontiers in Bioengineering and Biotechnology*, *7*. Available from https://doi.org/10.3389/fbioe.2019.00420, http://journal.frontiersin.org/journal/bioengineering-and-biotechnology#archive.

Tzimorotas, D., Afseth, N. K., Lindberg, D., Kjørlaug, O., Axelsson, L., & Shapaval, V. (2018). Pretreatment of different food rest materials for bioconversion into fungal lipid-rich biomass. *Bioprocess and Biosystems Engineering*, *41*(7), 1039–1049. Available from https://doi.org/10.1007/s00449-018-1933-0, http://link.springer.de/link/service/journals/00449/index.htm.

United Nations Department of Economic and Social Affairs PD. World Population Prospects. 2022 (2022).

van der Valk, J., Brunner, D., De Smet, K., Fex Svenningsen, A., Honegger, P., Knudsen, L. E., Lindl, T., Noraberg, J., Price, A., Scarino, M. L., & Gstraunthaler, G. (2010). Optimization of chemically defined cell culture media - Replacing fetal bovine serum in mammalian in vitro methods. *Toxicology In Vitro*, *24*(4), 1053–1063. Available from https://doi.org/10.1016/j.tiv.2010.030.016.

Venkatesan, M., Semper, C., Skrivergaard, S., Di Leo, R., Mesa, N., Rasmussen, M. K., Young, J. F., Therkildsen, M., Stogios, P. J., & Savchenko, A. (2022). Recombinant production of growth factors for application in cell culture. *iScience*, *25*(10), 105054. Available from https://doi.org/10.1016/j.isci.2022.105054.

Vergeer, R., Sinke, P., & Odegards, I. (2021). TEA of cultivated meat. Future projections of different scenarios. *CE Delft*.

White, Robin R., & Hall, M. B. (2017). Nutritional and greenhouse gas impacts of removing animals from US agriculture. *Proceedings of the National Academy of Sciences*, *114*(48). Available from https://doi.org/10.1073/pnas.1707322114.

Willett, W., Rockström, J., Loken, B., Springmann, M., Lang, T., Vermeulen, S., Garnett, T., Tilman, D., DeClerck, F., Wood, A., Jonell, M., Clark, M., Gordon, L. J., Fanzo, J., Hawkes, C., Zurayk, R., Rivera, J. A., De Vries, W., Majele Sibanda, L., … Murray, C. J. L. (2019). Food in the Anthropocene: The EAT–Lancet Commission on healthy diets from sustainable food systems. *The Lancet*, *393*(10170), 447–492. Available from https://doi.org/10.1016/S0140-6736(18)31788-4, http://www.journals.elsevier.com/the-lancet/.

Xu, Q. S., Yan, Y. S., & Feng, J. X. (2016). Efficient hydrolysis of raw starch and ethanol fermentation: A novel raw starch-digesting glucoamylase from penicillium oxalicum. *Biotechnology for Biofuels*, *9*(1). Available from https://doi.org/10.1186/s13068-016-0636-5, http://www.biotechnologyforbiofuels.com/.

Yamanaka, K., Haraguchi, Y., Takahashi, H., Kawashima, I., & Shimizu, T. (2023). Development of serum-free and grain-derived-nutrient-free medium using microalga-derived nutrients and mammalian cell-secreted growth factors for sustainable cultured meat production. *Scientific Reports*, *13*(1). Available from https://doi.org/10.1038/s41598-023-27629-w, https://www.nature.com/srep/.

Chapter 12

Emerging food trends: plant-based food revolution

Rui M.S. Cruz[1,2] and Fatma Boukid[3]

[1]Department of Food Engineering, Institute of Engineering, Universidade do Algarve, Campus da Penha, Faro, Portugal, [2]MED-Mediterranean Institute for Agriculture, Environment and Development and CHANGE-Global Change and Sustainability Institute, Faculty of Sciences and Technology, Campus de Gambelas, Universidade do Algarve, Faro, Portugal, [3]ClonBio Group, Ltd., Dublin, Ireland

12.1 Introduction

Plant-based food industry is expanding with no signs of slowing down and it is expected to reach a value of USD 35.9 billion by 2033 (Future Market Insights, 2023). The main reason behind this is that consumers increasingly associate plant-based diets with healthiness and well-being as recommended by public health organizations (Food and Agriculture Organization of the United Nations, F, 2016; World Health Organization, 2021). Even though meat and processed meat products are staple foods in the human diet, the overconsumption of meat might be related to health issues such as cardiovascular diseases, high cholesterol blood, and high blood pressure (Craig et al., 2021; Samtiya et al., 2021). Ethical concerns over animal welfare are also boosting the adoption of alternative sources such as cereals, pulses, algae, and insects. Niches like vegans, vegetarians, and flexitarians are increasing worldwide (Boukid, 2021b; Grossmann & McClements, 2021). Climate change and the rapid growth in the world's population pose a threat to food security (Ben Hassen & El Bilali, 2022).

The plant-based food market is segmented into meat, fish, egg, and dairy alternatives (Boukid & Castellari, 2021; Boukid, 2021a; Boukid et al., 2021). Despite the fast progress, plant-based products still encounter organoleptic challenges related to the presence of beany flavors, the lack of animal-like experience, and consumer acceptance (Alcorta, et al., 2021; Shanmugam et al., 2023). Innovative ingredients and technologies are being developed to boost the improvement of plant-based food quality for farm to fork (Fig. 12.1). Smart technologies, such as artificial intelligence, robotics, blockchain, 3D printing, and smart/intelligent packaging tools, are increasingly being tested to improve palatability, safety and quality, efficiency, traceability, and/or consumer's perception (Hassoun et al., 2022a). The implementation of these technologies is still in its early stages and many researchers are looking for the best way to benefit consumers and manufacturers. Therefore this chapter focuses on the current trends and challenges in the plant-based food sector and the role of smart technologies in boosting this sector.

12.2 Emerging trends in plant-based ingredients

12.2.1 Conventional and emerging sources of plant proteins

Oil seeds such as soy, canola, and sunflower are rich sources of proteins. Soy is the most widely used source of proteins due to its high protein content (around 40% protein per seed) (Van den Berg et al., 2022). Canola and sunflower proteins are receiving increased interest to valorize the high amounts of oil extraction by-products (oil cake) rich in proteins (40%–50%). Cereals are important sources of proteins that are used in several food applications including bakery products and meat products owing to the viscoelastic properties of gluten (Aoun & Boukid, 2022). Proteins that derive from rice and corn hold an important share in the gluten-free market (Lee et al., 2022; Linares-García et al., 2019). The use of pulses such as peas is rising as sustainable, genetically modified organism (GMO)-free, allergen-free, gluten-free, and lactose-free products. Faba bean, chickpea, lupine, lentil, and mung bean proteins are attracting interest to respond to the increased demand (Boukid & Pasqualone, 2021; Boukid, 2021; Shivakumar et al., 2019). Plant sources

Food Industry 4.0. DOI: https://doi.org/10.1016/B978-0-443-15516-1.00012-8

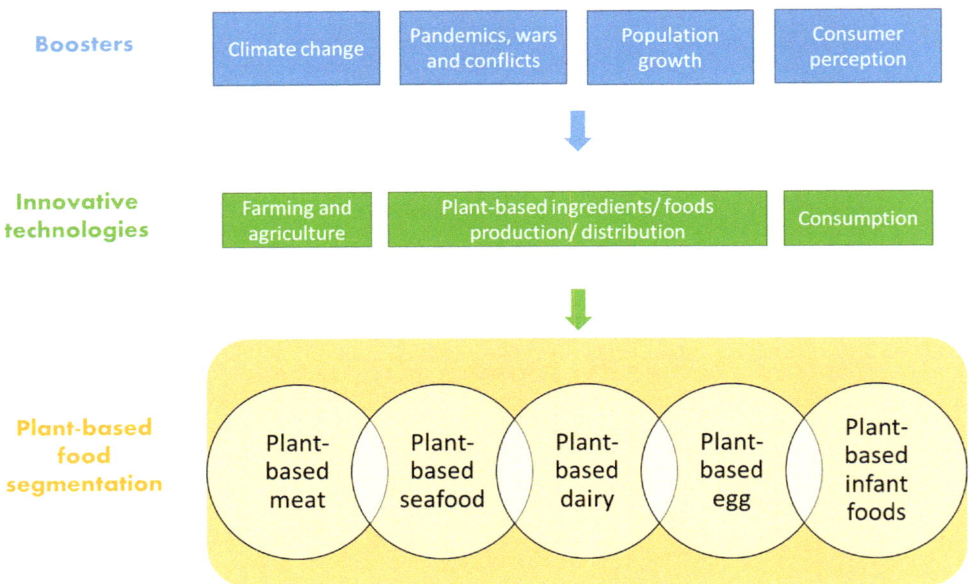

FIGURE 12.1 Application of innovative technologies to boost the growth of the plant-based food sector.

are sources of protein, carbohydrates, fibers, and minerals. However, some sources lack essential amino acids, which results in low protein digestibility-corrected amino acid score (PDCAAS) (Mathai et al., 2017). Soy protein, however, has a complete essential amino acid profile and PDCAAS of 1 (like animal sources) (Qin et al., 2022).

Seeking new sources is deemed necessary to respond to the increased demand for plant-based products and to offer consumers a large portfolio of products to fit their dietary needs (intolerance and allergies) and lifestyle (vegan and vegetarian) (Boukid & Pasqualone, 2021; Boukid, Lamri, et al., 2021). Furthermore, the interest in new crops could support the unstable plant-based market. This instability can be attributed to climate changes (e.g., rising global temperatures) unfavorable for plant crops, and the low productivity of some crops (Nadathur et al., 2017; Zhao et al., 2017). Furthermore, global crises such as the COVID-19 outbreak, and the Russian invasion of Ukraine have disrupted the food market and threatened food security (Galanakis, 2020). During the COVID-19 outbreak, consumers' dietary habits changed and the demand for plant-based food increased for health motives (Di Renzo et al., 2020; Mahwish et al., 2022). Considering that Ukraine and Russia are among the most important producers and exporters of cereals and oil seeds, prices have skyrocketed since the beginning of the war (Mottaleb et al., 2022). This alarmed countries to explore their underused national crops and reduce their dependency on exported crops such as wheat and soy. Furthermore, interest has increased in the bioconversion of agro-industrial by-products rich in proteins and other nutrients as substrates to produce fungal biomass (Gmoser et al., 2019; Souza Filho et al., 2019). Seaweed and microalgae are sustainable sources of proteins that are increasingly used in food applications. Moreover, they have a low impact on the environment, do not compete for water and land with terrestrial crops, and have a fast growth cycle (Boukid & Castellari, 2022).

12.2.2 Use of plant protein isolates and concentrates

Plant proteins are commercially available in different forms including flours (15%–40%), concentrates (65%–75%), and isolates (up to 90% protein) (Boukid & Pasqualone, 2021; Boukid, Lamri, et al., 2021). The degree of purity impacts the price and the functionality of plant proteins and thus offers the versatility of choice depending on the end-uses. Protein isolates and concentrates contribute to protein content enhancement in foods and differ in functionality compared to flours from the same source. Protein isolates and hydrolysates are rich in peptides and amino acids, which could improve protein digestibility and availability (Ding et al., 2020; Gavrilova et al., 2019). The process of purification and extraction contributes to the elimination/reduction of off-flavor and antinutrients as well as color improvement (Vatansever et al., 2021). Nevertheless, enzymatic hydrolysis of isolates might liberate peptides with a bitter taste (Cosson et al., 2022). Plant proteins can be further processed (extruded, electrospinned, or shear-structured) to obtain textured vegetable proteins (TVPs) with low and high moisture contents (Baune et al., 2022; Kyriakopoulou et al., 2021). These textured proteins are mainly used in meat alternatives to mimic meat fibrous structure (Bühler et al., 2020; Dekkers et al., 2018). From a sustainability perspective, processing flours to obtain isolates, concentrates, and textured proteins contributes to increasing product costs and might negatively impact the environment. Notably, plant proteins can be extracted from side streams of starch or oil extraction, thus reducing waste and carbon dioxide emissions (Ancuţa & Sonia, 2020; Boukid, 2021). The use of dry fractionation to recover protein concentrates can be also an approach to reduce the cost of energy and chemicals thus lowering the environmental impact (Lie-Piang et al., 2021; Vogelsang-O'Dwyer et al., 2021).

12.3 Emerging trends in plant-based food segments

12.3.1 Plant-based meat

Plant-based meat is available in different forms including burger patties, sausages, and meatballs (Saget et al., 2021). They are mainly made with plant proteins (isolates, concentrates, and/or TVP), flour and starches, vegetable fats, flavoring, and color ingredients (Alcorta et al., 2021). The products are cholesterol-free and were found to be lower in fat, saturated fatty acids, and protein but higher in carbohydrates and fibers compared to conventional meat products (Guess et al., 2022; Salomé et al., 2021). These alternative products also lack essential elements including vitamin B12 and iron (Curtain & Grafenauer, 2019; Harnack et al., 2021). Biofortification was used as a strategy to provide vitamins and minerals naturally lacking in plant sources (Chen et al., 2019; van Vliet et al., 2020). New product launches showed also an enhancement in protein content and essential amino acids composition (Boukid & Castellari, 2021; Guo et al., 2020). The use of TVP and high moisture extruded proteins contributes to improving the texture and enabling a meat-like bite (Boukid, Lamri, et al., 2021). Binding and thickening ingredients and additives are also added to reach the desired consistency and bite (Boukid, 2021). Traditionally, herbs and spices were used for flavoring, but currently vegan synthetic flavors are commercially available to mimic meat and fat flavors. Masking agents and protein blockers are available for mitigating the off-flavor of plant-based ingredients (Boukid & Gagaoua, 2022). Soy leghemoglobin (Impossible Foods, Inc), fat alternative (Melt & Marble), and Hemami (Motif Food Works), expressed by a genetically engineered yeast strain, were produced to deliver umami flavor and meaty aroma (Ismail et al., 2020; Jin et al., 2018).

12.3.2 Plant-based dairy

Plant-based alternatives to milk are water-soluble extracts of legumes, oil seeds, nuts, cereals, or pseudocereals (McClements, 2020; Paul et al., 2020). Conventional (soy, almond, coconut, oat, and rice) and new sources (e.g., quiona and faba) are being used in making these alternatives (Canaviri Paz et al., 2020; Cichońska & Ziarno, 2022). Compared to cow's milk, plant-based milk contains low protein, calcium, and vitamin content (Jeske et al., 2018; Silva et al., 2020). Lactic acid fermentation is increasingly used to improve the nutritional quality (e.g., amino acid profile, and protein digestibility, minerals, vitamins), health benefits (probiotic effect), organoleptic properties (texture and flavor), and palatability of dairy alternatives (Aydar et al., 2020; Gobbetti et al., 2020; Tangyu et al., 2019; Ziarno et al., 2020).

The formulation of plant-based yogurt alternatives consists generally of fermented plant-based aqueous extracts, and thickening and flavoring agents (Jiang et al., 2020; Mefleh, Faccia, et al., 2022; Pandey et al., 2021). *S. thermophilus* and *L. delbrueckii* subsp. *Bulgaricus* are the most common strains used to reach desirable texture, mouthfeel, and viscosity (Brückner-Gühmann et al., 2019; Lorusso et al., 2018; Ogundipe et al., 2021; Zannini et al., 2018). Compared to dairy yogurt, plant-based alternatives contain higher amounts of fats and carbohydrates resulting in higher calories (Boukid, Lamri, et al., 2021).

Plant-based cheeses are made essentially using vegetable oils and polysaccharides (Masiá et al., 2022; Mefleh, Pasqualone, et al., 2022). These ingredients are blended together with additional ingredients such as flavors, preservatives, and water to imitate cheese appearance, taste, and consistency (Boukid, Lamri, et al., 2021; Shabani et al., 2016). Nutritionally, due to the high incorporation of fat and starches, these products are characterized by a low protein content, high fat content, and high calories (Boukid, Lamri, et al., 2021; Harper et al., 2022). Therefore plant proteins such as soy proteins are being increasingly added to increase protein content and lower fat while preserving organoleptic properties (Mattice & Marangoni, 2020; Oyeyinka et al., 2019). Fermentation is also gaining interest as a sustainable process to improve the nutritional, organoleptic, and technological features as well as to prolong product shelf life (Jeewanthi & Paik, 2018; Li et al., 2020; Shirotani et al., 2021).

12.3.3 Plant-based seafood

Seafood alternatives or analogs are plant-based products designed with the objective to mimic the texture and organoleptic properties of fish and seafood products. This market is boosted by the rising awareness of overfishing, and the environmental impact of industrial fishery (Alcorta et al., 2021; Remize et al., 2021). There is currently a large variety of seafood alternatives such as tuna, shrimp, calamari, fish fingers, fish sticks, salmon, caviar, and fillet (Ran et al., 2022). Commercial products such as tuna, shrimp, caviar, and fillet alternatives showed lower protein content than conventional products, while calamari, fish fingers, fish sticks, and salmon alternatives showed similar amounts to those conventional (Boukid et al., 2022). More investigations are needed to improve and standardize the nutritional quality of these alternatives. To increase protein content and provide a fishy flavor, microalgae can play a crucial role in delivering both aspects (Coleman et al., 2022).

12.3.4 Plant-based egg

Plant-based egg is a complex concept due to the multifunctionality of the egg. Indeed, egg plays several functional roles in reaching the desired taste and texture depending on the food product (He, Meda, et al., 2021; He, Purdy, et al., 2021). Commercial plant-based eggs can be available in liquid, egg-shaped, and powder forms, and they mainly consist of starches, plant-based proteins, hydrocolloids, and/or emulsifiers (Boukid & Gagaoua, 2022; Rondoni et al., 2021). Compared to a whole egg, plant-based eggs are cholesterol-free and contain lower saturated fat contents but provide higher calories (Boukid & Gagaoua, 2022). This sector is still a niche, and more research is being conducted to improve the quality of egg alternative to deliver improved functionality and nutritional quality.

12.3.5 Plant-based infant foods

Plant-based infant food market is growing mainly due to the increased prevalence of cow's milk and dairy products allergies and intolerance in babies and children (Mintel, 2022). Furthermore, parents following vegan or vegetarian diets are also making the decision that their children follow a similar diet (Protudjer & Mikkelsen, 2020). In several cases, the nutritional composition of plant-based alternatives to milk does not provide the necessary micronutrients such as calcium and iron for the growth of infants and children (Escobar-Sáez et al., 2022; Le Roux et al., 2020; Wright & Smith, 2020). Similarly, a recent study showed high nutritional variability among vegan and vegetarian commercial infant products (such as baby formulas, baby cereals, baby fruit products, desserts and yogurts, baby biscuits and rusks, baby snacks, baby savory meals and dishes, and baby juices and drinks). This suggests that some products might be nutritionally deficient and not suit babies/toddlers' needs (Gasparre et al., 2022). Thus focusing on improving nutritional quality and enabling clear labeling would help parents to select the best option.

12.4 Role of smart technologies in the plant-based food sector from farm to fork

The current advances in technology are opening a wide spectrum of different applications, from agriculture to the food industry. Technologies such as artificial intelligence (AI), robotics, and smart sensors can be used to improve the efficiency of different processes and the quality of the final products, thus increasing the capabilities of the production of sustainable food for an increasing world population and reducing food waste and loss (Hassoun et al. 2022b). This section will focus on different applications of smart technologies from farm to fork.

12.4.1 Farming

The production of plant-based products can be improved by reducing human error with the use of digital systems to help control the growth needs of the plant itself (e.g., water and soil monitoring) and to predict and control external factors (e.g., weather forecasting, weeds, and insects) that may jeopardize the crop production (Javaid et al., 2022). On the other hand, harvesting systems using robotics can optimize labor work and increase productivity (Hassoun et al., 2022b).

Choi and Shin (2023) performed an input–output analysis and indicated that the networking of systems must be revised and updated to improve the organization and technical aspects and, thus, fully demonstrate the economic effects of smart farms. These aspects are very important since they will help the farmers make better decisions by using fewer resources (e.g., water and energy) in an environmentally and economically sustainable manner. In another study, Reinhardt (2022) showed that farm-to-fork (F2F) strategies can contribute to innovation in the agrifood sector; however,

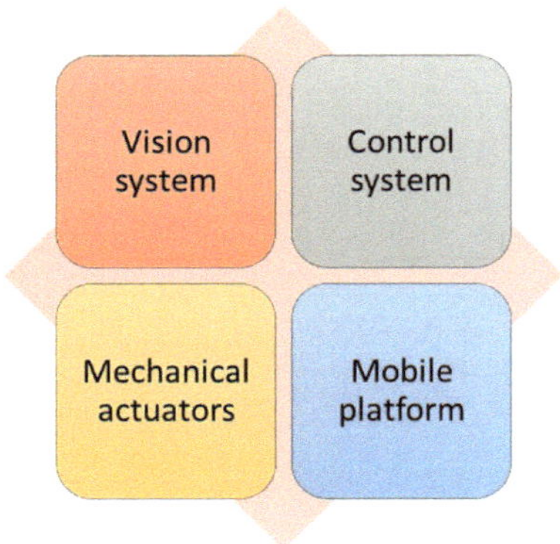

FIGURE 12.2 Different components of agricultural robots. *Adapted from Cheng, C., Fu, J., Su, H., & Ren, L. (2023). Recent advancements in agriculture robots: Benefits and challenges.* Machines, 11, *48. http://doi.org/10.3390/machines11010048.*

the market conditions for these technologies and the communication with farmers are some of the aspects that need to be improved.

Different technologies are available to constantly monitor and optimize crop production. One of those technologies is sensors. Smart sensors are devices that are connected to different equipment to collect data, inform (e.g., about the crop status), alert, and help make decisions. For example, from temperature readings in the air to pH, moisture, and nutrient levels in the soil, smart sensors offer key solutions in simple or more advanced farming systems, making the process more efficient and profitable (Dhanaraju et al., 2022).

Robotics are machines or equipment that, among other tasks, may replace the hand labor performed by humans. Robots may perform tasks in seeding and planting, spraying water or pesticides, and harvesting, increasing productivity compared with traditional methods (Cheng et al., 2023). Robots are mainly produced with four different parts that allow them to adapt to different farming conditions (Fig. 12.2).

Moreover, equipment like drones with sensors and cameras can also contribute to better decisions in land management (Dhanaraju et al., 2022). More advanced farming systems with AI allow for better and more robust decisions, and consequently better crop productivity and quality, since these systems will solve problems and think for themselves for a better solution (Addanki et al., 2022). Nevertheless, all these technologies require connectivity. All sensors and Wi-Fi and Global Positioning System (GPS) systems, if available, allow the farmers to make any decision.

12.4.2 Harnessing smart technologies for better plant-based ingredients/food production

In the next few decades, a great variety of new ingredients and food products will be developed, taking into consideration the needs of the world population and the environment, using new smart technologies to produce in a more efficient and sustainable manner. Consumer and environmental demands are key factors for new ideas and alternative formulations and products. Smart technologies, such as those of Industry 4.0 innovations, will be needed for plant-based ingredients to be used in healthier and more sustainable new products, compensating nutritional needs, mimicking other food products (as shown in Section 3), and improving flavor, texture, and taste. Moreover, new technologies for extending food shelf life and freshness will be fundamental for the next generation of food products (Hassoun et al., 2022c).

For instance, 3D food printing is a computer-aided technology that allows to use innovative ingredients and produce new food products for specific diets, new designs, colors, and structures that mimic existing food products, such as fish and meat. Furthermore, 3D printing can be used to valorize food by-products and food wastes (Varvara et al., 2021) using peels, seeds, and other materials. This technology can be also applied in the development of biodegradable packaging materials, and to a certain extent, it can also be used to design and print structures for 4D food. This printed food is a stimulus-responsive material and may change, for example, its structure or color due to stimulating agents (e.g., temperature, pH, and humidity). This change will inform the producer or consumer regarding the status of the food (Navaf et al., 2022).

In a study reported by Derossi et al. (2020), cereal-based snacks with different textures were produced using 3D printing technology. The study revealed that it is possible to control the generation of pores and thus produce cereal-based snacks with different textures. In another study, Keerthana et al. (2020) were able to develop healthy fiber-enriched snacks from button mushrooms, contributing to the use of a sustainable food source for personalized nutrition. Varghese et al. (2020) established a 3D printing methodology for cookie dough to produce structurally stable cookies using jackfruit seed powder and finger millet powder as base materials. Pant et al. (2021) developed 3D-printed fresh vegetables using food hydrocolloids for dysphagic patients. The results showed that this technology may also contribute to the development of food products suitable for people suffering from swallowing problems.

In what concerns 4D food, Phuhongsung et al. (2020a) studied the influence of surface pH on the color, texture, and flavor of a 3D-printed composite mixture of soy protein isolate, pumpkin, and beetroot. The results showed changes in color (at a pH below 6) and changes in flavor at higher pH values (8 and 10). More acidic pH values were revealed to increase the hardness of the samples. In another study, He et al. (2020) 4D printed a puree with mashed potatoes/purple sweet potatoes. The puree changed from red to green when the pH was modified from acidic to alkali. In another study reported by Phuhongsung et al. (2020b), several products based on soy protein isolate were developed with 4D printing technology. After microwave heating, the samples developed new flavor compounds and increased, among others, their bitterness and saltiness. Guo et al. (2021) developed 3D-printed buckwheat dough with yellow flesh peaches, enriched with complex coacervates microcapsules. Again, after microwave heating, the microcapsules in the dough were gradually destroyed, and with the embedded oil released, the samples changed their color and aroma.

Smart packaging is another technology that is fundamental to monitoring the environmental conditions of the packed food and the food itself in the supply chain to maintain quality and safety (Madhusudan et al., 2018). This type of packaging provides information regarding the status of the food product and thus communicates with the producer, retailer, and consumer, avoiding food waste and food loss (Li et al., 2020). In these systems, the interaction can be performed, for example, by quick response (QR) codes, other communication protocols, or even natural sensors produced from plant-based materials recovered from wastes or chips that give additional information regarding the product's temperature, integrity, or freshness (Vasuki et al., 2023).

In a study reported by Occhiuzzi et al. (2020), avocado fruits were packed with a radio-frequency identification (RFID) tag. The system was able to identify three ripening levels of the fruits and contributed to defining the right period to eat the fruit, helping with its display in the store, and consequently helping to reduce waste. In another study, anthocyanins from black carrots were incorporated into a starch matrix to develop a freshness/spoilage-indicating sensor (Boarca et al., 2019; Goodarzi et al., 2020) presented the use of a thermochromic ink for beer packed in metal or glass that allows the consumer to be informed about the right temperature for beer consumption. In a recent study reported by Romero et al. (2022), pectin films were developed using different berries from agro-industrial waste. The study showed a change in the color of each film for different pH values. The developed films were able to detect the freshness of salmon fillets by changing color with the pH variation of the fish. (Lei et al., 2023) developed packaging films based on pectin, sodium alginate, cellulose nanocrystals, and anthocyanins to be used as a colorimetric indicator for monitoring the freshness of shrimp. The developed films were able to provide real-time colorimetric changes due to pH variation associated with shrimp spoilage. (Zeng et al., 2023) reported the development of a pH indicator composite film based on pectin and chitosan incorporated with black rice anthocyanins for meat freshness monitoring. The study revealed that the presence of anthocyanins allows the film to change color from red to blue according to the degree of meat spoilage. In another study, a novel film based on pectin and betacyanins from the peel waste of pitaya (*Hylocereus undatus*) was developed. The results showed that adding color indicators from plant-based pigments to the pectin film allowed to monitor the shrimp freshness (Jiang et al., 2023).

12.4.3 Distribution and storage

During the distribution and storage of fruits and vegetables or other plant-based products, different technologies can be used to monitor the conditions and ensure product safety along the food supply chain. Technologies included in the food packaging, referred to in the previous section, using advanced blockchain technologies, big data analytics (BDAs), the internet of things (IoTs), and AI for real-time tracking and tracing purposes can be used, for example, during transport and storage of stocks (with information of the harvest date, pallet number, buyer and vendor identifiers, the quantity shipped, lot number, packaging date, unit of measure, and conditions such as temperature, humidity, oxygen, ethylene, food integrity, and freshness) in trucks and warehouses, respectively (Cicculo et al., 2022; Jin et al., 2020; Trollman et al., 2022; Yang et al., 2021). Along the food supply chain, a huge amount of data is generated by sensors and can be made available in real time. Using these technologies, the records are immutable, guaranteeing consumer confidence in the available data. There are several advantages, among them increased efficiency in shipping, tracking, and storage conditions, security and veracity in the food information, and the fact that the data are permanent and stored on a peer-to-peer network instead of a central location (DHL, 2018). In a study reported by Kamath (2018), blockchain technology was used in mangoes from South and Central America to North America. The results showed a huge reduction in the tracking time, from 7 days to 2.2 seconds. In another study, das Nair and Landani (2020) presented an example of a mobile-based platform for small and medium enterprises to be linked to fruit and vegetable farmers, allowing information and market access. This platform assured better products to consumers, made markets available to farmers with lower losses, and provided a trustworthy supply to companies. In a study reported by Yang et al. (2021), a blockchain technology traceability system was designed and implemented for the storage and query of product information in the supply chain of agricultural products. Several solutions were proposed to improve the query efficiency and the security of private information. The system guaranteed the validity and consistency of the data in supply chain management.

12.5 Conclusion

The market of plant-based ingredients/foods is witnessing a fast expansion. Consumers are continuously asking for improved nutritional quality, sustainability, and sensory experience as well as affordability, versatility, and availability. The ongoing progress of digital technologies is creating new possibilities in the plant-based food sector, adding value to the supply chain. The quality and safety status of the products in real time can be used for decision-making in the logistic control and the optimization of processes, avoiding unnecessary losses. Nevertheless, the implementation of all these new systems requires full understanding and perception from all stakeholders. The consumers will have the final say on the acceptance of these technologies, based on the benefits (e.g., more information, extended shelf life, healthier) and drawbacks (e.g., disparity of the 3D meat analogs, paying more, changing in purchasing habits) of these systems, and they may also contribute to the improvements in the existing systems and future ones, such as apps to suggest ideas or recipes before food expires and smartphone-integrated sensors to monitor the food freshness.

References

Addanki, M., Patra, P., & Kandra, P. (2022). Recent advances and applications of artificial intelligence and related technologies in the food industry. *Applied Food Research*, *2*(2). Available from https://doi.org/10.1016/j.afres.2022.100126.

Alcorta, A., Porta, A., Tárrega, A., Alvarez, M. D., & Vaquero, M. P. (2021). Foods for plant-based diets: Challenges and innovations. *Foods*, *10*, 293. Available from https://doi.org/10.3390/FOODS10020293.

Ancuța, P., & Sonia, A. (2020). Oil press-cakes and meals valorization through circular economy approaches: A review. *Applied Sciences (Switzerland)*, *10*(21), 1–31. Available from https://doi.org/10.3390/app10217432, https://www.mdpi.com/2076-3417/10/21/7432/pdf.

Aoun, M., & Boukid, F. (2022). Novel wheat quality features to expand durum wheat applications. *Journal of the Science of Food and Agriculture*. Available from https://doi.org/10.1002/JSFA.12374.

Aydar, E. F., Tutuncu, S., & Ozcelik, B. (2020). Plant-based milk substitutes: Bioactive compounds, conventional and novel processes, bioavailability studies, and health effects. *Journal of Functional Foods*, *70*, 103975. Available from https://doi.org/10.1016/J.JFF.2020.103975.

Baune, M. C., Terjung, N., Tülbek, M. Ç., & Boukid, F. (2022). Textured vegetable proteins (TVP): Future foods standing on their merits as meat alternatives. *Future Foods*, *6*. Available from https://doi.org/10.1016/j.fufo.2022.100181, http://www.journals.elsevier.com/future-foods.

Ben Hassen, T., & El Bilali, H. (2022). Impacts of the Russia-Ukraine war on global food security: Towards more sustainable and resilient food systems. *Foods*, *11*, 2301. Available from https://doi.org/10.3390/FOODS11152301.

Boarca, B., Lungu, I., & Holban, A. M. (2019). *Bioactive packaging for modern beverage industry trends in beverage packaging, . The science of beverages* (Volume 16, pp. 51–71). Romania: Elsevier. Available from https://www.sciencedirect.com/book/9780128166833.

Boukid, F. (2021). Peanut protein – an underutilised by-product with great potential: A review. *International Journal of Food Science & Technology*, *57*, 5585–5591. Available from https://doi.org/10.1111/IJFS.15495.

Boukid, F. (2021). Oat proteins as emerging ingredients for food formulation: where we stand. *European Food Research and Technology*, *247*, 535–544. Available from https://doi.org/10.1007/S00217-020-03661-2.

Boukid, F. (2021a). Plant-based meat analogues: from niche to mainstream. *European Food Research and Technology*, *247*, 297–308. Available from https://doi.org/10.1007/S00217-020-03630-9.

Boukid, F. (2021b). *The realm of plant proteins with focus on their application in developing new bakery products. Advances in food and nutrition research*. Academic Press. Available from https://doi.org/10.1016/BS.AFNR.2021.11.001.

Boukid, F., & Castellari, M. (2021). Veggie burgers in the EU market: a nutritional challenge. *European Food Research and Technology*, *1*, 1–9. Available from https://doi.org/10.1007/S00217-021-03808-9.

Boukid, F., & Castellari, M. (2022). Algae as nutritional and functional food sources. *Foods*, *12*, 122. Available from https://doi.org/10.3390/FOODS12010122.

Boukid, F., & Gagaoua, M. (2022). Vegan egg: A future-proof food ingredient. *Foods*, *11*, 161. Available from https://doi.org/10.3390/FOODS11020161.

Boukid, F., & Gagaoua, M. (2022). Meat alternatives: A proofed commodity? *Advances in Food and Nutrition Research*, *101*. Available from https://doi.org/10.1016/bs.afnr.2022.02.003, http://www.elsevier.com/wps/find/bookdescription.cws_home/704624/description#description.

Boukid, F., & Pasqualone, A. (2021). Lupine (*Lupinus* spp.) proteins: characteristics, safety and food applications. *European Food Research and Technology*, *1*, 3. Available from https://doi.org/10.1007/s00217-021-03909-5.

Boukid, F., Rosell, C. M., Rosene, S., Bover-Cid, S., & Castellari, M. (2021). Non-animal proteins as cutting-edge ingredients to reformulate animal-free foodstuffs: Present status and future perspectives. *Critical Reviews in Food Science and Nutrition*, *137*, 1–31. Available from https://doi.org/10.1080/10408398.2021.1901649.

Boukid, F., Lamri, M., Dar, B. N., Garron, M., & Castellari, M. (2021). Vegan alternatives to processed cheese and yogurt launched in the European market during 2020: A nutritional challenge. *Foods*, *10*, 2782. Available from https://doi.org/10.3390/FOODS10112782.

Boukid, F., Baune, M.-C., Gagaoua, M., & Castellari, M. (2022). Seafood alternatives: assessing the nutritional profile of products sold in the global market. *European Food Research and Technology*, *1*, 1–10. Available from https://doi.org/10.1007/S00217-022-04004-Z.

Brückner-Gühmann, M., Vasil'eva, E., Culetu, A., Duta, D., Sozer, N., & Drusch, S. (2019). Oat protein concentrate as alternative ingredient for non-dairy yoghurt-type product. *Journal of the Science of Food and Agriculture*, *99*, 5852–5857. Available from https://doi.org/10.1002/JSFA.9858.

Bühler, J. M., Dekkers, B. L., Bruins, M. E., & Van Der Goot, A. J. (2020). Modifying faba bean protein concentrate using dry heat to increase water holding capacity. *Foods*, *9*(8). Available from https://doi.org/10.3390/foods9081077, https://www.mdpi.com/2304-8158/9/8/1077.

Canaviri Paz, P., Janny, R. J., & Håkansson, Å. (2020). Safeguarding of quinoa beverage production by fermentation with *Lactobacillus plantarum* DSM 9843. *International Journal of Food Microbiology*, *324*. Available from https://doi.org/10.1016/j.ijfoodmicro.2020.108630, http://www.elsevier.com/locate/ijfoodmicro.

Chen, C., Chaudhary, A., & Mathys, A. (2019). Dietary change scenarios and implications for environmental, nutrition, human health and economic dimensions of food sustainability. *Nutrients*, *11*(4). Available from https://doi.org/10.3390/nu11040856.

Cheng, C., Fu, J., Su, H., & Ren, L. (2023). Recent advancements in agriculture robots: Benefits and challenges. *Machines*, *11*, 48. Available from https://doi.org/10.3390/machines11010048.

Choi, S.-W., & Shin, Y. J. (2023). Role of smart farm as a tool for sustainable economic growth of Korean agriculture: Using input–output analysis. *Sustainability*, *15*, 3450. Available from https://doi.org/10.3390/su15043450.

Ciccullo, F., Fabbri, M., Abdelkafi, N., & Pero, M. (2022). Exploring the potential of business models for sustainability and big data for food waste reduction. *Journal of Cleaner Production*, *340*, 130673. Available from https://doi.org/10.1016/j.jclepro.2022.130673.

Cichońska, P., & Ziarno, M. (2022). Legumes and legume-based beverages fermented with lactic acid bacteria as a potential carrier of probiotics and prebiotics. *Microorganisms*, *10*, 91. Available from https://doi.org/10.3390/MICROORGANISMS10010091.

Coleman, B., Van Poucke, C., Dewitte, B., Ruttens, A., Moerdijk-Poortvliet, T., Latsos, C., De Reu, K., Blommaert, L., Duquenne, B., Timmermans, K., van Houcke, J., Muylaert, K., & Robbens, J. (2022). Potential of microalgae as flavoring agents for plant-based seafood alternatives. *Future Foods*, *5*, 100139. Available from https://doi.org/10.1016/J.FUFO.2022.100139.

Cosson, A., Oliveira Correia, L., Descamps, N., Saint-Eve, A., & Souchon, I. (2022). Identification and characterization of the main peptides in pea protein isolates using ultra high-performance liquid chromatography coupled with mass spectrometry and bioinformatics tools. *Food Chemistry*, *367*. Available from https://doi.org/10.1016/j.foodchem.2021.130747.

Craig, W. J., Mangels, A. R., Fresán, U., Marsh, K., Miles, F. L., Saunders, A. V., Haddad, E. H., Heskey, C. E., Johnston, P., Larson-meyer, E., & Orlich, M. (2021). The safe and effective use of plant-based diets with guidelines for health professionals. *Nutrients*, *13*, 4144. Available from https://doi.org/10.3390/NU13114144.

Curtain, F., & Grafenauer, S. (2019). Plant-based meat substitutes in the flexitarian age: An audit of products on supermarket shelves. *Nutrients*, *11* (11). Available from https://doi.org/10.3390/nu11112603.

DHL. (2018). Blockchain in logistics—Perspectives on the upcoming impact of blockchain technology and use cases for the logistics industry. <https://www.dhl.com/content/dam/dhl/global/core/documents/pdf/glo-core-blockchain-trendreport.pdf/> Accessed 06.04.23.

Dekkers, B. L., Hamoen, R., Boom, R. M., & van der Goot, A. J. (2018). Understanding fiber formation in a concentrated soy protein isolate - Pectin blend. *Journal of Food Engineering*, *222*, 84–92. Available from https://doi.org/10.1016/J.JFOODENG.2017.11.014.

Derossi, A., Caporizzi, R., Paolillo, M., & Severini, C. (2020). Programmable texture properties of cereal-based snack mediated by 3D printing technology. *Journal of Food Engineering*, *289*, 110160. Available from https://doi.org/10.1016/j.jfoodeng.2020.110160.

das Nair, R., & Landani, N. (2020). Making agricultural value chains more inclusive through technology and innovation. WIDER working paper. UNU-WIDER. https://doi.org/10.35188/unu-wider/2020/795-8.

Dhanaraju, M., Chenniappan, P., Ramalingam, K., Pazhanivelan, S., & Kaliaperumal, R. (2022). Smart farming: Internet of things (IoT)-based sustainable agriculture. *Agriculture*, *12*(10). Available from https://doi.org/10.3390/agriculture12101745.

Di Renzo, L., Gualtieri, P., Pivari, F., Soldati, L., Attinà, A., Cinelli, G., Cinelli, G., Leggeri, C., Caparello, G., Barrea, L., Scerbo, F., Esposito, E., & De Lorenzo, A. (2020). Eating habits and lifestyle changes during COVID-19 lockdown: An Italian survey. *Journal of Translational Medicine*, *18*, 229. Available from https://doi.org/10.1186/S12967-020-02399-5.

Ding, J., Liang, R., Yang, Y., Sun, N., & Lin, S. (2020). Optimization of pea protein hydrolysate preparation and purification of antioxidant peptides based on an in silico analytical approach. *LWT*, *123*, 109126. Available from https://doi.org/10.1016/j.lwt.2020.109126.

Escobar-Sáez, D., Montero-Jiménez, L., García-Herrera, P., & Sánchez-Mata, M. C. (2022). Plant-based drinks for vegetarian or vegan toddlers: nutritional evaluation of commercial products, and review of health benefits and potential concerns. *Food Research International*, *160*, 111646. Available from https://doi.org/10.1016/J.FOODRES.2022.111646.

Food And Agriculture Organization of the United Nations, F. (2016). Pulses Nutritions SEEDS for sustainable future. FAO, 1–196.

Future Market Insights. (2023). Plant-based food market. https://www.futuremarketinsights.com/reports/plant-based-food-market.

Galanakis, C. M. (2020). The food systems in the era of the coronavirus (CoVID-19) pandemic crisis. *Foods*, *9*(4). Available from https://doi.org/10.3390/foods9040523, https://www.mdpi.com/2304-8158/9/4/523.

Gasparre, N., Mefleh, M., & Boukid, F. (2022). Nutritional facts and health/nutrition claims of commercial plant-based infant foods: Where do we stand. *Plants*, *11*(19). Available from https://doi.org/10.3390/plants11192531.

Gavrilova, K. V., Bychkov, A. L., Bychkova, E. S., Akimenko, Z. A., Chernonosov, A. A., Kalambet, Y. A., & Lomovskii, O. I. (2019). Mechanically activated hydrolysis of plant-derived proteins in food industry. *Foods and Raw Materials*, *7*(2), 255–263. Available from https://doi.org/10.21603/2308-4057-2019-2-255-263, http://jfrm.ru/files/archive/14/5.pdf.

Gmoser, R., Sintca, C., Taherzadeh, M. J., & Lennartsson, P. R. (2019). Combining submerged and solid state fermentation to convert waste bread into protein and pigment using the edible filamentous fungus *N. intermedia*. *Waste Management*, *97*, 63–70. Available from https://doi.org/10.1016/J.WASMAN.2019.07.039.

Gobbetti, M., De Angelis, M., Di Cagno, R., Polo, A., & Rizzello, C. G. (2020). The sourdough fermentation is the powerful process to exploit the potential of legumes, pseudo-cereals and milling by-products in baking industry. *Critical Reviews in Food Science and Nutrition*, *60*(13), 2158–2173. Available from https://doi.org/10.1080/10408398.2019.1631753, https://www.tandfonline.com/loi/bfsn20.

Goodarzi, M., Moradi, M., Tajik, H., Forough, M., Ezati, P., & Kuswandi, B. (2020). Development of an easy-to-use colorimetric pH label with starch and carrot anthocyanins for milk shelf life assessment. *International Journal of Biological Macromolecules*, *153*, 240–247. Available from https://doi.org/10.1016/j.ijbiomac.2020.03.014.

Grossmann, L., & McClements, D. J. (2021). The science of plant-based foods: Approaches to create nutritious and sustainable plant-based cheese analogs. *Trends in Food Science and Technology*, *118*, 207–229. Available from https://doi.org/10.1016/j.tifs.2021.10.004.

Guess, N., Klatt, K., Wei, D., Williamson, E., Ulgenalp, I., Trinidade, O., Kusaslan, E., Yilidrim, A., Gowers, C., Guard, R., & Mills, C. (2022). A cross-sectional study of the commercial plant-based landscape across the US, UK and Canada. *bioRxiv*. Available from https://doi.org/10.1101/2022.04.08.487708, https://www.biorxiv.org.

Guo, C., Zhang, M., & Devahastin, S. (2021). Color/aroma changes of 3D-Printed buckwheat dough with yellow flesh peach as triggered by microwave heating of gelatin-gum Arabic complex coacervates. *Food Hydrocolloids*, *112*, 106358. Available from https://doi.org/10.1016/j.foodhyd.2020.106358.

Guo, Z., Teng, F., Huang, Z., Lv, B., Lv, X., Babich, O., Yu, W., Li, Y., Wang, Z., & Jiang, L. (2020). Effects of material characteristics on the structural characteristics and flavor substances retention of meat analogs. *Food Hydrocolloids*, *105*, 105752. Available from https://doi.org/10.1016/j.foodhyd.2020.105752.

Harnack, L., Mork, S., Valluri, S., Weber, C., Schmitz, K., Stevenson, J., & Pettit, J. (2021). Nutrient composition of a selection of plant-based ground beef alternative products available in the United States. *Journal of the Academy of Nutrition and Dietetics*, *121*(12), 2401–2408. Available from https://doi.org/10.1016/j.jand.2021.05.002, e12.

Harper, A. R., Dobson, R. C. J., Morris, V. K., & Moggré, G. J. (2022). Fermentation of plant-based dairy alternatives by lactic acid bacteria. *Microbial Biotechnology*, *15*(5), 1404–1421. Available from https://doi.org/10.1111/1751-7915.14008, http://onlinelibrary.wiley.com/journal/10.1111/.

Hassoun, A, et al. (2022b). Exploring the Role of Green and Industry 4.0 Technologies in Achieving Sustainable Development Goals in Food Sectors. 112068. *Food Research International*. Available from https://doi.org/10.1016/j.foodres.2022.112068.

Hassoun, A., Aït-Kaddour, A., Abu-Mahfouz, A. M., Rathod, N. B., Bader, F., Barba, F. J., Biancolillo, A., Cropotova, J., Galanakis, C. M., Jambrak, A. R., Lorenzo, J. M., Måge, I., Ozogul, F., & Regenstein, J. (2022a). The fourth industrial revolution in the food industry—Part I: Industry 4.0 technologies. *Critical Reviews in Food Science and Nutrition*. Available from https://doi.org/10.1080/10408398.2022.2034735.

Hassoun, A., Boukid, F., Pasqualone, A., Bryant, C. J., García, G. G., Parra-López, C., Jagtap, S., Trollman, H., Cropotova, J., & Barba, F. J. (2022c). Emerging trends in the agri-food sector: Digitalisation and shift to plant-based diets. *Current Research in Food Science*, *5*, 2261–2269. Available from https://doi.org/10.1016/j.crfs.2022.11.010.

He, C., Zhang, M., & Guo, C. (2020). 4D printing of mashed potato/purple sweet potato puree with spontaneous color change. *Innovative Food Science & Emerging Technologies*, *59*, 102250. Available from https://doi.org/10.1016/j.ifset.2019.102250.

He, Y., Meda, V., Reaney, M. J. T., & Mustafa, R. (2021). Aquafaba, a new plant-based rheological additive for food applications. *Trends in Food Science and Technology*, *111*, 27–42. Available from https://doi.org/10.1016/J.TIFS.2021.02.035.

He, Y., Purdy, S. K., Tse, T. J., Tar'an, B., Meda, V., Reaney, M. J. T., & Mustafa, R. (2021). Standardization of aquafaba production and application in vegan mayonnaise analogs. *Foods*, *10*, 1978. Available from https://doi.org/10.3390/FOODS10091978.

Ismail, I., Hwang, Y. H., & Joo, S. T. (2020). Meat analog as future food: A review. *Journal of Animal Science and Technology*, *62*(2), 111–120. Available from https://doi.org/10.5187/jast.2020.62.2.111, http://www.ejast.org/archive/view_article?doi = 10.5187/jast.2020.62.2.111.

Javaid, M., Haleem, A., Singh, R. P., & Suman, R. (2022). Enhancing smart farming through the applications of Agriculture 4.0 technologies. *International Journal of Intelligent Networks*, *3*, 150–164. Available from https://doi.org/10.1016/j.ijin.2022.09.004.

Jeewanthi, R. K. C., & Paik, H. D. (2018). Modifications of nutritional, structural, and sensory characteristics of non-dairy soy cheese analogs to improve their quality attributes. *Journal of Food Science and Technology*, *55*(11), 4384–4394. Available from https://doi.org/10.1007/s13197-018-3408-3, http://www.springerlink.com/content/121580/.

Jeske, S., Zannini, E., & Arendt, E. K. (2018). Past, present and future: The strength of plant-based dairy substitutes based on gluten-free raw materials. *Food Research International*, *110*, 42–51. Available from https://doi.org/10.1016/J.FOODRES.2017.03.045.

Jiang, H., Zhang, W., Pu, Y., Chen, L., Cao, J., & Jiang, W. (2023). Development and characterization of a novel active and intelligent film based on pectin and betacyanins from peel waste of pitaya (*Hylocereus undatus*). *Food Chemistry*, *404*, 134444. Available from https://doi.org/10.1016/j.foodchem.2022.134444.

Jiang, Z. Q., Wang, J., Stoddard, F., Salovaara, H., & Sontag-Strohm, T. (2020). Preparation and characterization of emulsion gels from whole faba bean flour. *Foods*, *9*, 755. Available from https://doi.org/10.3390/foods9060755.

Jin, C., Bouzembrak, Y., Zhou, J., Liang, Q., van den Bulk, L. M., Gavai, A., Liu, N., van den Heuvel, L. J., Hoenderdaal, W., & Marvin, H. J. P. (2020). Big Data in food safety- A review. *Current Opinion in Food Science*, *36*, 24–32. Available from https://doi.org/10.1016/j.cofs.2020.11.006.

Jin, Y., He, X., Andoh-Kumi, K., Fraser, R. Z., Lu, M., & Goodman, R. E. (2018). Evaluating potential risks of food allergy and toxicity of Soy Leghemoglobin expressed in *Pichia pastoris*. *Molecular Nutrition and Food Research*, *62*, 1700297. Available from https://doi.org/10.1002/MNFR.201700297.

Kamath, R. (2018). Food traceability on blockchain: Walmart's pork and mango pilots with IBM. *The Journal of the British Blockchain Association*, 1. Available from https://doi.org/10.31585/jbba-1-1-(10)2018.

Keerthana, K., Anukiruthika, T., Moses, J. A., & Anandharamakrishnan, C. (2020). Development of fiber-enriched 3D printed snacks from alternative foods: A study on button mushroom. *Journal of Food Engineering*, *287*. Available from https://doi.org/10.1016/j.jfoodeng.2020.110116.

Kyriakopoulou, K., Keppler, J. K., & van der Goot, A. J. (2021). Functionality of ingredients and additives in plant-based meat analogues. *Foods*, *10*(3). Available from https://doi.org/10.3390/foods10030600, https://www.mdpi.com/2304-8158/10/3/600/pdf.

Le Roux, L., Ménard, O., Chacon, R., Dupont, D., Jeantet, R., Deglaire, A., & Nau, F. (2020). Are faba bean and pea proteins potential whey protein substitutes in infant formulas? An in vitro dynamic digestion approach. *Foods*, *9*, 362. Available from https://doi.org/10.3390/foods9030362.

Lee, J. S., Oh, H., Choi, I., Yoon, C. S., & Han, J. (2022). Physico-chemical characteristics of rice protein-based novel textured vegetable proteins as meat analogues produced by low-moisture extrusion cooking technology. *LWT*, *157*, 113056. Available from https://doi.org/10.1016/J.LWT.2021.113056.

Lei, Y., Yao, Q., Jin, Z., & Wang, Y.-C. (2023). Intelligent films based on pectin, sodium alginate, cellulose nanocrystals, and anthocyanins for monitoring food freshness. *Food Chemistry*, *404*, 134528. Available from https://doi.org/10.1016/j.foodchem.2022.134528.

Li, T., Lloyd, K., Birch, J., Wu, X., Mirosa, M., & Liao, X. (2020). A quantitative survey of consumer perceptions of smart food packaging in China. *Food Science & Nutrition*, *8*, 3977–3988. Available from https://doi.org/10.1002/fsn3.1563.

Li., Y., Zhang, X., Yang, J. J., Ma, X. Y., Jia, X. D., Du, P., & Li, A. li (2020). Influence of the addition of *Geotrichum candidum* on the microbial, chemical, textural, and sensory features of soft soy cheese. *Journal of Food Processing and Preservation*, *44*, e14823. Available from https://doi.org/10.1111/JFPP.14823.

Lie-Piang, A., Braconi, N., Boom, R. M., & van der Padt, A. (2021). Less refined ingredients have lower environmental impact — A life cycle assessment of protein-rich ingredients from oil- and starch-bearing crops. *Journal of Cleaner Production*, *292*. Available from https://doi.org/10.1016/j.jclepro.2021.126046.

Linares-García, L., Repo-Carrasco-Valencia, R., Paulet, P. G., & Schoenlechner, R. (2019). Development of gluten-free and egg-free pasta based on quinoa (*Chenopdium quinoa* Willd) with addition of lupine flour, vegetable proteins and the oxidizing enzyme POx. *European Food Research and Technology*, *245*, 2147–2156. Available from https://doi.org/10.1007/s00217-019-03320-1.

Lorusso, A., Coda, R., Montemurro, M., & Rizzello, C. G. (2018). Use of selected lactic acid bacteria and quinoa flour for manufacturing novel yogurt-like beverages. *Foods*, *7*(4). Available from https://doi.org/10.3390/foods7040051, https://www.mdpi.com/2304-8158/7/4.

Madhusudan, P., Chellukuri, N., & Shivakumar, N. (2018). Smart packaging of food for the 21st century — A review with futuristic trends, their feasibility and economics. *Materials Today: Proceedings*. Available from https://doi.org/10.1016/j.matpr.2018.06.494.

Mahwish, S. F., Afzaal, M., Hussain, M., Imran, M., Nawaz, T., & Siddeeg, A. (2022). Dietary guidelines to boost immunity during pre and post covid-19 conditions. *International Journal of Food Properties*, *25*, 1246–1265. Available from https://doi.org/10.1080/10942912.2022.2071287.

Masiá, C., Jensen, P. E., Petersen, I. L., & Buldo, P. (2022). Design of a functional pea protein matrix for fermented plant-based cheese. *Foods*, *11*(2). Available from https://doi.org/10.3390/foods11020178, https://www.mdpi.com/2304-8158/11/2/178/pdf.

Mathai, J. K., Liu, Y., & Stein, H. H. (2017). Values for digestible indispensable amino acid scores (DIAAS) for some dairy and plant proteins may better describe protein quality than values calculated using the concept for protein digestibility-corrected amino acid scores (PDCAAS. *British Journal of Nutrition*, *117*(4), 490–499. Available from https://doi.org/10.1017/S0007114517000125, http://journals.cambridge.org/BJN.

Mattice, K. D., & Marangoni, A. G. (2020). Evaluating the use of zein in structuring plant-based products. *Current Research in Food Science*, *3*, 59–66. Available from https://doi.org/10.1016/j.crfs.2020.03.004.

McClements, D. J. (2020). Development of next-generation nutritionally fortified plant-based milk substitutes: Structural design principles. *Foods*, *9*(4). Available from https://doi.org/10.3390/foods9040421, https://www.mdpi.com/2304-8158/9/4/421.

Mefleh, M., Pasqualone, A., Caponio, F., De Angelis, D., Natrella, G., Summo, C., & Faccia, M. (2022). Spreadable plant-based cheese analogue with dry-fractioned pea protein and inulin–olive oil emulsion-filled gel. *Journal of the Science of Food and Agriculture*, *102*(12), 5478–5487. Available from https://doi.org/10.1002/jsfa.11902.

Mefleh, M., Faccia, M., Natrella, G., De Angelis, D., Pasqualone, A., Caponio, F., & Summo, C. (2022). Development and chemical-sensory characterization of chickpeas-based beverages fermented with selected starters. *Foods*, *11*(22). Available from https://doi.org/10.3390/foods11223578.

Mintel. (2022). Driving success for plant-based drinks for kids - Mintel. https://clients.mintel.com/insight/driving-success-for-plant-based-drinks-for-kids?fromSearch = %3Ffreetext%3Dvegan%2520diet&resultPosition = 11.

Mottaleb, K. A., Kruseman, G., & Snapp, S. (2022). Potential impacts of Ukraine-Russia armed conflict on global wheat food security: A quantitative exploration. *Global Food Security*, *35*, 100659. Available from https://doi.org/10.1016/J.GFS.2022.100659.

Nadathur, S. R., Wanasundara, J. P. D., & Scanlin, L. (2017). *Proteins in the diet: Challenges in feeding the global population sustainable protein sources* (pp. 1–19). United States: Elsevier Inc. Available from http://www.sciencedirect.com/science/book/9780128027783.

Navaf, M., Sunooj, K. V., Aaliya, B., Akhila, P. P., Sudheesh, C., Mir, S. A., & George, J. (2022). 4D printing: a new approach for food printing; effect of various stimuli on 4D printed food properties. A comprehensive review. *Applied Food Research*, *2*(2). Available from https://doi.org/10.1016/j.afres.2022.100150, https://www.journals.elsevier.com/applied-food-research.

Occhiuzzi, C., D'Uva, N., Nappi, S., Amendola, S., Giallucca, C., Chiabrando, V., Garavaglia, L., Giacalone, G., & Marrocco, G. (2020). Radio-frequency-identification-based intelligent packaging: Electromagnetic classification of tropical fruit ripening. *IEEE Antennas and Propagation Magazine*, *62*(5), 64–75. Available from https://doi.org/10.1109/map.2020.3003212.

Ogundipe, O. O., Fasogbon, B. M., Ogundipe, F. O., Oredope, O., & Amaezenanbu, R. U. (2021). Nutritional composition of non-dairy yogurt from sprouted tigernut tubers. *Journal of Food Processing and Preservation*, *45*(11). Available from https://doi.org/10.1111/jfpp.15884, http://onlinelibrary.wiley.com/journal/10.1111/, ISSN: 1745-4549.

Oyeyinka, A. T., Odukoya, J. O., & Adebayo, Y. S. (2019). Nutritional composition and consumer acceptability of cheese analog from soy and cashew nut milk. *Journal of Food Processing and Preservation*, *43*(12). Available from https://doi.org/10.1111/jfpp.14285, http://onlinelibrary.wiley.com/journal/10.1111/, ISSN: 1745-4549.

Pandey, S., Ritz, C., & Perez-Cueto, F. J. A. (2021). An application of the theory of planned behaviour to predict intention to consume plant-based yogurt alternatives. *Foods*, *10*(1). Available from https://doi.org/10.3390/foods10010148, https://www.mdpi.com/2304-8158/10/1/148/pdf.

Pant, A., Lee, A. Y., Karyappa, R., Lee, C. P., An, J., Hashimoto, M., Tan, U. X., Wong, G., Chua, C. K., & Zhang, Y. (2021). 3D food printing of fresh vegetables using food hydrocolloids for dysphagic patients. *Food Hydrocolloids*, *114*. Available from https://doi.org/10.1016/j.foodhyd.2020.106546, https://www.journals.elsevier.com/food-hydrocolloids.

Paul, A. A., Kumar, S., Kumar, V., & Sharma, R. (2020). Milk analog: Plant based alternatives to conventional milk, production, potential and health concerns. *Critical Reviews in Food Science and Nutrition*, *60*, 3005–3023. Available from https://doi.org/10.1080/10408398.2019.1674243.

Phuhongsung, P., Zhang, M., & Devahastin, S. (2020a). Influence of surface pH on color, texture and flavor of 3D printed composite mixture of soy protein isolate, pumpkin, and beetroot. *Food and Bioprocess Technology*, *13*, 1600–1610. Available from https://doi.org/10.1007/s11947-020-02497-8.

Phuhongsung, P., Zhang, M., & Bhandari, B. (2020b). 4D printing of products based on soy protein isolate via microwave heating for flavor development. *Food Research International*, *137*, 109605. Available from https://doi.org/10.1016/j.foodres.2020.109605.

Protudjer, J. L. P., & Mikkelsen, A. (2020). Veganism and paediatric food allergy: two increasingly prevalent dietary issues that are challenging when co-occurring. *BMC Pediatrics*, *20*, 341. Available from https://doi.org/10.1186/S12887-020-02236-0.

Qin, P., Wang, T., & Luo, Y. (2022). A review on plant-based proteins from soybean: Health benefits and soy product development. *Journal of Agriculture and Food Research*, *7*, 100265. Available from https://doi.org/10.1016/J.JAFR.2021.100265.

Ran, X., Lou, X., Zheng, H., Gu, Q., & Yang, H. (2022). Improving the texture and rheological qualities of a plant-based fishball analogue by using konjac glucomannan to enhance crosslinks with soy protein. *Innovative Food Science & Emerging Technologies*, *75*, 102910. Available from https://doi.org/10.1016/J.IFSET.2021.102910.

Reinhardt, T. (2022). The farm to fork strategy and the digital transformation of the agrifood sector—An assessment from the perspective of innovation systems. *Applied Economic Perspectives and Policy*. Available from https://doi.org/10.1002/aepp.13246.

Remize, M., Brunel, Y., Silva, J. L., Berthon, J. Y., & Filaire, E. (2021). Microalgae n-3 PUFAs production and use in food and feed industries. *Marine Drugs*, *19*(2). Available from https://doi.org/10.3390/MD19020113, https://www.mdpi.com/1660-3397/19/2/113/pdf.

Romero, J., Cruz, R. M. S., Díez-Méndez, A., & Albertos, I. (2022). Valorization of berries' agro-industrial waste in the development of biodegradable pectin-based films for fresh salmon (Salmo salar) shelf-life monitoring. *International Journal of Molecular Sciences*, *23*(16). Available from https://doi.org/10.3390/ijms23168970, http://www.mdpi.com/journal/ijms.

Rondoni, A., Millan, E., & Asioli, D. (2021). Plant-based eggs: Views of industry practitioners and experts. *Journal of International Food & Agribusiness Marketing*, *34*, 564–587. Available from https://doi.org/10.1080/08974438.2021.1915222.

Saget, S., Costa, M., Santos, C. S., Vasconcelos, M. W., Gibbons, J., Styles, D., & Williams, M. (2021). Substitution of beef with pea protein reduces the environmental footprint of meat balls whilst supporting health and climate stabilisation goals. *Journal of Cleaner Production*, *297*, 126447. Available from https://doi.org/10.1016/J.JCLEPRO.2021.126447.

Salomé, M., Huneau, J. F., Le Baron, C., Kesse-Guyot, E., Fouillet, H., & Mariotti, F. (2021). Substituting meat or dairy products with plant-based substitutes has small and heterogeneous effects on diet quality and nutrient security: A simulation study in French adults (INCA3. *Journal of Nutrition*, *151*(8), 2435–2445. Available from https://doi.org/10.1093/jn/nxab146, http://jn.nutrition.org.

Samtiya, M., Aluko, R. E., Dhewa, T., & Moreno-Rojas, J. M. (2021). Potential health benefits of plant food-derived bioactive components: An overview. *Foods*, *10*(4). Available from https://doi.org/10.3390/foods10040839, https://www.mdpi.com/2304-8158/10/4/839/pdf.

Shabani, J., Sarfarazi, M., Mirzaei, H., & Jafari, S. M. (2016). Influence of the sunflower oil content, cooking temperature and cooking time on the physical and sensory properties of spreadable cheese analogues based on UF white-brined cheese. *International Journal of Dairy Technology*, *69*(4), 576–584. Available from https://doi.org/10.1111/1471-0307.12305, http://www.wiley.com/bw/journal.asp?ref = 1364-727x.

Shanmugam, K., Bryngelsson, S., Östergren, K., & Hallström, E. (2023). Climate impact of plant-based meat analogues: A review of life cycle assessments. *Sustainable Production and Consumption*, *36*, 328–337. Available from https://doi.org/10.1016/j.spc.2023.01.014.

Shirotani, N., Bygvraa Hougaard, A., Lametsch, R., Agerlin Petersen, M., Rattray, F. P., & Ipsen, R. (2021). Proteolytic activity of selected commercial Lactobacillus helveticus strains on soy protein isolates. *Food Chemistry*, *340*. Available from https://doi.org/10.1016/j.foodchem.2020.128152.

Shivakumar, N., Kashyap, S., Kishore, S., Thomas, T., Varkey, A., Devi, S., Preston, T., Jahoor, F., Sheshshayee, M. S., & Kurpad, A. V. (2019). Protein-quality evaluation of complementary foods in Indian children. *The American Journal of Clinical Nutrition*, *109*, 1319. Available from https://doi.org/10.1093/AJCN/NQY265.

Silva, A. R. A., Silva, M. M. N., & Ribeiro, B. D. (2020). Health issues and technological aspects of plant-based alternative milk. *Food Research International*, *131*, 108972. Available from https://doi.org/10.1016/j.foodres.2019.108972.

Souza Filho, P. F., Andersson, D., Ferreira, J. A., & Taherzadeh, M. J. (2019). Mycoprotein: environmental impact and health aspects. *World Journal of Microbiology and Biotechnology*, *35*(10). Available from https://doi.org/10.1007/s11274-019-2723-9, http://www.wkap.nl/journalhome.htm/0959-3993.

Tangyu, M., Muller, J., Bolten, C. J., & Wittmann, C. (2019). Fermentation of plant-based milk alternatives for improved flavour and nutritional value. *Applied Microbiology and Biotechnology*, *103*, 9263–9275. Available from https://doi.org/10.1007/S00253-019-10175-9.

Trollman, H., Garcia-Garcia, G., Jagtap, S., & Trollman, F. (2022). Blockchain for ecologically embedded coffee supply chains. *Logistics*, *6*, 43. Available from https://doi.org/10.3390/logistics6030043.

Varghese, C., Wolodko, J., Chen, L., Doschak, M., Srivastav, P. P., & Roopesh, M. S. (2020). Influence of selected product and process parameters on microstructure, rheological, and textural properties of 3D printed cookies. *Foods*, *9*, 907. Available from https://doi.org/10.3390/foods9070907, https://doi.

Varvara, R. A., Szabo, K., & Vodnar, D. C. (2021). 3D food printing: Principles of obtaining digitally-designed nourishment. *Nutrients*, *13*, 3617. Available from https://doi.org/10.3390/nu13103617.

Vasuki, M. T., Kadirvel, V., & Narayana, G. P. (2023). Smart packaging—An overview of concepts and applications in various food industries. *Food Bioengineering*, *2*, 25–41. Available from https://doi.org/10.1002/fbe2.12038.

Vatansever, S., Xu, M., Magallanes-López, A., Chen, B., & Hall, C. (2021). Supercritical carbon dioxide + ethanol extraction to improve organoleptic attributes of pea flour with applications of sensory evaluation, HS-SPME-GC, and GC-olfactory. *Processes*, *9*, 489. Available from https://doi.org/10.3390/PR9030489.

van Vliet, S., Kronberg, S. L., & Provenza, F. D. (2020). Plant-based meats, human health, and climate change. *Frontiers in Sustainable Food Systems*, *4*. Available from https://doi.org/10.3389/fsufs.2020.00128, http://www.frontiersin.org/journals/sustainable-food-systems#.

Vogelsang-O'Dwyer, M., Zannini, E., & Arendt, E. K. (2021). Production of pulse protein ingredients and their application in plant-based milk alternatives. *Trends in Food Science and Technology*, *110*, 364–374. Available from https://doi.org/10.1016/j.tifs.2021.01.090, http://www.elsevier.com/wps/find/journaldescription.cws_home/601278/description#description.

Van den Berg, L. A., Mes, J. J., Mensink, M., & Wanders, A. J. (2022). Protein quality of soy and the effect of processing: A quantitative review. *Frontiers in Nutrition*, *9*, 2148. Available from https://doi.org/10.3389/FNUT.2022.1004754/BIBTEX.

Wright, N. S., & Smith, M. (2020). Guidelines suggesting children avoid plant-based milks: A closer examination. *Maternal and Child Health Journal*, *24*(10), 1189–1192. Available from https://doi.org/10.1007/s10995-020-02970-y, http://www.kluweronline.com/issn/1092-7875/.

Yang, X., Li, M., Yu, H., Wang, M., Xu, D., & Sun, C. (2021). A trusted blockchain-based traceability system for fruit and vegetable agricultural products. *IEEE Access*, *9*, 36282–36293. Available from https://doi.org/10.1109/access.2021.3062845.

Zannini, E., Jeske, S., Lynch, K., & Arendt, E. K. (2018). Development of novel quinoa-based yoghurt fermented with dextran producer Weissella cibaria MG1. *International Journal of Food Microbiology*, *268*, 19–26. Available from https://doi.org/10.1016/j.ijfoodmicro.2018.01.001, http://www.elsevier.com/locate/ijfoodmicro.

Zeng, F., Ye, Y., Liu, J., & Fei, P. (2023). Intelligent pH indicator composite film based on pectin/chitosan incorporated with black rice anthocyanins for meat freshness monitoring. *Food Chemistry: X*, *17*, 100531. Available from https://doi.org/10.1016/j.fochx.2022.100531.

Zhao, C., Liu, B., Piao, S., Wang, X., Lobell, D. B., Huang, Y., Huang, M., Yao, Y., Bassu, S., Ciais, P., Durand, J. L., Elliott, J., Ewert, F., Janssens, I. A., Li, T., Lin, E., Liu, Q., Martre, P., Müller, C., ... Asseng, S. (2017). Temperature increase reduces global yields of major crops in four independent estimates. *Proceedings of the National Academy of Sciences of the United States of America*, *114*(35), 9326–9331. Available from https://doi.org/10.1073/pnas.1701762114, http://www.pnas.org/content/114/35/9326.full.pdf.

Ziarno, M., Bryś, J., Parzyszek, M., & Veber, A. (2020). Effect of lactic acid bacteria on the lipid profile of bean-based pant substitute of fermented milk. *Microorganisms*, *2020*(8), 1348. Available from https://doi.org/10.3390/MICROORGANISMS8091348.

Chapter 13

Toward Meat Industry 4.0: opportunities and challenges for digitalized red meat processing

Ruth M. Hamill[1], Alessandro Ferragina[1], Jyoti P. Mishra[1], Alan Kavanagh[2], Mike Hibbett[2], Mohammed Gagaoua[3], John Colreavy[4] and Ahmed Rady[1]

[1]*Food Quality and Sensory Science Department, Teagasc, Ashtown, Dublin, Ireland,* [2]*Irish Manufacturing Research, Mullingar, Co. Westmeath, Ireland,* [3]*PEGASE, INRAE, Institut Agro, Saint-Gilles, France,* [4]*Meat Technology Ireland, Teagasc, Ashtown, Dublin, Ireland*

13.1 Introduction

13.1.1 Industry 4.0 and the agrifood sector

Industry 4.0 (I4.0) is an established concept that relates to the transformation of the manufacturing sector through digitalization (McFarlane, 2018). The central vision of the Industry 4.0 movement is that future industrial businesses will build global, interconnected networks of machinery, products, and factories, which are represented as cyber-physical systems, seamlessly sharing digital information while being intelligently monitored and controlled to deliver significant improvements in industrial processes (Gilchrist, 2016). These improvements are made possible by the intrinsic business value residing in data that companies generate throughout their supply chains. These data are generated, in part, from the automation and information technologies arising from the third industrial revolution (e.g., automation and process control systems) but made available through a uniquely fourth industrial revolution technology, the Industrial Internet of Things (IIoT). Transforming these data to business-critical information is done using data analytics to generate insights, make predictions, and optimize processes. For a given industrial sector, progress towards achieving I4.0 is predicated on the incorporation of sectorially appropriate elements from a range of technologies (Mourtzis et al., 2022). At this time, a range of different industrial sectors have embraced digitalization and I4.0 solutions to greater or lesser extents and thus are characterized by a continuum of digital maturity levels. Semiconductors and pharmaceuticals are generally the most digitally mature sectors, while the food processing sector would be characterized by a lower digital maturity (Konur et al., 2021).

Global megatrends currently influencing the outlook for food production and processing, include the potential requirement to provide adequate nutrition to 9 billion people by 2050, increase in higher income middle-class consumers worldwide, who demand improved diets including more protein, while inequality and climate instability are increasing the risks associated with agriculture and providing new challenges for food security (OECD, 2022; UN, 2019). This places greater importance on new approaches such as I4.0, which offer the potential to stabilize and provide resilience to the agrifood sector. In recent years, there have been increasing efforts to increase the incorporation of digital technologies into food manufacturing. Approaches to I4.0 in the agriculture sector have led to a new area of the enterprise known as "AgTech", which includes incorporating farm management software, use of sensors in the field, drone-based imaging to detect stress or disease in crops, data transmission and management, temperature sensing of livestock for signs of estrus and with motion sensors, signs of parturition, for example, calving alerts, and automation in crop and fruit harvesting (Kumari et al., 2021a; O'Malley, 2019; Tingey-Holyoak et al., 2021; von Veltheim and Heise, 2020). These technologies hold great promise for efficient and resilient production, but I4.0 technologies also have the potential to beneficially impact and transform the food processing sector, including red meat processing.

Food Industry 4.0. DOI: https://doi.org/10.1016/B978-0-443-15516-1.00013-X

13.1.2 Potential of I4.0 for the red meat processing sector

The red meat sector not only provides income for the producers but also provides valuable nutrition for billions of people, and demand is forecast to remain strong into the future (OECD, 2022). Changing meat consumption patterns is closely linked with economic development (Andreoli et al., 2021). Worldwide meat consumption is predicted to increase by 14% by 2030 (OECD-FAO, 2022). Red meat is a source of high-quality protein and is a rich source of valuable micronutrients to support human growth and development and mitigate age-related conditions in an increasingly long-lived population (Chan, 2004; Mulvihill, 2004). Beyond these nutritive considerations, the red meat sector is of societal importance, with multifactorial impacts on the economy, culture, and gender equality, especially in the global south (Randolph et al., 2007), to support regenerative agriculture (Colley et al., 2020; Teague et al., 2016) and the conversion of marginal land, not capable of supporting other edible crops into human food (Steinfeld and Gerber, 2010).

Despite its societal importance, red meat processing is a low-margin sector, there are many drivers toward efficiency, for example, labor availability and the sustainability of raw materials. These challenges were exacerbated during COVID-19, and there is a recognition that similar crises will be present in the future (Ijaz et al., 2021). I4.0 technologies offer new opportunities to the red meat sector in the areas of managing automation and information flow in the supply chain (Barbut, 2020).

13.1.3 Objectives

In this chapter, we will provide an overview of enabling I4.0 technologies that have particular relevance in the red meat sector. We will describe ongoing research, opportunities, and barriers to uptake for these key technologies and the potential for I4.0 to support the red meat sector to meet the challenges it faces in the 21st century. We will also provide an overview of digital platforms that enable companies to self-assess their digital maturity and provide tailored guidance for companies on how they can best approach the opportunities in I4.0. Finally, we will offer our perspective on how the red meat sector can progress a step closer to embedding I4.0 and its next evolution, Industry 5.0.

13.1.4 Potential of I4.0 enabling technologies for the red meat industry

Key technologies in I4.0 for the meat sector include data analytics, sensors, IIoT, digital twin, automation, and traceability. These I4.0 technologies could support both visibility on plant systems, carcass, and meat attributes and increased automation in this labor-intensive sector, with benefits for worker and food safety, enhanced yield, and more consistent quality throughout the meat processing plant (Maselyne et al., 2017). The relevance of these technologies to the meat sector is described in the following sections.

13.2 Data analytics for the meat processing sector

Data analytics encompasses the processes of inspecting, cleaning, transforming, and modeling with the goal of discovering useful information, suggesting conclusions, and supporting decision-making to create business value (McFarlane, 2018). Data analytics is a significantly important area within I4.0 that is advancing at an unprecedented pace, largely due to rapid improvements in advanced Artificial Intelligence (AI). Deep learning algorithms and generative AI are examples of innovations in data analytics, that are improving in performance in speed and accuracy to enable rapid object recognition and are applied in diverse domains including monitoring crowds and facilitating self-driving cars (Redmon et al., 2016). The agrifood domain is assimilating deep learning at the research and development stage but is beginning to benefit from fruitful studies in various areas (Baglat et al., 2023; Elbeheiry and Balog, 2023; Iheonye et al., 2023; Serey et al., 2023).

Adoption rates of data analytics by companies across geographies, functions, and organizations of all sizes show a significant increase in usage uptake and a large drop in companies with no plans to adopt it (Addo-Tenkorang and Helo, 2016). This is an indication that data analytics is moving from an experimental or piecemeal endeavor to a critical pursuit within organizations. However, the opportunity to create business value and impact is being missed by many businesses, due to a number of factors. There is clear evidence indicating the adoption and real "on-the-ground" business impact of data analytics within the industry continues to pose major challenges. While the attention of many companies is currently around obtaining data—how to collect it; how to develop data infrastructures; how to store and access data—a lack of focus on creating value out of data, analyzing that data, and designing analytics strategies around it is leading to poor performance metrics with respect to data projects. A survey of over 1,300 manufacturing executives

(Betti et al., 2021), revealed just 39% had successfully scaled a data analytics use case beyond a pilot or single value stream. There are many studies reporting similar underwhelming outcomes. A recent survey of uptake of I4.0 technologies by 92 manufacturing companies revealed data analytics as one of the technologies lagging in terms of uptake and adoption despite the obvious benefits (Frank et al., 2019), and with much information stored on paper-based systems and a high reliance on domain knowledge of their most experienced staff, meat processing companies are no exception.

While several factors can be identified as potential causes, one key inhibitor to success in data analytics is the lack of understanding of how to solve problems using data. To address this, collaborative, cross-functional projects consisting of data scientists, IIoT, and domain experts will increase the likelihood of success; crucially not just in solving existing business problems but in unlocking the even greater potential inherent in business data to create new products, services, and business models that previously were not thought possible. To demonstrate this potential, one can look to the exemplar of Tesla. The electric car manufacturer's success in disrupting the industry is largely attributed to the value they leverage from their data with Artifical Intelligence including early adoption of an exclusive online platform, disrupting the traditional dealership model, and a holistic application of many aspects of I4.0 in the business.

There are reports emerging of capabilities and innovation within data analytics in research and in real meat processing environments. Computer vision coupled with AI has been successfully applied to classify different primal meat cuts (Huang et al., 2022) and different cut specifications from a single muscle (Prakash et al., 2021) and predict cut yield from images (Matthews et al., 2022). Thus computer vision systems and machine learning present attainable benefits for meat management systems and process analytics for meeting customer specifications. Offal yield prediction is a typical challenge in an industry where the incoming raw materials (i.e., animals) vary in shape, conformation, and size. For Silver Fern farms, a recent I4.0 data analytics and machine learning project, involving dashboard, analytics, and crucial integration of previously siloed data sets (multiyear yield reports, farm/animal-specific data, and climate data) led to benefits in more accurate yield forecasting of offal yield and deployment of user-friendly systems for data visualization for staff (Data-driven forecasting changing our meat industry | Callaghan Innovation, accessed 24/06/2023). Elsewhere, a commercial data analytics provider reports the benefits experienced by a client company (Voltigeurs Farm) in terms of productivity improvement of the order of 43% by introducing real-time monitoring and IIoT sensors and station-level visualization dashboards to allow for faster response to downtime leading to a significant increase in throughput per month through its Quebec abattoir (How Worximity helped Voltigeur Farms increase throughput by 15% in deboning and meat packing processes, accessed 24/06/2023). The focus on improving yield and productivity more generally is crucial as processors seek to increase margins in a highly competitive market.

13.3 Sensorization as an enabling technology in meat processing

Providing information on the shop floor product and processing equipment requires a means to record the relevant information and transfer it to enterprise resource planning infrastructure. Monitoring the raw material in a meat plant rapidly and without interfering with the product is a highly relevant challenge to the adoption of sensor-enabled technologies as inputs to data analytics and to the IIoT, discussed in the following section. Nondestructive sensors provide rapid, and relatively accurate evaluation for initial monitoring of various quality attributes of food materials. Among such sensors, optical sensors have been providing subjective quality evaluation for raw and processed food products over the last three decades (Singh and Jayas, 2013). Optical sensors (i.e., RGB), ultraviolet (UV), visible and/or near-infrared (Vis/NIR) spectroscopy, hyperspectral and multispectral imaging, Fourier-Transform (FT-IR) spectroscopy, X-ray, micro CT X-ray, nuclear magnetic resonance spectroscopy (NMR), TeraHertz, and Electronic Nose (E-Nose) provide high acquisition speed, minimal sample preparation, rapid predictions, versatility, and simultaneous prediction of multitraits (Osborne, 2007; Watson et al., 2021). Such technologies benefited from the development of efficient hardware, and high-performance computational power (Davies, 2013). Table 13.1 illustrates examples of various nondestructive systems used for assessing different quality traits of raw and processed meats especially those focused on in-line or on-line applications. Spectroscopic sensors including Vis/NIR, and RGB color imaging as subcategories of optical sensors are relatively low-cost compared to other sensors. Optical sensing relies on the interaction between the matter and the electromagnetic radiation as the incident radiation can be absorbed, transmitted, or reflected from the matter proportionally to the chemical composition and structure which results in a fingerprint of the material at a specific wavelength(s) (Osborne, 2007). Sensors based on microwave dielectric spectroscopy properties also provide noninvasive and rapid measurements of the quality attributes of food products (Meng et al., 2018). Measurements using such sensors rely on the correlation between the water content in biological materials and certain properties of the acquired signals in the microwave range such as relative permittivity, dielectric constant, and dielectric loss factor (Meng et al., 2018).

TABLE 13.1 Examples of nondestructive sensors utilized for quality evaluation of raw and processed meats.

Sensors group	Sensor subgroup/ acquisition mode	Meat product	Quality attribute measured	Modeling algorithm(s)	Advantages or drawbacks	References	Notes
Optical	NIR/Reflectance	Lamb, emu, camel, beef	Classification	PCA LDA	Handheld	Cozzolino et al. (2023)	(950–1600 nm) Class. Acc. = 87.8%
		Green ham	Composition (moisture, fat, protein)	PCA PLSR RF	Supports digitalization through smartphones	Kartakoullis et al. (2019)	740–1070 nm (handheld) 940–1700 nm $R^2 = 94.0\%–99.0\%$
	Raman spectroscopy	Beef	Quality (WBSF, IMF, drip loss, cook loss)	PLSR		Cama-Moncunill et al. (2020)	(250–3381 cm^{-1}) $R^2 = 50.0\%–90.0\%$
		Beef	Identifying meats from grass- and grain-fed cattle	PCA		Logan et al. (2020)	(600–1900 cm^{-1}) $R^2 = 50.0\%–90.0\%$
	Hyperspectral imaging	Beef striploin	Sensory	PLSR	High speed In-line	Wold et al. (2022)	(760–1047 nm) $R^2 = 81.0\%$ (slices) $R^2 = 82.0\%$ (whole)
		Cooked pork loin	Identifying brined samples	PCA PLS-DA PLSR	PAT	Achata et al. (2019)	(450–1664 nm) Sens. = 100% (raw) Sens. = 100% (cooked) $R^2 = 92.0\%$ (raw) $R^2 = 83.0\%$ (cooked)
	X-ray	Lamb carcass	Composition (fat%, lean%, bone%)	GLM analysis	On-line	Gardner et al. (2018)	Dual-energy X-ray absorptiometry

							$R^2 = 68.0\%$ (bone) $R^2 = 89.0\%$ (fat) $R^2 = 69.0\%$ (lean)
		Meats	Detecting metal needles	CNN		Ahn et al. (2020)	Classf Acc. = 99.8%
	CT X-ray	Pig	Developing a pig atlas	Parametric linear Lagrange or Cubic Hermite mesh		Ho et al. (2019)	CT scanner
	RGB color imaging	Pork loin	Color and marbling scores	SVMR	In-line	Sun et al. (2018)	$R^2 = 92.5$ (color score) $R^2 = 75.0$ (marbling score)
	FT-IR spectroscopy	Cured meats (hams, salami, dried lamb ribs, dried mutton legs)	Detection of molds (*Penicillium* spp.)	HotPLS		Schirmer et al. (2021)	FT-IR ($4000–500\ cm^{-1}$) Class. Acc. up to 100%
		Aged beef	Classification of samples based on aging/Prediction of tenderness	PCA PLSR		Beriain et al. (2022)	FT-IR-ATR ($4000–400\ cm^{-1}$) Class. Acc. up to 70.0% $R^2 = 7.71\%–38.54\%$
NMR spectroscopy	T2 NMR relaxometry	Dry-cured ham	Monitoring the quality at different processing times, measuring water content	Linear regression		García-García et al. (2019)	
Ultrasonic	Noncontact (pulse–echo mode) (1 MHz)	Dry-cured ham	Monitoring the salting process	Linear regression	On-line	Garcia-Perez et al. (2019)	$R^2 = 83\%–93\%$
Microwave spectroscopy	Two portable systems 100 MHz–5.4 GHz 100 MHz–6.5 GHz	Beef (Angus)	Assessing the depth of P8 fat and Rib fat	PLSR		Marimuthu and Gardner (2019)	$R^2 = 66.0\%–88.0\%$ for P8 $R^2 = 44.0.0\%–56.0\%$ for rib fat
E-Nose	An array of 10 gas sensors	Beef	Assessing beef adulteration with pork	1DCNN-RFR		Huang and Gu (2022)	$R^2 = 98.6\%–99.8\%$
Sensor fusion	Hyperspectral imaging + E-Nose	Frozen-thawed pork	Moisture content	PLSR		Cheng et al. (2022)	$R^2 = 81.4\%–95.3\%$

The general concept of developing predictive model(s) using nondestructive sensors involves acquiring spectral data, measurement of the trait(s) of interest (i.e., reference) for the training samples, developing training models of the trait(s) via machine learning, validation of the developed models using an external data set, testing the validated models on new samples with the unknown trait(s), and updating the models. Some of the key factors to achieving a successful application using sensors are the availability of sufficient spectral and phenotypic variability imposed by the samples, whether the trait(s) of interested is directly or indirectly correlated with the spectral data, the number of samples available for training, the applied chemometrics, and the validation process (Varmuza and Filzmoser, 2009). To calibrate the most powerful models, the ability to cover as much variability as possible has been one of the biggest challenges in the research and development of novel sensor applications for the prediction of traits or parameters of interest in the meat plant. Collecting large amounts of high-quality data that can then be modeled with the sensor data is labor-intensive and costly.

13.3.1 Current and potential applications of noninvasive and/or rapid sensors in meat factories

One of the most important applications of I4.0-related technologies is in the classification of animal carcasses for the purposes of grading. Meat carcass is graded before secondary processing to predict the meat yield and quality. Grading can be conducted via contact or noncontact probes for predicting meat yield. Optical sensors including computer vision systems (CVSs) provide noncontact scanning and are used for faster and more consistent grading. Extensive reviews of carcass grading and associated technologies have been conducted (Delgado-Pando et al., 2021; Gardner et al., 2021), and examples can be seen in Table 13.2. Some other key areas where sensing has potential application are related to meat quality through rapid noninvasive prediction of important postmortem characteristics such as sarcomere contractile status, pH profile, and glycogen depletion (Cama-Moncunill et al., 2020; Warriss, 2009) while food safety (Khaled et al., 2021) and authenticity (Monahan et al., 2018) are also areas with demonstrated potential. A series of studies were conducted by

TABLE 13.2 Examples of objective carcass grading technologies that can be integrated into Industry 4.0 systems.

Level of automation	Working principle	Important features	Examples	Link
Semi-automatic	Grading based on Rib eye area	2D and 3D images of rib eye are captured Instrument has on-board AI for QC and data analysis Grading parameters are displayed in a few seconds Can be integrated into plant production planning system RGB-D camera	Q-FOM from Frontmatec for beef carcass grading VGB-2000 from E + V	https://www.frontmatec.com/en/other/instruments/carcass-grading-traceability/q-fom-beef https://www.eplusv.com/en/products/beef/vbg-2000/
Fully automatic	Ultrasonic image based	Robust system Prediction of yield	Frontmatec fully automatic ultrasonic carcass grading system (Autofom III) for pig carcass grading	https://www.frontmatec.com/en/other/instruments/carcass-grading-traceability/autofom-iii
	3D reconstruction of carcass from images	Three-dimensional structure of carcass is captured using a camera array The depth parameter can be based on stereoscopy or structured light Uses point cloud data and volume The carcass is classified into different fat and conformation classes based on the EUROP standard	Frontmatec carcass grading station (BCC-3) for beef application E + V vision system (VBS-2000)	https://www.frontmatec.com/en/other/instruments/carcass-grading-traceability/beef-classification-center-bcc-3 https://www.eplusv.com/en/products/beef/vbs-2000
	X-ray grading system	X-ray image is captured by scanning the carcass	Scott X-ray grading system	https://scottautomation.com/en/products/meat/grading

Meat and Livestock Australia to investigate the feasibility of using microwave sensors coupled with regression algorithms for assessing different fats in cattle. The studies focused on using a portable microwave sensor that works in the range of 100 MHz–6.5 GHz to determine P8 fat depth and rib fat in Angus beef ($R^2 = 44\%–88\%$); and C-site fat depth, GR tissue depth, and eye muscle depth in live and preslaughtered lamb meats ($R^2 = 24\%–79\%$) (Marimuthu et al., 2021, 2022).

Applications of I4.0 entail automated data collection, input to appropriate cloud-based algorithms, and optimized meat management systems via data analytics. Some of the optical sensors have just such a potential, and computational modeling is becoming less of a barrier in recent years. Finally, the use of information recorded from various types of sensors can be used to assist with operations in the meat plant, for example, tasks guided by extended reality iterations, whether virtual or augmented reality (the latter involving overlay of virtual information on the real world). Although the uptake of extended reality in the food processing sector is not well advanced to date, concepts such as immersive scribing are being tested, for example, by the Australian Meat Processor Corporation, while augmented reality is under study for applications in hygiene (e.g., "CleanAssist"—Fraunhofer IVV at IFFA 2022—Fraunhofer IVV). Further applications are also possible in training (Christensen and Engell-Nørregård, 2016) and remote assistance by engineers for preventive maintenance and guiding specific operations (Vignali et al., 2018).

13.3.2 Future potential for online sensors in a Meat Industry 4.0

Meat processing plants can be considered as continuous processing lines with a high throughput of products, and in this context, the processing line should be considered as an integral part of a quality system optimizing the usage of the raw materials and the quality of the final product (Segtnan et al., 2009). Monitoring the production along the processing lines and the entire supply chain is important to guarantee standardized and high-quality products, traceability, and improve the sustainability of the overall process. The available literature produced in the last decades has proven the efficacy of optical imaging and spectroscopic systems for providing valuable information at every single step of processing (Cafferky et al., 2020; Cama-Moncunill et al., 2020; Kamruzzaman et al., 2012; Reis and Rosenvold, 2014). An area with great potential for the future is the development and implementation of multilocated spectroscopic sensors for autonomous data acquisition during meat processing that can be predictive of economically significant traits, generating and connecting valuable information.

Toohey et al. (2018) concluded that although, the benefits of using sensor technologies are obvious for the whole meat supply chain (on-farm, at the processors, and to the customers), the development and adoption are challenging and complex, including how to integrate these technologies at each stage of the supply chain to maximize and promote an equal benefit to all the sectors of the chain. They propose possible reasons for limited adoption including the need for complex and laborious calibration required as part of the vision of a reliable and efficient system for the meat sector (Antequera et al., 2021). Other reasons include insufficiently skilled labor, chain speed, accuracy, and resources (i.e., economic) to transfer the technology from the laboratory to the factory (Toohey et al., 2018). Translational research should now be prioritized regarding practical solutions to the rollout of sensorization in digitalized meat processing.

13.4 Industrial Internet of Things

An Internet of Things is simply a network of sensors, data storage, and processing tools that have the capability to exchange and analyze data, make a decision, and be accessed by users in a real-time scenario (Kim et al., 2020). IoT devices can communicate and interact with each other, as well as with other Internet-enabled devices and systems, without the need for human intervention. The global IoT market has been rapidly growing since 2018, with a total value of $750 billion in 2020 that is expected to reach $1.1 trillion in 2023. Such current and future growth is attributed to the increasing demand for connected devices, developing low-cost and user-friendly embedded systems, and the growing demand for AI in various industries (McFarlane, 2018). The IoT presents significant opportunities for companies in various industries, including manufacturing, healthcare, retail, and transportation. IIoT is simply about applying IoT technology specifically in the industrial, manufacturing, logistics, and utilities industries. IIoT allows industrial devices to collect and transmit data, enabling companies and staff to monitor and control their environment, automate processes, and make informed decisions based on real-time data (Kim et al., 2020). The investment by the industry in this technology is expected to exceed $40 billion by 2024.

13.4.1 Challenges in deploying IIoT technologies

Despite the advantages of adopting IIoT, challenges include (1) integrating IIoT systems with existing IT infrastructure can be complex and time consuming, requiring significant technical expertise and resources; (2) IIoT systems collect and transmit copious amounts of sensitive data, which can be vulnerable to hacking and cyber-attacks. Ensuring the security of these data is a major challenge for companies deploying IIoT technologies; (3) ensuring that different IIoT devices and systems can communicate with each other and exchange data is a significant challenge, requiring companies to invest in compatibility and interoperability testing; (4) the lack of standardization in the IIoT market makes it difficult for companies to integrate different devices and systems, leading to compatibility issues and increased complexity; (5) the deployment of IIoT technologies can be expensive, requiring significant investment in hardware, software, and infrastructure; (6) IIoT devices and systems collect vast amounts of personal and sensitive data, raising privacy concerns and the need for companies to comply with privacy regulations; (7) deployment of IIoT technologies requires a deep understanding of various technical and operational aspects, making it a complex process that requires significant technical expertise; and (8) growing from a small, localized proof of concept to an organizational-wide solution will involve significant costs and impact staff at all levels of the organization. For successful implementation, it is essential that long-term support is secured early in the deployment planning.

13.4.2 Industrial Internet of Things in the agrifood sector and potential in meat processing

The potential of IIoT technology, incorporating wireless sensor networks, in various Agri-Tech applications has been investigated to monitor plant health and yield (Gao et al., 2020; Khattab et al., 2019), soil conditions (Goap et al., 2018; Muangprathub et al., 2019), controlled environment conditions (Aafreen et al., 2019; Geng et al., 2019), and livestock behavior and welfare (Memon et al., 2016; Nadimi et al., 2012). IIoT systems also offer the potential to change how food factories operate—providing not only a basis for automation but possibilities for an unprecedented depth of visualization across all systems driven by the massive amounts of data that can be collected, analyzed, reported, and acted upon—often without the need for human interaction or involvement (Bigliardi et al., 2022). The potential for meat processors of this technology is huge as IoT systems would facilitate live monitoring of product performance within the meat plant and offer a means to dynamically assess the impact of changing conditions on product performance in real time. However, full implementation is dependent on the deployment of a plethora of industrial sensors, as referred to earlier.

IIoT systems in the meat processing industry could be applied to (1) product and environmental monitoring to improve the efficiency, safety, and quality through the meat processing stages, in addition to environmental conditions throughout the entire supply chain; (2) quality control by monitoring various aspects of the product, for example, pH, color, temperature, and parameters related to the production process; (3) traceability from farm, through slaughterhouses, factories, retailers, and finally consumers; (4) predictive maintenance: IIoT can monitor the performance of machines and equipment, reducing the risk of breakdowns and ensuring that maintenance is performed on a timely basis; (5) contamination detection which is a key safety factor in the meat processing industry; and (6) inventory management though monitoring the inventory levels of ingredients, raw materials, specific stock keeping units (SKUs) is an effective way for managing supplies and reducing food waste.

IIoT for meat is an area in which only limited research has commenced, but results are encouraging. Evaluation of postmortem meat quality has been studied using IoT tools in the research setting. Cama-Moncunill et al. (2022). developed an IoT-based system that provides real-time measurements of the temperature and relative humidity inside commercial dry-aging meat chambers. In another study, a low-cost IoT system comprising a computer vision system and gas-sensing measurements was studied to track pork freshness (Kim et al., 2022). The system can acquire and extract concentrations of total volatile organic compounds (TVOCs) and carbon dioxide (CO_2) data and wirelessly send them to a Raspberry Pi board, which uploads the data into the cloud. A one-dimensional convolutional neural network (1D-CNN) was used to develop the prediction model of pork freshness, and the results can be accessed through a mobile phone application. More research is required to test implementations of IIoT that may have relevance to the meat processing sector.

13.5 Digital twins

A digital twin is a core element of I4.0 consisting of creating a digital equivalent of a real-life object mirroring its behavior and states over its processing life (Verdouw et al., 2021). The first definition of a digital twin was introduced by NASA *as "an integrated multiphysics, multiscale, probabilistic simulation of a vehicle or system that uses the best available physical models, sensor updates, fleet history, etc., to mirror the life of its flying twin. It is ultra-realistic and may consider one or more important and interdependent vehicle systems*." The digital replica of an artifact (machine, line, product, etc.), process, or service is connected to the real-life twin by streams of data. These data can undergo analysis and be utilized for making faster, better decisions. Through the integration of real-time data, a digital twin is capable of providing a detailed model that can support decision-making and intervention at each step of the processing line (Krupitzer et al., 2022). The distinguishing digital-twin characteristic is continuous twinning, or mirroring of the operational conditions, of the physical twin whereby real-time data are continuously collected from sensors and transmitted to the virtual models to drive the simulation and prediction. Then, the parameters of the virtual models are fed back to optimize the physical entities. In the closed-loop interaction process, the physical entities and virtual models "coevolve." Digital twins can be used to examine and improve all manufacturing processes, from product design, manufacturing planning, and virtual equipment commissioning, to name a few. Digital twins can be seen as a culmination of several key industry 4.0 concepts often including IIoT, data analytics, simulation, and AI. Digital twins of a manufacturing system allows centralized analysis and—fundamentally—real-time control of the physical asset by leveraging the predictive capabilities of technologies from simulation to AI. Several digital twins of individual machines can form a virtualized production line or area, and ultimately a digital twin–enabled smart factory. The main attributes of such a smart factory are elaborated in Fig. 13.1. In practice, many digital-twin use cases are at a machine or production line level. There are only a small number of industries that can boast of having truly smart factories with all these attributes.

There are several examples of successful digital-twin deployments outside the factory setting. The aerospace giant Rolls Royce are digital-twin pioneers through a service offering advanced digital-twin capabilities. They have a comprehensive digital twin of their jet propulsion systems, which are constantly uploading actual in-flight performance data. When AI-

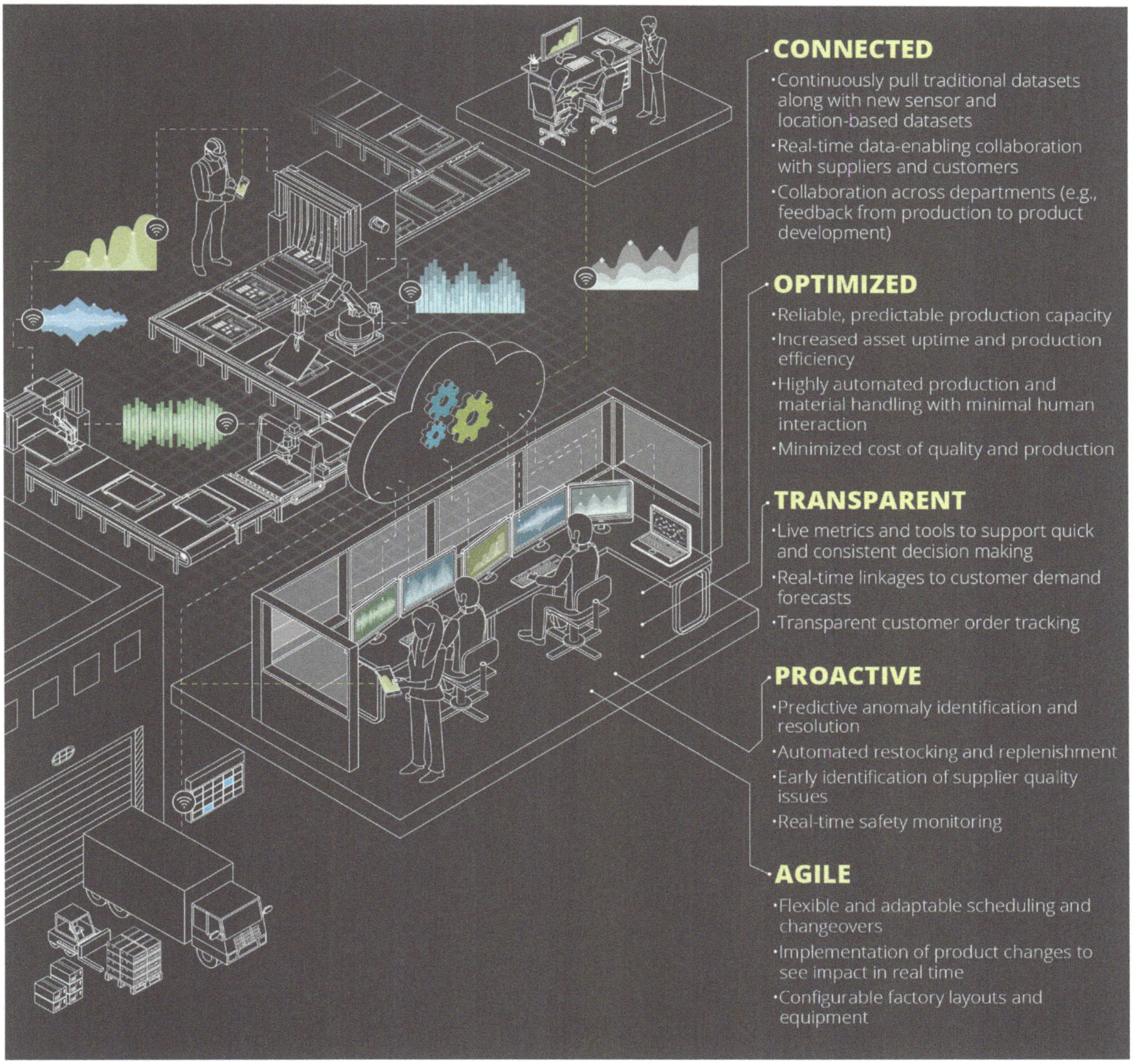

Source: Deloitte analysis. Deloitte University Press | dupress.deloitte.com

FIGURE 13.1 Five key characteristics of a smart factory. Smart factories should be connected, optimized, transparent, proactive, and agile. *Reproduced from Burke et al. Industry 4.0, smart factory, and connected manufacturing | Deloitte Insights.*

enabled, they can learn from data patterns and autonomously interact with their physical twins to make beneficial operational changes. Elsewhere tire manufacturers such as Bridgestone are investing in digital twins, where their sensor-equipped tires are detecting the road surface, capturing events in real time, and communicating continuously with vehicles to improve vehicle safety and performance, again utilizing AI. There are numerous other industry examples of digital twins. These innovations are part of a rapidly emerging "product as a service" market from hitherto traditional manufacturing companies. This refers to the trend of providing services packaged with or instead of a product for sale. More than 70% of EU-based manufacturers are expected to launch a new service in the next couple of years and digital twins will play a major role.

Within the meat sector, there would be a huge benefit to digital-twin installation in meat plants, for example, for tracking products and monitoring any potential disruption to cutting or boning operations. The linear nature of operations is such that any downtime represents significant economic losses, so timely information or even intervention would be extremely beneficial. Examples of digital-twin models and approaches for the food sector and farming have been proposed recently in several studies (Krupitzer et al., 2022; Purcell and Neubauer, 2023; Valero Hicks et al., 2023; Verdouw et al., 2021), and these guidelines could be useful for the development of such systems in the meat sector (Valero Hicks et al., 2023). Alongside applications in monitoring production and logistics, digital twins in simulation mode would provide an opportunity to explore process innovation without significant disruption to production. Some trials have commenced in the research setting in relation to digital twin−type implementations within meat processor facilities. The Australian Meat Processor Corporation (AMPC) developed a virtual model of a member company boning hall initially to support the onboarding and training of new staff during COVID-19 through an immersive virtual reality experience of the actual factory. To take its digital-twin concept further, the organization plans to incorporate sensor and machine data to view equipment status and even the movement of meat in real time via a 3D representation of the production line. From here, they want to move toward spatially anchored predictive maintenance of equipment utilizing advanced analytical techniques. Outside of the factory, there are initiatives to digitize all processing steps from "farm to fork" as part of environmental, social, and economic improvements associated with the processing of meat.

13.6 Automation

The degree of automation in the food supply chain is at a lower level compared to other industries, especially for meat deboning where the major operations are conducted manually (Romanov et al., 2022). The lack of trained operators and the need for efficient and sustainable production highlight the necessity for automation and digitalization in the meat sector (Echegaray et al., 2022). The main elements of the meat processing line can be classified into two steps: primary and secondary processing. Primary processing includes stunning, exsanguination, dehiding, evisceration, bung removal, carcass splitting, carcass dressing, and grading. Secondary processing includes cutting, deboning, trimming, and slicing. The level of automation in meat processing depends on the use of machinery and other cyber-physical systems. Romanov et al. (2022) stated that meat processing can be automated in either of three modes: (1) low automation (Level 1), which exists in most meat processing plants where limited automation and more manual laboring are utilized, especially in deboning and primal cutting; (2) semiautomation (Level 2), where some processes are automated (i.e., butchers use electric equipment) and manual labor is involved during or in between such processes; (3) full automation (Level 3), where all the processes are automated through robotics and data-driven tools. There are various equipment in the market for low- or semiautomated plants that are mainly used for primary processing in the hot process section (i.e., slaughterhouse). Skilled supervision and troubleshooting are required to achieve successful and continuous processing when adopting such standalone machines. Some examples of semi- and fully-automated equipment are listed in Table 13.3.

13.6.1 Meat processing with cyber-physical systems

Red meat animals are drawn from populations with natural biological variability in carcass size, tissue composition, and proportions of lean, fat, and bone (Warriss, 2009). The nature of meat processing and the potential for automation differ for the different meat animals (i.e., cattle, swine, sheep, and goats) in accordance with the nature of raw meat, which is inherently variable. Cyber-physical systems are intelligent systems that integrate sensing, computation, networking, and control of physical objects based on a set of algorithms embedded in a microprocessor or a robotic system. The variation in the input "raw material" of the meat processing industry poses a challenge even when it comes to implementing advanced robotics into meat production processes as, by and large, robots are normally deployed to perform operations on standard raw materials and components. Large carcass size poses additional challenges for beef processing automation. However, considering skilled and unskilled labor shortages in the meat industry and other drivers, it is increasingly necessary to develop intelligent robots that can perform complicated tasks as consistently and accurately as human

TABLE 13.3 Automated solutions across the slaughter line.

Processing step	Level of automation	Equipment name	Working principle	Features	Examples	Link
Stunning and exsanguination	Semiautomatic	Stunning box	Electrical/mechanical/CO_2 based	Fully enclosed chamber made up of GI or SS material. Electrodes or captive-bolt type guns. Blood collection and storage facility	Frontmatec stunning box for cattle	https://www.frontmatec.com/en/beef-solutions/unclean-line/stunning
					Frontmatec CO_2 stunning system	https://www.frontmatec.com/en/pork-solutions/unclean-line/stunning
					Marel Cattle stunning cage	https://marel.com/en/meat/cattle-beef/primary-processing/live-animal-handling
					Marel CO_2 and electrical stunning (Midas) system for pig	https://marel.com/en/meat/pig-pork/primary-processing/stunning
Bung removal	Fully automatic	Automatic bunger	Programmed machines	Synchronized movement along with carcass conveyor Hygienic design with self-cleaning after each operation	Frontmatec auto line bung dropper for automatic pig line	https://www.frontmatec.com/en/pork-solutions/clean-line-chill-room/autoline-machines/bunger-apb7
		Robotic bung dropper	Robot detects the tail position, drills, and removes the bung	The robot is provided with a protective cover and cutting tools, a vacuum suction unit, and a vision system.Sterilization facility in between cycles to avoid cross-contamination	AIRA RBD bung dropper for clean pig line	https://www.frontmatec.com/en/pork-solutions/clean-line-chill-room/aira-robots/aira-rbd-bung-dropper
					M-line bung remover (MBR) for pig line	https://marel.com/en/products/m-line-bung-remover-mbr
Carcass splitting	Fully automatic	Carcass splitting saw	Cutting blade with rolls move through the breast bones or aitch bone and back bones	Cutting machines are programmed and synchronized to the main line Machines are flexible, suitable for any line Cleaning of tools happens after each cycle	Frontmatec auto line carcass opener and splitting saws for pig processing	https://www.frontmatec.com/en/pork-solutions/clean-line-chill-room/autoline-machines/carcass-opener-apo7
		Robotic aitch bone cutter, breast and belly opener, carcass splitter	Robotic arm cuts at optimized location	Industrial robotic arm with good protection from contamination The arm is fitted with holding roll and circular cutting tools Laser scan-based 3D vision system gives feedback to the robots for moving the knife on the cutting trajectory	AIRA aitch bone cutter, breast and belly opener, neck clipper and splitter for clean pig line	https://www.frontmatec.com/en/pork-solutions/clean-line-chill-room/aira-robots/aira-rbo-belly-and-breast-openerhttps://www.frontmatec.com/en/pork-solutions/clean-line-chill-room/aira-robots/aira-rps-h-splitter-with-knives

operators (Barbar et al., 2022). Robotic systems are additionally preferred in harsh environments that can be dangerous for humans and for repetitive tasks, and where successfully deployed, benefits are expected to accrue in yield and throughput. The meat industry is thus slowly moving toward adopting more automated and robotic-incorporating systems (Xu et al., 2023). Due to the aforementioned variability, generalization across individuals is challenging, and sensing such as via X-ray, ultrasound, and RGB-D can be considered prerequisite enablers of automation but for each use case/point of installation of an automatic machine to conduct a task on the meat line, large amounts of underpinning research are required, for example, the Frontmatec automated loin trimmer technology is based on AutoFOM technology, refined based on the outputs of intensive academic research (Mosbech, Ersbøll et al., 2009a, Mosbech, Ersbøll et al., 2009b). A growing number of regional suppliers provide solutions for clean meat processing lines with robots considering the benefits in terms of hygiene and consistent efficiency. Some examples of commercially available automation systems and robotic solutions for red meat processing are listed below (Echegaray et al., 2022; Warriss, 2009):

AIRA Robotic suite for pig meat processing, which comprises the following units (https://www.frontmatec.com/en/pork-solutions/clean-line-chill-room/aira-robots):

- AIRA RBD bung dropper consists of a robot with a protective cover and cutting tools, a vacuum unit, and a vision system.
- AIRA RHC aitch bone cutter.
- AIRA RBO breast and belly opener cuts the breast bones and opens up the belly keeping the intestine intact.
- AIRA RNC neck clippers use a vision system for determining the optimal position for neck cutting.
- AIRA RPS splitter is equipped with a robotic arm, a holding roll, and circular cutting tools to split the carcass into two halves along the spine to the atlas. The laser scan-based 3D vision system gives feedback for moving the knife on the cutting trajectory. Three different combinations are available; (1) one robot and saw, (2) one robotic arm and knife combination, and (3) two robotic arms with knives.

M-Line (marel, Garðabær, Iceland) (https://marel.com/en/products/m-line-robots):

- Robotic solution for automatic pig processing.
- The system has a twin-tool technology, where each machine has double acting tools such that one tool operates, and the other tool is sterilized.
- The system contains a 3D camera coupled with software for precise movement with reference to the carcass coordinates.
- Products include M-line Precutter and Belly opener, M-Line Bung Remover, M-Line splitting saw, and M-Line neck cutter.

Robotic Beef Loin Drop (Frontmatec, Kolding, Denmark):

- The system provides a cutting solution to separate the sirloin from the round at a relatively fast rate. Such a system utilizes a computer vision system to scan the hindquarters and detects the position of aitch and tail bones. The band saw mounted on a robotic arm performs the cutting on the optimized path with high accuracy. The loin drops to the conveyor belt and the round continues on the rail.

Lamb Processing Solutions (Scott Automation, Dunedin, New Zealand):

- The system uses X-ray and 3D laser reconstruction to identify the bone structure and the robots give a precise cut at a specific location.
- The system predicts the height and angle of each cut to reduce wastage.
- The system operates at the stages of X-ray grading, X-ray primal cut, fore quarter, and middle and hindquarters.
- The knife in the middle system avoids cross-contamination of meat by sawdust and removes the spinal cord material by vacuum.
- The circular knives precisely split the loin from the rack.
- The system includes chine boning and splitting stations.
- The forequarter system uses a 3D vision camera, and the decision made by the software passes through a feedback loop to the robotic arm.

Automatic Pork processing solution (Mayekawa Mfg. co ltd., Tokyo, Japan) (http://www.mayekawa.eu/en/products/das).

- HAMDAS-RX is a robotic solution for pork ham. It has a capacity to debone 500 pork hams per hour. The ham is scanned with an X-ray and image processing software detects the thigh bone position and coordinates with other robots down the line for cutting and extracting the meat.
- WANDAS-RX is a robotic solution to debone pork shoulders following the same principle as HAMDAS-RX. The line is provided with multiple vertical multijointed robots, which are flexible enough to tackle irregularly shaped shoulder bones. This system has capacity to debone 600 shoulders per hour.

As exemplified by the Marel "Twin Tool" solution, the tooling that is associated with robotic systems, termed "soft robotics," is a specialist area in which underpinning research is required to stimulate the implementation of I4.0 level automation in meat plants. Grippers, saws, knives, and other handheld tools developed by the meat processing systems manufacturers for line operations are not necessarily directly transferable to robotic implementation and each requires detailed consideration (Takács et al., 2020) and robot tools for meat cutting are limited in commercial availability. Regarding a knife that can be used in conjunction with a robot, Mason et al. (2021a, 2022a, 2022b) have reviewed the requirements for a so-called "smart knife," for robotic meat applications, that is capable of making adjustments based on dynamic feedback to deliver the desired cut using sensors. The knife must not cause damage and must be fast enough to provide real-time information as the blade moves through the cut. The sensor should be appropriately sized to work well with the knife, be cost-effective, and robust and constructed of a hygienic and safe material for use in proximity to food products. With regard to special requirements for robotic automation (i.e., planning robot paths), the precise depth of the meat should be provided. The authors considered several sensing modalities, including electrical impedance measurements (latency and invasive of tissue), force sensing, optical methods (including infrared spectroscopy, optics embedded in the metal composing the knife, and polarization-sensitive optical coherence tomography), electromagnetic wave-based sensing, and rapid evaporative ionization mass spectrometry. The authors conclude that only optical and electromagnetic spectroscopies meet most of the criteria for smart knife integration, but this remains a challenging prospect.

13.6.2 Meat factory cell concept—a disruptive innovation in meat processing?

Meat processing is a labor-intensive sector, operating on a linear model. On the meat line, many operators are deployed each carrying out a single or small number of task(s) per carcass, as part of the sequence of carcass fabrication. Changes to this traditional slaughter set-up and associated processes are required to obtain feasibly robust, scalable, and sustainable robotic automation for the sector (Hinrichsen et al., 2022). The meat factory cell (MFC) is an innovative new concept, wherein several carcass fabrication tasks are carried out simultaneously by a team of robots (i.e., multiagent robot system).

The MFC is an alternative to line production that is highly conducive to scaling, which, depending on the embodiment, can theoretically match the throughput of line production. The concept for scaling is that multiple cells are placed parallel to each other and the number of cells follows the capacity of the production line. The system was initially devised as semi-automatic robotic cells, where the most complex activities are conducted by skilled butchers and carcass manipulation is performed by robots. Considering labor shortages during the recent pandemic, attention rapidly moved to systems with even greater automation (Mason et al., 2021c). MFCs have been used to examine the potential for robotic boning of both the cold carcass (Hinrichsen et al., 2022; Philipsen and Moeslund, 2020; Soendergaard and Christensen, 2022) and the hot carcass (Mason et al., 2021c, 2021b).

An innovative prototype for the removal of the main pork cuts from a hot carcass in a factory cell has been developed in Norway through the RoBUTCHER project (RoBUTCHER—A Robust, Flexible and Scalable Cognitive Robotics Platform) and a series of papers demonstrate the function and assess the potential for this concept have been published (Mason et al., 2021c; Sødring, et al., 2022; de Medeiros Esper et al., 2022; Alvseike et al. 2018, 2020). This prototype MFC (Fig. 13.2) consists of two industrial robots and a third essential element, which is a fixation frame to support the carcass. This custom-designed metal platform is called carcass handling unit (CHU). Activities like dressing and boning are merged in the cell, and the robots perform the key operations of butchery, that is, "stretching" of the carcass and "cutting" (Alvseike et al., 2017). A cutting robot has a load capacity of 40 kg and a reach of 2.55 m, and the manipulating robot has a load capacity of 60 kg and reach 2.05 m. The cutting robot is provided with cutting tools and the manipulating robot is provided with suitable clamps/grippers for holding and moving the carcass as required. The metallic clamps of the table hold the head and rear part of the carcass firmly. The cup-type vacuum grippers on the cutting table prevent the carcass from sagging or deformation when rotating the carcass horizontally. The table is designed to place the carcass at a horizontal or vertical position and can be rotated horizontally by as much as 180 degrees, which provides easy access to the carcass for the robots. The force sensed by the cutting tools and the efficiency of vacuum grippers are still under research (Mason et al., 2021c). The main cuts are removed while the carcass is hot, keeping offal removal toward the end. This approach keeps the carcass intact and firm during the automatic operation in addition to avoiding the cross-contamination of the external muscle by internal impurities (Alvseike et al., 2018; Sødring, et al., effect). A set of mechanical clamps stretch the legs during processing to support clean cutting with the robot knife or other cutting implements. One RGB-D camera is mounted on the bracket of the cutting arm to capture the carcass image (de Medeiros Esper et al., 2022). The system is loaded with a deep learning-enabled computer vision system, including modules that guide the manipulating robot to an optimal gripping location and direct the cutting robot's

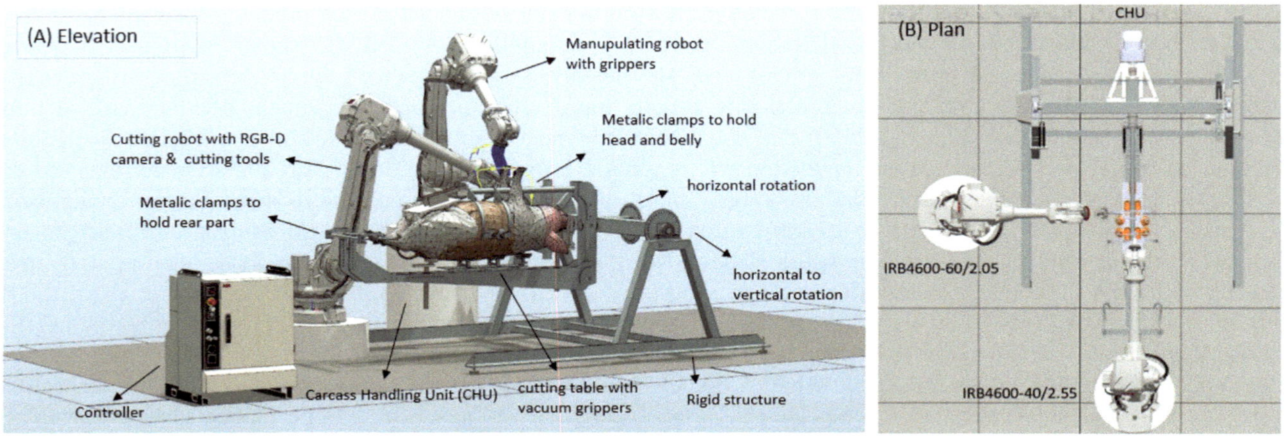

FIGURE 13.2 Robotic meat factory cell (MFC) for pig carcass dissection. The system (A) consists of two industrial robots (one with a robotic knife, and one with a robotic gripper tool) and a carcass handling unit (to hold and rotate the carcass). The overhead view (B) shows the relative locations of the robots and carcass handling unit. (A) *Adapted from Mason, A., et al. (2021c). The meat factory cell: A new way of thinking for meat producers. In 2021 IEEE 21st international symposium on computational intelligence and informatics (CINTI), Budapest, Hungary (pp. 000091–000096).* https://doi.org/10.1109/CINTI53070.2021.9668455. (B) Adapted from Mason, A., et al. (2021b). Meat factory cell: Assisting meat processors address sustainability in meat production. In 21st IEEE international symposium on computational intelligence and informatics, CINTI 2021 - Proceedings (pp. 000103-000108). https://doi.org/10.1109/ CINTI53070.2021.9668392.

cutting trajectory (Liubivyi et al., 2021). The algorithm is capable of predicting the work area, irrespective of biological variation, and this system is declared as cheaper compared to an X-ray-based solution (Mason et al., 2021a).

The MFC is an innovative concept that could be considered challenging for retrofitting into an existing plant, due to tight limitations in space in traditional meat plants (Fig. 13.3a). However, it is worth considering, as an alternative or replacement to the line, for many reasons. For some national sectors, there can be difficulties in transporting animals long distances and at times, excess capacity in meat plants. Additionally, automation of the standard line is extremely expensive, as an automation solution has to be found for every task, whereas, under a cell-based paradigm, the robots are set up to multitask. Furthermore, the overall footprint of the plant can be reduced and there is more flexibility in setting the plant layout, compared to the traditional line, which is illustrated in Fig. 13.3b. Future expansion is therefore easy as per capacity requirement. An unpredicted line stop and associated production loss due to mechanical failure of any critical machine can be minimized (Hinrichsen et al., 2022), because plants with multiple redundant cells have the advantage of addressing the breakdown of a single cell without halting production.

Codex Alimentarius codes like Good Hygienic Practice (GHP), Hazard Analysis, and Critical Control Point (HACCP) could be achieved to a high level in cell-based factories. Meat inspection in the MFC, while requiring consideration, could be straightforward as the removed carcass elements, including the primal cuts and visceral parts can be presented in an aligned arrangement, conducive to machine vision. This approach also permits cuts-based meat management systems, allowing flexibility to implement risk-based and quality-based interventions like customized chilling or targeted decontamination. Automation of meat inspection is an area of active research currently, and the MFC offers scope to automate inspection using sensors and diagnostic tools because the presentation mode as envisaged in the RoBUTCHER project, at least, is conducive to CVSs (Alvseike et al., 2018). The MFC concept is potentially disruptive and it is not surprising that regulation and legislation may need to be reexamined under the auspices of a cell-based operation. According to Alvseike et al. (2018), the language used in the legislation in European legislature is restrictive and is not conducive to the encouragement of innovation in the development of alternative approaches to carcass processing, relative to the classical line system of operation. They call for new approaches to the legislation, which demand functional outcomes, rather than being prescriptive with regard to specifying the system.

13.7 Tracking and tracing

In Europe, traceability of a lot, batch, or consignment one step forward and one step back in the meat value chain is a requirement under EU Law (EU Directive 178/2002) (Mc Carthy et al., 2011). However, complete traceability from retail back to individual animals is rarely implemented given the nature of meat processing and is usually applied only

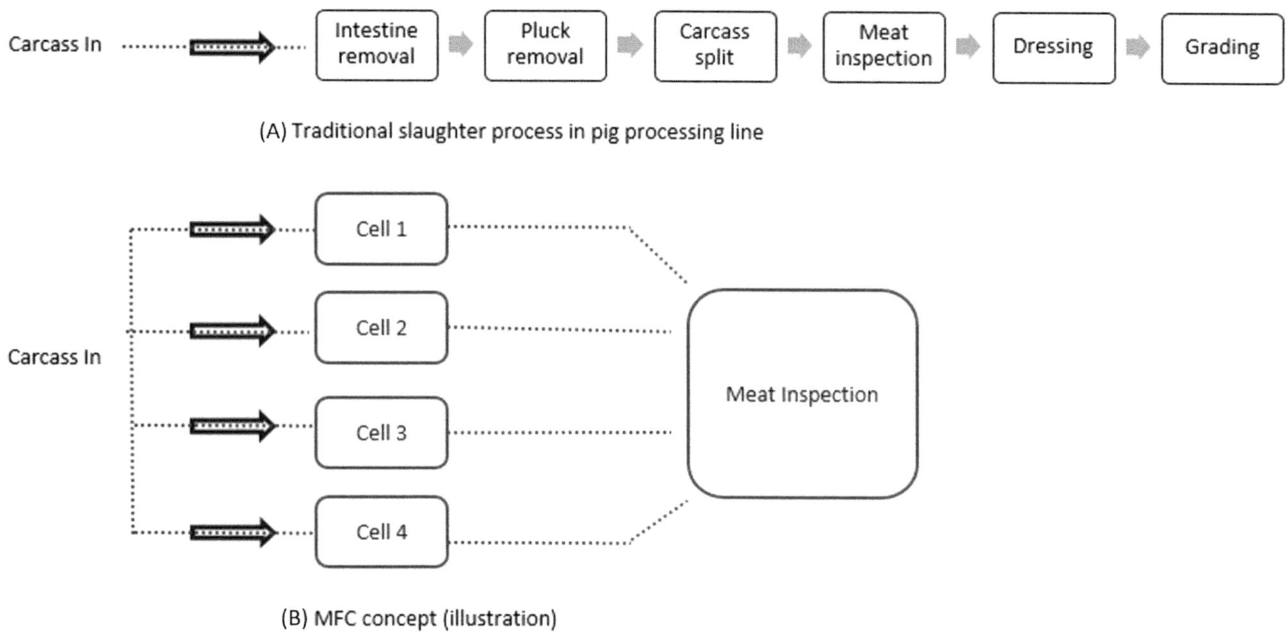

FIGURE 13.3 Contrast between (A) the traditional meat processing line and (B) the meat factory cell concept of carcass dressing. Parallel and multifunctional meat factory cell (MFC) units can complete the tasks of the processing line.

to higher-value niche products (Qian et al., 2020). In the current digital landscape, connected consumers increasingly expect to have access to information about their purchases at retail, which permits them to be confident that the product (s) they are about to purchase meets their expectations across a range of intrinsic and extrinsic attributes. Smartphone applications and QR code technology offer easy access to detailed on-pack information for many consumer goods and services. User-friendly tools that allow the consumer to instantly access information about the meat they are about to purchase could be possible and may be sought in the near future. Many technologies pertain to this future challenge to both track and reliably trace in the meat chain including barcodes, QR-codes, radiofrequency ID (RFID), wireless sensor networks (WSNs), and DNA-based traceability systems. Digital ledger technologies permit secure, immutable, and traceable data storage.

There are challenges in maintaining tracking and tracing in food chains, and not least, in tracking produce through meat processing plants. In some beef industries with relatively advanced traceability, individual animal's information is traceable from farm to factory via DNA samples, ear tags, RFID tags, and gambrels to meat grading and quartering. However, once the meat is boned from the quarter, individual traceability is vulnerable due to infrastructure gaps, though batch traceability is usually maintained. Tracking and tracing solutions include DNA archiving and RFID tags or labels. While RFID tags can, in theory, track meat as it is broken into individual cuts and portions through production, there is a risk of plastic contamination resulting from, for example, a broken tag with the meat. Thus RFID tags are placed on the box or tray, rather than on the meat, which limits individual traceability. RFID labels, which are re-encodable just like RFID tags and their printers, have recently become cheaper and smaller. Hence, it is anticipated that RFID labels have feasible applications in the meat value chain, alongside other enabling technologies, in bridging traceability gaps. One limitation of using RFID labels in food packages is the encoded signals from the labels can be dampened by the high moisture in food products (fresh, meats, etc.). Barcodes on labels or in conjunction with food grade inks (used to stamp some food products such as meats) also offer possibilities to improve the granularity of tracking in processing facilities, but these can fade or become illegible, limiting their value. Recent developments in machine vision and deep learning hold the potential to complement the applications of RFID tags and RFID labels for in-line tracking, to facilitate an improved traceability system, as has been demonstrated for exemplar parts in a learning factory (Kumar et al., 2021b). In conjunction with simple video cameras mounted strategically throughout the plant, deep learning algorithms have the potential to enhance the tracking of individual cuts/muscles through meat plants and research is ongoing into the potential to register and track meat primal movement paths on the conveyor belt with these technologies (D. Brandborg Nielsen, Pers. comm.).

The generation of reliable information on food movement, tracking, and tracing is an important challenge, but the transmission, preservation, security, and accessibility of that information to relevant stakeholders must also be addressed. Demestichas et al. (2020) provide a comprehensive review of traceability in the food chain and discuss the potential of digital ledger for example, blockchain, to provide validated, secure traceability in agricultural and food systems, while acknowledging the complexity of the landscape, that makes uptake a challenging prospect. In the case of meat, information that consumers could, in theory, have access to on-pack includes provenance/farm of origin, farming methods, feed type and source, antibiotic status, and metrics related to sustainability. However, regarding all these data that are possible to encode, certain attributes can be difficult for consumers to understand, the sheer variety of information can be overwhelming, and consumers find it challenging to decide what information they wish to have access to about the meat they are consuming (Sander et al., 2018). For red meat, Kamilaris et al. review efforts in meat, many of which are focused on premium produce, or particular markets, for example, a blockchain-based solution, "BeefLedger," to provide full traceability for Australian beef in the Chinese market, a Mongolian beef blockchain wherein information about nutrition, slaughtering and packaging dates and food safety data are provided through scanning a QR code and several US producer-led systems, providing blockchain systems to monitor traceability and validate raising information (Kamilaris et al., 2019).

Following further developments, available technologies, especially if applied in combination, have the potential to support reliable tracking of meat from the producer, through the processor, distribution, retail, and finally to the consumer and to support the deployment of I4.0 in the meat chain, but it will require multifactor stakeholder participation, in a sectoral response. Benefits would include enhanced consumer confidence, opportunities for feedback, and reductions in food waste, due to enhanced efficiency in the eventuality of recall.

13.8 Case study: digitalization journey joins up the Irish red meat chain

13.8.1 Irish government policy on digitalization

The agrifood sector is of strategic importance to Ireland, as the largest indigenous industry. While the presence of a relatively high number of companies in wider manufacturing sectors adopting AI and automation technologies places Ireland at the head of the digitization curve in today's manufacturing landscape, recent reports show concern that the food and drink processing sector lags back on the "digitalization journey", as is common across Europe (IFM, 2018), and there have been calls for State agencies to develop sectoral digitization roadmaps to maintain Ireland's leading position. The Harnessing Digital report (Anonymous, 2022) highlights the potential for digital technologies to tackle major challenges faced in the coming years—achieving climate change goals and enhancing productivity. In agriculture, digital technologies such as sensors and AI are recognized as already being deployed to address climate change mitigation. The policy report recognizes that precision farming generates a higher yield per animal and hectare through data-driven herd management.

13.8.2 Meat Technology in Ireland

In 2012 beef processors ABP, Ashbourne, Dawn Meats, Kepak, Slaney International, Liffey Meats, Irish Country Meats, and Hilton Foods initiated a cooperation with Enterprise Ireland (EI) to form a collaborative consortium. In 2016 Meat Technology Ireland (MTI), hosted by Teagasc, was launched to drive research focused on carcass and meat quality, shelf life, nutrition, the circular economy, and genetics, in an industry-led collaboration valued at over €8M (http://www.mti.ie). MTI Phase II commenced in 2021 with increased industry investment and a budget of €11m. The Phase II program addresses key topics such as sustainability (including methane) and digitalization, including digital maturity assessment. One MTI Phase II genetics project aims to support reducing lifetime methane emissions per animal by reducing both the age at slaughter and daily methane emissions. Data analytics will be applied to detailed records of inputs, performance, methane emissions, and so on across the animal's lifetime. This will permit modeling the precise time when the animal is ready to be slaughtered. MTI Phase II also involves the collection of individual animal methane data on a population genetically relevant to the Irish herd to deliver a national genetic breeding index for methane in Irish beef. The Irish Cattle Breeders Federation has been at the forefront globally in driving genetic improvement in dairy with impressive gains made on various traits over generations. The generated data will permit scientific Life Cycle Analysis (LCA) for different Irish beef production systems that support the most profit with the lowest emissions while reducing the age at slaughter. Meat Industry Ireland (MII—the beef and lamb sectors trade association) has published the Irish Beef Sustainability Report (https://youtu.be/OgJJj3Ze2Yk), which outlines four strategic pillars thereby facilitating its members' intent to support the agri-sector climate targets by 2030. The MTI Methane project brings the concept further from the farm gate to the factory gate.

13.8.3 Digitalization in support of meeting Ireland's climate action targets

Supported by the Department of Agriculture, Food and the Marine (DAFM), Teagasc in cooperation with the ICBF and Bord Bia (Irish Food Board), is building a new digital platform to facilitate a robust whole farm sustainability assessment and "counting" of carbon emissions and removals: the AgNav Sustainability Digital Platform https://www.agnav.ie/landing. The concept of mapping the carbon inputs from the farm, to the day the food product leaves the factory, relies on a digital chain that is connected, involving all of the actors working toward an agreed national strategy for the sector. The uniqueness of the AgNav platform is that it encompasses information from calf birth to the retail product leaving the factory gate.

13.9 Digital transformation strategies

The concept of digital transformation is centered around the use of technology to improve productivity and profitability. Business leaders have recognized the need to align their digitalization efforts with business strategy if they are to deliver the greatest benefits and keep up with their competition. This is due to the fact that digital transformation is not easy. According to McKinsey "more than 70% of companies find it difficult to implement and scale advanced technologies that deliver significant improvement in return on investment or operational KPIs" (Behrendt et al., 2021). It is clear that, although companies are aware of and implementing some digital transformation and I4.0 solutions, the results demonstrate that some are working, but others are not delivering the significant improvement that companies were hoping for. This is certainly a challenge for the entire manufacturing community.

The process of digital transformation is shaped and influenced by a large number of factors. The term digital transformation itself was coined by the consultancy firm Capgemini, working alongside MIT Center for Digital Business. They were among the first to look at how business and IT functions should work together as part of an effective digital transformation program and not something that can be achieved by working separately. In fact, all departments *should* cooperate together to boost productivity and profits. There is an onus across all employees, processes, and systems to communicate ubiquitous business and technical requirements when it comes to technology implementations. In many cases, however, digital transformation projects are often bottom-up, individually led efforts by well-intentioned employees. To be successful, change needs to be driven from the top, as part of a cohesive business and digital transformation strategy—with intense focus and buy-in from all stakeholders and departments.

The continuous process improvement adage "you cannot improve what you can't measure" is very applicable to the digital transformation of companies worldwide. This is the basis for the emergence of a microindustry that has sprung up to measure companies' digitalization state through the use of digital maturity models and assessments (DMAs). There are numerous companies, consultancies, and technology centeres around the world that have recognized the poor uptake and implementation challenges that have been experienced by companies of all sizes when it comes to I4.0. Some DMAs are quick self-assessment online questionnaires, while others include thorough audits of factories carried out by I4.0 consultants. Fundamentally, DMAs are formulated to answer the big question companies wanting to embark on digital transformation ask; where they are now in digitalization, and where they want to get to. Most companies do not want to and do not need to fully transform digitally. DMAs are intended to give companies tailored insights and recommendations into what technologies and processes they can reap the most benefit from and help shape a tailored digital transformation strategy, likely to generate impact.

The European Union is investing billions of euros in its own #DigitalEU programs. Launched in March 2021, the blocs' path to the "Digital Decade" sets out ambitious targets to make Europe a global leader in digital transformation. There is an abundance of EU-funded digital transformation initiatives and programs targeting the manufacturing sector. The World Economic Forum (WEF) has also acknowledged the problem in the uptake and impact of I4.0 technologies and the significant variation in the pace of adoption across industries. The WEF adopted a global Smart Industry Readiness Index (SIRI), launched in 2020, which champions a standardized DMA approach to help companies decide how and where to apply solutions to maximize benefits. The assessment focuses not just on technology adoption within companies, but looks at companies' readiness across a range of aspects in technology and organizational processes (Fig. 13.4A). This is because to maximize value from technology, the processes (i.e., operational, supply chain, and product lifecycle) must be effective, but so too, the organizational framework and behaviors should be in place to support technology integration into the business. Through DMA, companies can see where their factory compares to peers within their own industry category and industry groups can be compared against one another, to give further context.

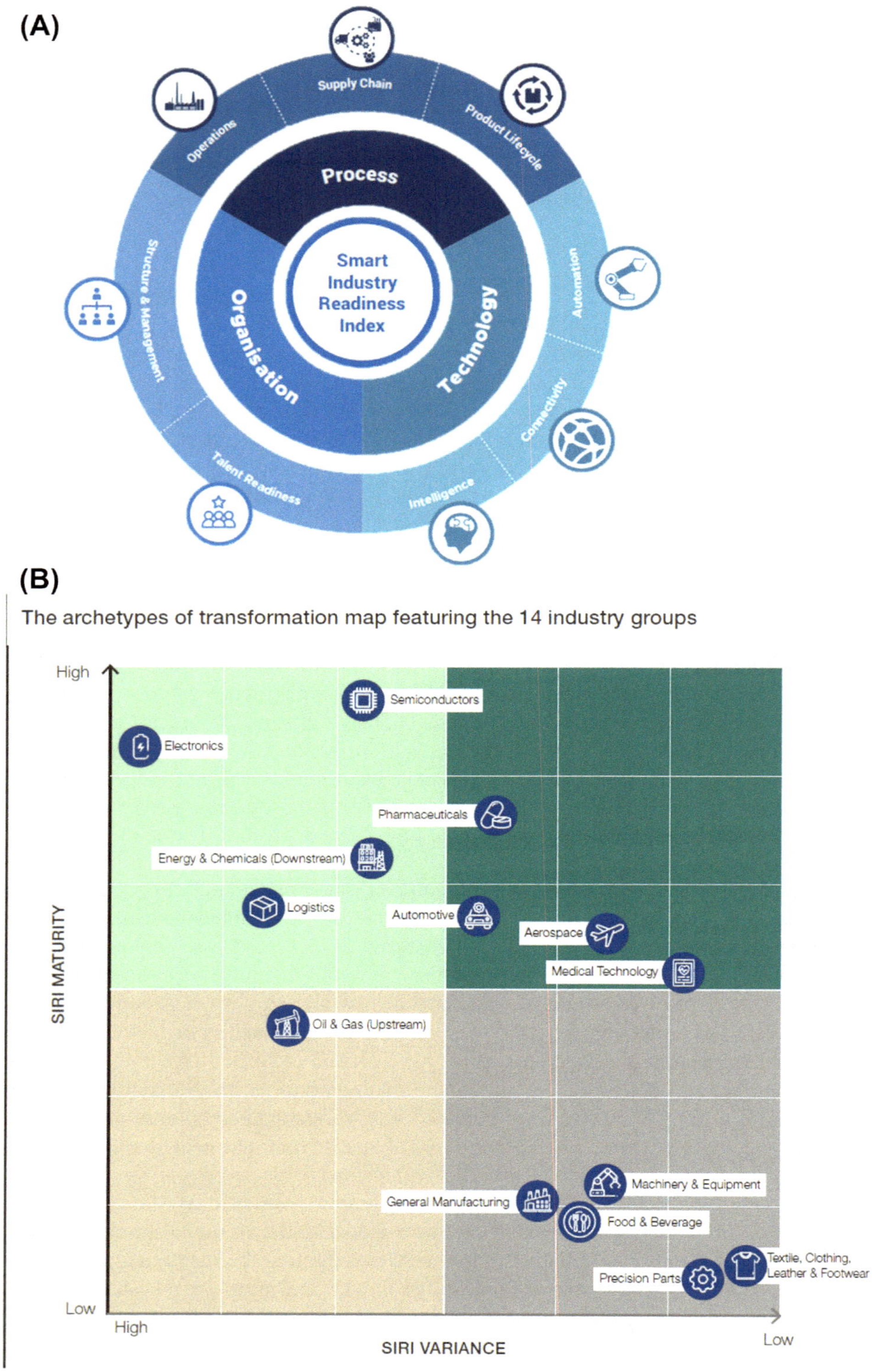

FIGURE 13.4 The Smart Industry Readiness Index (SIRI). (A) Digital maturity assessment wheel showing the 3 central pillars, 8 building blocks, and 16 dimensions used to determine a company's digital maturity as part of the SIRI process; (B) SIRI maturity and variation of 14 manufacturing industry collective groups.

Fig. 13.4B highlights the differing maturity levels across sectors. Multinational corporation (MNC) dominated industries like semiconductors, electronics, and pharmaceuticals sit well ahead of the more fragmented, small and medium-sized enterprise (SME) dominated sectors such as General Manufacturing and Food and Beverage.

While the meat processing sector is not listed separately in SIRI, it is captured under the Food and Beverage industry group. The Food and Beverage grouping is low down the digital maturity scoring as shown; however, there is considerable variation within the sector implying that some companies with relatively high digital maturity exist. In certain regions, targeted efforts are being made at benchmarking of the meat processing industry through such initiatives that will help shape the investment in technologies in line with business objectives and impact to the bottom line.

13.10 Benefits of adopting I4.0 in the meat industry

Impacts of I4.0 include greatly enhanced efficiencies, improved forecasting and logistics, provision of traceability, transparency, with associated reduction of food waste, and concomitant improvements in sustainability and consumer confidence. Improvements in connectivity and adoption of industry 4.0 technologies such as data analytics and AI will provide new ways of using information that companies already have access to to add value. The application of online sensors has huge potential both to monitor uptime in machinery and the attributes of the raw materials. Sensors also have important applications in automation. Automation in meat moving, tracking, and cutting processes is a key aspect of Industry 4.0, which offers the potential to benefit process efficiency, systems control, and specific metrics such as yield in valuable cuts. The application of cyber-physical systems and the MFC concept offers a viable prospect for high levels of automation, even for small or medium industries. Furthermore, the working day is not limited to a single shift, and indeed, automated systems are capable of operating up to 24 hours per day if required. With a reduction in worker density in the plant, and more limited contact with the carcass, microbial load will be reduced with benefits in downstream food safety and shelf life. Beyond the economic benefits, incorporation of Industry 4.0 technologies has important benefits for staff well-being, and quality of employment. Many of the procedures in carcass processing and deboning are manual and hazardous operations. It is noteworthy, that rather than a reduction in employment with automation and robotics being the norm, in industries associated with automation uptake, staff retention actually increases and likewise, a greater proportion of positions are high-quality technical roles.

13.11 Perspective

The nature of red meat processing has not altered significantly in 100 years. While automation has progressed, digitalization overall is lagging compared to other sectors. With consumer and regulatory requirements, to enhance transparency, carbon footprint, retention of staff, and resilience in the face of global shocks, the sector is required to re-evaluate and identify and take steps toward new ways of meeting these challenges. I4.0 offers many opportunities to support the industry meet the challenges it faces toward 2050. The accuracy, speed, and cost of modern I4.0 technologies present a promising opportunity to move toward a coherently digital meat processing chain from farm to fork with less impact on the environment and more authenticated products that meet consumers' requirements. Companies in the meat industry can significantly benefit from the utilization of I4.0 to promote their products into wider markets with affordable investment. However, with so many technologies (e.g., advanced sensors, AI, analytics, computing power) available and already deployed in other industries, the means of actually embracing, adopting, and delivering an impact as a result of the digitalization journey is daunting for many companies, more so given that the true benefits of I4.0 are realized when such technologies interact in a holistic system. To make continual progress, awareness of sectoral relative digital maturity and appropriate planning and risk management in deployment is essential to ensure maximum impact for the sector throughout the globe and a range of tools are available to facilitate this. The convergence of multiple technologies offers the potential for dramatic increases in productivity and innovation. Spurred on by the COVID-19 pandemic, companies will need to ensure that digital transformation is at the core of their operations, putting paid to the hitherto "wait and see" approach of many. Adoption of Industry 4.0, and its next iteration, Industry 5.0, which also encompasses worker well-being and environmental sustainability, is a realistic ambition for the meat sector and will be driven by companies with purpose, who will as a result, be more successful in the future.

Acknowledgments

This work was supported by Meat Technology Ireland (MTI) a co-funded Industry/Enterprise Project (TC 2021 031), a Research Leaders 2025 Fellowship Award to Dr. Ahmed Rady, under the European Union Horizon 2020 research and innovation program (Marie Skłodowska-Curie grant agreement no. 754380) and a Teagasc Walsh Scholarship Award (2021039) to Jyoti P. Mishra. We thank Giulia Romano, Teagasc and Kevin Dooley, IMR for helpful discussion.

References

Aafreen, R., Neyaz, S. Y., Shamim, R., & Beg, M. S. (2019). An IoT based system for telemetry and control of greenhouse environment. In 2019 International conference on electrical, electronics and computer engineering (UPCON), (pp. 1–6). Aligarh, India: IEEE.

Achata, E. M., Inguglia, E. S., Esquerre, C. A., Tiwari, B. K., & O'Donnell, C. P. (2019). Evaluation of Vis-NIR hyperspectral imaging as a process analytical tool to classify brined pork samples and predict brining salt concentration. *Journal of Food Engineering*, *246*, 134–140.

Addo-Tenkorang, R., & Helo, P. T. (2016). Big data applications in operations/supply-chain management: A literature review. *Computers and Industrial Engineering*, *101*, 528–543.

Ahn, J.-H., Jang, W.-J., Lee, W.-H., & Kim, J.-D. (2020). Detection of needles in meat using X-ray images and convolution neural networks. *Journal of Sensor Science and Technology*, *29*(6), 427–432.

Alvseike, O., Sverdvik, H., O'Farrell, M., & Berg, P. (2017). *Meat factory cell - A concept for the future. International congress of meat science and technology*. Cork, Ireland: Wageningen Academic Publishers.

Alvseike, O., Prieto, M., Torkveen, K., Ruud, C., & Nesbakken, T. (2018). Meat inspection and hygiene in a meat factory cell – An alternative concept. *Food Control*, *90*, 32–39.

Alvseike, O., Prieto, M., Bjørnstad, P. H., & Mason, A. (2020). Intact gastro-intestinal tract removal from pig carcasses in a novel meat factory cell approach. *Acta Veterinaria Scandinavica*, *62*(1). Available from https://doi.org/10.1186/s13028-020-00546-y.

Andreoli, V., Bagliani, M., Corsi, A., & Frontuto, V. (2021). Drivers of protein consumption: A cross-country analysis. *Sustainability*, *13*(13), 7399.

Anonymous. (2022). In E.a.I. Business (Ed.), Harnessing digital-2022 progress report, Government of Ireland, Ireland.

Antequera, T., Caballero, D., Grassi, S., Uttaro, B., & Perez-Palacios, T. (2021). Evaluation of fresh meat quality by hyperspectral imaging (HSI), nuclear magnetic resonance (NMR) and magnetic resonance imaging (MRI): A review. *Meat Science*, *172*, 108340.

Baglat, P., Hayat, A., Mendonça, F., Gupta, A., Mostafa, S. S., Morgado-Dias, F., et al. (2023). Non-destructive banana ripeness detection using shallow and deep learning: A systematic review. *Sensors*, *23*(2), 738. Available from https://doi.org/10.3390/s23020738.

Barbar, C., Bass, P. D., Barbar, R., Bader, J., & Wondercheck, B. (2022). Artificial intelligence-driven automation is how we achieve the next level of efficiency in meat processing. *Animal Frontiers*, *12*(2), 56–63.

Barbut, S. (2020). *Meat industry 4.0: A distant future? Animal Frontiers*, *10*(4), 38–47.

Behrendt, A., Kasah, E. d B. , T., & Koerber, B. (2021). Leveraging industrial IoT and advanced technologies for digital transformation. *McKinsey & Company*, 1–75.

Beriain, M. J., Lozano, M., Echeverría, J., Murillo-Arbizu, M. T., Insausti, K., & Beruete, M. (2022). Tenderness of PGI "Ternera de Navarra" Beef Samples Determined by FTIR-MIR Spectroscopy. *Foods*, *11*(21), 3426.

Betti, F., Bezamat, F., Bloempott, S., Fendri, M., & Kuepper, D. (2021). Data excellence: Transforming manufacturing and supply systems. World Economic Forum White Paper.

Bigliardi, B., Bottani, E., & Filippelli, S. (2022). A study on IoT application in the food industry using keywords analysis. *Procedia Computer Science*, *200*, 1826–1835.

Cafferky, J., Sweeney, T., Allen, P., Sahar, A., Downey, G., Cromie, A. R., & Hamill, R. M. (2020). Investigating the use of visible and near infrared spectroscopy to predict sensory and texture attributes of beef M. longissimus thoracis et lumborum. *Meat Science*, *159*, 107915. Available from https://doi.org/10.1016/j.meatsci.2019.107915, Jan;159.

Cama-Moncunill, R., Cafferky, J., Augier, C., Sweeney, T., Ferragina, A., Sullivan, C., Cromie, A., & Hamill, R. M. (2020). Prediction of Warner-Bratzler shear force, intramuscular fat, drip-loss and cook-loss in beef via Raman spectroscopy and chemometrics. *Meat Science*, *167*, 108157.

Cama-Moncunill, X., Da Costa, T., Murphy, F., Ward, S., Gillespie, J., & Ramanathan, R. (2022). *Reducing food waste during the dry ageing of beef using IoT technology*. https://vb.nweurope.eu/media/19114/burns-farm-meats-final.pdf.

Mc Carthy, U., Ayalew, G., Butler, F., McDonnell, K., & Ward, S. (2011). The case for UHF RFID application in the meat supply chain in the Irish context: A review perspective. *Agricultural Engineering International: CIGR Journal*, *13*(3).

Chan, W. (2004). Macronutrients in meat, in encyclopedia of meat sciencesIn W. K. J. , C. Devine, & M. Dikeman (Eds.), (pp. 614–618). Academic Press.

Cheng, J., Sun, J., Yao, K., Xu, M., Tian, Y., & Dai, C. (2022). A decision fusion method based on hyperspectral imaging and electronic nose techniques for moisture content prediction in frozen-thawed pork. *LWT*, *165*, 113778.

Christensen, L. B., & Engell-Nørregård, M. P. (2016). Augmented reality in the slaughterhouse - a future operation facility? *Cogent Food &. Agriculture*, *2*(1), 1188678.

Colley, T. A., Olsen, S. I., Birkved, M., & Hauschild, M. Z (2020). Delta life cycle assessment of regenerative agriculture in a sheep farming system. *Integrated Environmental Assessment and Management*, *16*(2), 282–290.

Cozzolino, D., Bureš, D., & Hoffman, L. C. (2023). Evaluating the use of a similarity index (SI) combined with near infrared (NIR) spectroscopy as method in meat species authenticity. *Foods*, *12*(1), 182.

Davies, E. (2013). *Machine vision in the food industry. Robotics and automation in the food industry* (pp. 75–110). Elsevier.

Delgado-Pando, G., Allen, P., Troy, D. J., & McDonnell, C. K. (2021). Objective carcass measurement technologies: Latest developments and future trends. *Trends in Food Science & Technology*, *111*, 771–782.

Demestichas, K., Peppes, N., Alexakis, T., & Adamopoulou, E. (2020). Blockchain in agriculture traceability systems: A review. *Applied Sciences*, *10*(12), 4113. Available from https://doi.org/10.3390/app10124113.

Echegaray, N., Hassoun, A., Jagtap, S., Tetteh-Caesar, M., Tomasevic, I., Goksen, G., & Lorenzo, J. M. (2022). Meat 4.0: Principles and applications of Industry 4.0 technologies in the meat industry. *Applied Sciences*, *12*(14), 6986. Available from https://doi.org/10.3390/app12146986.

Elbeheiry, N., & Balog, R. S. (2023). Technologies driving the shift to smart farming: A review. *IEEE Sensors Journal*, *23*(3), 1752–1769.

Frank, A. G., Dalenogare, L. S., & Ayala, N. F. (2019). Industry 4.0 technologies: Implementation patterns in manufacturing companies. *International Journal of Production Economics*, *210*, 15–26.

Gao, D., Sun, Q., Hu, B., & Zhang, S. (2020). A framework for agricultural pest and disease monitoring based on internet-of-things and unmanned aerial vehicles. *Sensors*, *20*(5), 1487.

Garcia-Perez, J., De Prados, M., Martinez, G., Alvarez-Arenas, T. G., & Benedito, J. (2019). Ultrasonic online monitoring of the ham salting process. Methods for signal analysis: Time of flight calculation. *Journal of Food Engineering*, *263*, 87–95.

García-García, A., Cambero, M., Castejón, D., Escudero, R., & Fernández-Valle, M. (2019). Dry cured-ham microestructure: A T2 NMR relaxometry, SEM and uniaxial tensile test combined study. *Food Structure (London, England: 1993)*, *19*, 100104.

Gardner, G., Starling, S., Charnley, J., Hocking-Edwards, J., Peterse, J., & Williams, A. (2018). Calibration of an on-line dual energy X-ray absorptiometer for estimating carcase composition in lamb at abattoir chain-speed. *Meat Science*, *144*, 91–99.

Gardner, G. E., Apps, R., McColl, R., & Craigie, C. R. (2021). Objective measurement technologies for transforming the Australian & New Zealand livestock industries. *Meat Science*, *179*, 108556.

Geng, X., Zhang, Q., Wei, Q., Zhang, T., Cai, Y., Liang, Y, & Sun, X (2019). A mobile greenhouse environment monitoring system based on the internet of things. *Ieee Access*, *7*, 135832–135844.

Gilchrist, A. (2016). *Industry 4.0: The industrial internet of things*. California, USA: Apress.

Goap, A., Sharma, D., Shukla, A. K., Krishna, C. R., et al. (2018). An IoT based smart irrigation management system using machine learning and open source technologies. *Computers and Electronics in Agriculture*, *155*, 41–49.

Hinrichsen, L. L., Wu, H., & Gregersen, K. D. (2022). We need to rethink production technology for meat-packers—the old cutting table is being revived. *Animal Frontiers*, *12*(2), 18–24.

Ho, H., Yu, H., Gangsei, L., & Kongsro, J. (2019). A CT-image based pig atlas model and its potential applications in the meat industry. *Meat Science*, *148*, 1–4.

Huang, C., & Gu, Y. (2022). A machine learning method for the quantitative detection of adulterated meat using a MOS-based E-nose. *Foods*, *11*(4), 602.

Huang, H., Zhan, W., Du, Z., Hong, S., Dong, T., She, J., & Min, C. (2022). Pork primal cuts recognition method via computer vision. *Meat Science*, *192*, 108898.

IFM. (2018). *A study on the digitalisation of the manufacturing sector and the policy implications for Ireland*, University of Cambridge, Department of Engineering: Cambridge, UK.

Iheonye, A. C., Raghavan, V., Ferrie, F. P., Orsat, V., & Gariepy, Y. (2023). Monitoring visual properties of food in real time during food drying. *Food Engineering Reviews*, *15*, 242–260.

Ijaz, M., Yar, M. K., Badar, I. H., Ali, S., Islam, M. S., Jaspal, M. H., Hayat, Z., Sardar, A., Ullah, S., & Guevara-Ruiz, D. (2021). Meat production and supply chain under COVID-19 scenario: Current trends and future prospects. *Frontiers in Veterinary Science*, *8*, 660736.

Kamilaris, A., Fonts, A., & Prenafeta-Boldύ, F. X. (2019). The rise of blockchain technology in agriculture and food supply chains. *Trends in Food Science & Technology*, *91*, 640–652.

Kamruzzaman, M., ElMasry, G., Sun, D. W., & Allen, P. (2012). Prediction of some quality attributes of lamb meat using near-infrared hyperspectral imaging and multivariate analysis. *Analytica Chimica Acta*, *714*, 57–67.

Kartakoullis, A., Comaposada, J., Cruz-Carrión, A., Serra, X., & Gou, P. (2019). Feasibility study of smartphone-based near infrared spectroscopy (NIRS) for salted minced meat composition diagnostics at different temperatures. *Food Chemistry*, *278*, 314–321.

Khaled, A. Y., Parrish, C. A., & Adedeji, A. (2021). Emerging nondestructive approaches for meat quality and safety evaluation—A review. *Comprehensive Reviews in Food Science and Food Safety*, *20*(4), 3438–3463.

Khattab, A., Habib, S. E., Ismail, H., Zayan, S., Fahmy, Y., & Khairy, M. M. (2019). An IoT-based cognitive monitoring system for early plant disease forecast. *Computers and Electronics in Agriculture*, *166*, 105028.

Kim, D.-E., Mai,N.-D., & Chung,W.-Y. (2022). IoT-based meat quality monitoring system using computer vision and air quality sensor. In 2022 IEEE sensors (pp. 1–4). doi:10.1109/SENSORS52175.2022.9967142.

Kim, W.-S., Lee, W.-S., & Kim, Y.-J. (2020). A review of the applications of the internet of things (IoT) for agricultural automation. *Journal of Biosystems Engineering*, *45*, 385–400.

Konur, S., Lan, Y., Thakker, D., Morkyani, G., Polovina, N., & Sharp, J. (2021). Towards design and implementation of Industry 4.0 for food manufacturing. *Neural Computing and Applications*, 1–13. Available from https://doi.org/10.1007/s00521-021-05726-z.

Krupitzer, C., Noack, T., & Borsum, C. (2022). Digital food twins combining data science and food science: System model, applications, and challenges. *Processes*, 10. Available from https://doi.org/10.3390/pr10091781.

Kumar, R., et al. (2021a). Machine vision and radio-frequency identification (RFID) based real-time part traceability in a learning factory. *in Procedia CIRP*, 630–635. Available from https://doi.org/10.1016/j.procir.2021.11.106.

Kumari, S., et al. (2021b). Digitization in agriculture: Insight into the networked world, in artificial intelligence and IoT-based technologies for sustainable farming and smart. *Agriculture.*, 1–24.

Liubivyi, A., Hopkins, D. L., Schmidtke, L., Morris, S., & Fowler, S. M. (2021). Computer vision for robotic butcher. In *Proceedings of the challenges in automated food processing workshop at the European Robotics Forum (ERF 2021).*

Logan, B. G., Hopkins, D. L., Schmidtke, L., Morris, S., & Fowler, S. M. (2020). Preliminary investigation into the use of Raman spectroscopy for the verification of Australian grass and grain fed beef. *Meat Science*, *160*, 107970.

Marimuthu, J., & Gardner,G.E. (2019). P8 and rib fat depth measurement on beef carcase using a portable microwave system. In *Proceedings of the international congress of meat science and technology*, Potsdam.

Marimuthu, J., Loudon, K., & Gardner, G. (2021). Prediction of lamb carcase C-site fat depth and GR tissue depth using a non-invasive portable microwave system. *Meat Science*, *181*, 108398.

Marimuthu, J., Loudon, K. M., & Gardner, G. (2022). Prediction of lamb carcase C-site fat depth and GR tissue depth using a non-invasive portable microwave system versus body condition scoring. *Meat Science*, *188*, 108764.

Maselyne, J., Mertens, K., Van Nuffel, A., Scholten, H., Athanasiadis, I., Larrañaga, M., & Fernández, I. (2017). IoF2020—The internet of meat: Towards applications of internet of things in the meat supply chain. In *Proceedings of the 8th European conference on precision livestock farming*.

Mason, A., et al. (2021b). Meat factory cell: Assisting meat processors address sustainability in meat production. In *21st IEEE international symposium on computational intelligence and informatics, CINTI 2021 - Proceedings* (pp. 000103–000108). https://doi.org/10.1109/CINTI53070.2021.9668392.

Mason, A., et al. (2021c). The meat factory cell: A new way of thinking for meat producers. In *2021 IEEE 21st international symposium on computational intelligence and informatics (CINTI)*, Budapest, Hungary (pp. 000091–000096). https://doi.org/10.1109/CINTI53070.2021.9668455.

Mason, A., et al. (2022a). Smart knife: Technological advances towards smart cutting tools in meat industry automation. *Sensor Review*, *42*(1), 155–163.

Mason, A., et al. (2022b). Smart knife: Integrated intelligence for robotic meat cutting. *IEEE Sensors Journal*, *22*(21), 20475–20483.

Mason, A., Romanov, D., Cordova-Lopez, L. E., & Korostynska, O. (2021a). *Smart knife for robotic meat cutting* (pp. 1–4). Sydney, Australia: IEEE Sensors. Available from https://doi.org/10.1109/SENSORS47087.2021.9639793.

Matthews, D., Pabiou, T., Evans, R. D., Beder, C., & Daly, A. (2022). Predicting carcass cut yields in cattle from digital images using artificial intelligence. *Meat Science*, *184*, 108671.

McFarlane, D. (2018). *Industrial internet of things applying IoT in the industrial context*. Engineering and Physical Sciences Research Council (EPSRC).

de Medeiros Esper, I., Cordova-Lopez, L. E., Romanov, D., Alvseike, O., From, P. J., & Mason, A. (2022). Pigs: A stepwise RGB-D novel pig carcass cutting dataset. *Data in Brief*, *41*, 107945. Available from https://doi.org/10.1016/j.dib.2022.107945, Feb 15:41.

Memon, M.H., Kumar, W., Memon, A., Chowdhry, B. S., Aamir, M., & Kumar, P. (2016). Internet of Things (IoT) enabled smart animal farm. In *2016 3rd International conference on computing for sustainable global development (INDIACom)*, New Delhi, India (pp. 2067–2072).

Meng, Z., Wu, Z., & Gray, J. (2018). Microwave sensor technologies for food evaluation and analysis: Methods, challenges and solutions. *Transactions of the Institute of Measurement and Control*, *40*(12), 3433–3448.

Monahan, F. J., Schmidt, O., & Moloney, A. P. (2018). Meat provenance: Authentication of geographical origin and dietary background of meat. *Meat Science*, *144*, 2–14.

Mosbech, T.H., Ersbøll,B.K., & Christensen,L.B. (2009a). Quantification and validation of soft tissue deformation. In *Progress in biomedical optics and imaging - Proceedings of SPIE*. doi:10.1117/12.811986.

Mosbech, T.H., Ersbøll,B.K., & Christensen,L.B. (2009b). An efficient data-driven tissue deformation model. In *2009 IEEE 12th international conference on computer vision workshops, ICCV workshops*.

Mourtzis, D., Angelopoulos, J., & Panopoulos, N. (2022). A literature review of the challenges and opportunities of the transition from Industry 4.0 to Society 5.0. *Energies*, *15*(17), 6276.

Muangprathub, J., Boonnam, N., Kajornkasirat, S., Lekbangpong, N., Wanichsombat, A., & Nillaor, P. (2019). IoT and agriculture data analysis for smart farm. *Computers and Electronics in Agriculture*, *156*, 467–474.

Mulvihill, B. (2004). In W. K. J. , C. Devine, & M. Dikeman (Eds.), *Micronutrients in meat*, in e*ncyclopedia of meat sciences* (pp. 618–623). Academic Press.

Nadimi, E. S., Jørgensen, R. N., Blanes-Vidal, V., & Christensen, S. (2012). Monitoring and classifying animal behavior using ZigBee-based mobile ad hoc wireless sensor networks and artificial neural networks. *Computers and Electronics in Agriculture*, *82*, 44–54.

O'Malley, M. (2019). *Securing the future of AgTech. Network Security*, *2019*(8), 12–14.

OECD, *OECD-FAO Agricultural Outlook 2022–2031*. (2022).

OECD-FAO, F. A. O. (2022). *Agricultural Outlook 2021–2030* (pp. 163–177). FAO.

Osborne, B. G. (2007). *Near-infrared spectroscopy in food analysis. Encyclopedia of analytical chemistry: Applications. Theory and Instrumentation*. Wiley, ISBN: 9780471976707.

Philipsen, M. P., & Moeslund, T. B. (2020). Cutting pose prediction from point clouds. *Sensors*, *20*(6), 1563.

Prakash, S., Berry, D. P., Roantree, M., Onibonoje, O., Gualano, L., Scriney, M., & McCarren, A. (2021). Using artificial intelligence to automate meat cut identification from the semimembranosus muscle on beef boning lines. *Journal of Animal Science*, *99*(12), skab319.

Purcell, W., & Neubauer, T. (2023). Digital twins in agriculture: A state-of-the-art review. smart agricultural. *Technology (Elmsford, N.Y.)*, *3*, 100094.

Qian, J., Ruiz-Garcia, L., Fan, B., Robla Villalba, J. I., McCarthy, U., Zhang, B., Yu, Q., & Wu, W. (2020). Food traceability system from governmental, corporate, and consumer perspectives in the European Union and China: A comparative review. *Trends in Food Science and Technology*, *99*, 402–412.

Randolph, T. F., Schelling, E., Grace, D., Nicholson, C. F., Leroy, J. L., Cole, D. C., Demment, M. W., Omore, A, Zinsstag, J., & Ruel, M. (2007). Invited review: Role of livestock in human nutrition and health for poverty reduction in developing countries. *Journal of Animal Science*, *85*(11), 2788–2800.

Redmon, J., et al. (2016). You only look once: Unified, real-time object detection. In IEEE conference on computer vision and pattern recognition (CVPR), Las Vegas, NV (pp. 779–788). https://doi.org/10.1109/CVPR.2016.91.

Reis, M. M., & Rosenvold, K. (2014). Early on-line classification of beef carcasses based on ultimate pH by near infrared spectroscopy. *Meat Science*, *96*(2), 862–869.

Romanov, D., Korostynska, O., Lekang, O., & Mason, A. (2022). Towards human–robot collaboration in meat processing: Challenges and possibilities. *Journal of Food Engineering*, *331*, 111117. Available from https://doi.org/10.1016/j.jfoodeng.2022.111117.

Sander, F., Semeijn, J., & Mahr, D. (2018). The acceptance of blockchain technology in meat traceability and transparency. *British Food Journal*, *120* (9), 2066–2079.

Schirmer, B. C., Liland, K. H., Øverby, L., Skaar, I., & Kure, C. F. (2021). Suitability of FTIR to distinguish pure cultures of problematic mould species from closely related species in the meat industry. *Journal of Applied Microbiology*, *131*(5), 2308–2316.

Segtnan, V. H., Hildrum, K. I., & Wold, J. P. (2009). 22 - New methods for analysis of factors affecting meat eating quality. In J. P. Kerry, & D. Ledward (Eds.), *Improving the sensory and nutritional quality of fresh meat* (pp. 519–538). Woodhead Publishing.

Serey, J., Alfaro, M., Fuertes, G., Vargas, M., Durán, C., Ternero, R., Rivera, R., & Sabattin, J. (2023). Pattern recognition and deep learning technologies, enablers of Industry 4.0, and their role in engineering research. *Symmetry*, *15*(2), 535. Available from https://doi.org/10.3390/sym15020535.

Singh, C., & Jayas, D. (2013). *Optical sensors and online spectroscopy for automated quality and safety inspection of food products. Robotics and automation in the food industry* (pp. 111–129). Elsevier.

Soendergaard, J., & Christensen,L.B. (2022). AI assisted cobot potential for meat cutting procedure. In *Proceedings, international congress of meat science and technology*, Kobe, Japan.

Steinfeld, H., & Gerber, P. (2010). Livestock production and the global environment: Consume less or produce better? *Proceedings of the National Academy of Sciences*, *107*(43), 18237–18238.

Sun, X., Young, J., Liu, J.-H., & Newman, D. (2018). Prediction of pork loin quality using online computer vision system and artificial intelligence model. *Meat Science*, *140*, 72–77.

Sødring, M., Thauland Håseth, T., Rasten Brunsdon, E., Bjørnstad, P. H., Sandnes, R., Røtterud, O. J., Mason, A., de Medeiros Esper, I., Hallenstvedt, E., Agerup, P., Kåsin, K., Egelandsdal, B., & Alvseike, O. (2022). Effects of meat factory cell on pork qualities, sensory characteristics and carcass hygiene: An exploratory study. *Acta Agriculturae Scandinavica, Section A—Animal Science*, *71*(1–4), 98–113. Available from https://doi.org/10.1080/09064702.2022.2113120.

Takács, K., Mason, A., Christensen, L. B., & Haidegger, T. (2020). Robotic grippers for large and soft object manipulation. In *2020 IEEE 20th international symposium on computational intelligence and informatics (CINTI)*.

Teague, W. R., Apfelbaum, S., Lal, R., Kreuter, U. P., Rowntree, J., Davies, C. A., Conser, R., Rasmussen, M., Hatfield, J., Wang, T., Wang, F., & Byck, P. (2016). The role of ruminants in reducing agriculture's carbon footprint in North America. *Journal of Soil and Water Conservation*, *71* (2), 156–164.

Tingey-Holyoak, J. L., Pisaniello, J. D., & Buss, P. (2021). Embedding smart technologies in accounting to meet global irrigation challenges. Meditari accountancy. *Research; a Journal of Science and its Applications*, *29*(5), 1146–1178.

Toohey, E. S., Van De Ven, R., & Hopkins, D. L. (2018). The value of objective online measurement technology: Australian red meat processor perspective. *Animal Production Science*, *58*(8), 1559–1565.

UN. (2019). How certain are the United Nations global population projections? In *Population facts*. Department of Economic and Social Affairs.

Valero, M. R., Hicks, B. J., & Nassehi, A. (2023a). *A conceptual framework of a digital-twin for a circular meat supply chain. Flexible automation and intelligent manufacturing: The human-data-technology nexus* (pp. 188–196). Cham: Springer International Publishing.

Varmuza, K., & Filzmoser, P. (2009). *Introduction to multivariate statistical analysis in chemometrics*. CRC Press. Available from https://doi.org/10.1201/9781420059496.

von Veltheim, F. R., & Heise, H. (2020). The agtech startup perspective to farmers ex ante acceptance process of autonomous field robots. *Sustainability (Switzerland)*, *12*(24), 1–18.

Verdouw, C., Tekinerdogan, B., Beulens, A., & Wolfert, S. (2021). Digital twins in smart farming. *Agricultural Systems*, 189.

Vignali, G., Bertolini, M., Bottani, E., Di Donato, L., Ferraro, A., & Longo, F. (2018). Design and testing of an augmented reality solution to enhance operator safety in the food industry. *International Journal of Food Engineering*, *14*(2), 20170122. Available from https://doi.org/10.1515/ijfe-2017-0122.

Warriss, P. D. (2009). *Meat Science*. CABI, ISBN: 978-1-84593-593-1.

Watson, N. J., Bowler, A. L., Rady, A., Fisher, O. J., Simeone, A., Escrig, J., Woolley, E., & Adedeji, A. A. (2021). Intelligent sensors for sustainable food and drink manufacturing. *Frontiers in Sustainable Food Systems*, *5*.

Wold, J. P., Solberg, L. E., Gaarder, M. Ø., Carlehøg, M., Sanden, K. W., & Rødbotten, R. (2022). In-line estimation of fat marbling in whole beef striploins (longissimus lumborum) by NIR hyperspectral imaging. A closer look at the role of myoglobin. *Foods*, *11*(9), 1219.

Xu, W., He, Y., Li, J., Zhou, J., Xu, E., Wang, W., & Liu, D. (2023). Robotization and intelligent digital systems in the meat cutting industry: From the perspectives of robotic cutting, perception, and digital development. *Trends in Food Science and Technology*, *135*, 234–251.

Chapter 14

From Food Industry 4.0 toward Food Industry 5.0: human-centric approach for enhanced sustainability and resilience in the food and agriculture sector

Abdo Hassoun
Sustainable AgriFoodtech Innovation & Research (SAFIR), Arras, France

14.1 Introduction

So much has happened over the last few years and disrupted economies and societies all over the world due to the emergence of a range of pressing issues including, but not limited to, climate change, biodiversity crisis, outbreak of armed conflicts and pandemics, problems of food waste and plastic pollution, and increased inequities and injustices, which threaten food security for millions of people and food sustainability for future generations. To mitigate these complex challenges, robust sustainable technologies and innovative ambitious solutions should be developed for "Great Food System Transformation" from farm to fork. These not only rely on technological innovations but also on the initiation of multistakeholder dialogue and build capacity at community, city, and country levels, along with ensuring open access to data and creating a forum for diplomacy, standards and target-setting, and policy making (Peydayesh et al., 2023; Singh et al., 2023; The EAT−Lancet 2.0 Commissioners & contributing authors, 2023).

From a technological viewpoint, significant advances and innovations have always emerged since the first industrial revolution to drive progress in every aspect of human life, including production and consumption patterns in agriculture and the food industry. Recent technological breakthroughs have enabled the emergence of new scientific paradigms, known as the fourth industrial revolution (or Industry 4.0), which is spurred by high automation and digitalization and characterized by the integration of information and communication technologies, offering new opportunities for the intelligentization of manufacturing, enhanced interconnection, and real-time data exchange (Birkel & Müller, 2021; Marvin et al., 2022; Sadeghi et al., 2022).

Nowadays, there is a lot of discussion around the concept of Industry 5.0 (the fifth industrial revolution) in various fields, such as supply chain (Frederico, 2021), healthcare environment (Javaid et al., 2020), logistics (Jafari et al., 2022), construction (Marinelli, 2023), engineering education (Gürdür et al., 2022), and others (Ghobakhloo et al., 2022; Khan et al., 2023). However, to the best of our knowledge, the concept of Food Industry 5.0 has not yet been discussed. Therefore this short chapter strives to address this knowledge gap by providing all-around popularization and discussion (i.e., a summary, not a comprehensive overview) of Industry 4.0 and Industry 5.0 in food sectors. Nevertheless, a detailed in-depth exploration of all valuable directions of these futurist concepts in food sectors will be the subject of our future publications. As a pioneer work, this chapter is expected to make important contributions to understanding the definition, values, and enabling technologies of Industry 5.0 in the agriculture and food industry sectors.

14.2 Overview of the Food Industry 4.0

Industry 4.0 can be described as a fusion of technologies across the physical, digital, and biological worlds (Callaghan, 2023; Hassoun, Aït-kaddour, et al., 2022; Mhlanga, 2021). Industry 4.0 puts the focus on developing cyber-physical systems (CPSs) and involves the digitalization and automation of industry (Aheleroff et al., 2022; Rana & Rathore, 2023).

Food Industry 4.0. DOI: https://doi.org/10.1016/B978-0-443-15516-1.00014-1

Although there has been a large number of reports and studies on Industry 4.0 in the literature, very little can be found on Food Industry 4.0. Recently published papers identified the main enabler of Industry 4.0 in agriculture and the food industry (Hassoun, Aït-kaddour, et al., 2022; Hassoun, Bekhit, et al., 2022; Hassoun, Siddiqui, et al., 2022; Javaid et al., 2022). Indeed, the main enablers of digital transformation in the agrifood sector (Fig. 14.1) include Artificial Intelligence (AI), big data, the Internet of Things (IoT), smart sensors, blockchain, robotics, digital twins, and virtual and augmented reality (Hassoun, Marvin, et al., 2023). Other advanced technologies such as additive manufacturing, CPS, cloud computing, cybersecurity, information and communication technologies (ICTs), as well as recent advances in biotechnology and nanotechnology have been widely reported to be in close association with the concept of Industry 4.0 (Alcácer & Cruz-Machado, 2019; Andronie et al., 2021; Bettiol et al., 2022; Ching et al., 2022; Khan et al., 2023).

In the food industry, the concepts of digital transformation, smart manufacturing, industrial IoT, and smart food factory are related in the sense that they all refer to the application of Industry 4.0 technologies in the food industry. On the other hand, digital agriculture, precision agriculture, smart farming, and Agriculture 4.0 are common terms used to refer to the implication of fourth industrial revolution technologies in agriculture (Araújo et al., 2021; Culot et al., 2020; Ferrag et al., 2021; Javaid et al., 2022; Liu et al., 2021; Müller, Buliga, et al., 2020).

Moreover, inspired by Industry 4.0, new concepts, such as Dairy 4.0 (Hassoun, Garcia-Garcia, et al., 2023), Meat 4.0 (Echegaray et al., 2022), Traceability 4.0 (Hassoun, Abdullah, et al., 2022; Hassoun, Kamiloglu, et al., 2023), Processing 4.0 (Hassoun, Jagtap, Trollman, et al., 2023), Quality 4.0 (Hassoun, Jagtap, et al., 2023), Flavor Analysis 4.0 (Zeng et al., 2023), and Packaging 4.0 (Sadeghi et al., 2022), have been recently introduced.

We have recently reviewed the application of Industry 4.0 technologies in the agrifood sector, highlighting the role of digitalization in accelerating the development of plant-based innovations and the shift to vegetal diets (Hassoun, Boukid, et al., 2022). The use of Industry 4.0 technologies (especially IoT) in the agrifood sector and their integration into the global value chain was also reviewed by Senturk et al. (2023). Industry 4.0 technologies have a big role to play in food sustainability to deliver sustainable development value. For example, the role of machine learning, blockchain, and other advanced technologies to achieve a digital and sustainable transition of the agrifood sector was highlighted by Abbate et al. (2023). In the same line, we argued in a recent publication that green and Industry 4.0 technologies have a promising potential to accelerate and enhance the achievement of sustainable development goals (SDGs), enabling a green and digital transition in food systems (Hassoun, Prieto, et al., 2022).

14.3 The fifth industrial revolution (Industry 5.0)

While several food-related sectors are still busy developing methods for interconnecting new technologies to improve efficiency and productivity in line with Industry 4.0, the guiding principles behind Industry 5.0 and its enabling

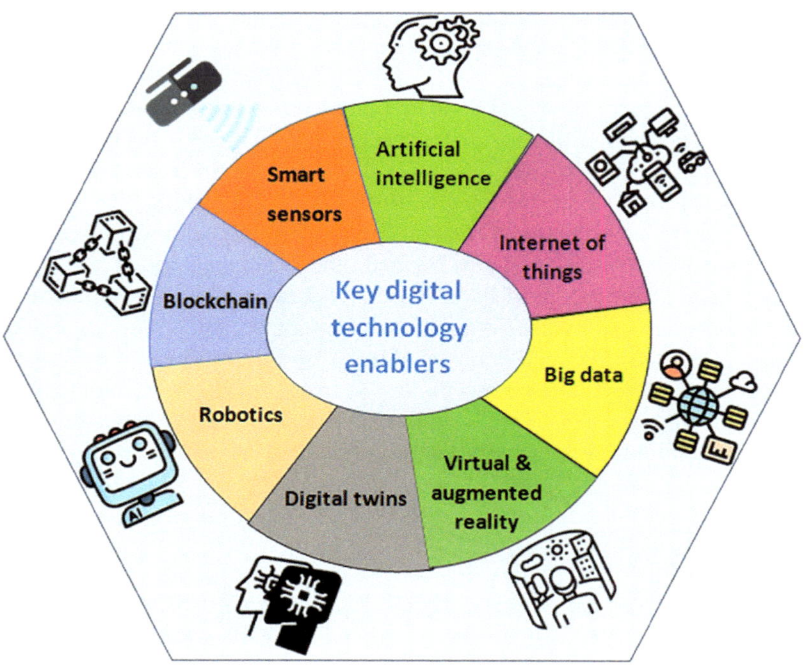

FIGURE 14.1 The main enabler of digital transformation in the agrifood sector.

technologies are being identified and used in various manufacturing sectors. This industrial interest in this new concept has been accompanied by similar academic and research focus. Indeed, a quick search inquiry on the Scopus database using the terms "Industry 5.0" or "fifth industrial revolution" in titles of publications showed 419 publications, with only 2 being published before 2015. The number of publications and citations on this topic has henceforth increased spectacularly, rising from 1 publication in 2015 to 178 in 2022, as shown in Fig. 14.2. Computer science, engineering, and business, management, and accounting are the main subject areas, followed by social and decision sciences, mathematics, economics, econometrics and finance, physics and astronomy, energy, materials science, environmental sciences, chemical engineering, and medicine, whereas only a few papers were published in the field of agricultural and biological sciences.

Despite the huge amount of publications and data on Industry 5.0, there is still a lack of consensus on its definition and scope, as well as its technological enablers (Ghobakhloo et al., 2022; Ghobakhloo, Iranmanesh, Tseng, et al., 2023). Indeed, various definitions and enabling technologies of Industry 5.0 can be found in the literature.

According to Maddikunta et al. (2022), Industry 5.0 is characterized by mass personalization (Fig. 14.3), and it combines human creativity with efficient, intelligent, and accurate machines. Nahavandi (2019) reported that Industry 5.0 is based on intelligent devices, intelligent systems, and intelligent automation, which are fully merged with the physical world in cooperation with human intelligence.

Leng et al. (2022) and Xu et al. (2021) argued that Industry 5.0 is value-driven (while Industry 4.0 is technology-driven) and characterized by human centricity, sustainability, and resiliency. Ordieres-meré et al. (2023) stated that it is the integration of human and machine data that fuels Industry 5.0 applications. Similarly, Zizic et al. (2022) acknowledged the human centricity of Industry 5.0, whereas Müller (2020) indicated that Industry 5.0 will support and foster socially and ecologically relevant values. Espina-romero et al. (2023) stated that the main intention of Industry 5.0 is to transform the industry into intelligent structures supported by cognitive computing and IoT. Industry 5.0 contribution to sustainable development has been underscored by Ghobakhloo et al. (2022) who revealed this vision of a future industry delivers sustainable development values through 16 functions related to resilience, environmental sustainability, and human centricity. Recently, the same authors referred to cognitive CPSs, cognitive AI, human interaction, and recognition technologies, extended reality, industrial smart wearable, and intelligent robots as the main enabling technologies of Industry 5.0 (Ghobakhloo, Iranmanesh, Foroughi, et al., 2023; Ghobakhloo et al., 2023). Another recent overview defined mutual-cognitive human–robot collaboration, digital twins and metaverse, extended reality and holography, big data analytics and cognitive computing, blockchain and decentralized computing, and the Internet of Everything (IoE) as principles enablers of Industry 5.0 (Leng et al., 2022).

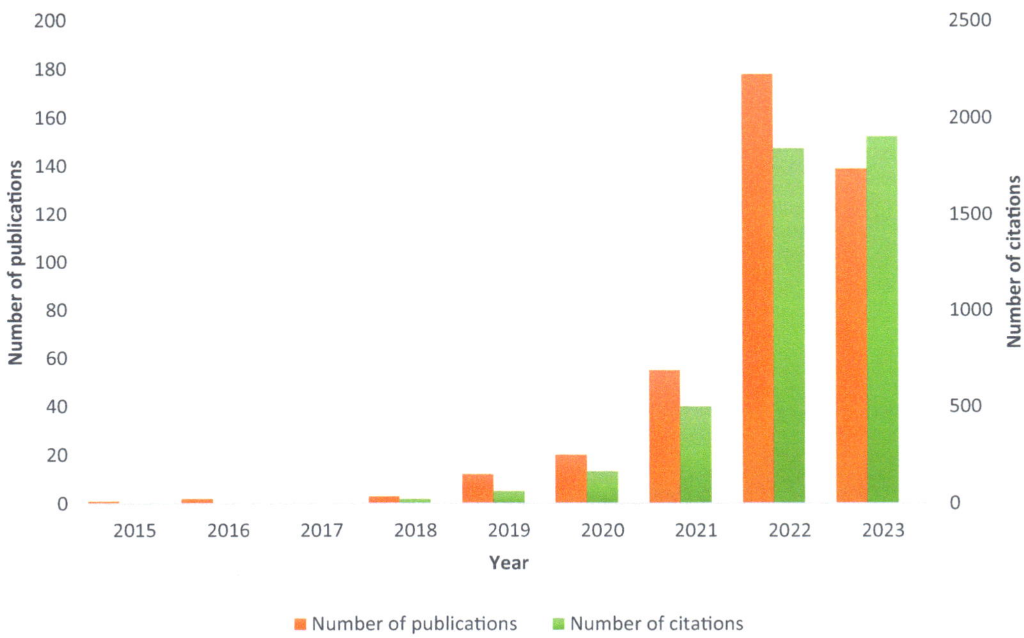

FIGURE 14.2 Evolution of publications and citations number per year on Industry 5.0 over the last few years (search query was done on July 15, 2023). the following keyword search query was used in the Scopus database: TITLE-ABS-KEY (Industry 5.0) or (fifth industrial revolution).

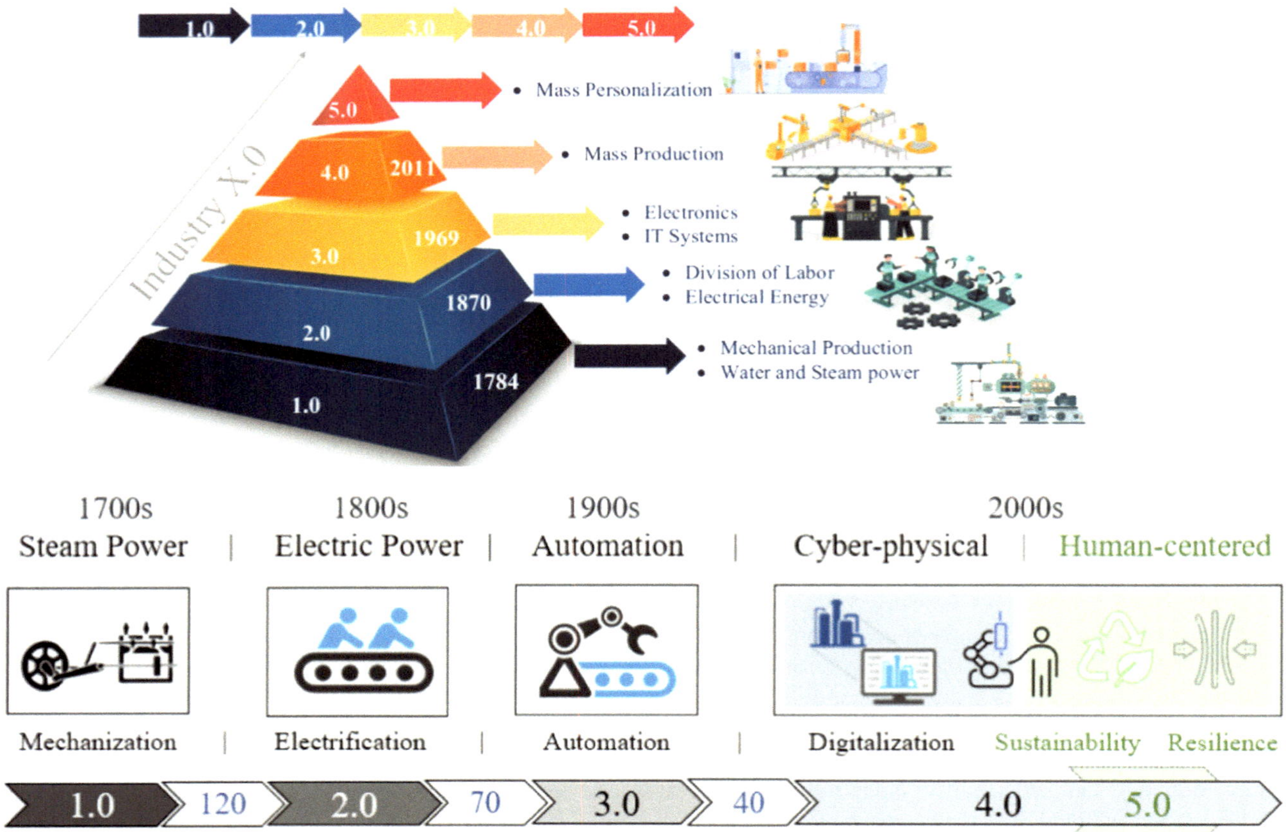

FIGURE 14.3 Evolution of the industrial revolution from 1.0 to 5.0 according to Aheleroff et al. (2022) and Maddikunta et al. (2022).

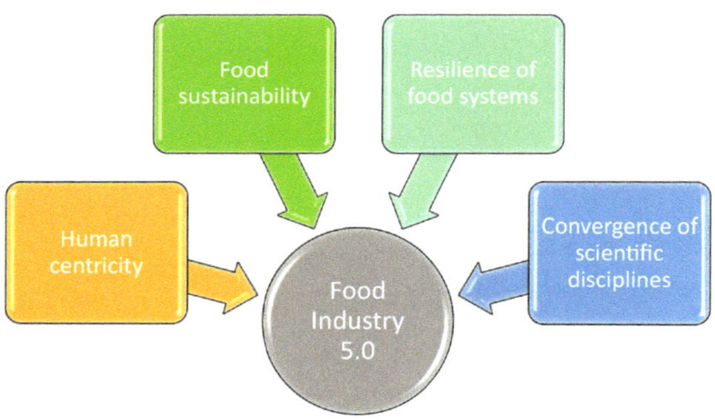

FIGURE 14.4 Food Industry 5.0.

14.4 Food Industry 5.0: Definition and a survey of enabling technologies

Food Industry 5.0 is inspired by the fifth industrial revolution and refers to the application of Industry 5.0 technologies in agriculture and the food industry. Consistent with the European Commission agenda (J. Müller, 2020) and recent publications in other sectors (Ghobakhloo et al., 2023; Leng et al., 2022; Sigov et al., 2022), we consider "Food Industry 5.0" a socio-ecological phenomenon, based on enhanced food sustainability and resilience and robust human–robot coworking and cooperation in food production and manufacturing sectors. Food Industry 5.0 can happen by the convergence phenomenon where the boundaries between different disciplines are eroding (Fig. 14.4).

Many technologies of Industry 5.0 can already be found in Industry 4.0 but with upgraded features along with exploring new breakthroughs in the field.

AI is one of the core Industry 4.0 technologies that has been widely used in many applications in the food and agriculture sector (Ben Ayed & Hanana, 2021; Siregar et al., 2022). Advanced AI-based solutions are expected to be at the heart of the Food Industry 5.0. For example, In 2022 OpenAI developed a chatbot called Chat Generative Pre-Trained Transformer (ChatGPT), whose creation has sparked enormous public and scientific interest and heralds a new era of AI. Indeed, this easy-to-implement AI-based tool is increasingly being discussed and adapted in various research and scientific fields, including the food and agriculture sector (Fusté-forné & Orea-giner, 2023; Loh, 2023). Recently, other innovations in the AI field such as cognitive and industrial AI have been reported (Ghobakhloo et al., 2023; Leng et al., 2022; Peres et al., 2020). IoT is another backbone of Industry 4.0 that has fundamentally reshaped several industries, including agriculture and the food industry (Cheng et al., 2022; Rejeb et al., 2022). A new related term; the IoE has been coined in the context of Industry 5.0, referring to the interconnection of people, process, information technology, and things (Guruswamy et al., 2022; Leng et al., 2022; Maddikunta et al., 2022).

One of the main features of Industry 5.0 is the human–machine collaboration and the use of collaborative robots (cobots) that are designed to work in close interaction with humans (Khan et al., 2023; Maddikunta et al., 2022). Other innovations in food printing (the evolution from 3D printing to 4D printing and beyond), communication technologies (passage from fifth-generation mobile networks 5G to 6G and beyond), and the use of smart materials are characteristics of the transition from Industry 4.0 to Industry 5.0 (Alwis et al., 2021; Khan et al., 2023; Ligon et al., 2017; Mena et al., 2023; Polymeni et al., 2023).

Other technologies, such as blockchain, big data analytics, miniaturized smart sensors, cloud computing, and augmented and extended reality (Ghobakhloo et al., 2023; Müller-Maatsch et al., 2021; Verma et al., 2022; Zeb et al., 2022) will continue to play a major role as enabling technologies of Food Industry 5.0.

There are only a few publications on Food Industry 5.0 available in the current scientific literature. For example, in its book entitled "Food 5.0," Saik (2019) reviewed technological advances in agriculture throughout history from "Agriculture 1.0" referring to the muscle era of farming to "Agriculture 5.0" denoting the convergence of new technologies such as sensors, robotics, and machine learning to make infinite sustainability achievable. Contreras-medina et al. (2022) provided a technological proposal for developing a coffee plant centered on indigenous producers' requirements from five localities in Mexico, using Industry 5.0 technologies. Guruswamy et al. (2022) highlighted the benefits of leveraging human creativity in collaboration with efficient, intelligent, and accurate machines to achieve better food security. Saptaningtyas and Rahayu (2020) proposed a process product innovation model, based on Industry 5.0 combined with business engineering, Society 5.0, and agile strategy for food manufacturing in small and medium enterprises.

In another context, AI has been used to analyze large amounts of data and generate insights to help improve gastronomy tourism offerings, such as optimizing menu prices and predicting food trends. For example, ChatGPT was used to explore the relationships between tourism and this language model from the perspective of gastronomy (Fusté-forné & Orea-giner, 2023). In another recent study, ChatGPT was used in nutritional sciences to investigate the safety and accuracy of diets, constructed for hypothetical individuals who are allergic to food allergens (Niszczota & Rybicka, 2023).

14.5 Conclusion

Although most food sectors are still grappling with the definition and implementation of principles and technologies of Industry 4.0, there are already many publications on theoretical framework and enabling technologies of the next industrial revolution in other manufacturing sectors. This underscores the need for more research and studies on both Industry 4.0 and Industry 5.0 in the food and agriculture sector.

Industry 5.0 can be understood as an upgrade (from the IoT to the IoE; from robots to collaborative robots: cobots; from AI to industrial and cognitive AI; from additive manufacturing: 3D printing to smart additive manufacturing, 4D printing and beyond, and use of smart materials) of Industry 4.0 technologies. In the context of this futurist vision, advanced robotics and intelligent machines would not be able to supersede humans but rather support farmers and food producers and manufacturers to take efficiency and food productivity a step further.

Growing evidence shows that Industry 4.0 and 5.0 innovations hold great potential to improve the environmental performance of food production systems worldwide. Moreover, leveraging these advanced technologies best could enhance food quality and safety, underpinning healthy consumption for all. Therefore there is everlasting interest in more adoption of Industry 4.0 and 5.0 technologies to explore new opportunities to help steer toward more just and equitable food systems for current and future generations.

To sum up, compared to Food Industry 4.0, Food Industry 5.0 is a more holistic approach, which is characterized by increased collaboration among humans and machines (with humans remaining the main actors) and enhanced focus on food sustainability, the resilience of the food supply chain, and the circularity of food systems. Today, the food and agriculture sector lives in Industry 4.0 but it is rapidly evolving toward Industry 5.0. Harnessing the full potential of Food Industry 4.0 and getting the most out of it will enable ground-breaking innovations in food sectors and pave the way for Food Industry 5.0.

References

Abbate, S., Centobelli, P., & Cerchione, R. (2023). The digital and sustainable transition of the agri-food sector. *Technological Forecasting and Social Change*, *187*(December 2022), 122222. Available from https://doi.org/10.1016/j.techfore.2022.122222.

Aheleroff, S., Huang, H., Xu, X., & Zhong, R. Y. (2022). Toward sustainability and resilience with Industry 4.0 and Industry 5.0. *Frontiers in Manufacturing Technology*, 1–20. Available from https://doi.org/10.3389/fmtec.2022.951643.

Alcácer, V., & Cruz-Machado, V. (2019). Scanning the Industry 4.0: A literature review on technologies for manufacturing systems. *Engineering Science and Technology, an International Journal*, *22*(3), 899–919. Available from https://doi.org/10.1016/J.JESTCH.2019.01.006.

Alwis, C. D., Kalla, A., Pham, Q. V., Kumar, P., Dev, K., Hwang, W. J., & Liyanage, M. (2021). Survey on 6G frontiers: Trends, applications, requirements, technologies and future research. *IEEE Open Journal of the Communications Society*, *2*(March), 836–886. Available from https://doi.org/10.1109/OJCOMS.2021.3071496.

Andronie, M., Lăzăroiu, G., Ştefănescu, R., Uţă, C., & Dijmărescu, I. (2021). Sustainable, smart, and sensing technologies for cyber-physical manufacturing systems: A systematic literature review. *Sustainability*, *13*(10), 5495. Available from https://doi.org/10.3390/su13105495.

Araújo, S. O., Peres, R. S., Barata, J., Lidon, F., & Ramalho, J. C. (2021). Characterising the Agriculture 4.0 landscape—Emerging trends, challenges and opportunities. *Agronomy*, *11*, 667.

Ben Ayed, R., & Hanana, M. (2021). Artificial intelligence to improve the food and agriculture sector. *Journal of Food Quality*, *2021*. Available from https://doi.org/10.1155/2021/5584754.

Bettiol, M., Capestro, M., Maria, E. D., & Micelli, S. (2022). Disentangling the link between ICT and Industry 4.0: Impacts on knowledge-related performance. *International Journal of Productivity and Performance Management*, *71*(4), 1076–1098. Available from https://doi.org/10.1108/IJPPM-10-2020-0573.

Birkel, H., & Müller, J. M. (2021). Potentials of industry 4.0 for supply chain management within the triple bottom line of sustainability – A systematic literature review. *Journal of Cleaner Production*, *289*, 125612. Available from https://doi.org/10.1016/j.jclepro.2020.125612.

Callaghan, C. (2023). Revisiting how scientific research drives technological change: The Fifth Industrial Revolution. *South African Journal of Science*, *119*(7–8), 8–10. Available from https://doi.org/10.17159/sajs.2023/15556.

Cheng, J., Van, F., En, P., Ren, H., Shiong, K., Chang, J., & Loke, P. (2022). Integration of internet-of-things as sustainable smart farming technology for the rearing of black soldier fly to mitigate food waste. *Journal of the Taiwan Institute of Chemical Engineers*, *137*, 104235. Available from https://doi.org/10.1016/j.jtice.2022.104235.

Ching, N. T., Ghobakhloo, M., Iranmanesh, M., Maroufkhani, P., & Asadi, S. (2022). Industry 4.0 applications for sustainable manufacturing: A systematic literature review and a roadmap to sustainable development. *Journal of Cleaner Production*, *334*, 130133. Available from https://doi.org/10.1016/j.jclepro.2021.130133.

Contreras-medina, D. I., Medina-cuéllar, S. E., & Rodríguez-garcía, J. M. (2022). Roadmapping 5.0 technologies in agriculture: A technological proposal for developing the coffee plant centered on indigenous producers' requirements from Mexico, via knowledge management. *Plants*, *11*(11), 1502. Available from https://doi.org/10.3390/PLANTS11111502.

Culot, G., Nassimbeni, G., Orzes, G., & Sartor, M. (2020). Behind the definition of Industry 4.0: Analysis and open questions. *International Journal of Production Economics*, *226*(August 2018), 107617. Available from https://doi.org/10.1016/j.ijpe.2020.107617.

Echegaray, N., Hassoun, A., Jagtap, S., Tetteh-Caesar, M., Kumar, M., Tomasevic, I., Goksen, G., & Lorenzo, J. M. (2022). Meat 4.0: Principles and applications of industry 4.0 technologies in the meat industry. *Applied Sciences*, *12*(14), 6986. Available from https://doi.org/10.3390/app12146986.

Espina-romero, L., Guerrero-alcedo, J., Avila, N. G., & Noroño, G. (2023). Industry 5.0: Tracking scientific activity on the most influential industries, associated topics, and future research agenda. *Sustainability*, *15*, 5554.

Ferrag, M. A., Shu, L., Djallel, H., & Choo, K.-K. R. (2021). Deep learning-based intrusion detection for distributed. *Electronics*, *10*, 1257.

Frederico, G. F. (2021). From Supply Chain 4.0 to Supply Chain 5.0: Findings from a systematic literature review and research directions. *Logistics*, *5*(3), 49. Available from https://doi.org/10.3390/LOGISTICS5030049.

Fusté-forné, F., & Orea-giner, A. (2023). Gastronomy in tourism management and marketing: An interview with ChatGPT. *The Journal of Automated Economy*, *4*(42), 1–13.

Ghobakhloo, M., Iranmanesh, M., Foroughi, B., & Base, L. (2023). Industry 5.0 implications for inclusive sustainable manufacturing: An evidence-knowledge-based strategic roadmap. *Journal of Cleaner Production*, *417*(June), 138023. Available from https://doi.org/10.1016/j.jclepro.2023.138023.

Ghobakhloo, M., Iranmanesh, M., Mubarak, M. F., Mubarik, M., Rejeb, A., & Nilashi, M. (2022). Identifying industry 5.0 contributions to sustainable development: A strategy roadmap for delivering sustainability values. *Sustainable Production and Consumption*, *33*, 716–737. Available from https://doi.org/10.1016/j.spc.2022.08.003.

Ghobakhloo, M., Iranmanesh, M., Tseng, M. L., Grybauskas, A., Stefanini, A., & Amran, A. (2023). Behind the definition of Industry 5.0: A systematic review of technologies, principles, components, and values. *Journal of Industrial and Production Engineering*, *40*(6), 432–447. Available from https://doi.org/10.1080/21681015.2023.2216701.

Gürdür, D., Kaynak, O., & Sait, S. M. (2022). Rethinking engineering education at the age of industry 5.0. *Journal of Industrial Information Integration*, *25*(October 2021), 100311. Available from https://doi.org/10.1016/j.jii.2021.100311.

Guruswamy, S., Subramanian, J., Sarang, S., Subbanagounder, A., & Jeoti, V. (2022). Toward better food security using concepts from Industry 5.0. *Sensors (Switzerland)*, *22*, 8377.

Hassoun, A., Abdullah, N. A., Aït-kaddour, A., Beşir, A., Zannou, O., Önal, B., Aadil, R. M., Lorenzo, J. M., Khaneghah, A. M., Regenstein, J. M., Beşir, A., Zannou, O., Önal, B., Aadil, R. M., & Lorenzo, J. M. (2022). Food traceability 4.0 as part of the fourth industrial revolution: Key enabling technologies. *Critical Reviews in Food Science and Nutrition*, 1−17. Available from https://doi.org/10.1080/10408398.2022.2110033.

Hassoun, A., Aït-kaddour, A., Abu-mahfouz, A. M., Rathod, N. B., Bader, F., Barba, F. J., Cropotova, J., Galanakis, C. M., Jambrak, A. R., Lorenzo, M., Måge, I., Ozogul, F., Regenstein, J., Rathod, B., Bader, F., Barba, F. J., Biancolillo, A., Cropotova, J., Galanakis, M., ... Regenstein, J. (2022). The fourth industrial revolution in the food industry—Part I: Industry 4.0 technologies. *Critical Reviews in Food Science and Nutrition*, 1−17. Available from https://doi.org/10.1080/10408398.2022.2034735.

Hassoun, A., Bekhit, A. E., Jambrak, A. R., Regenstein, J. M., Chemat, F., Morton, J. D., Gudjónsdóttir, M., Carpena, M., Prieto, A., Varela, P., Arshad, R. N., Aadil, R. M., & Ueland, Ø. (2022). The fourth industrial revolution in the food industry—part II : Emerging food trends. *Critical Reviews in Food Science and Nutrition*, 1−31. Available from https://doi.org/10.1080/10408398.2022.2106472.

Hassoun, A., Boukid, F., Pasqualone, A., Bryant, C. J., Jagtap, S., Trollman, H., Cropotova, J., García, G., Parra-l, C., & Barba, F. J. (2022). Emerging trends in the agri-food sector: Digitalisation and shift to plant-based diets. *Current Research in Food Science*, *5*, 2261−2269. Available from https://doi.org/10.1016/j.crfs.2022.11.010.

Hassoun, A., Garcia-Garcia, G., Trollman, H., Jagtap, S., Parra-López, C., Cropotova, J., Bhat, Z., Centobelli, P., & Aït-Kaddour, A. (2023). Birth of dairy 4.0: Opportunities and challenges in adoption of fourth industrial revolution technologies in the production of milk and itsderivatives. *Current Research in Food Science*, 100535. Available from https://doi.org/10.1016/J.CRFS.2023.100535.

Hassoun, A., Jagtap, S., Garcia-garcia, G., Trollman, H., Pateiro, M., Lorenzo, M., Trif, M., Vasile, A., Muhammad, R., Simat, V., Cropotova, J., & C, S. (2023). Food quality 4.0: From traditional approaches to digitalized automated analysis. *Journal of Food Engineering*, *337*, 111216. Available from https://doi.org/10.1016/j.jfoodeng.2022.111216.

Hassoun, A., Jagtap, S., Trollman, H., Garcia-garcia, G., Alhaj, N., Goksen, G., Bader, F., Ozogul, F., Barba, F. J., Cropotova, J., Munekata, P. E. S., & Lorenzo, M. (2023). Food processing 4.0: Current and future developments spurred by the fourth industrial revolution. *Food Control*, *145*, 109507. Available from https://doi.org/10.1016/j.foodcont.2022.109507.

Hassoun, A., Kamiloglu, S., Garcia-Garcia, G., Parra-Lopez, C., Trollman, H., Jagtap, S., Aadil, R. M., & Esatbeyoglu, T. (2023). Implementation of relevant fourth industrial revolution innovations across the supply chain of fruits and vegetables: A short update on Traceability 4.0. *Food Chemistry*, *409*, 135303. Available from https://doi.org/10.1016/j.foodchem.2022.135303.

Hassoun, A., Marvin, H. J. P., Bouzembrak, Y., Barba, F. J., Castagnini, J. M., Pallarés, N., Rabail, R., Aadil, R. M., Punia Bangar, S., Bhat, R., Cropotova, J., Maqsood, S., & Regenstein, J. M. (2023). Digital transformation in the agri-food industry: Recent applications and the role of the COVID-19 pandemic. *Frontiers in Sustainable Food Systems*, *7*, 1217813. Available from https://doi.org/10.3389/FSUFS.2023.1217813.

Hassoun, A., Prieto, M. A., Carpena, M., Bouzembrak, Y., Marvin, H. J. P., Pallar, N., Barba, F. J., Punia, S., Chaudhary, V., Ibrahim, S., & Bono, G. (2022). Exploring the role of green and Industry 4.0 technologies in achieving sustainable development goals in food sectors. *Food Research International*, *162*, 112068. Available from https://doi.org/10.1016/j.foodres.2022.112068.

Hassoun, A., Siddiqui, S. A., Smaoui, S., Ucak, İ., Arshad, N., Bhat, Z. F., Bhat, H. F., Carpena, M., Miguel, A., Aït-kaddour, A., Pereira, J. A. M., Zacometti, C., Tata, A., Ibrahim, S. A., Ozogul, F., Camara, J. S., Arshad, N., Bhat, Z. F., Bhat, H. F., ... Ibrahim, S. A. (2022). Emerging technological advances in improving the safety of muscle foods: Framing in the context of the food revolution 4.0. *Food Research International*, 0−42. Available from https://doi.org/10.1080/87559129.2022.2149776.

Jafari, N., Azarian, M., & Yu, H. (2022). Moving from Industry 4.0 to Industry 5.0: What are the implications for smart logistics? *Logistics*, *6*(2), 26. Available from https://doi.org/10.3390/logistics6020026.

Javaid, M., Haleem, A., Singh, R. P., & Suman, R. (2022). Enhancing smart farming through the applications of Agriculture 4.0 technologies. *International Journal of Intelligent Networks*, *3*, 150−164. Available from https://doi.org/10.1016/j.ijin.2022.09.004.

Javaid, M., Haleem, A., Singh, R. P., Ul Haq, M. I., Raina, A., & Suman, R. (2020). Industry 5.0: Potential applications in covid-19. *Journal of Industrial Integration and Management*, *5*(4), 507−530. Available from https://doi.org/10.1142/S2424862220500220.

Khan, M., Haleem, A., & Javaid, M. (2023). Changes and improvements in Industry 5.0: A strategic approach to overcome the challenges of Industry 4.0. *Green Technologies and Sustainability*, *1*(2), 100020. Available from https://doi.org/10.1016/j.grets.2023.100020.

Leng, J., Sha, W., Wang, B., Zheng, P., Zhuang, C., Liu, Q., Wuest, T., Mourtzis, D., & Wang, L. (2022). Industry 5.0: Prospect and retrospect. *Journal of Manufacturing Systems*, *65*(September), 279−295. Available from https://doi.org/10.1016/j.jmsy.2022.09.017.

Ligon, S. C., Liska, R., Stampfl, J., Gurr, M., & Mülhaupt, R. (2017). Polymers for 3D printing and customized additive manufacturing. *Chemical Reviews*, *117*(15), 10212−10290. Available from https://doi.org/10.1021/acs.chemrev.7b00074.

Liu, Y., Ma, X., Shu, L., Hancke, G. P., & Abu-Mahfouz, A. M. (2021). From Industry 4.0 to Agriculture 4.0: Current status, enabling technologies, and research challenges. *IEEE Transactions on Industrial Informatics*, *17*(6), 4322−4334. Available from https://doi.org/10.1109/TII.2020.3003910.

Loh, E. (2023). ChatGPT and generative AI chatbots: Challenges and opportunities for science, medicine and medical leaders. *BMJ Leader*, 1−4. Available from https://doi.org/10.1136/leader-2023-000797.

Maddikunta, P. K. R., Pham, Q. V., B, P., Deepa, N., Dev, K., Gadekallu, T. R., Ruby, R., & Liyanage, M. (2022). Industry 5.0: A survey on enabling technologies and potential applications. *Journal of Industrial Information Integration*, *26*, 100257. Available from https://doi.org/10.1016/j.jii.2021.100257.

Marinelli, M. (2023). From Industry 4.0 to Construction 5.0: Exploring the path towards human−robot collaboration in construction. *Systems*, *11*, 152.

Marvin, H. J. P., Bouzembrak, Y., van der Fels-Klerx, H. J., Kempenaar, C., Veerkamp, R., Chauhan, A., Stroosnijder, S., Top, J., Simsek-Senel, G., Vrolijk, H., Knibbe, W. J., Zhang, L., Boom, R., & Tekinerdogan, B. (2022). Digitalisation and artificial intelligence for sustainable food systems. *Trends in Food Science & Technology*, *120*, 344–348. Available from https://doi.org/10.1016/J.TIFS.2022.01.020.

Mena, B., Littlefair, G., & Singamneni, S. (2023). From 3D to 4D printing: A review. *Materials Today: Proceedings*. Available from https://doi.org/10.1016/j.matpr.2023.05.707.

Mhlanga, D. (2021). Artificial intelligence in the industry 4.0, and its impact on poverty, innovation, infrastructure development, and the sustainable development goals: Lessons from emerging economies? *Sustainability (Switzerland)*, *13*(11), 5788. Available from https://doi.org/10.3390/su13115788.

Müller, J. (2020). *Enabling technologies for Industry 5.0: Results of a workshop with Europe's technology leaders*. European Commission *September*, 19. Available from https://doi.org/10.2777/082634.

Müller, J. M., Buliga, O., & Voigt, K. (2020). The role of absorptive capacity and innovation strategy in the design of industry 4.0 business models - A comparison between SMEs and large enterprises. *European Management Journal*, *39*(3), 333–343. Available from https://doi.org/10.1016/j.emj.2020.01.002.

Müller-Maatsch, J., Bertani, F. R., Mencattini, A., Gerardino, A., Martinelli, E., Weesepoel, Y., & van Ruth, S. (2021). The spectral treasure house of miniaturized instruments for food safety, quality and authenticity applications: A perspective. *Trends in Food Science and Technology*, *110* (February), 841–848. Available from https://doi.org/10.1016/j.tifs.2021.01.091.

Nahavandi, S. (2019). Industry 5.0—A human-centric solution. *Sustainability*, *11*(16), 4371. Available from https://doi.org/10.3390/SU11164371.

Niszczota, P., & Rybicka, I. (2023). The credibility of dietary advice formulated by ChatGPT: Robo-diets for people with food allergies. *Nutrition*, *112*, 112076. Available from https://doi.org/10.1016/j.nut.2023.112076.

Ordieres-meré, J., Gutierrez, M., & Villalba-díez, J. (2023). Toward the industry 5.0 paradigm: Increasing value creation through the robust integration of humans and machines. *Computers in Industry*, *150*(April), 103947. Available from https://doi.org/10.1016/j.compind.2023.103947.

Peres, R. S., Jia, X., Lee, J. A. Y., Sun, K., Colombo, A. W., & Barata, J. (2020). Industrial artificial intelligence in Industry 4.0 - Systematic review, challenges and outlook. *IEEE Access*, *8*, 220121–220139.

Peydayesh, M., Bagnani, M., Soon, W. L., & Mezzenga, R. (2023). Turning food protein waste into sustainable technologies. *Chemical Reviews*, *123* (5), 2112–2154. Available from https://doi.org/10.1021/acs.chemrev.2c00236.

Polymeni, S., Plastras, S., & Kormentzas, G. (2023). The impact of 6G-IoT technologies on the development of Agriculture 5.0: A review. *Electronics*, *12*, 2651.

Rana, B., & Rathore, S. S. (2023). Industry 4.0 - Applications, challenges and opportunities in industries and academia: A review. *Materials Today: Proceedings*, *79*, 389–394. Available from https://doi.org/10.1016/j.matpr.2022.12.162.

Rejeb, A., Rejeb, K., Abdollahi, A., Al-Turjman, F., & Treiblmaier, H. (2022). The Interplay between the Internet of Things and agriculture: A bibliometric analysis and research agenda. *Internet of Things (Netherlands)*, *19*(July), 100580. Available from https://doi.org/10.1016/j.iot.2022.100580.

Sadeghi, K., Kim, J., & Seo, J. (2022). Packaging 4.0: The threshold of an intelligent approach. *Comprehensive Reviews in Food Science and Food Safety*, *21*(3), 2615–2638. Available from https://doi.org/10.1111/1541-4337.12932.

Saik, R. (2019). *Food 5.0: How we feed the future*. Lioncrest Publishing.

Saptaningtyas, W. W. E., & Rahayu, D. K. (2020, August). A proposed model for food manufacturing in smes: Facing Industry 5.0. In *Proceedings of the international conference on industrial engineering and operations management* (pp. 1653–1661).

Senturk, S., Senturk, F., & Karaca, H. (2023). Industry 4.0 technologies in agri-food sector and their integration in the global value chain: A review. *Journal of Cleaner Production*, *408*(April), 137096. Available from https://doi.org/10.1016/j.jclepro.2023.137096.

Sigov, A., Ratkin, L., Ivanov, L. A., & Da, L. (2022). Emerging enabling technologies for Industry 4.0 and beyond. *Information Systems Frontiers, September*, 2021. Available from https://doi.org/10.1007/s10796-021-10213-w.

Singh, B. K., Fraser, E. D. G., Arnold, T., Biermayr-Jenzano, P., Broerse, J. E. W., Brunori, G., Caron, P., De Schutter, O., Fabbri, K., Fan, S., Fanzo, J., Gajdzinska, M., Gurinovic, M., Hugas, M., McGlade, J., Nellemann, C., Njuki, J., Tuomisto, H. L., Tutundjian, S., ... Webb, P. (2023). Ensuring societal considerations are met when translating science into policy for sustainable food system transformation. *Trends in Food Science and Technology*, *137*(October 2022), 104–108. Available from https://doi.org/10.1016/j.tifs.2023.04.021.

Siregar, R. R. A., Seminar, K. B., Wahjuni, S., & Santosa, E. (2022). Vertical farming perspectives in support of precision agriculture using artificial intelligence: A review. *Computers*, *11*, 135. Available from https://doi.org/10.3390/COMPUTERS11090135.

The EAT–Lancet 2.0 Commissioners and contributing authors. (2023). EAT–Lancet Commission 2.0: Securing a just transition to healthy, environmentally sustainable diets for all. *The Lancet*, *6736*(23), 1–3. Available from https://doi.org/10.1016/S0140-6736(23)01290-4.

Verma, A., Bhattacharya, P., & Madhani, N. (2022). Blockchain for Industry 5.0: Vision, opportunities, key enablers, and future directions. *IEEE Access*, *10*(July), 69160–69199. Available from https://doi.org/10.1109/ACCESS.2022.3186892.

Xu, X., Lu, Y., Vogel-Heuser, B., & Wang, L. (2021). Industry 4.0 and Industry 5.0—Inception, conception and perception. *Journal of Manufacturing Systems*, *61*, 530–535. Available from https://doi.org/10.1016/J.JMSY.2021.10.006.

Zeb, S., Member, S., Mahmood, A., Member, S., Khowaja, S. A., Member, S., Dev, K., Member, S., Hassan, S. A., Member, S., Muhammad, N., Qureshi, F., Member, S., Gidlund, M., Member, S., Bellavista, I. P., & Member, S. (2022). Industry 5.0 is coming: A survey on intelligent NextG wireless networks as technological enablers. *ArXiv Preprint* (Ci), 1–42.

Zeng, X., Cao, R., Xi, Y., Li, X., Yu, M., & Zhao, J. (2023). Food flavor analysis 4.0: A cross-domain application of machine learning. *Trends in Food Science & Technology*, *138*(June), 116–125. Available from https://doi.org/10.1016/j.tifs.2023.06.011.

Zizic, M. C., Mladineo, M., Gjeldum, N., & Celent, L. (2022). From Industry 4.0 towards Industry 5.0: A review and analysis of paradigm shift for the people, organization and technology. *Energies*, *15*, 5221.

Index

Note: Page numbers followed by "*f*" and "*t*" refer o figures and tables, respectively.

L

M

N

O